Introduction to Finite Element Analysis and Design

Introduction to Finite Element Analysis and Design

Nam-Ho Kim and Bhavani V. Sankar

*Department of Mechanical & Aerospace Engineering,
University of Florida, Gainesville, FL 32611*

WILEY

John Wiley & Sons, Inc.
New York / Chichester / Weinheim / Brisbane / Singapore / Toronto

ACQUISITIONS EDITOR	Michael McDonald
EDITORIAL ASSISTANT	Rachael Leblond
MARKETING MANAGER	Chris Ruel
DESIGN DIRECTOR	Harry Nolan
SENIOR DESIGNER	Hope Miller
SENIOR PRODUCTION EDITOR	Patricia McFadden
PRODUCTION MANAGEMENT SERVICES	Thomson Digital

This book was set in by 10/12 Times New Roman and printed and bound by Hamilton Printing. The cover was printed by Phoenix Color.

This book is printed on acid free paper. ∞

To order books or for customer service please, call 1-800-CALL WILEY (225-5945).

ISBN- 978-0-470-12539-7

Printed in the United States of America
10 9 8 7 6 5 4 3

To our mothers, Sookyung and Rajeswari

Preface

Finite Element Method (FEM) is a numerical method for solving differential equations that describe many engineering problems. One of the reasons for FEM's popularity is that the method results in computer programs versatile in nature that can solve many practical problems with a small amount of training. Obviously, there is a danger in using computer programs without proper understanding of the theory behind them, and that is one of the reasons to have a thorough understanding of the theory behind FEM.

Many universities teach FEM to students at the junior/senior level. One of the biggest challenges to the instructor is finding a textbook appropriate to the level of students. In the past, FEM was taught only to graduate students who would carry out research in that field. Accordingly, many textbooks focus on theoretical development and numerical implementation of the method. However, the goal of an undergraduate FEM course is to introduce the basic concepts so that the students can use the method efficiently and interpret the results properly. Furthermore, the theoretical aspects of FEM must be presented without too many mathematical niceties. Practical applications through several design projects can help students to understand the method clearly.

This book is suitable for junior/senior level undergraduate students and beginning graduate students in mechanical, civil, aerospace, biomedical and industrial engineering, and engineering mechanics; researchers and design engineers in the above fields.

The textbook is organized into three parts and ten chapters. Part 1 reviews some concepts in mathematics and mechanics of materials that are prerequisite for learning finite element analysis. The objective of Part 1 is to establish a common ground before learning the main topics of FEM. Depending on the prerequisite courses, some portions of Part 1 can be skipped. Chapter 0 summarizes most mathematical preliminaries that will be repeatedly used in the text. The purpose of the chapter is by no means to provide a comprehensive mathematical treatment of the subject. Rather, it provides a common notation and the minimum amount of mathematical knowledge that will be required in future chapters, including matrix algebra and minimization of quadratic functions. In Chapter 1, the concepts of stress and strain are presented followed by constitutive relations and equilibrium equations. We limit our interest to linear, isotropic materials in order to make the concepts simple and clear. However, advanced concepts such as transformation of stress and strain, and the eigen value problem for calculating the principal values, are also included. Since in practice FEM is used mostly for designing a structure or a mechanical system, design criteria are also introduced in Chapter 1. These design criteria will be used in conjunction with FEM to determine whether a structure is safe or not.

Part 2 introduces one-dimensional finite elements, including truss and beam elements. This is the major part of the text that will teach the fundamental aspects of the FEM. We take an approach that provides students with the concepts of FEM incrementally, rather than providing all of them at the same time. Chapter 2 first introduces the direct stiffness method using spring elements. The concepts of nodes, elements, internal forces, equilibrium, assembly, and applying boundary conditions are presented in detail. The spring element is then extended to the uniaxial bar element without introducing interpolation. The concept of local (elemental) and global coordinates and their transformations are introduced via two– and three–dimensional truss elements. Four design projects are provided at the end of the chapter, so that students can apply the method to real life problems. The direct method in Chapter 2 provides a clear physical insight into FEM and is preferred in the beginning stages of learning the principles. However, it is limited in its application in that it can be used to solve one-dimensional problems only. The variational method is akin to the methods of calculus of variations and is a powerful tool for deriving the finite element equations. However, it requires the existence of a functional, minimization which results in the solution of the differential equations. We include a simple 1–D variational formulation in Chapter 3 using boundary value problems. The concept of polynomial approximation and domain discretization is introduced. The formal procedure of finite element analysis is also presented in this chapter. We focus on making a smooth connection between the discrete formulation in Chapter 2 and the continuum formulation in Chapter 3. The 1–D formulation is further extended to beams and plane frames in Chapter 4. At this point, the direct method is not useful because the stiffness matrix generated from the direct method cannot provide a clear physical interpretation. Accordingly, we use the principle of minimum potential energy to derive the matrix equation at the element level. The 1–D beam element is extended to 2–D frame element by using coordinate transformation. A 2–D bicycle frame design project is included at the end of the chapter. The finite element formulation is extended to the steady–state heat transfer problem in Chapter 5. This chapter is a special application of Chapter 3, including the convection boundary condition as a special case.

Part 3 is focused on two– and three–dimensional finite elements. Unlike 1–D elements, finite element modeling (domain discretization) becomes an important aspect for 2–D and 3–D problems. In Chapter 6, we introduce 2–D isoparametric solid elements. First, we introduce plane-stress and plane-strain approximation of 3–D problems. The governing variational equation is developed using the principle of minimum potential energy. Different types of elements are introduced, including triangular, rectangular, and quadrilateral elements. Numerical performance of each element is discussed through examples. We also emphasize the concept of isoparametric mapping and numerical integration. Three design projects are provided at the end of the chapter, so that students can apply the method to real life problems. In Chapter 7, we discuss traditional finite element analysis procedures, including preliminary analysis, pre-processing, solving matrix equations, and post-processing. Emphasis is on selection of element types, approximating the part geometry, different types of meshing, convergence, and taking advantage of symmetry. A design project involving 2–D analysis is provided at the end of the chapter. Since one of the important goals of FEM is to use the tool for engineering design, Chapter 8 briefly introduces structural design using FEM. The basic concept of design parameterization and the standard design problem formulation are presented. This chapter can be skipped depending on the schedule and content of the course.

Depending on the students' background one can leave out certain chapters. For example, if the course is offered at a junior level, one can leave out Chapters 3 and 8 and include Chapters 0 and 1. On the otherhand, Chapters 0,1 and 8 can be left out and Chapter 3 included at senior/beginning graduate level class.

Usage of commercial FEA programs is summarized in the Appendix. It includes various examples in the text using Pro/Engineer, Nastran, ANSYS, and the MATLAB toolbox developed at the Lund University in Sweden. Depending on availability and experience of the instructor, any program can be used as part of homework assignments and design projects. The textbook website will maintain up-to-date examples with the most recent version of the commercial programs.

Each chapter contains a comprehensive set of homework problems, some of which require commercial FEA programs. A total of nine design projects are provided in the book. We have included access to the NEi Nastran software which can be downloaded from www.wiley.com/college/kim.

We are thankful to the students who took our course and used the course package that had the same material as in this book. We are grateful for their valuable suggestions especially regarding the example and exercise problems.

January, 2008
Nam-Ho Kim and Bhavani V. Sankar

Contents

Preface   vii

Chapter 0. Mathematical Preliminaries   1

0.1. Vectors and Matrices   1
0.2. Vector-Matrix Calculus   3
0.3. Matrix Equations   8
0.4. Eigen Values and Eigen Vectors   8
0.5. Quadratic Forms   12
0.6. Maxima and Minima of Functions   13
0.7. Exercise   14

Chapter 1. Stress-Strain Analysis   17

1.1. Stress   17
1.2. Strain   30
1.3. Stress-Strain Relations   35
1.4. Boundary Value Problems   39
1.5. Failure Theories   43
1.6. Safety Factor   49
1.7. Exercise   52

Chapter 2. Uniaxial Bar and Truss Elements–Direct Method   60

2.1. Illustration of the Direct Method   61
2.2. Uniaxial Bar Element   66
2.3. Plane Truss Elements   73
2.4. Three-Dimensional Truss Elements (Space Truss)   83
2.5. Thermal Stresses   87
2.6. Projects   94
2.7. Exercise   98

Chapter 3. Weighted Residual and Energy Methods for One-Dimensional Problems   108

3.1. Exact vs. Approximate Solution   108
3.2. Galerkin Method   109
3.3. Higher-Order Differential Equations   114
3.4. Finite Element Approximation   117
3.5. Formal Procedure   124
3.6. Energy Methods   129
3.7. Exercise   138

Chapter 4. Finite Element Analysis of Beams and Frames   143

4.1. Review of Elementary Beam Theory   143
4.2. Rayleigh–Ritz Method   148
4.3. Finite Element Interpolation   153
4.4. Finite Element Equation for the Beam Element   158
4.5. Bending Moment and Shear Force Distribution   166
4.6. Plane Frame   171
4.7. Project   175
4.8. Exercise   176

Chapter 5. Finite Elements for Heat Transfer Problems   185

5.1. Introduction   185
5.2. Fourier Heat Conduction Equation   186
5.3. Finite Element Anlaysis – Direct Method   188
5.4. Galerkin Method for Heat Conduction Problems   194
5.5. Convection Boundary Conditions   200
5.6. Exercise   207

Chapter 6. Finite Elements for Plane Solids   211

6.1. Introduction   211
6.2. Types of Two-Dimensional Problems   211
6.3. Principle of Minimum Potential Energy   214
6.4. Constant Strain Triangular (CST) Element   216
6.5. Four-Node Rectangular Element   229
6.6. Four-Node Iso-Parametric Quadrilateral Element   237
6.7. Numerical Integration   248
6.8. Project   253
6.9. Exercise   254

Chapter 7. Finite Element Procedures and Modeling   261

7.1. Finite Element Analysis Procedures   261
7.2. Finite Element Modeling Techniques   281

7.3. Project   291
7.4. Exercise   292

Chapter 8. Structural Design Using Finite Elements   297

8.1. Introduction   297
8.2. Safety Margin in Design   298
8.3. Intuitive Design: Fully-Stressed Design   301
8.4. Design Parameterization   304
8.5. Parameter Study – Sensitivity Analysis   307
8.6. Structural Optimization   313
8.7. Project: Design Optimization of a Bracket   325
8.8. Exercise   327

Appendix A. Finite Element Analysis Using Pro/Engineer   333

A.1. Introduction   333
A.2. Getting Start   333
A.3. Plate with a Hole Analysis   334
A.4. Design Sensitivity Analysis/Parameter Study   339
A.5. Design Optimization   341

Appendix B. Finite Element Analysis Using NEi Nastran   343

B.1. Introduction   343
B.2. Getting Start   343
B.3. Plate with a Hole Analysis   343
B.4. Static Analysis of Beams   350
B.5. Examples in the text   356

Appendix C. Finite Element Analysis Using ANSYS   363

C.1. Introduction   363
C.2. Getting Start   364
C.3. Static Analysis of a Corner Bracket   365
C.4. Examples in the Text   381

Appendix D. Finite Element Analysis Using MATLAB Toolbox   391

D.1. Finite Element Analysis of Bar and Truss   391
D.2. Finite Element Analysis Using Frame Elements   403
D.3. Finite Element Analysis Using Plane Solid Elements   408

Index   417

Chapter 0

Mathematical Preliminaries

Since vector calculus and linear algebra are used extensively in finite element analysis, it is worth reviewing some fundamental concepts and recalling some important results that will be used in this book. A brief summary of concepts and results pertinent to the development of the subject are provided for the convenience of students. For a thorough understanding of the mathematical concepts, readers are advised to refer to any standard textbook, e.g., Kreyszig[1] and Strang.[2]

0.1 VECTORS AND MATRICES

0.1.1 Vector

A *vector* is a collection of scalars and is defined using a bold typeface[3] inside a pair of braces, such as

$$\{\mathbf{a}\} = \begin{Bmatrix} a_1 \\ a_2 \\ \vdots \\ a_N \end{Bmatrix} \tag{0.1}$$

In Eq. (0.1), $\{\mathbf{a}\}$ is an N-dimensional column vector. When the context is clear, we will remove the braces and simply use the letter "a" to denote the vector. The transpose of $\mathbf{a}$ above will be a row vector and will be denoted by $\mathbf{a}^T$.

$$\{\mathbf{a}\}^T = \{ a_1 \quad a_2 \quad \cdots \quad a_N \} \tag{0.2}$$

By default, in this text all vectors are considered as column vectors unless specified. For simplicity of notation, a geometric vector in two- or three-dimensional space is denoted by a bold typeface without braces:

$$\mathbf{a} = \begin{Bmatrix} a_1 \\ a_2 \\ a_3 \end{Bmatrix}, \qquad \text{or} \qquad \mathbf{a} = \begin{Bmatrix} a_1 \\ a_2 \end{Bmatrix} \tag{0.3}$$

where a_1, a_2, and a_3 are components of the vector $\mathbf{a}$ in the x-, y-, and z-direction, respectively, as shown in Figure (0.1). To save space, the above column vector $\mathbf{a}$ can be written as $\mathbf{a} = \{a_1, a_2, a_3\}^T$, in which $\{ \bullet \}^T$ denotes the *transpose*. The above three-dimensional geometric vector can also be denoted using a unit vector in each coordinate direction. Let

---

[1] E. Kreyszig, *Advanced Engineering Mathematics,* 5th ed., John Wiley & Sons, New York, 1983.

[2] G. Strang, *Linear Algebra and its Applications,* 2nd ed., Academic Press, New York, 1980.

[3] In the classroom one can use an underscore ($\underline{a}$) to denote vectors on the blackboard.

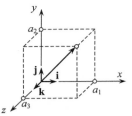

Figure 0.1 Three-dimensional geometric vector

$\mathbf{i} = \{1, 0, 0\}^T$, $\mathbf{j} = \{0, 1, 0\}^T$, and $\mathbf{k} = \{0, 0, 1\}^T$ be the unit vectors in the x-, y-, and z-directions, respectively. Then,

$$\mathbf{a} = a_1\mathbf{i} + a_2\mathbf{j} + a_3\mathbf{k} \tag{0.4}$$

The magnitude of the vector $\mathbf{a}$, $\|\mathbf{a}\|$, is given by

$$\|\mathbf{a}\| = \sqrt{a_1^2 + a_2^2 + a_3^2} \tag{0.5}$$

0.1.2   Matrix

A *matrix* is a collection of vectors and is defined using a bold typeface within square brackets. For example, let the matrix $[\mathbf{M}]$ be a collection of K number of column vectors $\{\mathbf{m}^i\}$, $i = 1, 2, \ldots, K$. Then, the matrix $[\mathbf{M}]$ is denoted by

$$[\mathbf{M}] = [\{\mathbf{m}^1\} \quad \{\mathbf{m}^2\} \quad \cdots \quad \{\mathbf{m}^K\}] \tag{0.6}$$

where

$$\{\mathbf{m}^i\} = \begin{Bmatrix} m_1^i \\ m_2^i \\ \vdots \\ m_N^i \end{Bmatrix}, \qquad i = 1, \cdots, K \tag{0.7}$$

By expanding each component of $\{\mathbf{m}^i\}$, the matrix $[\mathbf{M}]$ can be denoted using the $N \times K$ number of components as

$$[\mathbf{M}] = \begin{bmatrix} M_{11} & M_{12} & \cdots & M_{1K} \\ M_{21} & M_{22} & \cdots & M_{2K} \\ \vdots & \vdots & \ddots & \vdots \\ M_{N1} & M_{N2} & \cdots & M_{NK} \end{bmatrix} \tag{0.8}$$

where $M_{ij} = m_i^j$ is a component of the matrix. The notation for the subscripts M_{ij} is such that the first index denotes the position in the row, while the second index denotes the position in the column. The components that have the same indices are called diagonal components, e.g., M_{11}, M_{22}, etc. In Eq. (0.8), the dimensions of the matrix $[\mathbf{M}]$ are $N \times K$. When $N = K$, the matrix is called a *square* matrix.

A column vector can be considered as a matrix containing only one column. The column vector $\{\mathbf{m}^i\}$ in Eq. (0.7) is then an $N \times 1$ matrix.

0.1.3   Transpose of a Matrix

The *transpose* of a matrix can be obtained by switching the rows and columns of the matrix. For example, transpose of the matrix [**M**] in Eq. (0.8) can be written as

$$[\mathbf{M}]^T = \begin{bmatrix} M_{11} & M_{21} & \cdots & M_{N1} \\ M_{12} & M_{22} & \cdots & M_{N2} \\ \vdots & \vdots & \ddots & \vdots \\ M_{1K} & M_{2K} & \cdots & M_{NK} \end{bmatrix} \tag{0.9}$$

which is now a matrix of size $K \times N$.

0.1.4   Symmetric Matrix

A matrix is called *symmetric* when the matrix and its transpose are identical. It is clear from the definition that only a square matrix can be a symmetric matrix. For example, if [**S**] is symmetric, then

$$[\mathbf{S}] = [\mathbf{S}]^T = \begin{bmatrix} S_{11} & S_{12} & \cdots & S_{1N} \\ S_{12} & S_{22} & \cdots & S_{2N} \\ \vdots & \vdots & \ddots & \vdots \\ S_{1N} & S_{2N} & \cdots & S_{NN} \end{bmatrix} \tag{0.10}$$

Matrix [**A**] is called *skew-symmetric* when $[\mathbf{A}]^T = -[\mathbf{A}]$. It is clear that the diagonal components of a skew-symmetric matrix are zero. A typical skew-symmetric matrix can be defined as

$$[\mathbf{A}] = \begin{bmatrix} 0 & A_{12} & \cdots & A_{1N} \\ -A_{12} & 0 & \cdots & A_{2N} \\ \vdots & \vdots & \ddots & \vdots \\ -A_{1N} & -A_{2N} & \cdots & 0 \end{bmatrix} \tag{0.11}$$

0.1.5   Diagonal and Identity Matrices

A *diagonal matrix* is a special case of symmetric matrix in which all off-diagonal components are zero. An identity matrix is a diagonal matrix in which all diagonal components are equal to unity. For example, the (3×3) *identity matrix* is given by

$$[\mathbf{I}_3] = \begin{bmatrix} 1 & 0 & 0 \\ 0 & 1 & 0 \\ 0 & 0 & 1 \end{bmatrix} \tag{0.12}$$

0.2   VECTOR-MATRIX CALCULUS

0.2.1   Vector and Matrix Operations

Addition and subtraction of vectors and matrices are possible when their dimensions are the same. Let $\{\mathbf{a}\}$ and $\{\mathbf{b}\}$ be two N-dimensional vectors. Then, the addition and subtraction of these two vectors are defined as

$$\{\mathbf{c}\} = \{\mathbf{a}\} + \{\mathbf{b}\}, \quad \Rightarrow \quad c_i = a_i + b_i, \quad i = 1, \cdots, N$$
$$\{\mathbf{d}\} = \{\mathbf{a}\} - \{\mathbf{b}\}, \quad \Rightarrow \quad d_i = a_i - b_i, \quad i = 1, \cdots, N$$
(0.13)

Note that the dimensions of the resulting vectors $\{\mathbf{c}\}$ and $\{\mathbf{d}\}$ are the same as those of $\{\mathbf{a}\}$ and $\{\mathbf{b}\}$.

A scalar multiple of a vector is obtained by multiplying all of its components by a constant. For example, k times a vector $\{\mathbf{a}\}$ is obtained by multiplying each component of the vector by the constant k:

$$k\{\mathbf{a}\} = \{ka_1 \quad ka_2 \quad \cdots \quad ka_N\}^T$$
(0.14)

Similar operations can be defined for matrices. Let $[\mathbf{A}]$ and $[\mathbf{B}]$ be $N \times K$ matrices. Then, the addition and subtraction of these two matrices are defined as

$$[\mathbf{C}] = [\mathbf{A}] + [\mathbf{B}], \quad \Rightarrow \quad C_{ij} = A_{ij} + B_{ij}, \quad i = 1, \cdots, N, \quad j = 1, \cdots, K$$
$$[\mathbf{D}] = [\mathbf{A}] - [\mathbf{B}], \quad \Rightarrow \quad D_{ij} = A_{ij} - B_{ij}, \quad i = 1, \cdots, N, \quad j = 1, \cdots, K$$
(0.15)

Note that the dimensions of matrices $[\mathbf{C}]$ and $[\mathbf{D}]$ are the same as those of matrices $[\mathbf{A}]$ and $[\mathbf{B}]$. Similarly, one also can define the scalar multiple of a matrix.

Although the above matrix addition and subtraction are very similar to those of scalars, the multiplication and division of vectors and matrices are quite different from those of scalars.

0.2.2 Scalar Product

Since scalar products between two vectors will frequently appear in this text, it is necessary to clearly understand their definitions and notations used. Let $\mathbf{a}$ and $\mathbf{b}$ be two three-dimensional geometric vectors defined by

$$\mathbf{a} = \{ a_1 \quad a_2 \quad a_3 \}^T, \quad \text{and} \quad \mathbf{b} = \{ b_1 \quad b_2 \quad b_3 \}^T$$
(0.16)

The scalar product of $\mathbf{a}$ and $\mathbf{b}$ is defined as

$$\mathbf{a} \cdot \mathbf{b} = a_1 b_1 + a_2 b_2 + a_3 b_3$$
(0.17)

which is the summation of component-by-component products. Often notations in matrix product can be used such that $\mathbf{a} \cdot \mathbf{b} = \mathbf{a}^T \mathbf{b} = \mathbf{b}^T \mathbf{a}$. If $\mathbf{a}$ and $\mathbf{b}$ are two geometric vectors, then the scalar product can be written as

$$\mathbf{a} \cdot \mathbf{b} = \|a\|\|b\|\cos\theta$$
(0.18)

where θ is the angle between two vectors. Note that the scalar product of two vectors is a scalar, and hence the name *scalar product*. A scalar product can also be defined as the matrix product of one of the vectors and transpose of the other. In order for the scalar product to exist, the dimensions of the two vectors must be the same.

0.2.3 Norm

The *norm* or the magnitude of a vector [see Eq. (0.5)] can also be defined using the scalar product. For example, the norm of a three-dimensional vector $\mathbf{a}$ can be defined as

$$\|\mathbf{a}\| = \sqrt{\mathbf{a} \cdot \mathbf{a}}$$
(0.19)

Note that the norm is always a non-negative scalar and is the length of the geometric vector. When $\|\mathbf{a}\| = 1$, the vector $\mathbf{a}$ is called the *unit vector*.

0.2.4   Determinant of a Matrix

Determinant is an important concept, and it is useful in solving a linear system of equations. If the determinant of a matrix is zero, then it is not invertible and it is called a singular matrix. The determinant is defined only for square matrices. The formula for calculating the *determinant* of any square matrix can be easily understood by considering a 2×2 or 3×3 matrix. The determinant of a 2×2 matrix is defined as

$$|\mathbf{A}| = \begin{vmatrix} a_{11} & a_{12} \\ a_{21} & a_{22} \end{vmatrix} = a_{11}a_{22} - a_{12}a_{21} \tag{0.20}$$

The determinant of a 3×3 matrix is defined as

$$|\mathbf{A}| = \begin{vmatrix} a_{11} & a_{12} & a_{13} \\ a_{21} & a_{22} & a_{23} \\ a_{31} & a_{32} & a_{33} \end{vmatrix}$$

$$= a_{11} \begin{vmatrix} a_{22} & a_{23} \\ a_{32} & a_{33} \end{vmatrix} - a_{12} \begin{vmatrix} a_{21} & a_{23} \\ a_{31} & a_{33} \end{vmatrix} + a_{13} \begin{vmatrix} a_{21} & a_{22} \\ a_{31} & a_{32} \end{vmatrix} \tag{0.21}$$

$$= a_{11}(a_{22}a_{33} - a_{23}a_{32}) - a_{12}(a_{21}a_{33} - a_{23}a_{31}) + a_{13}(a_{21}a_{32} - a_{22}a_{31})$$

A matrix is called *singular* when its determinant is zero.

EXAMPLE 0.1   *Determinant*

The reader is encouraged to derive the following results using Eq. (0.20).

$$\begin{vmatrix} a & b \\ 0 & 0 \end{vmatrix} = 0$$

$$\begin{vmatrix} ka & kb \\ c & d \end{vmatrix} = k \begin{vmatrix} a & b \\ c & d \end{vmatrix}$$

$$\begin{vmatrix} a & b \\ c & d \end{vmatrix} = - \begin{vmatrix} c & d \\ a & b \end{vmatrix} = - \begin{vmatrix} b & a \\ d & c \end{vmatrix}$$

$$\begin{vmatrix} a & b \\ ka & kb \end{vmatrix} = 0, \qquad \begin{vmatrix} a & ka \\ b & kb \end{vmatrix} = 0$$

$$\begin{vmatrix} a+e & b+f \\ c & d \end{vmatrix} = \begin{vmatrix} a & b \\ c & d \end{vmatrix} + \begin{vmatrix} e & f \\ c & d \end{vmatrix} = (ad - bc) + (ed - cf)$$

0.2.5   Vector Product

Different from the scalar product, the result of the *vector product* is another vector. In the three-dimensional space, the vector product of two vectors $\mathbf{a}$ and $\mathbf{b}$ can be defined

Figure 0.2 Illustration of vector product

by the determinant as

$$
\mathbf{a} \times \mathbf{b} = \begin{vmatrix} \mathbf{i} & \mathbf{j} & \mathbf{k} \\ a_1 & a_2 & a_3 \\ b_1 & b_2 & b_3 \end{vmatrix}
$$
$$
= (a_2 b_3 - a_3 b_2)\mathbf{i} + (a_3 b_1 - a_1 b_3)\mathbf{j} + (a_1 b_2 - a_2 b_1)\mathbf{k} \qquad (0.22)
$$
$$
= \begin{Bmatrix} a_2 b_3 - a_3 b_2 \\ a_3 b_1 - a_1 b_3 \\ a_1 b_2 - a_2 b_1 \end{Bmatrix}
$$

In Eq. (0.22), we consider unit vectors $\mathbf{i}$, $\mathbf{j}$, and $\mathbf{k}$ as components of a matrix. As with the scalar product, the vector product can be defined only when the dimensions of two vectors are the same.

In the conventional notation, the vector product of two geometric vectors is defined by

$$
\mathbf{a} \times \mathbf{b} = \|\mathbf{a}\| \|\mathbf{b}\| \sin \theta \, \mathbf{n} \qquad (0.23)
$$

where θ is the angle between two vectors and $\mathbf{n}$ is the unit vector that is perpendicular to the plane that contains both vectors $\mathbf{a}$ and $\mathbf{b}$. The right-hand rule is used to determine the positive direction of vector $\mathbf{n}$ as shown in Figure 0.2. It is clear from its definitions in Eqs. (0.22) and (0.23), $\mathbf{a} \times \mathbf{a} = \mathbf{0}$, and $\mathbf{b} \times \mathbf{a} = -\mathbf{a} \times \mathbf{b}$.

0.2.6 Matrix-Vector Multiplication

The matrix-vector multiplication often appears in the finite element analysis. Let $[\mathbf{M}]$ be a 3×3 matrix defined by

$$
[\mathbf{M}] = \begin{bmatrix} m_{11} & m_{12} & m_{13} \\ m_{21} & m_{22} & m_{23} \\ m_{31} & m_{32} & m_{33} \end{bmatrix}
$$

The multiplication between a matrix $[\mathbf{M}]$ and a vector $\mathbf{a}$ is defined by

$$
\mathbf{c} = [\mathbf{M}] \cdot \mathbf{a} = \begin{bmatrix} m_{11} & m_{12} & m_{13} \\ m_{21} & m_{22} & m_{23} \\ m_{31} & m_{32} & m_{33} \end{bmatrix} \cdot \begin{Bmatrix} a_1 \\ a_2 \\ a_3 \end{Bmatrix} = \begin{Bmatrix} m_{11}a_1 + m_{12}a_2 + m_{13}a_3 \\ m_{21}a_1 + m_{22}a_2 + m_{23}a_3 \\ m_{31}a_1 + m_{32}a_2 + m_{33}a_3 \end{Bmatrix} \qquad (0.24)
$$

where $\mathbf{c}$ is a 3×1 column vector. Using a conventional summation notation, Eq. (0.24) can be written as

$$
c_i = \sum_{j=1}^{3} m_{ij} a_j, \quad i = 1, 2, 3 \qquad (0.25)
$$

Since the result of Eq. (0.24) is a vector, it is possible to obtain the scalar product of $\mathbf{c}$ with a vector $\mathbf{b}$, yielding

$$\mathbf{b} \cdot [\mathbf{M}] \cdot \mathbf{a} = b_1(m_{11}a_1 + m_{12}a_2 + m_{13}a_3) \\ + b_2(m_{21}a_1 + m_{22}a_2 + m_{23}a_3) \\ + b_3(m_{31}a_1 + m_{32}a_2 + m_{33}a_3) \tag{0.26}$$

which is a scalar.

The above matrix-vector multiplication can be generalized to arbitrary dimensions. For example, let $[\mathbf{M}]$ be an $N \times K$ matrix and $\{\mathbf{a}\}$ be an $L \times 1$ vector. The multiplication of $[\mathbf{M}]$ and $\{\mathbf{a}\}$ can be defined if and only if $K = L$. In addition, the result $\{\mathbf{c}\}$ will be a vector of $N \times 1$ dimension.

$$\{\mathbf{c}\}_{N \times 1} = [\mathbf{M}]_{N \times K} \{\mathbf{a}\}_{K \times 1} \\ c_i = \sum_{j=1}^{K} m_{ij}a_j, \quad i = 1, \ldots N \tag{0.27}$$

0.2.7   Matrix-Matrix Multiplication

The matrix-matrix multiplication is a more general case of Eq. (0.24). For 3×3 matrices, the matrix-matrix multiplication can be defined as

$$[\mathbf{C}] = [\mathbf{A}][\mathbf{B}] \tag{0.28}$$

where $[\mathbf{C}]$ is also a 3×3 matrix. Using the component notation, Eq. (0.28) is equivalent to

$$C_{ij} = \sum_{k=1}^{3} A_{ik}B_{kj}, \quad i = 1, 2, 3, \quad j = 1, 2, 3 \tag{0.29}$$

The above matrix-matrix multiplication can be generalized to arbitrary dimensions. For example, let the dimensions of matrices $[\mathbf{A}]$ and $[\mathbf{B}]$ be $N \times K$ and $L \times M$, respectively. The multiplication of $[\mathbf{A}]$ and $[\mathbf{B}]$ can be defined if and only if $K = L$, i.e., the number of columns in the first matrix must be equal to the number of rows in the second matrix. In addition, the dimension of the resulting matrix $[\mathbf{C}]$ will be $N \times M$.

$$C_{ij} = \sum_{k=1}^{K} A_{ik}B_{kj}, \quad i = 1, \ldots, N, \quad j = 1, \ldots, M \tag{0.30}$$

0.2.8   Inverse of a Matrix

If a square matrix $[\mathbf{A}]$ is invertible, then one can find another square matrix $[\mathbf{B}]$ such that $[\mathbf{A}][\mathbf{B}] = [\mathbf{B}][\mathbf{A}] = [\mathbf{I}]$, and then $[\mathbf{B}]$ is called the inverse of $[\mathbf{A}]$ and vice versa. A simple expression can be obtained for the *inverse* of a matrix when the dimension is 2×2, as

$$[\mathbf{A}]^{-1} = \begin{bmatrix} a_{11} & a_{12} \\ a_{21} & a_{22} \end{bmatrix}^{-1} = \frac{1}{|\mathbf{A}|} \begin{bmatrix} a_{22} & -a_{12} \\ -a_{21} & a_{11} \end{bmatrix} \tag{0.31}$$

For procedures of inverting a general $N \times N$ matrix, the reader should refer to textbooks such as Kreyszig[4] or Strang.[5] If a matrix is *singular* ($|\mathbf{A}| = 0$), then the inverse does not exist.

---

[4] E. Kreyszig, *Advanced Engineering Mathematics,* 5th ed., John Wiley & Sons, New York, 1983.

[5] G. Strang, *Linear Algebra and its Applications,* 2nd ed., Academic Press, New York, 1980.

0.2.9   Rules of Matrix Multiplication

The following rules of matrix multiplication will be useful in manipulating matrices and their functions. We present some results without proof.

Associative rule: $(\mathbf{AB})\mathbf{C} = \mathbf{A}(\mathbf{BC})$ $\qquad\qquad$ (0.32)

Distributive rule: $\mathbf{A}(\mathbf{B} + \mathbf{C}) = \mathbf{AB} + \mathbf{AC}$ $\qquad\qquad$ (0.33)

Non-commutative: $\mathbf{AB} \neq \mathbf{BA}$ $\qquad\qquad$ (0.34)

Transpose of product: $(\mathbf{AB})^T = \mathbf{B}^T\mathbf{A}^T, \ (\mathbf{ABC})^T = \mathbf{C}^T\mathbf{B}^T\mathbf{A}^T$ $\qquad\qquad$ (0.35)

Inverse of product: $(\mathbf{AB})^{-1} = \mathbf{B}^{-1}\mathbf{A}^{-1}, \ (\mathbf{ABC})^{-1} = \mathbf{C}^{-1}\mathbf{B}^{-1}\mathbf{A}^{-1}$ $\qquad\qquad$ (0.36)

0.3   MATRIX EQUATIONS

Consider the following simultaneous linear equations:

$$\begin{aligned} a_{11}x_1 + a_{12}x_2 + \cdots + a_{1N}x_N &= b_1 \\ a_{21}x_1 + a_{22}x_2 + \cdots + a_{2N}x_N &= b_2 \\ &\vdots \\ a_{N1}x_1 + a_{N2}x_2 + \cdots + a_{NN}x_N &= b_N \end{aligned} \qquad (0.37)$$

Equation (0.37) has N number of unknowns ($x_1, x_2, \ldots, x_N$), and there are N number of equations. If all equations are independent, then Eq. (0.37) has a unique solution. Equation (0.37) can be equivalently denoted using the matrix notation, as

$$[\mathbf{A}] \cdot \{\mathbf{x}\} = \{\mathbf{b}\} \qquad (0.38)$$

where

$$[\mathbf{A}] = \begin{bmatrix} a_{11} & a_{12} & \cdots & a_{1N} \\ a_{21} & a_{22} & \cdots & a_{2N} \\ \vdots & \vdots & \ddots & \vdots \\ a_{N1} & a_{N2} & \cdots & a_{NN} \end{bmatrix}, \quad \{\mathbf{x}\} = \begin{Bmatrix} x_1 \\ x_2 \\ \vdots \\ x_N \end{Bmatrix}, \quad \{\mathbf{b}\} = \begin{Bmatrix} b_1 \\ b_2 \\ \vdots \\ b_N \end{Bmatrix}$$

When the matrix $[\mathbf{A}]$ and the vector $\{\mathbf{b}\}$ are known, the solution $\{\mathbf{x}\}$ can be obtained by multiplying both sides of the equation by $[\mathbf{A}]^{-1}$ to obtain

$$\begin{aligned} [\mathbf{A}]^{-1}[\mathbf{A}] \cdot \{\mathbf{x}\} &= [\mathbf{A}]^{-1} \cdot \{\mathbf{b}\} \\ [\mathbf{I}] \cdot \{\mathbf{x}\} &= [\mathbf{A}]^{-1} \cdot \{\mathbf{b}\} \\ \{\mathbf{x}\} &= [\mathbf{A}]^{-1} \cdot \{\mathbf{b}\} \end{aligned} \qquad (0.39)$$

Note that $[\mathbf{I}]\{\mathbf{x}\} = \{\mathbf{x}\}$. Thus, a unique solution can be obtained if $[\mathbf{A}]^{-1}$ exists or, equivalently, if the matrix $[\mathbf{A}]$ is not singular.

0.4   EIGEN VALUES AND EIGEN VECTORS

Consider the equation shown below for a square matrix $[\mathbf{A}]$

$$[\mathbf{A}] \cdot \{\mathbf{x}\} = \lambda\{\mathbf{x}\} \qquad (0.40)$$

where λ is a scalar. The above equation can be thought of as a matrix equation similar to that in Eq. (0.38) except that $\{b\}$ is replaced by a scalar multiple of $\{x\}$ itself. Such equations, which arise in many engineering applications, are interesting and have physical significance. There are only certain special values for λ that will satisfy Eq. (0.40), and they are called the *eigen values*[6] of the square matrix $[A]$. For each eigen value there will be a corresponding vector $\{x\}$ called the *eigen vector*. Of course the null vector $\{x\} = \{0\}$ is a solution of Eq. (0.40) and we do not consider it, as it is trivial. It can be shown that the maximum number of eigen values will be equal to the number of rows (or columns) of $[A]$.

The procedure for computing the eigen values and eigen vectors is as follows. Equation (0.40) can be written as

$$[A] \cdot \{x\} - \lambda\{x\} = \{0\} \tag{0.41}$$

or

$$[A - \lambda I] \cdot \{x\} = \{0\} \tag{0.42}$$

where $[I]$ is the identity matrix of same dimensions as $[A]$. There are two possibilities for Eq. (0.42). Obviously $\{x\} = \{0\}$ is a solution of Eq. (0.42), but we have already declared it as trivial. If we want a nontrivial solution for $\{x\}$, then the matrix $[A - \lambda I]$ must be singular. Otherwise one can invert that matrix and multiply with $\{0\}$ on the RHS[7] to obtain the trivial solution $\{x\} = \{0\}$. Letting $[A - \lambda I]$ singular will open up new possibilities for $\{x\}$. In order for the coefficient matrix to be singular, its determinant must be equal to zero, i.e.,

$$|A - \lambda I| = 0 \tag{0.43}$$

or

$$\begin{vmatrix} a_{11} - \lambda & a_{12} & \cdots & a_{1n} \\ a_{21} & a_{22} - \lambda & \cdots & a_{2n} \\ \vdots & \vdots & \ddots & \vdots \\ a_{n1} & \cdots & \cdots & a_{nn} - \lambda \end{vmatrix} = 0 \tag{0.44}$$

The determinant in Eq. (0.44) can be expanded to obtain a polynomial equation in λ as

$$\lambda^n + C_1\lambda^{n-1} + \ldots\ldots C_{n-1}\lambda + C_n = 0 \tag{0.45}$$

The n-th degree polynomial on LHS of Eq. (0.45) is called the characteristic polynomial of matrix $[A]$. The n roots of the polynomial equation are the n eigen values of $[A]$.

Each one of the n eigen values can be substituted back in Eq. (0.42) to obtain a set of simultaneous equations for the unknown $\{x\}$. The solutions are called eigen vectors. The eigen vector corresponding the i-th eigen value is denoted by $\{x_i\}$. One cannot obtain a unique solution for $\{x_i\}$, as the set of equations are not linearly independent (remember that the determinant of the coefficient matrix was set to zero in order to solve for the eigen values). We will discuss this further as we find applications for the concepts of eigen values and eigen vectors.

The numerical method for solving the eigen value problem can be found in the literature[8]. An analytical method is available to solve Eq. (0.42) when $n = 2$ or 3. Here we

---

[6] **eigen** means "own" and Eigenschaft means "characteristic" in German.

[7] In this book, RHS and LHS mean the right-hand side and left-hand side of an equation, respectively.

[8] W. H. Press, B. P. Flannery, S. A. Teukolsky, and W. T. Vetterling, *Numerical Recipes,* Cambridge University Press, Cambridge, 1986.

introduce an analytical method when $n = 3$ and the coefficient matrix is symmetric. In such a case, the characteristic polynomial in Eq. (0.45) is cubic and can be written as

$$\lambda^3 + C_1\lambda^2 + C_2\lambda + C_3 = 0 \tag{0.46}$$

where

$$\begin{aligned}
C_1 &= -(a_{11} + a_{22} + a_{33}) \\
C_2 &= a_{11}a_{22} + a_{22}a_{33} + a_{33}a_{11} - a_{12}^2 - a_{23}^2 - a_{13}^2 \\
C_3 &= -(a_{11}a_{22}a_{33} + 2a_{12}a_{23}a_{13} - a_{11}a_{23}^2 - a_{22}a_{13}^2 - a_{33}a_{12}^2)
\end{aligned} \tag{0.47}$$

A general analytical solution for the above cubic equation can be written as

$$\begin{aligned}
\lambda_1 &= g\cos\frac{\phi}{3} - \frac{C_1}{3} \\
\lambda_2 &= g\cos\left(\frac{\phi + 2\pi}{3}\right) - \frac{C_1}{3} \\
\lambda_3 &= g\cos\left(\frac{\phi + 4\pi}{3}\right) - \frac{C_1}{3}
\end{aligned} \tag{0.48}$$

where

$$\begin{cases}
\phi = \cos^{-1}\left[-\dfrac{b}{2\sqrt{-a^3/27}}\right] \\
g = 2\sqrt{-a/3} \\
a = \dfrac{1}{3}(3C_2 - C_1^2) \\
b = \dfrac{1}{27}(2C_1^3 - 9C_1C_2 + 27C_3)
\end{cases} \tag{0.49}$$

It can be shown that eigen values of a real symmetric matrix are always real, and hence one can always compute three (not necessarily different) eigen values.

EXAMPLE 0.2   *Eigen Values and Eigen Vectors*

Find the eigen values and eigen vectors of the 3×3 matrix **A** given below.

$$A = \begin{bmatrix} 1 & 0 & 2 \\ 0 & 1 & 0 \\ 2 & 0 & 4 \end{bmatrix}$$

SOLUTION   The first step is to derive the characteristic equation for the matrix **A** similar to that shown in Eq. (0.46). It can be derived as

$$\lambda^3 - 6\lambda^2 + 5\lambda = 0$$

The solutions of the above cubic equation are the eigen values, and they are $\lambda = 0$, 1, and 5. The eigen vectors for each of the above eigen values are calculated using Eq. (0.42).
For $\lambda = 0$, we obtain

$$\begin{bmatrix} (1-0) & 0 & 2 \\ 0 & (1-0) & 0 \\ 2 & 0 & (4-0) \end{bmatrix} \begin{Bmatrix} x_1 \\ x_2 \\ x_3 \end{Bmatrix} = \begin{Bmatrix} 0 \\ 0 \\ 0 \end{Bmatrix}$$

The above equation yields three simultaneous equations for x_1, x_2, and x_3, as follows:

$$x_1 + 2x_3 = 0$$
$$x_2 = 0$$
$$2x_1 + 4x_3 = 0$$

As we mentioned earlier, there is no unique solution to the above set of equations. The solution can be written as $x_2 = 0$ and $x_1 = -2x_3$. One possible solution is $x_1 = -2$, $x_2 = 0$, and $x_3 = 1$. Thus, the eigen vector corresponding to the eigen value 0 can be written as $(-2, 0, 1)$. Usually the eigen vector is normalized such that its norm is equal to unity. Then the above eigen vector takes the form $\mathbf{x}^{(1)} = (-2, 0, 1)/\sqrt{5}$, where the superscript denotes that this is the first eigen vector.

For $\lambda = 1$, we obtain

$$\begin{bmatrix} 0 & 0 & 2 \\ 0 & 0 & 0 \\ 2 & 0 & 3 \end{bmatrix} \begin{Bmatrix} x_1 \\ x_2 \\ x_3 \end{Bmatrix} = \begin{Bmatrix} 0 \\ 0 \\ 0 \end{Bmatrix}$$

First we note that the second equation is not useful as it will be satisfied by any set of $\mathbf{x}$. The first equation clearly yields $x_1 = 0$. Substituting for x_1 in the third equation, we obtain $x_3 = 0$. We note that x_2 is arbitrary, and hence the eigen vector can be taken as $\mathbf{x}^{(2)} = (0, 1, 0)$.

Next, consider $\lambda = 5$. Following the same procedure as for the other eigen values, we obtain

$$\begin{bmatrix} -4 & 0 & 2 \\ 0 & -4 & 0 \\ 2 & 0 & -1 \end{bmatrix} \begin{Bmatrix} x_1 \\ x_2 \\ x_3 \end{Bmatrix} = \begin{Bmatrix} 0 \\ 0 \\ 0 \end{Bmatrix}$$

The solution can be derived as $x_2 = 0$ and $x_3 = 2x_1$. After normalizing, the eigen vector takes the form $\mathbf{x}^{(3)} = (1, 0, 2)/\sqrt{5}$.

It may be noted that the scalar product of any two eigen vectors of a symmetric matrix is equal to zero. That is,

$$\mathbf{x}^{(1)} \cdot \mathbf{x}^{(2)} = \mathbf{x}^{(2)} \cdot \mathbf{x}^{(3)} = \mathbf{x}^{(3)} \cdot \mathbf{x}^{(1)} = 0$$

Physically it means that the eigen vectors are orthogonal to each other. In geometric terms, the three directions represented by the eigen vectors are perpendicular to each other.

EXAMPLE 0.3   *Repeated Eigen Values*

Find the eigen values and eigen vectors of the 3×3 matrix $\mathbf{A}$ given below.

$$A = \begin{bmatrix} 9 & 4 & 0 \\ 4 & 3 & 0 \\ 0 & 0 & 1 \end{bmatrix}$$

SOLUTION   The characteristic equation is

$$\lambda^3 - 13\lambda^2 + 23\lambda - 11 = 0$$

The roots of the above cubic equation are $\lambda = 1$, 1, and 11. Thus, we note that the equation has repeated roots or the matrix has repeated eigen values. Let us first determine the eigen vector for $\lambda = 11$. The set of simultaneous equations are

$$\begin{bmatrix} -2 & 4 & 0 \\ 4 & -8 & 0 \\ 0 & 0 & -10 \end{bmatrix} \begin{Bmatrix} x_1 \\ x_2 \\ x_3 \end{Bmatrix} = \begin{Bmatrix} 0 \\ 0 \\ 0 \end{Bmatrix}$$

The solution for x_3 is uniquely obtained as $x_3 = 0$, and we also obtain $x_1 = 2x_2$. Thus we can write the eigen vector as $\mathbf{x}^{(1)} = (2, 1, 0)/\sqrt{5}$. Next we consider the repeating eigen value $\lambda_2 = \lambda_3 = 1$.

The set of simultaneous equations are

$$\begin{bmatrix} 8 & 4 & 0 \\ 4 & 2 & 0 \\ 0 & 0 & 0 \end{bmatrix} \begin{Bmatrix} x_1 \\ x_2 \\ x_3 \end{Bmatrix} = \begin{Bmatrix} 0 \\ 0 \\ 0 \end{Bmatrix}$$

We note that the third equation is not useful, and the first two equations are essentially the same, i.e., they are not linearly independent. The first (or the second) equation yields $x_2 = -2x_1$. We do not have any information to determine x_3, and hence it can be considered arbitrary. Thus, the eigen vector can be written as $\mathbf{x}^{(2)} = \mathbf{x}^{(3)} = (1, -2, \alpha)$, where α is an arbitrary number. Thus, there is an infinite number of eigen vectors. Such a situation arises whenever there are repeating eigen values. It may be noted that $\mathbf{x}^{(1)} \cdot \mathbf{x}^{(2)} = 0$ is satisfied for any value of α. Thus, we can state that any direction perpendicular to $\mathbf{x}^{(1)}$ is also an eigen vector.

0.5   QUADRATIC FORMS

The sum of products of variables x_i of the form

$$\begin{aligned} F \equiv a_{11}x_1^2 + a_{22}x_2^2 + \cdots + a_{nn}x_n^2 \\ + a_{12}x_1x_2 + a_{13}x_1x_3 + \cdots + a_{n,n-1}x_nx_{n-1} \end{aligned} \tag{0.50}$$

is called the *quadratic form* in $x_1, x_2, \ldots, x_n$, where a_{ij} are real constants. The quadratic form can be written in matrix form as

$$\begin{aligned} F &= \{\mathbf{x}\}^T[\mathbf{A}]\{\mathbf{x}\} \\ &= \{x_1, x_2, \ldots, x_n\} \begin{bmatrix} a_{11} & a_{12} & \cdots & a_{1n} \\ a_{21} & a_{22} & \cdots & a_{2n} \\ \vdots & \vdots & \ddots & \vdots \\ a_{n1} & a_{n2} & \cdots & a_{nn} \end{bmatrix} \begin{Bmatrix} x_1 \\ x_2 \\ \vdots \\ x_n \end{Bmatrix} \\ &= \sum_{i=1}^{n}\sum_{j=1}^{n} a_{ij}x_ix_j \end{aligned} \tag{0.51}$$

where $\{\mathbf{x}\} = \{x_1, x_2, \ldots, x_n\}^T$ and $[\mathbf{A}] = [a_{ij}]$ is the coefficient matrix of the quadratic form. The quadratic form appears often in engineering applications, such as strain energy of a solid or structure.

For a general $n \times n$ matrix $[\mathbf{B}]$ that is not necessarily symmetric, consider the quadratic form

$$F = \{\mathbf{x}\}^T[\mathbf{B}]\{\mathbf{x}\} \tag{0.52}$$

The matrix $[\mathbf{B}]$ can be decomposed into symmetric ($\mathbf{B}_S$) and skew-symmetric ($\mathbf{B}_A$) parts as

$$\begin{aligned} [\mathbf{B}] &= \frac{1}{2}[\mathbf{B} + \mathbf{B}^T] + \frac{1}{2}[\mathbf{B} - \mathbf{B}^T] \\ &= [\mathbf{B}_S] + [\mathbf{B}_A] \end{aligned} \tag{0.53}$$

The quadratic form using the symmetric part can be obtained as

$$F_S = \{\mathbf{x}\}^T[\mathbf{B}_S]\{\mathbf{x}\} \tag{0.54}$$

It can easily be shown that F and F_S are identical; i.e.,

$$F = \{\mathbf{x}\}^T[\mathbf{B}]\{\mathbf{x}\} = \{\mathbf{x}\}^T[\mathbf{B}_S]\{\mathbf{x}\} = F_S \tag{0.55}$$

The reader can show that the quadratic form of the skew-symmetric part is identically equal to zero. Thus, a non-symmetric matrix $[\mathbf{B}]$ in a quadratic form can always be replaced by the symmetric part of the matrix without affecting the value of the quadratic form.

0.5.1 Positive Definite Quadratic Form

If $\{\mathbf{x}\}^T[\mathbf{A}]\{\mathbf{x}\} \geq 0$ for all real vectors $\{\mathbf{x}\}$ and if $\{\mathbf{x}\}^T[\mathbf{A}]\{\mathbf{x}\} = 0$ only if $\{\mathbf{x}\} = \{\mathbf{0}\}$, then the quadratic form, hence the symmetric matrix $[\mathbf{A}]$, is said to be *positive definite*; i.e.,

$$\begin{aligned} \{\mathbf{x}\}^T[\mathbf{A}]\{\mathbf{x}\} &\geq 0, \quad \text{for all } \{\mathbf{x}\} \text{ in } R^n \\ \{\mathbf{x}\}^T[\mathbf{A}]\{\mathbf{x}\} &= 0, \quad \text{only if } \{\mathbf{x}\} = \{\mathbf{0}\} \end{aligned} \tag{0.56}$$

A quadratic form with the symmetric matrix $[\mathbf{A}]$ is said to be *positive semidefinite* when it takes on only nonnegative values for all values of the variables x but vanishes for some nonzero value of the variables; i.e.,

$$\begin{aligned} \{\mathbf{x}\}^T[\mathbf{A}]\{\mathbf{x}\} &\geq 0, \text{ for all } \{\mathbf{x}\} \text{ in } R^n \\ \{\mathbf{x}\}^T[\mathbf{A}]\{\mathbf{x}\} &= 0, \text{ for some } \{\mathbf{x}\} \neq \{\mathbf{0}\} \end{aligned} \tag{0.57}$$

Positive definiteness is an important property in structural analysis. When a matrix is positive definite, each column of the matrix is linearly independent, and as discussed in Section 0.3, the matrix can be inverted.

EXAMPLE 1.4 *Quadratic Form*

Consider the following quadratic form:

$$F(x,y) = \{x \quad y\}[\mathbf{A}]\begin{Bmatrix} x \\ y \end{Bmatrix} = \{x \quad y\}\begin{bmatrix} 1 & -1 \\ -1 & 2 \end{bmatrix}\begin{Bmatrix} x \\ y \end{Bmatrix}$$

F can be expanded as $F(x,y) = x^2 - 2xy + 2y^2 = (x-y)^2 + y^2$. Since F is the sum of two squared quantities, it is always positive, and $F = 0$ only if $x = y = 0$. Thus, $[\mathbf{A}]$ is positive definite.

Now consider different matrix $[\mathbf{A}]$ defined as

$$[\mathbf{A}] = \begin{bmatrix} 1 & -1 \\ -1 & 1 \end{bmatrix}$$

Then, $F(x,y) = x^2 - 2xy + y^2 = (x-y)^2$. One can note that F is always positive except when $x = y$. Hence $[\mathbf{A}]$ is positive semidefinite.

0.6 MAXIMA AND MINIMA OF FUNCTIONS

Consider the function $F(\mathbf{x})$ given by $F(\mathbf{x}) = \frac{1}{2}\{\mathbf{x}\}^T[\mathbf{A}]\{\mathbf{x}\} - \{\mathbf{x}\}^T\{\mathbf{b}\}$, where $\{\mathbf{x}\} = \{x_1, x_2, \ldots, x_n\}^T$ and $[\mathbf{A}]$ is a symmetric positive definite matrix of size $n \times n$. Then, we can show that the function $F(x) = F(x_1, \quad x_2, \ldots, x_n)$ has its minimum value at the point where $[\mathbf{A}]\{\mathbf{x}\} = \{\mathbf{b}\}$.

PROOF[9]

Let $\{\mathbf{y}\}$ be an arbitrary vector of size n. Consider the following expression:

$$F(\mathbf{y}) - F(\mathbf{x}) = \frac{1}{2}\{\mathbf{y}\}^T[\mathbf{A}]\{\mathbf{y}\} - \{\mathbf{y}\}^T\{\mathbf{b}\} - \frac{1}{2}\{\mathbf{x}\}^T[\mathbf{A}]\{\mathbf{x}\} + \{\mathbf{x}\}^T\{\mathbf{b}\} \tag{0.58}$$

---

[9] G. Strang, *Linear Algebra and its Applications,* Academic Press, New York, 1976.

Substituting $\{\mathbf{b}\} = [\mathbf{A}]\{\mathbf{x}\}$ in the above equation, we obtain

$$F(\mathbf{y}) - F(\mathbf{x}) = \frac{1}{2}\{\mathbf{y}\}^T[\mathbf{A}]\{\mathbf{y}\} - \{\mathbf{y}\}^T[\mathbf{A}]\{\mathbf{x}\} + \frac{1}{2}\{\mathbf{x}\}^T[\mathbf{A}]\{\mathbf{x}\}$$

$$= \frac{1}{2}\{\mathbf{y} - \mathbf{x}\}^T[\mathbf{A}]\{\mathbf{y} - \mathbf{x}\}$$

(0.59)

In deriving the above, we have used the relation $\{\mathbf{x}\}^T[\mathbf{A}]\{\mathbf{y}\} = \{\mathbf{y}\}^T[\mathbf{A}]\{\mathbf{x}\}$, which is due to the fact that $[\mathbf{A}]$ is symmetric, i.e., $[\mathbf{A}]^T = [\mathbf{A}]$. Since $[\mathbf{A}]$ is positive definite, the quantity on the RHS of the above equation is always positive except when $\{\mathbf{x}\} = \{\mathbf{y}\}$. That is, $F(\mathbf{y})$ is always greater than $F(\mathbf{x})$ except at $\{\mathbf{y}\} = \{\mathbf{x}\}$. Hence, the minimum value of F occurs at $\{\mathbf{x}\}$.

0.7   EXERCISE

1. Consider the following 3×3 matrix $[\mathbf{T}]$:

$$[\mathbf{T}] = \begin{bmatrix} 1 & 7 & 2 \\ 3 & 4 & 3 \\ 6 & 5 & 7 \end{bmatrix}$$

   (a) Write the transpose $\mathbf{T}^T$.
   (b) Show that the matrix $[\mathbf{S}] = [\mathbf{T}] + [\mathbf{T}]^T$ is a symmetric matrix.
   (c) Show that the matrix $[\mathbf{A}] = [\mathbf{T}] - [\mathbf{T}]^T$ is a skew-symmetric matrix. What are the diagonal components of the matrix $[\mathbf{A}]$?

2. Consider the following two 3×3 matrices $[\mathbf{A}]$ and $[\mathbf{B}]$:

$$[\mathbf{A}] = \begin{bmatrix} 1 & 7 & 2 \\ 3 & 4 & 3 \\ 6 & 5 & 7 \end{bmatrix}, \quad [\mathbf{B}] = \begin{bmatrix} 3 & 7 & 2 \\ 2 & 1 & 8 \\ 7 & 4 & 5 \end{bmatrix}$$

   (a) Calculate $[\mathbf{C}] = [\mathbf{A}] + [\mathbf{B}]$.
   (b) Calculate $[\mathbf{D}] = [\mathbf{A}] - [\mathbf{B}]$.
   (c) Calculate the scalar multiple $[\mathbf{D}] = 3[\mathbf{A}]$.

3. Consider the following two three-dimensional vectors $\mathbf{a}$ and $\mathbf{b}$:

$$\mathbf{a} = \{1, \ 4, \ 6\}^T \quad \text{and} \quad \mathbf{b} = \{4, \ 7, \ 2\}^T$$

   (a) Calculate the scalar product $c = \mathbf{a} \cdot \mathbf{b}$.
   (b) Calculate the norm of vector $\mathbf{a}$.
   (c) Calculate the vector product of $\mathbf{a}$ and $\mathbf{b}$.

4. For the matrix $[\mathbf{T}]$ in Problem 1 and the two vectors $\mathbf{a}$ and $\mathbf{b}$ in Problem 3, answer the following questions.

   (a) Calculate the product of the matrix-vector multiplication $[\mathbf{T}] \cdot \mathbf{a}$.
   (b) Calculate $\mathbf{b} \cdot [\mathbf{T}] \cdot \mathbf{a}$.

5. For the two matrices $[\mathbf{A}]$ and $[\mathbf{B}]$ in Problem 2, answer the following questions.

   (a) Evaluate the matrix-matrix multiplication $[\mathbf{C}] = [\mathbf{A}][\mathbf{B}]$.
   (b) Evaluate the matrix-matrix multiplication $[\mathbf{D}] = [\mathbf{B}][\mathbf{A}]$.

6. Calculate the determinant of the following matrices:

$$[\mathbf{A}] = \begin{bmatrix} 4 & 2 \\ 3 & 7 \end{bmatrix}, \quad [\mathbf{B}] = \begin{bmatrix} 1 & 3 & 2 \\ 1 & 4 & 5 \\ 2 & 6 & 7 \end{bmatrix}$$

7. Calculate the inverse of the matrix [A] in Problem 6.

8. Matrices [A] and [B] are defined below. If $\mathbf{B} = \mathbf{A}^{-1}$, determine the values of p, q, r, and s.

$$[\mathbf{A}] = \begin{bmatrix} 2 & 4 & p \\ 1 & 2 & 3 \\ q & 5 & 6 \end{bmatrix}, \quad [\mathbf{B}] = \begin{bmatrix} -3 & 1 & 2 \\ 3 & -3 & -1 \\ -1 & r & s \end{bmatrix}.$$

9. Solve the following simultaneous system of equations using the matrix method:

$$4x_1 + 3x_2 = 3$$
$$x_1 + 3x_2 = 3$$

10. Consider the following row vectors and matrices

$$\mathbf{a} = \begin{bmatrix} 5 & 12 & 3 \end{bmatrix} \quad \mathbf{b} = \begin{bmatrix} 1 & 0 & 4 \end{bmatrix}$$
$$\mathbf{A} = \begin{bmatrix} 1 & 6 & 3 \\ 2 & 8 & 4 \end{bmatrix} \quad \mathbf{B} = \begin{bmatrix} \cdot & 5 & 4 \\ 7 & 2 & 0 \end{bmatrix}$$

Using MATLAB, calculate $\mathbf{A}^T$, $\mathbf{a}^T$, $\mathbf{A} + \mathbf{B}$, $\mathbf{A} - \mathbf{B}$, $\mathbf{ab}^T$, $\mathbf{a}^T\mathbf{b}$, $\mathbf{A}^T\mathbf{B}$ $\mathbf{C} = \mathbf{BA}^T$, $\mathbf{AB}$, $\mathbf{C}^{-1}$, $\det(\mathbf{C})$. Test the commands $\mathbf{a.*b}$, $\mathbf{A.*B}$ and explain the difference between them and $\mathbf{a*b}$, $\mathbf{A*B}$, respectively.

11. Find the eigen values and eigen vectors of the following matrices:

(a) $[\mathbf{A}] = \begin{bmatrix} 5 & 2 & 0 \\ 2 & 2 & 0 \\ 0 & 0 & -1 \end{bmatrix}$

(b) $[\mathbf{B}] = \begin{bmatrix} 2 & 0 & 2 \\ 0 & 1 & 0 \\ 2 & 0 & 5 \end{bmatrix}$

12. Construct the quadratic form for the matrix [A] in Problem 2, i.e., $\{\mathbf{x}\}^T[\mathbf{A}]\{\mathbf{x}\}$, and compare with the quadratic form that is calculated using symmetric part $[\mathbf{A}_S]$.

13. Consider the matrix equation $[\mathbf{A}]\{\mathbf{x}\} = \{\mathbf{b}\}$ given by

$$\begin{bmatrix} 2 & -1 & 0 \\ -1 & 2 & -1 \\ 0 & -1 & 2 \end{bmatrix} \begin{Bmatrix} x_1 \\ x_2 \\ x_3 \end{Bmatrix} = \begin{Bmatrix} 4 \\ 0 \\ 4 \end{Bmatrix}$$

(a) Construct the quadratic form $F(\mathbf{x}) = \{\mathbf{x}\}^T[\mathbf{A}]\{\mathbf{x}\} - 2\{\mathbf{x}\}^T\{\mathbf{b}\}$.

(b) Find $\{\mathbf{x}\} = \{\mathbf{x}^*\}$ by minimizing $F(\mathbf{x})$.

(c) Verify that the vector $\{\mathbf{x}^*\}$ satisfies $[\mathbf{A}]\{\mathbf{x}\} = \{\mathbf{b}\}$.

14. A function $f(x_1, x_2)$ of two variables x_1 and x_2 is given by

$$f(x_1, x_2) = \frac{1}{2}\{x_1 \quad x_2\}\begin{bmatrix} 1 & -1 \\ -1 & 2 \end{bmatrix}\begin{Bmatrix} x_1 \\ x_2 \end{Bmatrix} - \{x_1 \quad x_2\}\begin{Bmatrix} 0 \\ 2 \end{Bmatrix} - 1$$

(a) Multiply the matrices and express f as a polynomial in x_1 and x_2.

(b) Determine the extreme (maximum or minimum) value of the function and corresponding x_1 and x_2.

(c) Is this a maxima or minima?

15. A function $f(x, y, z)$ of x, y, and z is defined as

$$f(x, y, z) = \frac{1}{2}\{x \quad y \quad z\}[\mathbf{K}]\begin{Bmatrix} x \\ y \\ z \end{Bmatrix} - \{x \quad y \quad z\}\{\mathbf{R}\} + 10$$

where

$$[\mathbf{K}] = \begin{bmatrix} 1 & 2 & 3 \\ 2 & 4 & 5 \\ 3 & 5 & 6 \end{bmatrix} \quad \text{and} \quad \{\mathbf{R}\} = \begin{Bmatrix} 1 \\ 2 \\ 1 \end{Bmatrix}$$

(a) Multiply the matrices and express f as a polynomial in x, y, and z.

(b) Write down the three equations necessary to find the extreme value of the function in the form

$$\underset{3\times3}{[\mathbf{A}]} \underset{3\times1}{\begin{Bmatrix} x \\ y \\ z \end{Bmatrix}} = \underset{3\times1}{\{\mathbf{b}\}} .$$

(c) Solve the equations in (b) to determine x, y, and z corresponding to the extreme value of f.

(d) Compute the extreme value of f.

(e) Is this a maxima or minima?

(f) Compute the determinant of $[\mathbf{K}]$.

Chapter 1

Stress-Strain Analysis

Stress analysis is a major step, and in fact it can be considered the most important one in the mechanical design process. There are many considerations that influence the design of a machine element or structure. The most important design considerations are the following:[1] (i) the stress at every point should be below a certain limit for the material; (ii) the deflection should not exceed the maximum allowable for proper functioning of the system; (iii) the structure should be stable; and (iv) the structure or machine element should not fail by fatigue. The failure mode corresponding to instability is also referred to as buckling. The failure due to excessive stress can take different forms, such as brittle fracture, yielding of the material causing inelastic deformations, and fatigue failure. Stress analysis of structures plays a crucial role in predicting failure types (i), (ii), and (iv) above. The stability analysis of a structure requires a slightly different treatment but in general uses most of the methods of stress analysis. Often, the results from the stress analysis can be used to predict buckling. Thus, stress is often used as a criterion for mechanical design.

In the elementary mechanics of materials or physics courses, stress is defined as force per unit area. While such a notion is useful and sufficient to analyze one-dimensional structures under a uniaxial state of stress, a complete understanding of the state of stress in a three-dimensional body requires a thorough understanding of the concept of stress at a point. Similarly, strain is defined as the change in length per original length of a one-dimensional body. However, the concept of strain at a point in a three-dimensional body is quite interesting and is required for complete understanding of deformation a solid undergoes. While stresses and strains are concepts developed by engineers for better understanding of the physics of deformation of a solid, the relation between stresses and strains is phenomenological in the sense that it is something observed and described as a simplified theory. Robert Hooke[2] was the first to establish the linear relation between stresses and strains in an elastic body. Although he explained his theory for one-dimensional objects, later his theory became the generalized Hooke's law that relates the stresses and strains in three-dimensional elastic bodies.

1.1 STRESS

1.1.1 Surface Traction

Consider a solid subjected to external forces and in static equilibrium, as shown in Figure 1.1. We are interested in the state of stress at a point P in the interior of the solid. We cut the body into two halves by passing an imaginary plane through P. The unit vector normal

---

[1] J. E. Shigley, C.R. Mischke, and R.G. Budynas, *Mechanical Engineering Design*, McGraw-Hill, New York, 2004

[2] R. Hooke, *De Potentia Restitutiva*, London, 1678

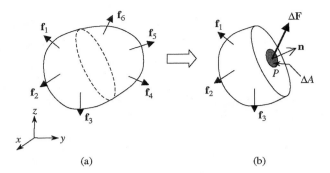

Figure 1.1 Surface traction acting on a plane at a point

(a)   (b)

to the plane is denoted by **n** [see Figure 1.1(b)]. The portion of the body on the left side is in equilibrium because of the external forces $\mathbf{f}_1$, $\mathbf{f}_2$, and $\mathbf{f}_3$ and also the internal forces acting on the cut surface. "Surface traction" is defined as the internal force per unit area or the force intensity acting on the cut plane. In order to measure the intensity or traction specifically at P, we consider the force $\Delta\mathbf{F}$ acting over a small area ΔA that contains point P. Then the surface traction $\mathbf{T}^{(\mathbf{n})}$ acting at the point P is defined as

$$\mathbf{T}^{(\mathbf{n})} = \lim_{\Delta A \to 0} \frac{\Delta\mathbf{F}}{\Delta A} \tag{1.1}$$

In Eq. (1.1), the right superscript $(\mathbf{n})$ is used to denote the fact that this surface traction is defined on a plane whose normal is **n**. It should be noted that at the same point P, the traction vector **T** would be different on a different plane passing through P. It is clear from Eq. (1.1) that the dimension of the traction vector is the same as that of pressure, force per unit area.

Since $\mathbf{T}^{(\mathbf{n})}$ is a vector, one can resolve it into components and write it as

$$\mathbf{T}^{(\mathbf{n})} = T_x\mathbf{i} + T_y\mathbf{j} + T_z\mathbf{k} \tag{1.2}$$

and its magnitude can be computed from

$$\|\mathbf{T}^{(\mathbf{n})}\| = T = \sqrt{T_x^2 + T_y^2 + T_z^2} \tag{1.3}$$

EXAMPLE 1.1 *Stress in an Inclined Plane*

Consider a uniaxial bar with the cross-sectional area $A = 2 \times 10^{-4}\,\mathrm{m}^2$, as shown in Figure 1.2. If an axial force $F = 100\,\mathrm{N}$ is applied to the bar, determine the surface traction on the plane whose normal is at an angle θ from the axial direction.

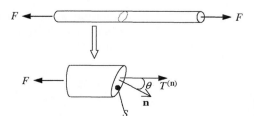

Figure 1.2 Equilibrium of a uniaxial bar under axial force

SOLUTION To simplify the analysis, let us assume that the traction on the plane is uniform, i.e., the stresses are equally distributed over the cross-section of the bar. In fact, this is the fundamental assumption in the analysis of uniaxial bars. The force on the inclined plane S can be obtained by integrating the constant surface traction $T^{(\mathbf{n})}$ over the plane S. In this simple example, direction of the surface traction $T^{(\mathbf{n})}$ must be opposite that of the force F. Since the member is in static equilibrium, the integral of the surface traction must be equal to the magnitude of the force F.

$$F = \iint_S T^{(\mathrm{n})}\, dS = T \iint_S dS = T \frac{A}{\cos\theta}$$

$$\therefore T = \frac{F}{A}\cos\theta = 5 \times 10^5 \cos\theta \, \frac{\mathrm{N}}{m^2} = 5 \times 10^5 \cos\theta \, \mathrm{Pa}$$

Note that the unit of traction is Pascal (Pa or N/m^2). It is clear that the surface traction depends on the direction of the normal to the plane.

1.1.2 Normal Stresses and Shear Stresses

The surface traction $\mathbf{T}^{(\mathbf{n})}$ defined by Eq. (1.1) does not act in general in the direction of $\mathbf{n}$; i.e., $\mathbf{T}$ and $\mathbf{n}$ are not necessarily parallel to each other. Thus, we can decompose the surface traction into two components, one parallel to $\mathbf{n}$ and the other perpendicular to $\mathbf{n}$, which will lie on the plane. The component normal to the plane or parallel to $\mathbf{n}$ is called the normal stress and denoted by σ_n. The other component parallel to the plane or perpendicular to $\mathbf{n}$ is called the shear stress and is denoted by τ_n.

The normal stress can be obtained from the scalar product of $\mathbf{T}^{(\mathbf{n})}$ and $\mathbf{n}$, as (see Figure 1.3)

$$\sigma_n = \mathbf{T}^{(\mathbf{n})} \cdot \mathbf{n} \tag{1.4}$$

and shear stress can be calculated from the relation

$$\tau_n = \sqrt{\left\|\mathbf{T}^{(\mathbf{n})}\right\|^2 - \sigma_n^2} \tag{1.5}$$

The angle between $\mathbf{T}^{(\mathbf{n})}$ and $\mathbf{n}$ can be obtained from the definition of scalar product given in Eq. (0.18)

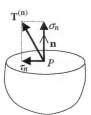

Figure 1.3 Normal and shear stresses at a point P

EXAMPLE 1.2 *Normal and Shear Stresses*

The surface traction at a point is $\mathbf{T}^{(\mathbf{n})} = \{3,\ 4,\ 5\}^T$ on a plane whose normal vector is parallel to the z-axis. Calculate the normal and shear stresses on this plane. What is the angle between $\mathbf{T}^{(\mathbf{n})}$ and $\mathbf{n}$?

SOLUTION   Note that a direction parallel to the z-axis is given by $\mathbf{n} = \{0, 0, 1\}^T$.

$$\sigma_n = \mathbf{T}^{(\mathbf{n})} \cdot \mathbf{n} = 3 \times 0 + 4 \times 0 + 5 \times 1 = 5$$

$$\left\| \mathbf{T}^{(\mathbf{n})} \right\|^2 = 3^2 + 4^2 + 5^2 = 50$$

$$\tau_n = \sqrt{50 - 5^2} = 5$$

The angle θ between $\mathbf{T}^{(\mathbf{n})}$ and $\mathbf{n}$ can be calculated using the relation $\sigma_n = \mathbf{T}^{(\mathbf{n})} \cdot \mathbf{n} = T \cos \theta$. Thus,

$$\theta = \cos^{-1}(\sigma_n / T) = \cos^{-1}(5 / \sqrt{50}) = 45 \text{ deg}.$$

1.1.3 Rectangular or Cartesian Stress Components

Since the surface traction at a point varies depending on the direction of the normal to the plane, one can obtain an infinite number of traction vectors $\mathbf{T}^{(\mathbf{n})}$ and corresponding normal and shear stresses for a given state of stress at a point. However, one might be interested in the maximum values of these stresses and the corresponding plane. Fortunately, the state of stress at a point can be completely characterized by defining the traction vectors on three mutually perpendicular planes passing through the point. That is, from the knowledge of $\mathbf{T}^{(\mathbf{n})}$ acting on three orthogonal planes, one can determine $\mathbf{T}^{(\mathbf{n})}$ on any arbitrary plane passing through the same point. For convenience, these planes are taken as the three planes that are normal to the x, y, and z axes.

Let us denote the traction vector on the yz-plane, which is normal to the x-axis, as $\mathbf{T}^{(x)}$. Instead of decomposing the traction vector into its normal and shear components, we will use its components parallel to the coordinate directions and denote them as $T_x^{(x)}$, $T_y^{(x)}$, and $T_z^{(x)}$. That is,

$$\mathbf{T}^{(x)} = T_x^{(x)} \mathbf{i} + T_y^{(x)} \mathbf{j} + T_z^{(x)} \mathbf{k} \tag{1.6}$$

It may be noted that $T_x^{(x)}$ in Eq. (1.6) is the normal stress, and $T_y^{(x)}$ and $T_z^{(x)}$ are the shear stresses in the y- and z-directions, respectively. In contemporary solid mechanics, the stress components in Eq. (1.6) are denoted by σ_{xx}, τ_{xy}, and τ_{xz}, where σ_{xx} is the normal stress and τ_{xy} and τ_{xz} are components of shear stress. In this notation, the first subscript denotes the plane on which the stress component acts—in this case, the plane normal to the x-axis or simply the x-plane—and the second subscript denotes the direction of the stress component. We can repeat this exercise by passing two more planes, normal to y- and z-axes, respectively, through the point P. Thus, the surface tractions acting on the plane normal to y will be τ_{yx}, σ_{yy} and τ_{yz}. The stresses acting on the z-plane can be written as τ_{zx}, τ_{zy} and σ_{zz}.

The stress components acting on the three planes can be depicted using a cube, as shown in Figure 1.4. It must be noted that this cube is not a physical cube and hence has

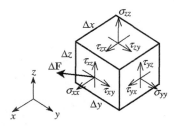

Figure 1.4 Stress components in Cartesian coordinate system

Table 1.1   Description of Stress Components

Stress component	Description
σ_{xx}	Normal stress on the x face in the x direction
σ_{yy}	Normal stress on the y face in the y direction
σ_{zz}	Normal stress on the z face in the z direction
τ_{xy}	Shear stress on the x face in the y direction
τ_{yx}	Shear stress on the y face in the x direction
τ_{yz}	Shear stress on the y face in the z direction
τ_{zy}	Shear stress on the z face in the y direction
τ_{xz}	Shear stress on the x face in the z direction
τ_{zx}	Shear stress on the z face in the x direction

no dimensions. The six faces of the cube represent the three pairs of planes normal to the coordinate axes. The top face, for example, is the $+z$ plane, and then the bottom face is the $-z$ plane or whose normal is in the $-z$ direction. Note that the three visible faces of the cube in Figure 1.4 represent the three positive planes, i.e., planes whose normal are the positive x-, y-, and z-axes. On these faces, all tractions are shown in the positive direction. For example, τ_{yz} is the traction on the y-plane acting in the positive z direction. The description of all stress components is summarized in Table 1.1.

Knowledge of the nine stress components is necessary to determine the components of the surface traction $\mathbf{T}^{(\mathbf{n})}$ acting on an arbitrary plane with normal $\mathbf{n}$.

Stresses are second-order tensors, and their sign convention is different from that of regular force vectors. Stress components, in addition to the direction of the force, contain information of the surface on which they are defined. A stress component is positive when both the surface normal and the stress component are either in the positive or in the negative coordinate direction. If the surface normal is in the positive direction and the stress component is in the negative direction, then the stress component has a negative sign.

Normal stress is positive when it is a tensile stress and negative when it is compressive. Shear stress acting on the positive face is positive when it is acting in the positive coordinate direction. The positive directions of all the stress components are shown in Figure 1.4.

1.1.4   Traction on an Arbitrary Plane Through a Point

If the components of stress at a point, say P, are known, it is possible to determine the surface traction acting on any plane passing through that point. Let $\mathbf{n}$ be the unit normal to the plane on which we want to determine the surface traction. The normal vector can be represented as

$$\mathbf{n} = n_x\mathbf{i} + n_y\mathbf{j} + n_z\mathbf{k} = \left\{ \begin{array}{c} n_x \\ n_y \\ n_z \end{array} \right\} \tag{1.7}$$

For convenience, we choose P as the origin of the coordinate system, as shown in Figure 1.5, and consider a plane parallel to the intended plane but passing at an infinitesimally small distance h away from P. Note that the normal to the face ABC is also $\mathbf{n}$. We will calculate the tractions on this plane first and then take the limit, as h approaches zero. We will consider the equilibrium of the tetrahedron PABC. If A is the area of the triangle ABC,

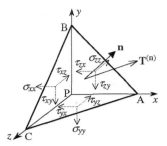

Figure 1.5 Surface traction and stress components acting on faces of an infinitesimal tetrahedron at a given point P

then the areas of triangles PAB, PBC, and PAC are given by An_z, An_x, An_y, respectively. Let $\mathbf{T}^{(\mathbf{n})} = T_x^{(\mathbf{n})}\mathbf{i} + T_y^{(\mathbf{n})}\mathbf{j} + T_z^{(\mathbf{n})}\mathbf{k}$ be the surface traction acting on the face ABC.

From the definition of surface traction in Eq. (1.1), the force on the surface can be calculated by multiplying the stresses with the surface area. Since the tetrahedron should be in equilibrium, the sum of the forces acting on its surfaces should be equal to zero. Force balance in the x-direction yields

$$\sum F_x = T_x^{(\mathbf{n})}A - \sigma_{xx}An_x - \tau_{yx}An_y - \tau_{zx}An_z = 0$$

In the above equation, we have assumed that the stresses acting on a surface are uniform. This will not be true if the size of the tetrahedron is not small. However, the tetrahedron is infinitesimally small, which is the case as h approaches zero. Dividing the above equation by A, we obtain the following relation:

$$T_x^{(\mathbf{n})} = \sigma_{xx}n_x + \tau_{yx}n_y + \tau_{zx}n_z \tag{1.8}$$

Similarly, force balance in the y- and z-directions yields

$$\begin{aligned} T_y^{(\mathbf{n})} &= \tau_{xy}n_x + \sigma_{yy}n_y + \tau_{zy}n_z \\ T_z^{(\mathbf{n})} &= \tau_{xz}n_x + \tau_{yz}n_y + \sigma_{zz}n_z \end{aligned} \tag{1.9}$$

From Eqs. (1.8) and (1.9), it is clear that the surface traction acting on the surface whose normal is $\mathbf{n}$ can be determined if the nine stress components are available. Using matrix notation, we can write these equations as

$$\mathbf{T}^{(\mathbf{n})} = [\boldsymbol{\sigma}] \cdot \mathbf{n} \tag{1.10}$$

where

$$[\boldsymbol{\sigma}] = \begin{bmatrix} \sigma_{xx} & \tau_{yx} & \tau_{zx} \\ \tau_{xy} & \sigma_{yy} & \tau_{zy} \\ \tau_{xz} & \tau_{yz} & \sigma_{zz} \end{bmatrix} \tag{1.11}$$

$[\boldsymbol{\sigma}]$ is called the stress matrix and it completely characterizes the state of stress at a given point.

EXAMPLE 1.3   *Normal and Shear Stresses on a Plane*

The state of stress at a particular point in the xyz coordinate system is given by the following stress matrix:

$$[\boldsymbol{\sigma}] = \begin{bmatrix} 3 & 7 & -7 \\ 7 & 4 & 0 \\ -7 & 0 & 2 \end{bmatrix}$$

Determine the normal and shear stresses on a surface passing through the point and parallel to the plane given by the equation $4x - 4y + 2z = 2$.

SOLUTION   To determine the surface traction $\mathbf{T}^{(n)}$, it is necessary to determine the unit normal vector to the plane. From solid geometry, the normal to the plane is found to be in the direction $\mathbf{d} = \{4, -4, 2\}^T$ and $\|\mathbf{d}\| = 6$. Thus, the unit normal vector becomes

$$\mathbf{n} = \left\{ \frac{2}{3}, -\frac{2}{3}, \frac{1}{3} \right\}^T$$

The surface traction can be obtained as

$$\mathbf{T}^{(n)} = [\boldsymbol{\sigma}] \cdot \mathbf{n} = \frac{1}{3}\begin{bmatrix} 3 & 7 & -7 \\ 7 & 4 & 0 \\ -7 & 0 & 2 \end{bmatrix} \cdot \left\{ \begin{array}{c} 2 \\ -2 \\ 1 \end{array} \right\} = \left\{ \begin{array}{c} -5 \\ 2 \\ -4 \end{array} \right\}$$

Remaining calculations are similar to those of Example 1.2:

$$\sigma_n = \mathbf{T}^{(n)} \cdot \mathbf{n} = -5 \times \frac{2}{3} - 2 \times \frac{2}{3} - 4 \times \frac{1}{3} = -6$$

$$\left\| \mathbf{T}^{(n)} \right\|^2 = 5^2 + 2^2 + 4^2 = 45$$

$$\tau_n = \sqrt{\left\| \mathbf{T}^{(n)} \right\|^2 - 6^2} = 3$$

1.1.5   Symmetry of Stress Matrix and Vector Notation

The nine components of the stress matrix can be reduced to six using the symmetry property of the stress matrix. Figure 1.6 shows an infinitesimal portion ($\Delta l \times \Delta l$) of a solid with shear stresses acting on its surface. The dimension in the z-direction is taken as unity. The direction of the shear stress τ_{xy} on surface BD is in the positive y-direction, while on surface AC it is in the negative y-direction. As the body is in static equilibrium, the sum of the moments about the z-axis must be equal to zero, which implies that the shear stresses τ_{xy} and τ_{yx} must be equal to each other, as shown below.

$$\sum M_z = \Delta l(\tau_{xy} - \tau_{yx}) = 0$$
$$\Rightarrow \tau_{xy} = \tau_{yx}$$

Similarly, we can derive the following relations:

$$\tau_{yz} = \tau_{zy}$$
$$\tau_{xz} = \tau_{zx}$$

(1.12)

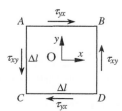

Figure 1.6 Equilibrium of a square element subjected to shear stresses

Thus, we need only six components to fully represent the stress at a point. In some occasions, stress at a point is written as a 6×1 pseudo vector, as shown below:

$$\{\boldsymbol{\sigma}\} = \begin{Bmatrix} \sigma_{xx} \\ \sigma_{yy} \\ \sigma_{zz} \\ \tau_{yz} \\ \tau_{zx} \\ \tau_{xy} \end{Bmatrix} \tag{1.13}$$

Some textbooks use a single subscript for normal stresses when the stress is written in a vector form.

1.1.6 Principal Stresses

As shown in the previous section, the normal and shear stresses acting on a plane passing through a given point in a solid change as the orientation of the plane is changed. Then a natural question is: Is there a plane on which the normal stress is the maximum? Similarly, we would also like to find the plane on which the shear stress attains a maximum. These questions are not only academic but also have significance in predicting the failure of the material at that point. In the following, we will provide some answers to the above questions without furnishing the proofs. The interested reader is referred to books on continuum mechanics, e.g., L.E. Malvern[3] or advanced solid mechanics,[4] for a more detailed treatment of the subject.

It can be shown that at every point in a solid there are at least three mutually perpendicular planes on which the normal stress attains extremum (maximum or minimum) values. On all these planes, the shear stresses vanish. Thus, the traction vector $\mathbf{T}^{(\mathbf{n})}$ will be parallel to the normal vector $\mathbf{n}$, i.e., $\mathbf{T}^{(\mathbf{n})} = \sigma_n \mathbf{n}$, on these planes. Of these three planes, one plane corresponds to the global maximum value of the normal stress and the other to the global minimum. The third plane will carry the intermediate normal stress. These special normal stresses are called the *principal stresses* at that point; the planes on which they act are the principal stress planes and the corresponding normal vectors are the principal stress directions. The principal stresses are denoted by σ_1, σ_2, and σ_3 such that $\sigma_1 \geq \sigma_2 \geq \sigma_3$.

Based on the above observations, the principal stresses can be calculated as follows. When the normal direction to a plane is the principal direction, the surface normal and the surface traction are in the same direction ($\mathbf{T}^{(\mathbf{n})} \| \mathbf{n}$). Thus, the surface traction on a plane can be represented by the product of the normal stress σ_n and the normal vector $\mathbf{n}$, as

$$\mathbf{T}^{(\mathbf{n})} = \sigma_n \mathbf{n} \tag{1.14}$$

Combining Eq. (1.14) with Eq. (1.10) for the surface traction, we obtain

$$[\boldsymbol{\sigma}] \cdot \mathbf{n} = \sigma_n \mathbf{n} \tag{1.15}$$

---

[3] L. E. Malvern, *Introduction to the Mechanics of a Continuous Medium*, Prentice-Hall, Englewood Cliffs, New Jersey, 1969

[4] Boresi et al., *Advanced Mechanics of Materials*, John Wiley & Sons, New York, 2003

Equation (1.15) represents the eigen value problem, where σ_n is the eigen value and $\mathbf{n}$ is the corresponding eigen vector (see Section 0.4 and Eq. 0.40). Equation (1.15) can be rearranged as

$$([\boldsymbol{\sigma}] - \sigma_n[\mathbf{I}]) \cdot \mathbf{n} = 0 \tag{1.16}$$

where $[\mathbf{I}]$ is a 3×3 identity matrix. In the component form, the above equation can be written as

$$\begin{bmatrix} \sigma_{xx} - \sigma_n & \tau_{yx} & \tau_{zx} \\ \tau_{xy} & \sigma_{yy} - \sigma_n & \tau_{zy} \\ \tau_{xz} & \tau_{yz} & \sigma_{zz} - \sigma_n \end{bmatrix} \begin{Bmatrix} n_x \\ n_y \\ n_z \end{Bmatrix} = \begin{Bmatrix} 0 \\ 0 \\ 0 \end{Bmatrix} \tag{1.17}$$

Note that $\mathbf{n} = 0$ satisfies the above equation, which is not only a trivial solution but also physically not possible, as $\|\mathbf{n}\|$ must be equal to unity. The above set of linear simultaneous equations will have a non-trivial, physically meaningful solution if and only if the determinant of the coefficient matrix is zero, i.e.,

$$\begin{vmatrix} \sigma_{xx} - \sigma_n & \tau_{yx} & \tau_{zx} \\ \tau_{xy} & \sigma_{yy} - \sigma_n & \tau_{zy} \\ \tau_{xz} & \tau_{yz} & \sigma_{zz} - \sigma_n \end{vmatrix} = 0 \tag{1.18}$$

Expanding this determinant, we obtain the following cubic equation in σ_n:

$$\sigma_n^3 - I_1 \sigma_n^2 + I_2 \sigma_n - I_3 = 0 \tag{1.19}$$

where

$$I_1 = \sigma_{xx} + \sigma_{yy} + \sigma_{zz}$$

$$I_2 = \begin{vmatrix} \sigma_{xx} & \tau_{xy} \\ \tau_{xy} & \sigma_{yy} \end{vmatrix} + \begin{vmatrix} \sigma_{yy} & \tau_{yz} \\ \tau_{yz} & \sigma_{zz} \end{vmatrix} + \begin{vmatrix} \sigma_{xx} & \tau_{zx} \\ \tau_{zx} & \sigma_{zz} \end{vmatrix} \tag{1.20}$$

$$= \sigma_{xx}\sigma_{yy} + \sigma_{yy}\sigma_{zz} + \sigma_{zz}\sigma_{xx} - \tau_{xy}^2 - \tau_{yz}^2 - \tau_{zx}^2$$

$$I_3 = |\sigma| = \sigma_{xx}\sigma_{yy}\sigma_{zz} + 2\tau_{xy}\tau_{yz}\tau_{zx} - \sigma_{xx}\tau_{yz}^2 - \sigma_{yy}\tau_{zx}^2 - \sigma_{zz}\tau_{xy}^2$$

In the above equation, I_1, I_2, and I_3 are the three invariants of the stress matrix $[\sigma]$, which can be shown to be independent of the coordinate system. The three roots of the cubic equation (1.19) correspond to the three principal stresses. We will denote them by σ_1, σ_2, and σ_3 in the order of $\sigma_1 \geq \sigma_2 \geq \sigma_3$. A method to solve the cubic equation was described in Chapter 0.

EXAMPLE 1.4   *Principal Stresses*

For the stress matrix given below, determine the principal stresses.

$$[\boldsymbol{\sigma}] = \begin{bmatrix} 3 & 1 & 1 \\ 1 & 0 & 2 \\ 1 & 2 & 0 \end{bmatrix}$$

SOLUTION   Setting the determinant of the coefficient matrix to zero [see Eq. (1.18)] yields

$$\begin{vmatrix} 3 - \sigma_n & 1 & 1 \\ 1 & -\sigma_n & 2 \\ 1 & 2 & -\sigma_n \end{vmatrix} = 0$$

By expanding the determinant, we obtain

$$(3 - \sigma_n)(\sigma_n^2 - 4) - (-\sigma_n - 2) + (2 + \sigma_n) = -(\sigma_n + 2)(\sigma_n - 1)(\sigma_n - 4) = 0$$

Three roots of the above equation are the principal stresses. They are

$$\sigma_n = 4, \quad \sigma_2 = 1, \quad \sigma_3 = -2$$

1.1.7   Principal Directions

Once the principal stresses have been computed, we can substitute them one at a time into Eq. (1.17) to obtain the linear system of equations that have three unknowns, n_x, n_y, and n_z. Corresponding to each principal value, we will get a principal direction that will be denoted as $\mathbf{n}^1$, $\mathbf{n}^2$, and $\mathbf{n}^3$. For example, if σ_n is replaced by the first principal stress σ_1, then we obtain

$$\begin{bmatrix} \sigma_{xx} - \sigma_1 & \tau_{yx} & \tau_{zx} \\ \tau_{xy} & \sigma_{yy} - \sigma_1 & \tau_{zy} \\ \tau_{xz} & \tau_{yz} & \sigma_{zz} - \sigma_1 \end{bmatrix} \begin{Bmatrix} n_x^1 \\ n_y^1 \\ n_z^1 \end{Bmatrix} = \begin{Bmatrix} 0 \\ 0 \\ 0 \end{Bmatrix} \tag{1.21}$$

In Eq. (1.21), n_x^1, n_y^1, and n_z^1 are components of $\mathbf{n}^1$. The above equations are three linear simultaneous equations in three unknowns. However, since the determinant of the matrix is zero (i.e., the matrix is singular), they are not independent. Thus, an infinite number of solutions exist. We need one more equation to find a unique value for the principal directions $\mathbf{n}^i$. Note that $\mathbf{n}$ is a unit vector, and hence its components must satisfy the following relation:

$$\|\mathbf{n}^i\|^2 = (n_x^i)^2 + (n_y^i)^2 + (n_z^i)^2 = 1, \quad i = 1, 2, 3 \tag{1.22}$$

It can be shown that the planes on which the principal stresses act are mutually perpendicular. Let us consider any two principal directions $\mathbf{n}^i$ and $\mathbf{n}^j$, with $i \neq j$. If σ_i and σ_j are the corresponding principal stresses, then they satisfy the following equations:

$$[\sigma] \cdot \mathbf{n}^i = \sigma_i \mathbf{n}^i$$
$$[\sigma] \cdot \mathbf{n}^j = \sigma_j \mathbf{n}^j \tag{1.23}$$

Multiplying the first equation by $\mathbf{n}^j$ and the second equation by $\mathbf{n}^i$, we obtain

$$\mathbf{n}^j \cdot [\sigma] \cdot \mathbf{n}^i = \sigma_i \mathbf{n}^j \cdot \mathbf{n}^i$$
$$\mathbf{n}^i \cdot [\sigma] \cdot \mathbf{n}^j = \sigma_j \mathbf{n}^i \cdot \mathbf{n}^j \tag{1.24}$$

Considering the symmetry of $[\sigma]$, i.e., $[\sigma] = [\sigma]^T$, and the rule for transpose of matrix products [Eq. (0.35)], one can show that $\mathbf{n}^j \cdot [\sigma] \cdot \mathbf{n}^i = \mathbf{n}^i \cdot [\sigma] \cdot \mathbf{n}^j$. Then, subtracting the first equation from the second in Eq. (1.24), we obtain

$$(\sigma_i - \sigma_j)\mathbf{n}^i \cdot \mathbf{n}^j = 0 \tag{1.25}$$

This implies that if the principal stresses are distinct, i.e., $\sigma_i \neq \sigma_j$, then

$$\mathbf{n}^i \cdot \mathbf{n}^j = 0 \tag{1.26}$$

which means that $\mathbf{n}^i$ and $\mathbf{n}^j$ are orthogonal. The three planes on which the principal stresses act are mutually perpendicular.

There are three different possibilities for principal stresses and directions:

(a) σ_1, σ_2, and σ_3 are distinct $\Rightarrow$ principal stress directions are three unique mutually orthogonal unit vectors.

(b) $\sigma_1 = \sigma_2 \neq \sigma_3 \Rightarrow \mathbf{n}^3$ is a unique principal stress direction, and any two orthogonal directions on the plane that is perpendicular to $\mathbf{n}^3$ are the other principal directions.

(c) $\sigma_1 = \sigma_2 = \sigma_3 \Rightarrow$ any three orthogonal directions are principal stress directions. This state of stress is called hydrostatic or isotropic state of stress.

EXAMPLE 1.5 *Principal Directions*

Calculate the principal direction corresponding to $\sigma_3 = -2$ in Example 1.4.

SOLUTION By substituting $\sigma_n = -2$ into Eq. (1.17), we obtain the following simultaneous equations:

$$5n_x + n_y + n_z = 0$$

$$n_x + 2n_y + 2n_z = 0$$

$$n_x + 2n_y + 2n_z = 0$$

We note that the three equations are not independent. In fact, the second and the third equations are identical. From the first two equations we obtain the following ratios between components:

$$n_x : n_y : n_z = 0 : -1 : 1$$

A unique solution can be obtained using Eq. (1.22) as

$$\mathbf{n}^{(3)} = \frac{1}{\sqrt{2}} \left\{ \begin{array}{c} 0 \\ -1 \\ 1 \end{array} \right\}$$

1.1.8 Transformation of Stress

Let the state of stress at a point be given by the stress matrix $[\boldsymbol{\sigma}]_{xyz}$, where the components are expressed with reference to the xyz coordinate as shown in Figure 1.7. The question is how the components will look like in a different coordinate system, say $x'y'z'$, or how to determine $[\boldsymbol{\sigma}]_{x'y'z'}$. It will be shown that the stress matrix $[\boldsymbol{\sigma}]_{x'y'z'}$ can be obtained by rotating the stress matrix $[\boldsymbol{\sigma}]_{xyz}$ using a transformation matrix of direction cosines.

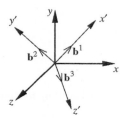

Figure 1.7 Coordinate transformation of stress

Let us define the $x'y'z'$ coordinate system whose unit vectors $\mathbf{b}^1$, $\mathbf{b}^2$, and $\mathbf{b}^3$, respectively, are given in the xyz coordinate system as

$$\mathbf{b}^1 = \left\{ \begin{array}{c} b_1^1 \\ b_2^1 \\ b_3^1 \end{array} \right\}, \ \mathbf{b}^2 = \left\{ \begin{array}{c} b_1^2 \\ b_2^2 \\ b_3^2 \end{array} \right\}, \ \mathbf{b}^3 = \left\{ \begin{array}{c} b_1^3 \\ b_2^3 \\ b_3^3 \end{array} \right\}$$

For example, $\mathbf{b}^1 = \{1, 0, 0\}^T$ in the $x'y'z'$ coordinate system, while $\mathbf{b}^1 = \{b_1^1, b_2^1, b_3^1\}$ in the xyz coordinates. Using these vectors, let us define a transformation matrix as

$$[\mathbf{N}] = [\, \mathbf{b}^1 \quad \mathbf{b}^2 \quad \mathbf{b}^3 \,] = \begin{bmatrix} b_1^1 & b_1^2 & b_1^3 \\ b_2^1 & b_2^2 & b_2^3 \\ b_3^1 & b_3^2 & b_3^3 \end{bmatrix} \tag{1.27}$$

The matrix $[\mathbf{N}]$ transforms a vector in the $x'y'z'$ coordinates into a vector in the xyz coordinates, while its transpose $[\mathbf{N}]^T$ transforms a vector in the xyz coordinates into a vector in $x'y'z'$ coordinates. For example, the unit vector $\mathbf{b}_{x'y'z'}^1 = \{1, 0, 0\}^T$ in $x'y'z'$ coordinates can be represented in the xyz coordinates as

$$\mathbf{b}_{xyz}^1 = [\mathbf{N}] \cdot \mathbf{b}_{x'y'z'}^1 = \left\{ \begin{array}{c} b_1^1 \\ b_2^1 \\ b_3^1 \end{array} \right\} \tag{1.28}$$

Transformation of a stress matrix is more complicated than transformation of a vector. We will perform the transformation in two steps. First we will determine the traction vectors on the three planes normal to the x'-, y'-, and z'-axes. This can be accomplished by multiplying the stress matrix and the corresponding direction cosine vector as described in Eq. (1.10). The three sets of traction vectors are written as columns of a square matrix, as shown below:

$$[\, \mathbf{T}^{(\mathbf{b}^1)} \quad \mathbf{T}^{(\mathbf{b}^2)} \quad \mathbf{T}^{(\mathbf{b}^3)} \,]_{xyz} = [\sigma]_{xyz}[\, \mathbf{b}^1 \quad \mathbf{b}^2 \quad \mathbf{b}^3 \,] = [\sigma]_{xyz}[\mathbf{N}] \tag{1.29}$$

Equation (1.29) represents the three surface traction vectors on planes perpendicular to $\mathbf{b}^1$, $\mathbf{b}^2$, and $\mathbf{b}^3$ in the xyz coordinates system. In the next step, we would like to transform the traction vectors to the $x'y'z'$ coordinate system. This transformation can be accomplished by multiplying $[\mathbf{N}]^T$ in front of Eq. (1.29). It may be recognized that the traction vectors $\mathbf{T}^{(\mathbf{b}^1)}$, $\mathbf{T}^{(\mathbf{b}^2)}$, and $\mathbf{T}^{(\mathbf{b}^3)}$ represented in the $x'y'z'$ coordinate system will be the transformed stress matrix. Thus, the stress matrix in the new coordinate system can be obtained by

$$\boxed{[\boldsymbol{\sigma}]_{x'y'z'} = [\mathbf{N}]^T[\sigma]_{xyz}[\mathbf{N}]} \tag{1.30}$$

EXAMPLE 1.6   *Coordinate Transformation*

The state of stress at a point in the xyz coordinates is

$$[\boldsymbol{\sigma}] = \begin{bmatrix} 2 & 1 & 0 \\ 1 & 2 & 0 \\ 0 & 0 & 2 \end{bmatrix}$$

Determine the stress matrix relative to the $x'y'z'$ coordinates, which is obtained by rotating the xyz coordinates by 45° about the z–axis, as shown in Figure 1.8.

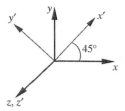

Figure 1.8 Coordinate transformation of Example 1.6

SOLUTION   Since the z'-direction is parallel to the z-direction, the coordinate transformation is basically a two-dimensional rotation. If $a = \cos 45° = \sin 45°$, then the transformation matrix becomes

$$[\mathbf{N}] = \begin{bmatrix} a & -a & 0 \\ a & a & 0 \\ 0 & 0 & 1 \end{bmatrix}$$

Using Eq. (1.30), the stress matrix in the transformed coordinates becomes

$$[\mathbf{N}]^T[\boldsymbol{\sigma}][\mathbf{N}] = \begin{bmatrix} a & a & 0 \\ -a & a & 0 \\ 0 & 0 & 1 \end{bmatrix} \begin{bmatrix} 2 & 1 & 0 \\ 1 & 2 & 0 \\ 0 & 0 & 2 \end{bmatrix} \begin{bmatrix} a & -a & 0 \\ a & a & 0 \\ 0 & 0 & 1 \end{bmatrix} = \begin{bmatrix} 6a^2 & 0 & 0 \\ 0 & 2a^2 & 0 \\ 0 & 0 & 2 \end{bmatrix} = \begin{bmatrix} 3 & 0 & 0 \\ 0 & 1 & 0 \\ 0 & 0 & 2 \end{bmatrix}$$

Note that the stress matrix is diagonal after transformation, which means that the $x'y'z'$ coordinates are the principal stress directions and the diagonal terms are the principal stresses.

1.1.9   Maximum Shear Stress

Maximum shear stress plays an important role in the failure of ductile materials. Let σ_1, σ_2, and σ_3 be the three principal stresses at a point such that $\sigma_1 \geq \sigma_2 \geq \sigma_3$. The stress state at this point can be described using three Mohr's circles as shown in Figure 1.9.

In Figure 1.9, any point (σ, τ) located in the shaded area represents the normal and shear stresses, σ_n and τ_n, on a plane through the point. One can note that the maximum shear stress is given by the radius of the largest Mohr's circle as

$$\tau_{\max} = \frac{\sigma_1 - \sigma_3}{2} \tag{1.31}$$

The plane on which the shear stress attains maximum bisects the first and third principal stress planes. The normal stress on this plane is given by

$$\sigma_n = \frac{\sigma_1 + \sigma_3}{2} \tag{1.32}$$

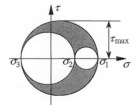

Figure 1.9 Mohr's circles and maximum shear stress

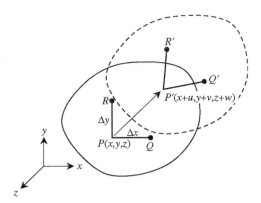

Figure 1.10 Deformation of line segments

1.2 STRAIN

When a solid is subjected to forces, it deforms. A quantitative measure of the deformation is provided by strains. Imagine an infinitesimal line segment in an arbitrary direction at a point in a solid. After deformation, the length of the line segment changes. Strain, specifically the normal strain, in the original direction of the line segment is defined as the change in length divided by the original length. However, this strain will be different in different directions at the same point. In the following discussion, we develop the concept of strain in a three-dimensional body.

Figure 1.10 shows a body before and after deformation. Let the points P, Q, and R in the undeformed body move to P', Q', and R', respectively, after deformation. The displacement of P can be represented by three displacement components, u, v, and w in the x, y, and z directions. Thus, the coordinates of P' are $(x + u, y + v, z + w)$. The functions $u(x,y,z)$, $v(x,y,z)$, and $w(x,y,z)$ are components of a vector field that are referred to as the deformation field or the displacement field. The displacements of point Q will be slightly different from those of P. They can be written as

$$u_Q = u + \frac{\partial u}{\partial x}\Delta x$$

$$v_Q = v + \frac{\partial v}{\partial x}\Delta x \tag{1.33}$$

$$w_Q = w + \frac{\partial w}{\partial x}\Delta x$$

Similarly, displacements of point R are

$$u_R = u + \frac{\partial u}{\partial y}\Delta y$$

$$v_R = v + \frac{\partial v}{\partial y}\Delta y \tag{1.34}$$

$$w_R = w + \frac{\partial w}{\partial y}\Delta y$$

The coordinates of P, Q, and R before and after deformation are as follows:

$$P : (x, y, z)$$
$$Q : (x + \Delta x, y, z)$$
$$R : (x, y + \Delta y, z)$$
$$P' : (x + u_P, \, y + v_P, \, z + w_P) = (x + u, \, y + v, \, z + w)$$
$$Q' : (x + \Delta x + u_Q, \, y + v_Q, \, z + w_Q)$$
$$= \left(x + \Delta x + u + \frac{\partial u}{\partial x} \Delta x, \; y + v + \frac{\partial v}{\partial x} \Delta x, \; z + w + \frac{\partial w}{\partial x} \Delta x \right)$$
$$R' : (x + u_R, \, y + \Delta y + v_R, \, z + w_R)$$
$$= \left(x + u + \frac{\partial u}{\partial y} \Delta y, \; y + \Delta y + v + \frac{\partial v}{\partial y} \Delta y, \; z + w + \frac{\partial w}{\partial y} \Delta y \right)$$

$$(1.35)$$

Length of the line segment $P'Q'$ can be calculated as

$$P'Q' = \sqrt{(x_{P'} - x_{Q'})^2 + (y_{P'} - y_{Q'})^2 + (z_{P'} - z_{Q'})^2} \tag{1.36}$$

Substituting for the coordinates of P' and Q' from Eq. (1.35), we obtain

$$
\begin{aligned}
P'Q' &= \Delta x \sqrt{\left(1 + \frac{\partial u}{\partial x}\right)^2 + \left(\frac{\partial v}{\partial x}\right)^2 + \left(\frac{\partial w}{\partial x}\right)^2} \\
&= \Delta x \left(1 + 2\frac{\partial u}{\partial x} + \left(\frac{\partial u}{\partial x}\right)^2 + \left(\frac{\partial v}{\partial x}\right)^2 + \left(\frac{\partial w}{\partial x}\right)^2 \right)^{1/2} \\
&\approx \Delta x \left(1 + \frac{\partial u}{\partial x} + \frac{1}{2}\left(\frac{\partial u}{\partial x}\right)^2 + \frac{1}{2}\left(\frac{\partial v}{\partial x}\right)^2 + \frac{1}{2}\left(\frac{\partial w}{\partial x}\right)^2 \right)
\end{aligned}
\tag{1.37}
$$

It may be noted that we have used two-term binomial expansion in deriving an approximate expression for change in length. In this book we will consider only small deformations such that all deformation gradients are very small compared to unity, e.g., $\partial u/\partial x \ll 1$, $\partial v/\partial x \ll 1$. Then we can neglect the higher order terms in Eq. (1.37) to obtain

$$P'Q' \approx \Delta x \left(1 + \frac{\partial u}{\partial x} \right) \tag{1.38}$$

Now we invoke the definition of normal strain as the ratio of change in length to original length to derive the expression for strain as

$$\varepsilon_{xx} = \frac{P'Q' - PQ}{PQ} = \frac{\partial u}{\partial x} \tag{1.39}$$

Thus, the normal strain ε_{xx} at a point can be defined as the change in length per unit length of an infinitesimally long line segment originally parallel to the x-axis. Similarly, we can derive normal strains in the y and z directions as

$$\varepsilon_{yy} = \frac{\partial v}{\partial y}, \quad \varepsilon_{zz} = \frac{\partial w}{\partial z} \tag{1.40}$$

The shear strain, say γ_{xy}, is defined as the change in angle between a pair of infinitesimal line segments that were originally parallel to x- and y-axes. From Figure 1.10, the angle between PQ and $P'Q'$ can be derived as

$$\theta_1 = \frac{y_{Q'} - y_{P'}}{\Delta x} = \frac{\partial v}{\partial x} \tag{1.41}$$

Similarly, the angle between PR and $P'R'$ is

$$\theta_2 = \frac{x_{R'} - x_{P'}}{\Delta y} = \frac{\partial u}{\partial y} \tag{1.42}$$

Using the aforementioned definition of shear strain,

$$\gamma_{xy} = \theta_1 + \theta_2 = \frac{\partial u}{\partial y} + \frac{\partial v}{\partial x} \tag{1.43}$$

Similarly, we can derive shear strains in the yz- and zx-planes as

$$\gamma_{yz} = \frac{\partial v}{\partial z} + \frac{\partial w}{\partial y}$$
$$\gamma_{zx} = \frac{\partial w}{\partial x} + \frac{\partial u}{\partial z} \tag{1.44}$$

The shear strains, γ_{xy}, γ_{yz} and γ_{zx}, are called engineering shear strains. We define tensorial shear strains as

$$\varepsilon_{xy} = \frac{1}{2}\left(\frac{\partial u}{\partial y} + \frac{\partial v}{\partial x}\right)$$
$$\varepsilon_{yz} = \frac{1}{2}\left(\frac{\partial v}{\partial z} + \frac{\partial w}{\partial y}\right) \tag{1.45}$$
$$\varepsilon_{zx} = \frac{1}{2}\left(\frac{\partial w}{\partial x} + \frac{\partial u}{\partial z}\right)$$

It may be noted that the tensorial shear strains are one-half of the corresponding engineering shear strains. It can be shown that the normal strains and the tensorial shear strains transform from one coordinate system to another following tensor transformation rules.

In general three-dimensional case, the strain matrix is defined as

$$[\varepsilon] = \begin{bmatrix} \varepsilon_{xx} & \varepsilon_{xy} & \varepsilon_{xz} \\ \varepsilon_{yx} & \varepsilon_{yy} & \varepsilon_{yz} \\ \varepsilon_{zx} & \varepsilon_{zy} & \varepsilon_{zz} \end{bmatrix} \tag{1.46}$$

As is clear from the definition in Eq. (1.45), the strain matrix is symmetric. Like the stress vector, the symmetric strain matrix can be represented as a pseudo vector:

$$\{\varepsilon\} = \begin{Bmatrix} \varepsilon_{xx} \\ \varepsilon_{yy} \\ \varepsilon_{zz} \\ \gamma_{yz} \\ \gamma_{zx} \\ \gamma_{xy} \end{Bmatrix} \tag{1.47}$$

where $\gamma_{yz}, \gamma_{zx},$ and γ_{xy} are used instead of $\varepsilon_{yz}, \varepsilon_{zx},$ and ε_{xy}. The six components of strain completely define the deformation at a point. The normal strain in any arbitrary direction at that point and also the shear strain in any arbitrary plane passing through the point can be calculated using the above strain components, as explained in the following section. Since strain is a tensor, it has properties similar to stress tensor. For example, transformation of strain, principal strains, and corresponding principal strain directions can be determined using the procedures we described for stresses.

1.2.1 Transformation of Strain

Let the state of strain at a point be given by the strain matrix $[\varepsilon]_{xyz}$, where the components are expressed in the xyz coordinates shown in Figure 1.7. As for the case of stresses, the question again is: What are the components of strain at the same point in a different coordinates system, $x'y'z'$ (i.e., $[\varepsilon]_{x'y'z'}$)?

Let $\mathbf{b}^1, \mathbf{b}^2,$ and $\mathbf{b}^3$ be unit vectors along the axes $x', y',$ and z', respectively, represented in xyz coordinates. The strain matrix expressed with respect to the $x'y'z'$ coordinates system is

$$[\varepsilon]_{x'y'z'} = [\mathbf{N}]^T [\varepsilon]_{xyz} [\mathbf{N}] \tag{1.48}$$

which is the same transformation used for stresses in Eq. (1.30). It should be emphasized that tensorial shear strains must be used in using the transformation in Eq. (1.48).

Let us consider the term $\varepsilon_{x'x'}$ in Eq. (1.48). It can be written as

$$\varepsilon_{x'x'} = \{\mathbf{b}^1\}^T [\varepsilon]_{xyz} \{\mathbf{b}^1\} \tag{1.49}$$

We note that the normal strain in the x'-direction depends only on the strain tensor and the direction cosines of the x'-axis. In fact, this relation can be generalized to any arbitrary direction $\mathbf{n}$ and written as

$$\begin{aligned} \varepsilon_{nn} &= \{\mathbf{n}\}^T [\varepsilon]_{xyz} \{\mathbf{n}\} \\ &= \varepsilon_{xx} n_x^2 + \varepsilon_{yy} n_y^2 + \varepsilon_{zz} n_z^2 + 2\varepsilon_{xy} n_x n_y + 2\varepsilon_{yz} n_y n_z + 2\varepsilon_{zx} n_z n_x \\ &= \varepsilon_{xx} n_x^2 + \varepsilon_{yy} n_y^2 + \varepsilon_{zz} n_z^2 + \gamma_{xy} n_x n_y + \gamma_{yz} n_y n_z + \gamma_{zx} n_z n_x \end{aligned} \tag{1.50}$$

Similarly, one can derive the shear strain in a plane containing two mutually perpendicular vectors $\mathbf{m}$ and $\mathbf{n}$ as

$$\begin{aligned} \varepsilon_{mn} &= \{\mathbf{m}\}^T [\varepsilon]_{xyz} \{\mathbf{n}\} \\ &= \varepsilon_{xx} m_x n_x + \varepsilon_{yy} m_y n_y + \varepsilon_{zz} m_z n_z + \varepsilon_{xy} (m_x n_y + m_y n_x) \\ &\quad + \varepsilon_{yz} (m_y n_z + m_z n_y) + \varepsilon_{zx} (m_z n_x + m_x n_z) \end{aligned} \tag{1.51}$$

1.2.2 Principal Strains

Since strains, like stresses, are second-order tensors, one can compute eigen values and eigen vectors for a given strain matrix. The eigen values are the three principal strains, and the eigen vectors are the corresponding principal strain directions. The principal strain directions represent the directions in which the normal strain ε_{nn} in Eq. (1.50) takes an extreme value. The maximum and minimum of the three principal strains are the global maximum and minimum, respectively. The three principal planes intersect along

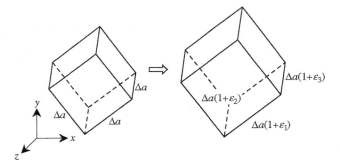

Figure 1.11 Deformation in the principal directions

the principal strain directions. The shear strain vanishes on the principal strain planes. That is, a pair of infinitesimally small line segments parallel to any two principal strain directions remains perpendicular to each other after deformation. That is, the angle between them does not change after deformation. In other words, the shear strain on this plane is equal to zero.

The physical meaning of strains can be explained as follows. Consider an infinitesimal cube of size $\Delta a \times \Delta a \times \Delta a$ whose sides are parallel to the coordinate axes before deformation. After deformation the rectangular parallelepiped will become an oblique parallelepiped of size $\Delta a(1 + \varepsilon_{xx}) \times \Delta a(1 + \varepsilon_{yy}) \times \Delta a(1 + \varepsilon_{zz})$. The angles between the edges of the parallelepiped will reduce by $\gamma_{xy}, \gamma_{yz},$ and γ_{zx}. If the same parallelepiped is oriented such that its edges are parallel to the principal strain directions, then, after deformation, the parallelepiped will remain as a rectangular parallelepiped and its dimensions will be $\Delta a(1 + \varepsilon_1) \times \Delta a(1 + \varepsilon_2) \times \Delta a(1 + \varepsilon_3)$. Since the shear strains in the principal strain planes are zero, the angle between the edges will remain as 90° (see Figure 1.11).

Principal strain is the extensional strain on a plane, where there is no shear strain. Principal strains and their directions are the eigen values and eigen vectors of the following eigen value problem:

$$[\varepsilon] \cdot \mathbf{n} = \varepsilon_n \mathbf{n} \tag{1.52}$$

where $\mathbf{n}$ is the eigen vector of matrix $[\varepsilon]$. The above eigen value problem has three solutions, $\varepsilon_n = \varepsilon_1, \varepsilon_2,$ and ε_3, which are the principal strains.

If the principal strains $\varepsilon_1, \varepsilon_2,$ and ε_3 are known, then the maximum engineering shear strain $\gamma_{\max}$ can be computed as

$$\frac{\gamma_{\max}}{2} = \frac{\varepsilon_1 - \varepsilon_3}{2} \tag{1.53}$$

where ε_1 and ε_3 are the maximum and minimum principal strains, respectively.

1.2.3 Stress Vs. Strain

Stresses and strains defined in the previous two sections are second-order tensors and hence share some common properties, as shown in Table 1.2. When the material is isotropic, which means the material properties are the same in all directions or independent of the coordinates system, the principal stress directions and principal strain directions coincide. If the material is anisotropic, the principal stress and strain directions are in general different. In this textbook, we will focus on isotropic materials only.

Table 1.2   Comparison of Stress and Strain

$[\sigma]$ is a symmetric 3×3 matrix	$[\varepsilon]$ is a symmetric 3×3 matrix
Normal stress in the direction **n** is $$\sigma_{nn} = \mathbf{n}^T[\sigma]\mathbf{n}$$	Normal strain in the direction **n** is $$\varepsilon_{nn} = \mathbf{n}^T[\varepsilon]\mathbf{n}$$
Shear stress in a plane containing two mutually perpendicular unit vectors **m** and **n** is $$\tau_{mn} = \mathbf{m}^T[\sigma]\mathbf{n}$$	Shear strain in a plane containing two mutually perpendicular unit vectors **m** and **n** is $$\varepsilon_{mn} = \frac{\gamma_{mn}}{2} = \mathbf{m}^T[\varepsilon]\mathbf{n}$$
Transformation of stresses $$[\sigma]_{x'y'z'} = [\mathbf{N}]^T[\sigma]_{xyz}[\mathbf{N}]$$	Transformation of strains $$[\varepsilon]_{x'y'z'} = [\mathbf{N}]^T[\varepsilon]_{xyz}[\mathbf{N}]$$
Three mutually perpendicular principal directions and principal stresses can be computed as eigen vectors and eigen values of the stress matrix: $$[\sigma]\,\mathbf{n} = \sigma_n\mathbf{n}$$	Three mutually perpendicular principal directions and principal strains can be computed as eigen vectors and eigen values of the strain matrix: $$[\varepsilon]\mathbf{n} = \varepsilon_n\mathbf{n}$$

1.3   STRESS-STRAIN RELATIONS

Finding a relationship between the loads acting on a structure and its deflection has been of great interest to scientists since the 17th century.[5] Robert Hooke, Jacob Bernoulli, and Leonard Euler are some of the pioneers who developed various theories to explain the bending of beams and stretching of bars. Forces applied to a solid create stresses within the body in order to satisfy equilibrium. These stresses also cause deformation or strains. Accumulation of strains over the volume of a body manifests as deflections or gross deformation of the body. Hence, it is clear that a fundamental knowledge of relationship between stresses and strains is necessary in order to understand the global behavior. Navier tried to explain deformations considering the forces between neighboring particles in a body, as they tend to separate and come closer. Later this approach was abandoned in favor of Cauchy's stresses and strains. Robert Hooke was the first one to propose the linear uniaxial stress-strain relation, which states that the stress is proportional to strain. Later, the general relation between the six components of strains and stresses called the generalized Hooke's law was developed. The generalized Hooke's law states that each component of stress is a linear combination of strains. It should be mentioned that stress-strain relations are called phenomenological models or theories as they are based on commonly observed behavior of materials and verified by experiments. Only recently, with the advancement of computers and computational techniques, behavior of materials based on first principles or fundamental atomistic behavior is being developed. This new field of study is called computational materials and involves techniques such as molecular dynamic simulations and multiscale modeling. Stress-strain relations are also called constitutive relations as they describe the constitution of the material.

A cylindrical test specimen is loaded along its axis as shown Figure 1.12. This type of loading ensures that the specimen is subjected to uniaxial state of stress. If the

F  F **Figure 1.12** Uniaxial tension test

---

[5] Timoshenko, S.P., *History of Strength of Materials*, Dover Publications, New York, 1983

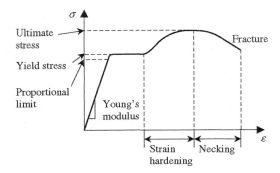

Figure 1.13 Stress-strain diagram for a typical ductile material in tension

Table 1.3   Explanations of Uniaxial Tension Test

Terms	Explanation
Proportional limit	The greatest stress for which the stress is still proportional to the strain
Elastic limit	The greatest stress without resulting in any permanent strain on release of stress
Young's Modulus	Slope of the linear portion of the stress-strain curve
Yield stress	The stress required to produce 0.2% plastic strain
Strain hardening	A region where more stress is required to deform the material
Ultimate stress	The maximum stress the material can resist
Necking	Cross section of the specimen reduces during deformation

stress-strain relation of the uniaxial tension test in Figure 1.12 is plotted, then a typical ductile material may show a behavior as in Figure 1.13. The explanation of terms in the figure is summarized in Table 1.3.

After the material yields, the shape of the structure permanently changes. Hence, many engineering structures are designed such that the maximum stress is smaller than the yield stress of the material. Under this range of the stress, the stress-strain relation can be approximated by a linear relation.

1.3.1   Linear Elastic Relationship (Generalized Hooke's Law)

The one-dimensional stress-strain relation described in the previous section can be extended to the three-dimensional state of stress. When the stress-strain relation is linear, the relationship between stress and strain can be written as

$$\{\boldsymbol{\sigma}\} = [\mathbf{C}] \cdot \{\boldsymbol{\varepsilon}\} \qquad (1.54)$$

where

$$\{\boldsymbol{\sigma}\} = \begin{Bmatrix} \sigma_{xx} \\ \sigma_{yy} \\ \sigma_{zz} \\ \tau_{yz} \\ \tau_{zx} \\ \tau_{xy} \end{Bmatrix}, [\mathbf{C}] = \begin{bmatrix} C_{11} & C_{12} & C_{13} & C_{14} & C_{15} & C_{16} \\ C_{21} & C_{22} & C_{23} & C_{24} & C_{25} & C_{26} \\ C_{31} & C_{32} & C_{33} & C_{34} & C_{35} & C_{36} \\ C_{41} & C_{42} & C_{43} & C_{44} & C_{45} & C_{46} \\ C_{51} & C_{52} & C_{53} & C_{54} & C_{55} & C_{56} \\ C_{61} & C_{62} & C_{63} & C_{64} & C_{65} & C_{66} \end{bmatrix}, \text{ and } \{\boldsymbol{\varepsilon}\} = \begin{Bmatrix} \varepsilon_{xx} \\ \varepsilon_{yy} \\ \varepsilon_{zz} \\ \gamma_{yz} \\ \gamma_{zx} \\ \gamma_{xy} \end{Bmatrix}$$

The matrix $[\mathbf{C}]$ is called the stress-strain matrix, or elasticity matrix. It can be shown that $[\mathbf{C}]$ must be a symmetric matrix and hence the number of independent coefficients or elastic constants for an anisotropic material is only 21. Many composite materials, naturally occurring composites such as wood or bone and man-made materials such as fiber-reinforced composites, can be modeled as an orthotropic material with nine independent elastic constants. Some composites are transversely isotropic and require only five independent elastic constants. Most materials are isotropic, and for such materials, the 21 constants in the symmetric matrix [**C**] can be expressed in terms of two independent constants called engineering elastic constants.

For isotropic materials, the relation between stress and strain can be written as

$$\left\{\begin{array}{c} \varepsilon_{xx} \\ \varepsilon_{yy} \\ \varepsilon_{zz} \end{array}\right\} = \frac{1}{E} \begin{bmatrix} 1 & -\nu & -\nu \\ -\nu & 1 & -\nu \\ -\nu & -\nu & 1 \end{bmatrix} \left\{\begin{array}{c} \sigma_{xx} \\ \sigma_{yy} \\ \sigma_{zz} \end{array}\right\} \tag{1.55}$$

$$\gamma_{xy} = \frac{\tau_{xy}}{G}, \ \gamma_{yz} = \frac{\tau_{yz}}{G}, \ \gamma_{zx} = \frac{\tau_{zx}}{G}$$

where E the Young's modulus and ν the Poisson's ratio are the two independent elastic constants, and G is the shear modulus, defined by

$$G = \frac{E}{2(1+\nu)} \tag{1.56}$$

Note that there are only two independent constants for isotropic materials.

Alternately, stresses can be written as a function of strains by inverting the relations in Eq. (1.55), as

$$\left\{\begin{array}{c} \sigma_{xx} \\ \sigma_{yy} \\ \sigma_{zz} \end{array}\right\} = \frac{E}{(1+\nu)(1-2\nu)} \begin{bmatrix} 1-\nu & \nu & \nu \\ \nu & 1-\nu & \nu \\ \nu & \nu & 1-\nu \end{bmatrix} \left\{\begin{array}{c} \varepsilon_{xx} \\ \varepsilon_{yy} \\ \varepsilon_{zz} \end{array}\right\} \tag{1.57}$$

$$\tau_{xy} = G\gamma_{xy}, \quad \tau_{yz} = G\gamma_{yz}, \quad \tau_{zx} = G\gamma_{zx}$$

The elasticity matrix [**C**] in Eq. (1.54) can be written as

$$[\mathbf{C}] = \frac{E}{(1+\nu)(1-2\nu)} \begin{bmatrix} 1-\nu & \nu & \nu & 0 & 0 & 0 \\ \nu & 1-\nu & \nu & 0 & 0 & 0 \\ \nu & \nu & 1-\nu & 0 & 0 & 0 \\ 0 & 0 & 0 & \frac{1}{2}-\nu & 0 & 0 \\ 0 & 0 & 0 & 0 & \frac{1}{2}-\nu & 0 \\ 0 & 0 & 0 & 0 & 0 & \frac{1}{2}-\nu \end{bmatrix} \tag{1.58}$$

EXAMPLE 1.7   *Stress-Strain Relationship*

The stress at a point in a body is

$$[\boldsymbol{\sigma}] = \begin{bmatrix} 5 & 3 & 2 \\ 3 & -1 & 0 \\ 2 & 0 & 4 \end{bmatrix} \times 10^3 \text{ psi}$$

Determine the strain components for an isotropic material when $E = 10^7$ psi and $\nu = 0.3$.

SOLUTION   From Eq. (1.55),

$$\varepsilon_{xx} = \frac{1}{10^7}[5 - 0.3(-1+4)] \times 10^3 = 4.1 \times 10^{-4}$$

$$\varepsilon_{yy} = \frac{1}{10^7}[-1 - 0.3(5+4)] \times 10^3 = -3.7 \times 10^{-4}$$

$$\varepsilon_{zz} = \frac{1}{10^7}[4 - 0.3(5-1)] \times 10^3 = 2.8 \times 10^{-4}$$

$$\gamma_{xy} = \frac{2(1+0.3)}{10^7} 3000 = 7.8 \times 10^{-4}$$

$$\gamma_{yz} = \frac{2(1+0.3)}{10^7} 0 = 0$$

$$\gamma_{xz} = \frac{2(1+0.3)}{10^7} 2000 = 5.2 \times 10^{-4}$$

1.3.2   Simplified Laws for Two-Dimensional Analysis

The general three-dimensional stress-strain relations in Eq. (1.54) can be simplified for certain special situations that often occur in practice. The two-dimensional stress-strain relations can be categorized into two cases: plane stress and plane strain.

Most practical structures consist of thin plate-like components in order to be efficient. Assume a thin plate that is parallel to the xy-plane. If we assume that the top and bottom surfaces of the plate are not subjected to any significant forces, i.e., the plate is subjected to forces in its plane only, in the x and y directions, then the transverse stresses (stresses with a z subscript) vanish on the top and bottom surfaces, i.e., $\sigma_{zz} = \tau_{xz} = \tau_{yz} = 0$ on the top and bottom surfaces. If the thickness is much smaller compared to the lateral dimensions of the plate, then we can assume that the aforementioned transverse stresses are approximately zero through the entire thickness. Then the plate is said to be in a state of plane stress parallel to the xy-plane or normal to the z-axis.

In order to derive the stress-strain relations for the state of plane stress, we start with Eq. (1.55). We set $\sigma_{zz} = \tau_{xz} = \tau_{yz} = 0$ on the right–hand side of the equations to obtain

$$\left\{ \begin{array}{c} \varepsilon_{xx} \\ \varepsilon_{yy} \end{array} \right\} = \frac{1}{E} \begin{bmatrix} 1 & -\nu \\ -\nu & 1 \end{bmatrix} \left\{ \begin{array}{c} \sigma_{xx} \\ \sigma_{yy} \end{array} \right\}$$

$$\varepsilon_{zz} = -\frac{\nu}{E}(\sigma_{xx} + \sigma_{yy}) \tag{1.59}$$

$$\gamma_{xy} = \frac{\tau_{xy}}{G}$$

$$\gamma_{yz} = \gamma_{zx} = 0$$

Inverting the above relations, we obtain

$$\{\boldsymbol{\sigma}\} = \left\{ \begin{array}{c} \sigma_{xx} \\ \sigma_{yy} \\ \tau_{xy} \end{array} \right\} = \frac{E}{1-\nu^2} \begin{bmatrix} 1 & \nu & 0 \\ \nu & 1 & 0 \\ 0 & 0 & \frac{1}{2}(1-\nu) \end{bmatrix} \left\{ \begin{array}{c} \varepsilon_{xx} \\ \varepsilon_{yy} \\ \gamma_{xy} \end{array} \right\} \tag{1.60}$$

Similar to plane stress, one can define a state of plane strain in which strains with a z subscript are all equal to zero. This situation corresponds to a structure whose deformation in the z-direction is constrained (i.e., $w = 0$), so that the following relation holds:

$$\varepsilon_{zz} = 0, \ \varepsilon_{xz} = 0, \ \varepsilon_{yz} = 0 \tag{1.61}$$

Plane strain can also be used if the structure is infinitely long in the z-direction. The stress-strain relations for the case of plane strain can be derived by starting with Eq. (1.54) with the stress-strain matrix $[\mathbf{C}]$ in Eq. (1.58) and setting the strains with a z subscript, ε_{zz}, γ_{xz}, and γ_{yz}, equal to zero to obtain

$$\{\boldsymbol{\sigma}\} = \begin{Bmatrix} \sigma_{xx} \\ \sigma_{yy} \\ \tau_{xy} \end{Bmatrix} = \frac{E}{(1+v)(1-2v)} \begin{bmatrix} 1-v & v & 0 \\ v & 1-v & 0 \\ 0 & 0 & \frac{1}{2}-v \end{bmatrix} \begin{Bmatrix} \varepsilon_{xx} \\ \varepsilon_{yy} \\ \gamma_{xy} \end{Bmatrix} \tag{1.62}$$

The inverse relation is given by

$$\begin{Bmatrix} \varepsilon_{xx} \\ \varepsilon_{yy} \\ \gamma_{xy} \end{Bmatrix} = \frac{(1+v)}{E} \begin{bmatrix} 1-v & v & 0 \\ v & 1-v & 0 \\ 0 & 0 & 2 \end{bmatrix} \begin{Bmatrix} \sigma_{xx} \\ \sigma_{yy} \\ \tau_{xy} \end{Bmatrix} \tag{1.63}$$

Note that the normal stress σ_{zz} is not zero in the plane strain problem but can be calculated from ε_{xx} and ε_{yy}:

$$\sigma_{zz} = \frac{Ev}{(1+v)(1-2v)} (\varepsilon_{xx} + \varepsilon_{yy}) \tag{1.64}$$

1.4   BOUNDARY VALUE PROBLEMS

1.4.1   Equilibrium Equations

As we discussed earlier, the state of stress at a point is defined by the six stress components, three normal and three shear stresses. These components in general vary within the solid. In static problems, the stresses can be represented by six functions of the spatial coordinates x, y, and z, $\sigma_{xx}(x, y, z)$, $\sigma_{yy}(x, y, z)$, etc. These six functions cannot be arbitrary and must satisfy certain relations called stress equilibrium equations. The functions are also called the stress field in the solid.

Consider the equilibrium of a differential element represented in two-dimensions by a square (see Figure 1.14) whose center is (x, y) and sides are dx and dy,

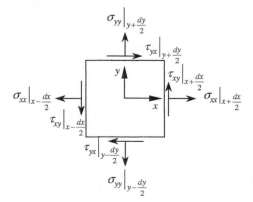

Figure 1.14 Stress variations in infinitesimal components

respectively. The dimension in the z-direction is taken as unity. The stresses are assumed to be independent of z, and the solid is in a state of plane stress or plane strain normal to the z-axis.

Equilibrium in the x-direction yields the following equation:

$$\left(\sigma_{xx}\big|_{x+\frac{dx}{2}}\right)dy - \left(\sigma_{xx}\big|_{x-\frac{dx}{2}}\right)dy + \left(\tau_{yx}\big|_{y+\frac{dy}{2}}\right)dx - \left(\tau_{yx}\big|_{y-\frac{dy}{2}}\right)dx = 0 \tag{1.65}$$

If the first-order Taylor series expansion is used to represent stresses on the surfaces of the rectangle in terms of stresses at the center, the first two terms in Eq. (1.65) can be approximated by

$$\left(\sigma_{xx}\big|_{x+\frac{dx}{2}}\right)dy - \left(\sigma_{xx}\big|_{x-\frac{dx}{2}}\right)dy$$

$$= \left(\sigma_{xx}\big|_x + \frac{\partial\sigma_{xx}}{\partial x}\frac{dx}{2}\right)dy - \left(\sigma_{xx}\big|_x - \frac{\partial\sigma_{xx}}{\partial x}\frac{dx}{2}\right)dy = \frac{\partial\sigma_{xx}}{\partial x}dxdy$$

Similarly, the last two terms can be approximated by

$$\left(\tau_{yx}\big|_{y+\frac{dy}{2}}\right)dx - \left(\tau_{yx}\big|_{y-\frac{dy}{2}}\right)dx$$

$$= \left(\tau_{yx}\big|_y + \frac{\partial\tau_{yx}}{\partial y}\frac{dy}{2}\right)dx - \left(\tau_{yx}\big|_y - \frac{\partial\tau_{yx}}{\partial y}\frac{dy}{2}\right)dx = \frac{\partial\tau_{yx}}{\partial y}dxdy$$

By substituting these two equations into Eq. (1.65), we obtain an equilibrium equation in the x-direction as

$$\frac{\partial\sigma_{xx}}{\partial x} + \frac{\partial\tau_{yx}}{\partial y} = 0 \tag{1.66}$$

Similarly, equilibrium in the y-direction yields the following equation:

$$\frac{\partial\tau_{xy}}{\partial x} + \frac{\partial\sigma_{yy}}{\partial y} = 0 \tag{1.67}$$

Equations (1.66) and (1.67) are the equilibrium equations for a solid subjected to two-dimensional state of stress. We can similarly derive the equations for a three-dimensional state of stress by considering the equilibrium of a three-dimensional differential element to obtain

$$\begin{cases} \dfrac{\partial\sigma_{xx}}{\partial x} + \dfrac{\partial\tau_{yx}}{\partial y} + \dfrac{\partial\tau_{zx}}{\partial z} = 0 \\[2mm] \dfrac{\partial\tau_{xy}}{\partial x} + \dfrac{\partial\sigma_{yy}}{\partial y} + \dfrac{\partial\tau_{zy}}{\partial z} = 0 \\[2mm] \dfrac{\partial\tau_{xz}}{\partial x} + \dfrac{\partial\tau_{yz}}{\partial y} + \dfrac{\partial\sigma_{zz}}{\partial z} = 0 \end{cases} \tag{1.68}$$

Equation (1.68) is obtained by considering force equilibrium in the x-, y-, and z-directions. As has been shown in Eq. (1.12), moment equilibrium yields symmetry of stress matrix.

1.4.2 Traction or Stress Boundary Conditions

While the stress field must satisfy the differential equations of equilibrium (1.68), there are other conditions the stress field must satisfy on the boundaries of a solid. These are called traction or stress boundary conditions. Consider the surface of a solid subjected to distributed forces such that the tractions in the x, y and z directions are, t_x, t_y, and t_z, respectively (see Figure 1.15). Let the direction cosines of the normal to the surface be n_x, n_y, and n_z. Then the state of stress at a point on the surface of the body must satisfy the boundary conditions shown below:

$$\sigma_{xx}n_x + \tau_{yx}n_y + \tau_{zx}n_z = t_x$$
$$\tau_{xy}n_x + \sigma_{yy}n_y + \tau_{zy}n_z = t_y \qquad (1.69)$$
$$\tau_{xz}n_x + \tau_{yz}n_y + \sigma_{zz}n_z = t_z$$

The derivation of the above boundary conditions is similar to that of Eq. (1.10).

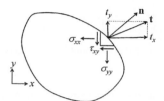

Figure 1.15 Traction boundary condition of a plane solid

1.4.3 Boundary Value Problems

Consider an arbitrary body subjected to external forces and displacement constraints as shown in Figure 1.16. Our goal is to determine the displacement field $\mathbf{u}(x,y,z)$. Once the displacements are determined, strains can be found using the strain displacement relations, e.g., Eqs. (1.39), (1.40), and (1.45), and then stress field can be determined using the constitutive relations, e.g., Eq. (1.54). This problem, typical of solid and structural mechanics, is called the elasto-static boundary value problem. Techniques for solving the boundary value problems in two- and three-dimensions are considered in advanced courses in solid mechanics and theory of elasticity.[6] The elasticity problems can be simplified by making certain approximations in the displacement field, e.g., Euler-Bernoulli beam theory or Kirchhoff plate theory. Finite element method is a numerical method that can solve complex two- and three-dimensional elasticity problems and will be discussed in later chapters.

In general, solving an elasticity problem involves solving the three equilibrium equations in Eq. (1.68) in conjunction with the strain-displacement relations in Eqs. (1.39),

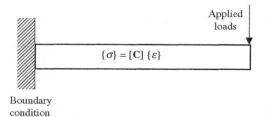

Boundary condition

Figure 1.16 Boundary value problem

---

[6] Boresi, A., Schmidt, R., and Sidebottom, O. J., *Advanced Mechanics of Materials*, 6th ed., Wiley, New York, 2003

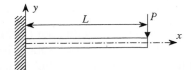

Figure 1.17 Cantilever beam bending problem

(1.40), and (1.45) and constitutive relations in Eq. (1.54). The boundary conditions also need to be provided in terms of given displacements or prescribed traction forces.

EXAMPLE 1.8 *Stress Distribution of a Cantilever Beam*

The displacement field for the thin beam shown in Figure 1.17 considering bending only is

$$u(x, y) = \frac{P}{EI}\left(Lx - \frac{x^2}{2}\right)y - \frac{\nu P}{6EI}y^3$$

$$v(x, y) = \frac{-\nu P}{2EI}(L - x)y^2 - \frac{P}{EI}\left(\frac{Lx^2}{2} - \frac{x^3}{6}\right)$$

where P is the applied force at the tip, I is the area moment of inertia about bending axis, and L is the length of the beam. Determine the entire stress field.

SOLUTION   Since the thickness of the beam is small, we can assume the plane stress condition along the z-direction. From the definition of strain,

$$\varepsilon_{xx} = \frac{\partial u}{\partial x} = \frac{P}{EI}(L - x)y$$

$$\varepsilon_{yy} = \frac{\partial v}{\partial y} = \frac{-\nu P}{EI}(L - x)y$$

$$\gamma_{xy} = \frac{\partial v}{\partial x} + \frac{\partial u}{\partial y} = \left[\frac{\nu P y^2}{2EI} - \frac{P}{EI}\left(Lx - \frac{x^2}{2}\right)\right]$$

$$+ \left[\frac{P}{EI}\left(Lx - \frac{x^2}{2}\right)y - \frac{\nu P y^2}{2EI}\right] = 0$$

Substituting into Eq. (1.60) yields the following stress field:

$$\sigma_{xx} = \frac{E}{1 - \nu^2}\left[\frac{P}{EI}(L - x)y - \frac{\nu^2 P}{EI}(L - x)y\right] = \frac{P}{I}(L - x)y$$

$$\sigma_{yy} = \frac{E}{1 - \nu^2}\left[-\frac{\nu P}{EI}(L - x)y + \frac{\nu P}{EI}(L - x)y\right] = 0$$

$$\tau_{xy} = 0$$

Since the normal stress σ_{xx} changes linearly in the y-direction, the stress field represents bending of a beam.

1.4.4 Compatibility Conditions

In the previous section, we talked about the displacement field $\mathbf{u}(x, y, z)$. Usually solutions to the boundary value problems are obtained by trial and error or inverse methods. We first assume a physically suitable displacement field and adjust the terms in the solution to satisfy the equilibrium equations and boundary conditions. More often one needs to start with a physically possible displacement field. In fact, a set of any three non-singular functions will represent a possible displacement field. Of course, the forces required to produce such a displacement field may be complex and difficult in practice but still physically possible. However, the same thing cannot be said about the strain field or stress field. That is, one cannot choose six arbitrary functions in spatial coordinates and claim the set to be a possible strain field. The particular strain field may not be physically possible. For example, a valid strain field should not create a discontinuous deformation or overlapping of certain points. The six strain functions must satisfy three relations called compatibility equations. The derivation of the three-dimensional compatibility equation is beyond the scope of this book. The interested reader is referred to more advanced elasticity books, e.g., Timoshenko.[7] For two-dimensional (plane) problems, the compatibility equation is given as

$$\frac{\partial^2 \gamma_{xy}}{\partial x \partial y} = \frac{\partial^2 \varepsilon_{xx}}{\partial y^2} + \frac{\partial^2 \varepsilon_{yy}}{\partial x^2} \tag{1.70}$$

EXAMPLE 1.9   *Compatibility Relation*

Choose two functions in x and y to represent the two displacements $u(x,y,z)$ and $v(x,y,z)$. Derive the strain field from the assumed displacement field. Show that the strains satisfy the compatibility equation in Eq. (1.70).

SOLUTION   Let the displacement field be given by

$$u(x,y) = x^4 + y^4, \ v(x,y) = x^2 y^2$$

Then the strains can be derived as

$$\varepsilon_{xx} = 4x^3, \quad \varepsilon_{yy} = 2x^2 y, \quad \gamma_{xy} = 4y^3 + 2xy^2$$

It is trivial to show that the above strains satisfy the compatibility requirement in Eq. (1.70).

1.5   FAILURE THEORIES

In the previous section, we introduced the concepts of stress, strain, and the relationship between stresses and strains. We also discussed failure of materials under uniaxial state of stress. Failure of engineering materials can be broadly classified into ductile and brittle failure. Most metals are ductile and fail due to yielding. Hence, the yield strength characterizes their failure. Ceramics and some polymers are brittle, and rupture or fracture when the stress exceeds certain maximum value. Their stress-strain behavior is linear up to the point of failure, and they fail abruptly.

The stress required to break the atomic bond and separate the atoms is called the theoretical strength of the material. It can be shown that the theoretical strength is approximately equal to $E/3$, where E is Young's modulus.[8] However, most materials fail at a stress about one-hundredth or even one-thousandth of the theoretical strength. For example, the theoretical strength of aluminum is about 22 GPa. However, the yield strength of aluminum is in the order of 100 MPa, which is 1/220th of the theoretical strength. This enormous discrepancy could be explained as follows.

[7] S. P. Timoshenko and J. N. Goodier, *Theory of Elasticity*, 3rd ed., McGraw-Hill, New York, 1970

[8] T.L. Anderson, *Fracture Mechanics—Fundamentals and Applications*, 3rd ed., CRC Press, Boca Raton, FL, 2006

Figure 1.18 Material failure due to relative sliding of atomic planes

In ductile material, yielding occurs not due to separation of atoms but due to sliding of atoms (movement of dislocations), as depicted in Figure 1.18. Thus, the stress or energy required for yielding is much less than that required for separating the atomic planes. Hence, in a ductile material, the maximum shear stress causes yielding of the material.

In brittle materials, the failure or rupture still occurs due to separation of atomic planes. However, the high value of stress required is provided locally by stress concentration caused by small pre-existing cracks or flaws in the material. The stress concentration factors can be in the order of 100 to 1,000. That is, the applied stress is amplified by an enormous amount due to the presence of cracks, and it is sufficient to separate the atoms. When this process becomes unstable, the material separates over a large area, causing brittle failure of the material.

Although research is underway not only to explain but also quantify the strength of materials in terms of its atomic structure and properties, it is still not practical to design machines and structures based on such atomistic models. Hence, we resort to phenomenological failure theories, which are based on observations and testing over a period of time. The purpose of failure theories is to extend the strength values obtained from uniaxial tests to multi-axial states of stress that exists in practical structures. It is not practical to test a material under all possible combinations of stress states. In the following, we describe some well-established phenomenological failure theories for both ductile and brittle materials.

1.5.1 Strain Energy

When a force is applied to a solid, it deforms. Then, we can say that work is done on the solid, which is proportional to the force and deformation. The work done by applied force is stored in the solid as potential energy, which is called the *strain energy*. The strain energy in the solid may not be distributed uniformly throughout the solid. We introduce the concept of strain energy density, which is strain energy per unit volume, and we denote it by U_0. Then the strain energy in the body can be obtained by integration as follows:

$$U = \iiint_V U_o(x,\, y,\, z)dV \qquad (1.71)$$

where the integration is performed over the volume V of the solid. In the case of uniaxial stress state, strain energy density is equal to the area under the stress-strain curve (see Figure 1.19). Thus, it can be written as

$$U_0 = \frac{1}{2}\sigma\varepsilon \qquad (1.72)$$

For the general three-dimensional case, the stain energy density is expressed as

$$U_0 = \frac{1}{2}\left(\sigma_x\varepsilon_x + \sigma_y\varepsilon_y + \sigma_z\varepsilon_z + \tau_{yz}\gamma_{yz} + \tau_{zx}\gamma_{zx} + \tau_{xy}\gamma_{xy}\right) \qquad (1.73)$$

If the material is elastic, then the strain energy can be completely recovered by unloading the body.

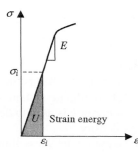

Figure 1.19 Stress-strain curve and the strain energy

The strain energy density in Eq. (1.73) can be further simplified. Consider a coordinate system that is parallel to the principal stress directions. In this coordinate system, no shear components exist. Extending Eq. (1.73) to this stress states yields

$$U_0 = \frac{1}{2}(\sigma_1\varepsilon_1 + \sigma_2\varepsilon_2 + \sigma_3\varepsilon_3) \tag{1.74}$$

From Section 1.3, we know that stresses and strains are related through the linear elastic relations. For example, in cases of principal stresses and strains,

$$\begin{cases} \varepsilon_1 = \frac{1}{E}(\sigma_1 - \nu\sigma_2 - \nu\sigma_3) \\[2mm] \varepsilon_2 = \frac{1}{E}(\sigma_2 - \nu\sigma_1 - \nu\sigma_3) \\[2mm] \varepsilon_3 = \frac{1}{E}(\sigma_3 - \nu\sigma_1 - \nu\sigma_2) \end{cases} \tag{1.75}$$

Substituting from Eq. (1.75) into Eq. (1.74), we can write the strain energy density in terms of principal stresses as

$$U_0 = \frac{1}{2E}[\sigma_1^2 + \sigma_2^2 + \sigma_3^2 - 2\nu(\sigma_1\sigma_2 + \sigma_2\sigma_3 + \sigma_1\sigma_3)] \tag{1.76}$$

The strain energy density can be thought of as consisting of two components: one due to dilation or change in volume and the other due to distortion or change in shape. The former is called dilatational strain energy, and the latter distortional energy. Many experiments have shown that ductile materials can be hydrostatically stressed to levels beyond their ultimate strength in compression without failure. This is because the hydrostatic state of stress reduces the volume of the specimen without changing its shape.

1.5.2   Decomposition of Strain Energy

The strain energy density at a point in a solid can be divided into two parts: dilatational strain energy density, U_h, which is due to change in volume, and distortional strain energy density, U_d, which is responsible for change in shape. To compute these components, we divide the stress matrix also into similar components, dilatational stress matrix, σ_h, and deviatoric stress matrix, σ_d. For convenience we will refer the stresses to the principal stress coordinates. Then, the aforementioned stress components can be derived as follows:

$$\begin{bmatrix} \sigma_1 & 0 & 0 \\ 0 & \sigma_2 & 0 \\ 0 & 0 & \sigma_3 \end{bmatrix} = \begin{bmatrix} \sigma_h & 0 & 0 \\ 0 & \sigma_h & 0 \\ 0 & 0 & \sigma_h \end{bmatrix} + \begin{bmatrix} \sigma_{1d} & 0 & 0 \\ 0 & \sigma_{2d} & 0 \\ 0 & 0 & \sigma_{3d} \end{bmatrix} \tag{1.77}$$

The dilatational component σ_h is defined as

$$\sigma_h = \frac{\sigma_1 + \sigma_2 + \sigma_3}{3} = \frac{\sigma_{xx} + \sigma_{yy} + \sigma_{zz}}{3} \tag{1.78}$$

which is also called the volumetric stress. Note that $3\sigma_h$ is nothing but the first invariant I_1 of the stress matrix in Eq. (1.20). Thus, it is independent of coordinate system. One can note that σ_h is a state of hydrostatic stress, and hence the subscript h is used to denote the dilatational stress component as well as dilatational energy density.

The dilatational energy density can be obtained by substituting the stress components of the hydrostatic stress state in Eq. (1.78) into the expression for strain energy density in Eq. (1.76),

$$U_h = \frac{1}{2E}[\sigma_h^2 + \sigma_h^2 + \sigma_h^2 - 2\nu(\sigma_h\sigma_h + \sigma_h\sigma_h + \sigma_h\sigma_h)]$$
$$= \frac{3}{2}\frac{(1 - 2\nu)}{E}\sigma_h^2 \tag{1.79}$$

and using the relation in Eq. (1.78),

$$U_h = \frac{3}{2}\frac{(1 - 2\nu)}{E}\left(\frac{\sigma_1 + \sigma_2 + \sigma_3}{3}\right)^2$$
$$= \frac{1 - 2\nu}{6E}[\sigma_1^2 + \sigma_2^2 + \sigma_3^2 + 2(\sigma_1\sigma_2 + \sigma_2\sigma_3 + \sigma_1\sigma_3)] \tag{1.80}$$

1.5.3   Distortion Energy

The distortion part of the strain energy is now found by subtracting Eq. (1.80) from Eq. (1.76), as

$$U_d = U_0 - U_h$$
$$= \frac{1 + \nu}{3E}[\sigma_1^2 + \sigma_2^2 + \sigma_3^2 - \sigma_1\sigma_2 - \sigma_2\sigma_3 - \sigma_1\sigma_3] \tag{1.81}$$
$$= \frac{1 + \nu}{3E}\frac{(\sigma_1 - \sigma_2)^2 + (\sigma_2 - \sigma_3)^2 + (\sigma_3 - \sigma_1)^2}{2}$$

It is customary to write U_d in terms of an equivalent stress called von Mises stress σ_{VM} as

$$U_d = \frac{1 + \nu}{3E}\sigma_{VM}^2 \tag{1.82}$$

The von Mises stress is defined in terms of principal stresses as

$$\sigma_{VM} = \sqrt{\frac{(\sigma_1 - \sigma_2)^2 + (\sigma_2 - \sigma_3)^2 + (\sigma_3 - \sigma_1)^2}{2}} \tag{1.83}$$

1.5.4   Distortion Energy Theory (von Mises)

According to von Mises theory, a ductile solid will yield when the distortion energy density reaches a critical value for that material. Since this should be true for uniaxial stress state also, the critical value of the distortional energy can be estimated from the uniaxial test. At the instance of yielding in a uniaxial tensile test, the state of stress in terms of principal stress is given by: $\sigma_1 = \sigma_Y$ (yield stress) and $\sigma_2 = \sigma_3 = 0$. The distortion energy density associated with yielding is

$$U_d = \frac{1 + \nu}{3E} \sigma_Y^2 \tag{1.84}$$

Thus, the energy density given in Eq. (1.84) is the critical value of the distortional energy density for the material. Then, according to von Mises failure criterion, the material under multi-axial loading will yield when the distortional energy is equal to or greater than the critical value for the material:

$$\frac{1 + \nu}{3E} \sigma_{VM}^2 \geq \frac{1 + \nu}{3E} \sigma_Y^2$$

$$\therefore \sigma_{VM} \geq \sigma_Y \tag{1.85}$$

Thus, the distortion energy theory can be stated that material yields when the von Mises stress exceeds the yield stress obtained in a uniaxial tensile test.

The von Mises stress in Eq. (1.81) can be rewritten in terms of stress components as

$$\sigma_{VM} = \sqrt{\frac{(\sigma_{xx} - \sigma_{yy})^2 + (\sigma_{yy} - \sigma_{zz})^2 + (\sigma_{zz} - \sigma_{xx})^2 + 6(\tau_{xy}^2 + \tau_{yz}^2 + \tau_{zx}^2)}{2}} \tag{1.86}$$

For a two-dimensional plane stress state, $\sigma_3 = 0$, the von Mises stress can be defined in terms of principal stresses as

$$\sigma_{VM} = \sqrt{\sigma_1^2 - \sigma_1 \sigma_2 + \sigma_2^2} \tag{1.87}$$

and in terms of general stress components as

$$\sigma_{VM} = \sqrt{\sigma_{xx}^2 + \sigma_{yy}^2 - \sigma_{xx}\sigma_{yy} + 3\tau_{xy}^2} \tag{1.88}$$

The two-dimensional distortion energy equation (1.87) describes an ellipse, which when plotted on the $\sigma_1 - \sigma_2$ plane is shown in Figure 1.20. The interior of this ellipse defines the region of combined biaxial stress where the material is safe against yielding under static loading.

Consider a situation in which only a shear stress exists, such that $\sigma_x = \sigma_y = 0$, and $\tau_{xy} = \tau$. For this stress state, the principal stresses are $\sigma_1 = -\sigma_2 = \tau$ and $\sigma_3 = 0$. On the $\sigma_1 - \sigma_2$ plane, this pure shear state is represented as a straight line through the origin at $-45°$ as shown in Figure 1.20. The line intersects the von Mises failure envelope at two points, A and B. The magnitude of σ_1 and σ_2 at these points can be found from Eq. (1.87) as

$$\sigma_Y^2 = \sigma_1^2 + \sigma_1 \sigma_1 + \sigma_1^2 = 3\sigma_1^2 = 3\tau_{\max}^2$$

$$\tau_{\max} = \sigma_1 = \frac{\sigma_Y}{\sqrt{3}} = 0.577\sigma_Y \tag{1.89}$$

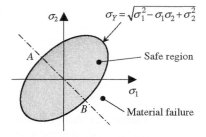

Figure 1.20 Failure envelope of the distortion energy theory

Thus, in pure shear stress state, the material yields when the shear stress reaches $0.577\,\sigma_Y$. This value will be compared to the maximum shear stress theory described below.

1.5.5   The Maximum Shear Stress Theory (Tresca)

According to the maximum shear stress theory, the material yields when the maximum shear stress at a point equals the critical shear stress value for that material. Since this should be true for uniaxial stress state, we can use the results from uniaxial tension test to determine the maximum allowable shear stress. The stress state in a tensile specimen at the point of yielding is given by $\sigma_1 = \sigma_Y$, $\sigma_2 = \sigma_3 = 0$. The maximum shear stress is calculated as

$$\tau_{\mathrm{max}} = \frac{\sigma_1 - \sigma_3}{2} \geq \tau_Y = \frac{\sigma_Y}{2} \tag{1.90}$$

This value of maximum shear stress is also called the yield shear stress of the material and is denoted by τ_Y. Note that $\tau_Y = \sigma_Y/2$. Thus, the Tresca's yield criterion is that yielding will occur in a material when the maximum shear stress equals the yield shear strength, τ_Y, of the material.

The hexagon in Figure 1.21 represents the two-dimensional failure envelope according to maximum shear stress theory. The ellipse corresponding to von Mises theory is also shown in the same figure. The hexagon is inscribed within the ellipse and contacts it at six vertices. Combinations of principal stresses σ_1 and σ_2 that lie within this hexagon are considered safe based on the maximum shear stress theory, and failure is considered to occur when the combined stress state reaches the hexagonal boundary. This is obviously a more conservative failure theory than distortion energy theory as it is contained within the latter. In the pure shear stress state, the shear stress at the points C and D correspond to $0.5\sigma_Y$, which is smaller than $0.577\sigma_Y$ according to the distortion energy theory.

1.5.6   Maximum Principal Stress Theory (Rankine)

According to the maximum principal stress theory, a brittle material ruptures when the maximum principal stress in the specimen reaches some limiting value for the material. Again, this critical value can be inferred as the tensile strength measured using a uniaxial tension test. In practice, this theory is simple, but it can only be used for brittle materials. Some practitioners have modified this theory for ductile materials as

$$\sigma_1 \geq \sigma_U \tag{1.91}$$

where σ_1 is the maximum principal stress and σ_U the ultimate strength described in Table 1.3. Figure 1.22 shows the failure envelope based on the maximum principal stress

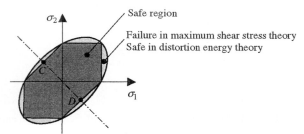

Figure 1.21 Failure envelope of the maximum shear stress theory

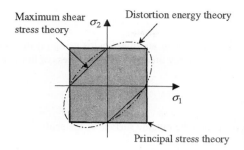

Figure 1.22 Failure envelope of the maximum principal stress theory

theory. Note that the failure envelopes in the first and third quadrants are coincident with those of the maximum shear stress theory and contained within the distortion energy theory. However, the envelopes in the second and fourth quadrants are well outside of the other two theories. Hence, the maximum principal stress theory is not considered suitable for ductile materials. However, it can be used to predict failure in brittle materials.

1.6   SAFETY FACTOR

One can notice that all of the aforementioned failure theories are of the following form:

$$A \text{ function of stress} \geq \text{strength.} \tag{1.92}$$

The term on the LHS of a failure criterion depends on the state of stress at a point. Various stress analysis methods, including the finite element method discussed in this book, are used to evaluate the stress term. The RHS of the equation is a material property usually determined from material tests. There are many uncertainties in calculating the state of stress at a point. These include uncertainties in the loads, material properties such as Young's modulus, dimensions, and geometry of the solid. Similarly, there are uncertainties in the strength of a material depending on the tests used to measure the strength, manufacturing process, etc. In order to account for the uncertainties, engineers use a factor of safety in the design of a solid or structural component. Thus, the failure criteria are modified as

$$N \times \text{stress} = \text{strength}, \tag{1.93}$$

where N is the safety factor. That is, we assume the stress is N times the calculated or estimated state of stress. Another interpretation is that the strength is reduced by a factor N to account for uncertainties. Thus, the space of allowable stresses is contained well within the failure envelope shown in Figure 1.20. For example, the safety factor in the von Mises theory is defined as

$$N_{VM} = \frac{\sigma_Y}{\sigma_{VM}} \tag{1.94}$$

In many engineering applications, N is in the range of 1.1–1.5.

The safety factor in the maximum shear stress theory is defined as

$$N_\tau = \frac{\tau_Y}{\tau_{\max}} = \frac{\sigma_Y/2}{\tau_{\max}} \tag{1.95}$$

It can be shown that for any two-dimensional loading, $N_{VM} \geq N_\tau$

$$N_{VM} \geq N_\tau$$
$$\frac{\sigma_Y}{\sigma_{VM}} \geq \frac{\tau_Y}{\tau_{\max}} \tag{1.96}$$

EXAMPLE 1.10 *Yield Criteria of a Shaft*

Estimate the torque on a 10-mm-diameter steel shaft when yielding begins using (a) the maximum shear stress theory and (b) the maximum distortion energy theory. The yield stress of the steel is 140 MPa.

SOLUTION

(a) For torsion, the maximum shear stress occurs on the outside surface of the shaft:

$$T_{max} = \frac{\tau_{max} J}{d/2} \tag{1.97}$$

where $J = \pi d^4 / 32$ is the polar moment of inertia for a solid circular shaft. Using the relation that the maximum shear stress is half of the yield stress, the shear yield stress for torsion can be obtained from Eq. (1.97), as

$$\tau_{max} = \frac{16T}{\pi d^3} = \frac{1}{2}\sigma_Y.$$

Solving for the torque yields

$$T = \frac{\pi d^3}{32}\sigma_Y = \frac{\pi(0.01)^3}{32}140 \times 10^6 = 13.74\,\text{N} \cdot \text{m}$$

(b) The principal stresses for the torsion problem can be obtained from the Mohr's circle,

$$\sigma_1 = \tau, \quad \sigma_2 = 0, \quad \sigma_3 = -\tau.$$

From Eq. (1.89),

$$\tau = 0.577\sigma_Y$$

Substituting into Eq. (1.97) and solving for the torque results in

$$T = \frac{0.577\pi d^3}{16}\sigma_Y = \frac{0.577\pi(0.01)^3}{16}140 \times 10^6 = 15.86\,\text{N} \cdot \text{m}$$

Thus, it can be seen that for yielding in pure torsion, the distortion energy theory predicts a torque, which is 15% greater than the prediction of the maximum shear stress theory. Tests on ductile materials have shown that the distortion energy theory is more accurate for predicting yield, but in design the more conservative answer predicted by maximum shear stress theory is commonly used.

EXAMPLE 1.11 *Safety Factor of a Bracket*

The bracket shown in Figure 1.23 consists of a rigid arm and a flexible rod. The latter has the following properties: moment of inertia $I = 1.0$, polar moment of inertia $J = 0.5$, radius $r = 0.1$, length $l = 10.0$, and yield stress $\sigma_Y = 2.8$. When a vertical force of $F = 1.0$ is applied at the end of the rigid

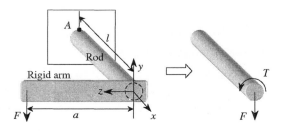

Figure 1.23 Bracket structure

arm $(a = 5\sqrt{2})$, calculate the stress matrix at point A and evaluate the safety factors at that point using both the distortion energy theory and maximum shear stress theory.

SOLUTION   As shown in Figure 1.23, the torque and the transverse shear force is applied to the rod. Since the transverse shear stress caused by F is zero at point A, only the bending moment and the torque contribute to the stress in the rod. Thus, stress components are

$$\sigma_{xx} = \frac{M \cdot r}{I} = \frac{F \cdot l \cdot r}{I} = 1$$

$$\tau_{xz} = \frac{T \cdot r}{J} = \frac{5\sqrt{2} \cdot 0.1}{0.5} = \sqrt{2}$$

$$\sigma_{yy} = \sigma_{zz} = \tau_{xy} = \tau_{yz} = 0.0$$

Thus, the symmetric stress matrix can be written as

$$[\boldsymbol{\sigma}] = \begin{bmatrix} 1 & 0 & \sqrt{2} \\ 0 & 0 & 0 \\ \sqrt{2} & 0 & 0 \end{bmatrix}$$

The principal stresses are found by solving the following eigen value problem:

$$([\boldsymbol{\sigma}] - \lambda[\mathbf{I}]) \cdot \mathbf{n} = 0$$

The solution to the above eigen value problem is non-trivial only if the coefficient matrix is singular or, equivalently, its determinant is equal to zero.

$$\begin{vmatrix} 1 - \lambda & 0 & \sqrt{2} \\ 0 & -\lambda & 0 \\ \sqrt{2} & 0 & -\lambda \end{vmatrix} = 0$$

After expanding the Jacobian, the following three solutions are obtained:

$$-\lambda[-\lambda(1 - \lambda) - 2] = -\lambda(\lambda^2 - \lambda - 2) = 0$$

$$\Rightarrow -\lambda(\lambda - 2)(\lambda + 1) = 0, \qquad \therefore \lambda = 2, 0, -1$$

which are the three principal stresses. Thus, we obtain

$$\sigma_1 = 2, \quad \sigma_2 = 0, \quad \sigma_3 = -1$$

To apply for the failure theory, the maximum shear stress and von Mises stress are calculated by

$$\tau_{max} = \frac{\sigma_1 - \sigma_3}{2} = 1.5$$

$$\sigma_{VM} = \sqrt{4 + 2 + 1} = \sqrt{7}$$

The safety factor from the distortion energy theory is

$$N = \frac{\sigma_Y}{\sigma_{VM}} = \frac{2.8}{\sqrt{7}} = 1.0583$$

which means that the structure is safe.

The safety factor from the maximum shear stress theory is

$$N = \frac{\tau_Y}{\tau_{max}} = \frac{1.4}{1.5} = 0.9333$$

which means that the structure will yield.

1.7   EXERCISE

1. A vertical force F is applied to a two-bar truss as shown in the figure. Let cross-sectional areas of the members 1 and 2 be A_1 and A_2, respectively. Determine the area ratio A_1/A_2 in order to have the same magnitude of stress in both members.

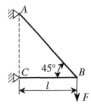

2. The stress at a point P is given below. The direction cosines of the normal $\mathbf{n}$ to a plane that passes through P have the ratio $n_x : n_y : n_z = 3 : 4 : 12$. Determine (a) the traction vector $\mathbf{T^{(n)}}$; (b) the magnitude T of $\mathbf{T^{(n)}}$; (c) the normal stress σ_n; (d) the shear stress τ_n; and (e) the angle between $\mathbf{T^{(n)}}$ and $\mathbf{n}$.
   Hint: Use $n_x^2 + n_y^2 + n_z^2 = 1$.

$$[\sigma] = \begin{bmatrix} 13 & 13 & 0 \\ 13 & 26 & -13 \\ 0 & -13 & -39 \end{bmatrix}$$

3. At a point P in a body, Cartesian stress components are given by $\sigma_{xx} = 80\,\text{MPa}$, $\sigma_{yy} = -40\,\text{MPa}$, $\sigma_{zz} = -40\,\text{MPa}$, and $\tau_{xy} = \tau_{yz} = \tau_{zx} = 80\,\text{MPa}$. Determine the traction vector, its normal component, and its shear component on a plane that is equally inclined to all three coordinate axes.
   Hint: When a plane is equally inclined to all three coordinate axes, the direction cosines of the normal are equal to each other.

4. If $\sigma_{xx} = 90\,\text{MPa}$, $\sigma_{yy} = -45\,\text{MPa}$, $\tau_{xy} = 30\,\text{MPa}$, and $\sigma_{zz} = \tau_{xz} = \tau_{yz} = 0$, compute the surface traction $\mathbf{T^{(n)}}$ on the plane shown in the figure, which makes an angle of $\theta = 40°$ with the vertical axis. What are the normal and shear components of stress on this plane?

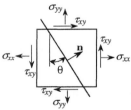

5. Find the principal stresses and the corresponding principal directions stress for the following cases of plane stress.

   (a) $\sigma_{xx} = 40\,\text{MPa}$, $\sigma_{yy} = 0\,\text{MPa}$, $\tau_{xy} = 80\,\text{MPa}$
   (b) $\sigma_{xx} = 140\,\text{MPa}$, $\sigma_{yy} = 20\,\text{MPa}$, $\tau_{xy} = -60\,\text{MPa}$
   (c) $\sigma_{xx} = -120\,\text{MPa}$, $\sigma_{yy} = 50\,\text{MPa}$, $\tau_{xy} = 100\,\text{MPa}$

6. If the minimum principal stress is $-7\,\text{MPa}$, find σ_{xx} and the angle that the principal stress axes make with the xy axes for the case of plane stress illustrated.

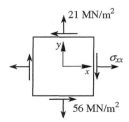

7. Determine the principal stresses and their associated directions when the stress matrix at a point is given by

$$[\sigma] = \begin{bmatrix} 1 & 1 & 1 \\ 1 & 1 & 2 \\ 1 & 2 & 1 \end{bmatrix} \text{MPa}$$

8. Let $x'y'z'$ coordinate system be defined using the three principal directions obtained from Problem 7. Determine the transformed stress matrix $[\sigma]_{x'y'z'}$ in the new coordinates system.

9. For the stress matrix below, the two principal stresses are given as $\sigma_3 = -3$ and $\sigma_1 = 2$, respectively. In addition, the two principal stress directions corresponding to the two principal stresses are also given below.

$$[\sigma] = \begin{bmatrix} 1 & 0 & 2 \\ 0 & 1 & 0 \\ 2 & 0 & -2 \end{bmatrix}, \ \mathbf{n}^1 = \begin{bmatrix} \dfrac{2}{\sqrt{5}} \\ 0 \\ \dfrac{1}{\sqrt{5}} \end{bmatrix} \text{and } \mathbf{n}^3 = \begin{bmatrix} \dfrac{1}{\sqrt{5}} \\ 0 \\ \dfrac{-2}{\sqrt{5}} \end{bmatrix}$$

   (a) What is the normal and shear stress on a plane whose normal vector is parallel to (2, 1, 2)?
   (b) Calculate the missing principal stress σ_2 and the principal direction $\mathbf{n}^2$.
   (c) Write stress matrix in the new coordinates system that is aligned with $\mathbf{n}^1$, $\mathbf{n}^2$, and $\mathbf{n}^3$.

10. With respect to the coordinate system xyz, the state of stress at a point P in a solid is

$$[\sigma] = \begin{bmatrix} -20 & 0 & 0 \\ 0 & 50 & 0 \\ 0 & 0 & 50 \end{bmatrix} \text{MPa}$$

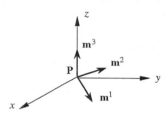

   (a) $\mathbf{m}^1$, $\mathbf{m}^2$, and $\mathbf{m}^3$ are three mutually perpendicular vectors such that $\mathbf{m}^1$ makes 45° with both x- and y-axes and $\mathbf{m}^3$ is aligned with the z-axis. Compute the normal stresses on planes normal to $\mathbf{m}^1$, $\mathbf{m}^2$, and $\mathbf{m}^3$.
   (b) Compute two components of shear stress on the plane normal to $\mathbf{m}^1$ in the directions $\mathbf{m}^2$ and $\mathbf{m}^3$.
   (c) Is the vector $\mathbf{n} = \{0, 1, 1\}^T$ a principal direction of stress? Explain. What is the normal stress in the direction $\mathbf{n}$?
   (d) Draw an infinitesimal cube with faces normal to $\mathbf{m}^1$, $\mathbf{m}^2$, and $\mathbf{m}^3$ and display the stresses on the positive faces of this cube.
   (e) Express the state of stress at the point P with respect to the $x'y'z'$ coordinates system that is aligned with the vectors $\mathbf{m}^1$, $\mathbf{m}^2$, and $\mathbf{m}^3$.
   (f) What are the principal stress and principal directions of stress at the point P with respect to the $x'y'z'$ coordinates system? Explain.
   (g) Compute the maximum shear stress at the point P. Which plane(s) does this maximum shear stress act on?

11. A solid shaft of diameter $d = 5$ cm, as shown in the figure, is subjected to tensile force $P = 13,000$ N and a torque $T = 6,000$ N · cm. At point A on the surface, what is the state of stress (write in matrix form), the principal stresses, and the maximum shear stress? Show the coordinate system you are using.

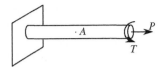

12. If the displacement field is given by

$$\begin{cases} u = x^2 + 2y^2 \\ v = -y^2 - 2x(y - z) \\ w = -z^2 - 2xy \end{cases}$$

(a) Write down 3×3 strain matrix.

(b) What is the normal strain component in the direction of $(1,1,1)$ at point $(1, -3, 1)$?

13. Consider the following displacement field in a plane solid:

$$u(x, y) = 0.04 - 0.01x + 0.006y$$
$$v(x, y) = 0.06 + 0.009x + 0.012y$$

(a) Compute the strain components ε_{xx}, ε_{yy}, and γ_{xy}. Is this a state of uniform strain?

(b) Determine the principal strains and their corresponding directions. Express the principal strain directions in terms of *angles* the directions make with the x-axis.

(c) What is the normal strain at Point O in a direction $45°$ to the x-axis?

14. The displacement field in a solid is given by

$$\begin{cases} u = kx^2 \\ v = 2kxy^2 \\ w = k(x + y)z \end{cases}$$

where k is a constant.

(a) Write down the strain matrix.

(b) What is the normal strain in the direction of $\mathbf{n} = \{1,\ 1,\ 1\}^T$?

15. Draw a 2×2-inch square OABC on the engineering paper. The coordinates of O are $(0, 0)$ and of B are $(2, 2)$. Using the displacement field in Problem 13, determine the u and v displacements of the corners of the square. Let the deformed square be denoted as O'A'B'C'.

(a) Determine the change in lengths of OA and OC. Relate the changes to the strain components.

(b) Determine the change in $\angle AOC$. Relate the change to the shear strain.

(c) Determine the change in length in the diagonal OB. How is it related to the strain(s)?

(d) Show that the relative change in the area of the square (change in area/original area) is given by $\Delta A/A = \varepsilon_{xx} + \varepsilon_{yy} = \varepsilon_1 + \varepsilon_2$.

Hint: You can use the old-fashioned method of using set-squares (triangles) and protractor or use spreadsheet to do the calculations. Place the origin somewhere in the bottom middle of the paper so that you have enough room to the left of the origin.

16. Draw a 2×2 -inch square $OPQR$ such that OP makes $+73°$ to the x-axis. Repeat questions (a) through (d) in Problem 15 for $OPQR$. Give physical interpretations to your results.

Note: The principal strains and the principal strain directions are given by

$$\varepsilon_{1,2} = \frac{(\varepsilon_{xx} + \varepsilon_{yy})}{2} \pm \sqrt{\left(\frac{\varepsilon_{xx} - \varepsilon_{yy}}{2}\right)^2 + \left(\frac{\gamma_{xy}}{2}\right)^2}$$

$$\tan 2\theta = \frac{\gamma_{xy}}{\varepsilon_{xx} - \varepsilon_{yy}}$$

17. For steel, the following material data are applicable: Young's modulus $E = 207$ GPa and shear modulus $G = 80$ GPa. For the strain matrix at a point shown below, determine the symmetric 3×3 stress matrix.

$$[\varepsilon] = \begin{bmatrix} 0.003 & 0 & -0.006 \\ 0 & -0.001 & 0.003 \\ -0.006 & 0.003 & 0.0015 \end{bmatrix}$$

18. Strain at a point is such that $\varepsilon_{xx} = \varepsilon_{yy} = 0$, $\varepsilon_{zz} = -0.001$, $\varepsilon_{xy} = 0.006$, and $\varepsilon_{xz} = \varepsilon_{yz} = 0$. Note: You need not solve the eigen value problem for this question.

   (a) Show that $\mathbf{n}^1 = \mathbf{i} + \mathbf{j}$ and $\mathbf{n}^2 = -\mathbf{i} + \mathbf{j}$ are principal directions of strain at this point.

   (b) What is the third principal direction?

   (c) Compute the three principal strains.

19. Derive the stress-strain relationship in Eq. (1.60) from Eq. (1.55) and the plane stress conditions.

20. A thin plate of width b, thickness t, and length L is placed between two frictionless rigid walls a distance b apart and is acted on by an axial force P. The material properties are Young's modulus E and Poisson's ratio ν?.

   (a) Find the stress and strain components in the xyz coordinate system.

   (b) Find the displacement field.

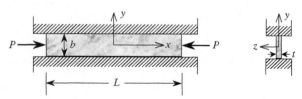

21. A solid with Young's modulus $E = 70$ GPa and Poisson's ratio $= 0.3$ is in a state of **plane strain** parallel to the xy-plane. The in-plane strain components are measured as follows: $\varepsilon_{xx} = 0.007$, $\varepsilon_{yy} = -0.008$, and $\gamma_{xy} = 0.02$.

   (a) Compute the principal strains and corresponding principal strain directions.

   (b) Compute the stresses including σ_{zz}, corresponding to the above strains.

   (c) Determine the principal stresses and corresponding principal stress directions. Are the principal stress and principal strain directions the same?

   (d) Show that the principal stresses could have been obtained from the principal strains using the stress-strain relations.

   (e) Compute the strain energy density using the stress and strain components in xy-coordinate system.

   (f) Compute the strain energy density using the principal stresses and principal strains.

22. Assume that the solid in Problem 21 is under a state of plane stress. Repeat (b) through (f).

23. A strain rosette consisting of three strain gages was used to measure the strains at a point in a thin-walled plate. The measured strains in the three gages are $\varepsilon_A = 0.001$, $\varepsilon_B = -0.0006$, and $\varepsilon_C = 0.0007$. Note that Gage C is at 45° with respect to x-axis. Assume the plane stress state.

   (a) Determine the complete state of strains and stresses (all six components) at that point. Assume $E = 70$ GPa, and $\nu = 0.3$.

   (b) What are the principal strains and their directions?

   (c) What are the principal stresses and their directions?

   (d) Show that the principal strains and stresses satisfy the stress-strain relations.

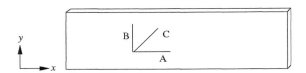

24. A strain rosette consisting of three strain gages was used to measure the strains at a point in a thin-walled plate. The measured strains in the three gages are: $\varepsilon_A = 0.016$, $\varepsilon_B = 0.004$, and $\varepsilon_C = 0.016$. Determine the complete state of strains and stresses (all six components) at that point. Assume $E = 100$ GPa and $\nu = 0.3$.

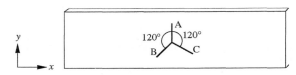

25. A strain rosette consisting of three strain gages was used to measure the strains at a point in a thin-walled plate. The measured strains in the three gages are: $\varepsilon_A = 0.008$, $\varepsilon_B = 0.002$, and $\varepsilon_C = 0.008$. Determine the complete state of strains and stresses (all six components) at that point. Assume $E = 100$ GPa and $\nu = 0.3$.

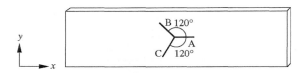

26. The figure below illustrates a thin plate of thickness t. An approximate displacement field, which accounts for displacements due to the weight of the plate, is given by

$$u(x, y) = \frac{\rho}{2E}(2bx - x^2 - \nu y^2)$$

$$v(x, y) = -\frac{\nu\rho}{E} y(b - x)$$

   (a) Determine the corresponding plane stress field.

   (b) Qualitatively draw the deformed shape of the plate.

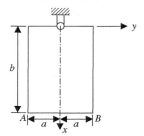

27. The stress matrix at a particular point in a body is

$$[\boldsymbol{\sigma}] = \begin{bmatrix} -2 & 1 & -3 \\ 1 & 0 & 4 \\ -3 & 4 & 5 \end{bmatrix} \times 10^7 \, Pa$$

Determine the corresponding strain if $E = 20 \times 10^{10}$ Pa and $\nu = 0.3$.

28. For a *plane stress* problem, the strain components in the x-y plane at a point P are computed as

$$\varepsilon_{xx} = \varepsilon_{yy} = .125 \times 10^{-2}, \quad \varepsilon_{xy} = .25 \times 10^{-2}$$

   (a) Compute the state of stress at this point if Young's modulus $E = 2 \times 10^{11}$ Pa and Poisson's ratio $\nu = 0.3$.
   (b) What is the normal strain in the z-direction?
   (c) Compute the normal strain in the direction of $n = \{1,\ 1,\ 1\}^T$.

29. The state of stress at a point is given by

$$[\boldsymbol{\sigma}] = \begin{bmatrix} 80 & 20 & 40 \\ 20 & 60 & 10 \\ 40 & 10 & 20 \end{bmatrix} MPa$$

   (a) Determine the strains using Young's modulus of 100 GPa and Poisson's ratio of 0.25.
   (b) Compute the strain energy density using these stresses and strains.
   (c) Calculate the principal stresses.
   (d) Calculate the principal strains from the strains calculated in (a).
   (e) Show that the principal stresses and principal strains satisfy the constitutive relations.
   (f) Calculate the strain energy density using the principal stresses and strains.

30. Consider the state of stress in problem 29 above. The yield strength of the material is 100 MPa. Determine the safety factors according to the following: (a) maximum principal stress criterion, (b) Tresca Criterion, and (c) von Mises criterion.

31. A thin-walled tube is subject to a torque T. The only non-zero stress component is the shear stress τ_{xy}, which is given by $\tau_{xy} = 10{,}000\,T$(Pa), where T is the torque in N.m. When the yield strength $\sigma_Y = 300$ MPa and the safety factor $N = 2$, calculate the maximum torque that can be applied using

   (a) Maximum principal stress criterion (Rankine)
   (b) Maximum shear stress criterion (Tresca)
   (c) Distortion energy criterion (Von Mises)

32. A thin-walled cylindrical pressure vessel with closed ends is subjected to an internal pressure $p = 100$ psi and also a torque T around its axis of symmetry. Determine T that will cause yielding according to von Mises' yield criterion. The design requires a safety factor of 2. The nominal diameter D of the pressure vessel $= 20$ inches, wall thickness $t = 0.1$ inch, and yield strength of the material $= 30$ ksi. (1 ksi $= 1000$ psi). Stresses in a thin-walled cylinder are longitudinal stress σ_l, hoop stress σ_h, and shear stress τ due to torsion. They are given by

$$\sigma_l = \frac{pD}{4t}, \quad \sigma_h = \frac{pD}{2t}, \quad \tau = \frac{2T}{\pi D^2 t}$$

33. A cold–rolled steel shaft is used to transmit 60 kW at 500 rpm from a motor. What should be the diameter of the shaft if the shaft is 6 m long and is simply supported at both ends? The shaft also experiences bending due to a distributed transverse load of 200 N/m. Ignore bending due to

the weight of the shaft. Use a factor of safety 2. The tensile yield limit is 280 MPa. Find the diameter using both maximum shear stress theory and von Mises criterion for yielding.

34. For the stress matrix below, the two principal stresses are given as $\sigma_1 = 2$ and $\sigma_3 = -3$, respectively. In addition, two principal directions corresponding to the two principal stresses are also given below. The yield stress of the structure is given as $\sigma_Y = 4.5$.

$$[\sigma] = \begin{bmatrix} 1 & 0 & 2 \\ 0 & 1 & 0 \\ 2 & 0 & -2 \end{bmatrix}, \quad \mathbf{n}^1 = \begin{bmatrix} \dfrac{2}{\sqrt{5}} \\ 0 \\ \dfrac{1}{\sqrt{5}} \end{bmatrix} \quad \text{and} \quad \mathbf{n}^3 = \begin{bmatrix} \dfrac{1}{\sqrt{5}} \\ 0 \\ \dfrac{-2}{\sqrt{5}} \end{bmatrix}$$

(a) Calculate the safety factor based on the maximum shear stress theory and determine whether the structure is safe.

(b) Calculate the safety factor based on the distortion energy theory and determine whether the structure is safe.

35. The figure below shows a shaft of 1.5-in diameter loaded by a bending moment $M_z = 5,000\ \text{lb} \cdot \text{in}$, a torque $T = 8,000\ \text{lb} \cdot \text{in}$, and an axial tensile force $N = 6,000\ \text{lb}$. If the material is ductile with the yielding stress $\sigma_Y = 40,000\ \text{psi}$, determine the safety factor using (a) the maximum shear stress theory and (b) the maximum distortion energy theory.

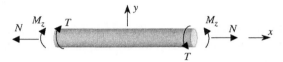

36. A 20-mm-diameter rod made of a ductile material with a yield strength of 350 MPa is subject to a torque of $T = 100\ \text{N} \cdot \text{m}$ and a bending moment of $M = 150\ \text{N} \cdot \text{m}$. An axial tensile force P is then gradually applied. What is the value of the axial force when yielding of the rod occurs? Solve the problem in two ways using: (a) the maximum shear stress theory and (b) the maximum distortional energy theory.

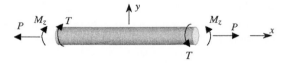

37. A circular shaft of radius r in the figure has a moment of inertia I and polar moment of inertia J. The shaft is under torsion T_z in the positive z-axis and bending moment M_x in the positive x-axis. The material is mild steel with yield strength of 2.8 MPa. Use only the given coordinate system for your calculations.

(a) If T_z and M_x are gradually increased, which point (or points) will fail first among four points (A, B, C, and D)? Identify all.

(b) Construct stress matrix $[\sigma]_A$ at point A in xyz-coordinates in terms of given parameters (i.e., T_z, M_x, I, J, and r).

(c) Calculate three principal stresses at point B in terms of given parameters.

(d) When the principal stresses at point C are $\sigma_1 = 1$, $\sigma_2 = 0$, and $\sigma_3 = -2\ \text{MPa}$, calculate safety factors **(1)** from maximum shear stress theory and **(2)** from distortion energy theory.

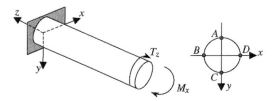

38. A rectangular plastic specimen of size $100 \times 100 \times 10\,\text{mm}^3$ is placed in a rectangular metal cavity. The dimensions of the cavity are $101 \times 101 \times 9\,\text{mm}^3$. The plastic is compressed by a rigid punch until it is completely inside the cavity. Due to Poisson effect, the plastic also expands in the x and y directions and fills the cavity. Calculate all stress and strain components and the force exerted by the punch. Assume there is no friction between all contacting surfaces. The metal cavity is rigid. Elastic constants of the plastic are $E = 10\,\text{GPa}$, $\nu = 0.3$.

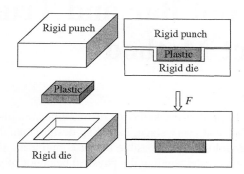

39. Repeat Problem 38 with elastic constants of the plastic as $E = 10\,\text{GPa}$ and $\nu = 0.485$.

40. Repeat Problem 38 with the specimen of size $100 \times 100 \times 10\,\text{mm}^3$ and the dimensions of the cavity $104 \times 104 \times 9\,\text{mm}^3$. Elastic constants of the plastic are $E = 10\,\text{GPa}$, $\nu = 0.3$.

Chapter 2

Uniaxial Bar and Truss Elements: Direct Method

An ability to predict the behavior of machines, engineering systems in general, is of great importance at every stage, including design, manufacture, and operation. Such predictive methodologies are possible because engineers and scientists have made tremendous progress in understanding the physical behavior of materials and structures and have developed mathematical models, albeit approximate, in order to describe their physical behavior. Most often the mathematical models result in algebraic, differential, or integral equations or combinations thereof. Seldom can these equations be solved in a closed form, and hence numerical methods are used to obtain solutions. Finite difference method is a classical method that provides approximate solutions to differential equations with reasonable accuracy. There are other methods of solving mathematical equations that are covered in traditional numerical methods courses.[1]

Finite Element Method (FEM) is one of the numerical methods of solving differential equations. The FEM, originated in the area of structural mechanics, has been extended to other areas of solid mechanics and later to other fields such as heat transfer, fluid dynamics, and electromagnetism. In fact, FEM has been recognized as a powerful tool for solving partial differential equations and integro-differential equations, and in the near future, it may become the numerical method of choice in many engineering and applied science areas. One of the reasons for FEM's popularity is that the method results in computer programs versatile in nature that can solve many practical problems with the least amount of training. Obviously, there is a danger in using computer programs without proper understanding of the theory behind them, and that is one of the reasons to have a thorough understanding of the theory behind the FEM.

The basic principle of FEM is to divide or *discretize* the system into a number of smaller elements called finite elements (FEs); to identify the degrees of freedom (DOFs) that describe its behavior; and then to write down the equations that describe the behavior of each element and its interaction with neighboring elements. The element-level equations are assembled to obtain global equations, often a linear system of equations, which are solved for the unknown DOFs. The phrase *finite element* refers to the fact that the elements are of finite size as opposed to the infinitesimal or differential element considered in deriving the governing equations of the system. Another interpretation is that the FE equations deal with a finite number of DOFs, as opposed to the infinite number of DOFs of a continuous system.

In general, solutions of practical engineering problems are quite complex, and they cannot be represented using simple expressions. An important concept of FEM is that the solution is approximated using simple polynomials, often linear or quadratic, within each

---

[1] Atkinson, K. E., *An Introduction to Numerical Analysis*, Wiley, New York, 1978

element. Since elements are connected throughout the system, the solution of the system is approximated using piecewise polynomials. Such approximation may contain errors when the size of element is large. As the size of element reduces, however, the approximated solution will converge to the exact solution.

There are three methods that can be used to derive the FE equations of a problem: (a) direct method; (b) variational method; and (c) weighted residual method. The direct method provides a clear physical insight into FEM and is preferred in the beginning stages of learning the principles. However, it is limited in its application in that it can be used to solve one-dimensional problems only. The variational method is akin to the methods of calculus of variations and is a powerful tool for deriving the FE equations. However, it requires the existence of a functional, minimization of which results in the solution of the differential equations. The Galerkin method is one of the popular weighted residual methods and is applicable to most problems. If a variational function exists for the problem, then the variational and Galerkin methods yield identical solutions.

In this chapter, we will illustrate the direct method of FE analysis using one-dimensional elements, such as linear spring, uniaxial bar, and truss elements. The emphasis is on construction and solution of the finite element equations and interpretation of the results, rather than the rigorous development of the general principles of the FEM.

2.1   ILLUSTRATION OF THE DIRECT METHOD

Consider a system of rigid bodies connected by springs, as shown in Figure 2.1. The bodies move only in the horizontal direction. Furthermore, we consider only the static, problem, and hence the mass effects (inertia) will be ignored. External forces, F_2, F_3, and F_4, are applied on the rigid bodies as shown. The objectives are to determine the displacement of each body, forces in the springs, and support reactions.

We will introduce the principles involved in the FEM through this example. Notice that there is no need to discretize the system, as it already consists of discrete elements, namely, the springs. The elements are connected at the nodes. In this case, the rigid bodies are the nodes. Of course, the two walls are also the nodes, as they connect to the elements. Numbers inside the little circles mark the nodes.

Consider the free body diagram of a typical Element e as shown in Figure 2.2. It has two nodes, Node i and Node j. They will also be referred to as the first and second node or Local Node 1 and Local Node 2, respectively. Assume a coordinate system going from left to right. The convention for first and second nodes is that $x_j > x_i$. The forces acting at the nodes are denoted by $f_i^{(e)}$ and $f_j^{(e)}$. In this notation, the subscripts denote the node numbers and the superscript the element number. Furthermore, the forces are shown in the positive direction. The unknown displacements of the Nodes i and j are u_i and u_j, respectively. Note that there is no superscript for u, as the displacement is unique to the node denoted by the subscript. We would like to develop a relationship between the nodal displacements u_i and u_j and the element forces $f_i^{(e)}$ and $f_j^{(e)}$.

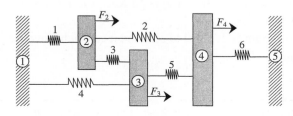

Figure 2.1 Rigid bodies connected by springs

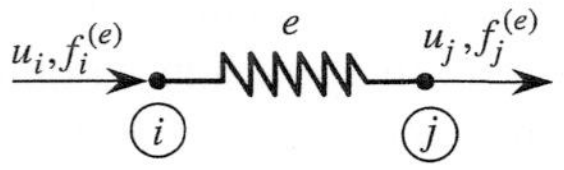

Figure 2.2 Spring Element (e) connected by Node i and Node j

The elongation of the spring is denoted by $\Delta^{(e)} = u_j - u_i$. Then the force in the spring is given by

$$P^{(e)} = k^{(e)}\Delta^{(e)} = k^{(e)}(u_j - u_i) \tag{2.1}$$

where $k^{(e)}$ is the spring rate or *stiffness* of Element e. If $u_j > u_i$, then the spring is elongated and the force in the spring is positive (tension). Otherwise, the spring is in compression. The spring force is related to the nodal force by

$$f_j^{(e)} = P^{(e)} \tag{2.2}$$

For equilibrium, the sum of the forces acting on Element e must be equal to zero, i.e.,

$$f_i^{(e)} + f_j^{(e)} = 0 \quad \text{or} \quad f_i^{(e)} = -f_j^{(e)} \tag{2.3}$$

From Eqs. (2.1)–(2.3), we can obtain a relation between the element forces and the displacements as

$$\begin{aligned} f_i^{(e)} &= k^{(e)}(u_i - u_j) \\ f_j^{(e)} &= k^{(e)}(-u_i + u_j) \end{aligned} \tag{2.4}$$

Equation (2.4) can be written in matrix forms as

$$k^{(e)}\begin{bmatrix} 1 & -1 \\ -1 & 1 \end{bmatrix}\begin{Bmatrix} u_i \\ u_j \end{Bmatrix} = \begin{Bmatrix} f_i^{(e)} \\ f_j^{(e)} \end{Bmatrix} \tag{2.5}$$

We also write Eq. (2.5) in a shorthand notation as

$$[\mathbf{k}^{(e)}]\begin{Bmatrix} u_i \\ u_j \end{Bmatrix} = \begin{Bmatrix} f_i^{(e)} \\ f_j^{(e)} \end{Bmatrix}$$

or,

$$\boxed{[\mathbf{k}^{(e)}]\{\mathbf{q}^{(e)}\} = \{\mathbf{f}^{(e)}\}} \tag{2.6}$$

where $[\mathbf{k}^{(e)}]$ is the element stiffness matrix, $\{\mathbf{q}^{(e)}\}$ is the vector of DOFs associated with Element e, and $\{\mathbf{f}^{(e)}\}$ is the vector of element forces. Sometimes we will omit the superscript e with the understanding that we are dealing with a generic element. Equation (2.6) is called the *element equilibrium equation*.

The element stiffness matrix $[\mathbf{k}^{(e)}]$ has the following properties:

1. It is square, as it relates to the same number of forces as the displacements.
2. It is symmetric (a consequence of Betti–Rayleigh Reciprocal Theorem in solid and structural mechanics[2]).
3. It is singular, i.e., the determinant is equal to zero and it cannot be inverted;
4. It is positive semi-definite.

---

[2] Y. C. Fung, *Foundations of Solid Mechanics,* Prentice-Hall, Englewood Cliffs, New Jersey, 1965.

Figure 2.3 Free body diagram of Node 3 in the example shown in Figure 2.1. The external force F_3 and the forces $f_3^{(e)}$ exerted by the springs attached to the node are shown. Note that the forces $f_3^{(e)}$ act in the negative direction

Properties 3 and 4 are related to each other, and they have physical significance. Consider Eq. (2.6). If the nodal displacements u_i and u_j of a spring element in a system are given, then it should be possible to predict the force P in the spring from its change in length $(u_j - u_i)$, and hence the forces $\{f^{(e)}\}$ acting at its nodes can be predicted. In fact, the element forces can be computed by performing the matrix multiplication $[k^{(e)}]\{q^{(e)}\}$. On the other hand, if the spring force is given, the nodal displacements cannot be determined uniquely, as a rigid body displacement (equal u_i and u_j) can be added without affecting the spring force. If $[\mathbf{k}^{(e)}]$ were to have an inverse, then it would have been possible to solve for $\{\mathbf{q}^{(e)}\} = [\mathbf{k}^{(e)}]^{-1}\{\mathbf{f}^{(e)}\}$ uniquely in violation of the physics. Property 4 also has a physical interpretation, which will be discussed in conjunction with energy methods.

In the next step, we develop a relation between the internal forces f and the known external forces F. For example, consider the free body diagram of Node 3 (or the rigid body in this case) in Figure 2.1. The forces acting on the node are the external force F_3 and the forces from the springs connected to Node 3 as shown in Figure 2.3.

For equilibrium of the node, the sum of the forces acting on the node should be equal to zero:

$$F_i - \sum_{e=1}^{i_e} f_i^{(e)} = 0$$

or,

$$F_i = \sum_{e=1}^{i_e} f_i^{(e)}, \quad i = 1, \ldots ND \tag{2.7}$$

where i_e is the number of elements connected to Node i, and ND is the total number of nodes in the model. Such equations can be written for each node including the boundary nodes, e.g., Nodes 1 and 5 in Figure 2.1. The element forces $f_i^{(e)}$ in Eq. (2.7) can be replaced by the unknown DOFs $\mathbf{q}$ by using Eq. (2.6). For example, the force equilibrium for the springs in Figure 2.1 can be written as

$$\begin{cases} F_1 = f_1^{(1)} + f_1^{(4)} = k^{(1)}(u_1 - u_2) + k^{(4)}(u_1 - u_3) \\ F_2 = f_2^{(1)} + f_2^{(3)} + f_2^{(2)} = k^{(1)}(u_2 - u_1) + k^{(3)}(u_2 - u_3) + k^{(2)}(u_2 - u_4) \\ F_3 = f_3^{(3)} + f_3^{(4)} + f_3^{(5)} = k^{(3)}(u_3 - u_2) + k^{(4)}(u_3 - u_1) + k^{(5)}(u_3 - u_4) \\ F_4 = f_4^{(2)} + f_4^{(5)} + f_4^{(6)} = k^{(2)}(u_4 - u_2) + k^{(5)}(u_4 - u_3) + k^{(6)}(u_4 - u_5) \\ F_5 = f_5^{(6)} = k^{(6)}(u_5 - u_4) \end{cases} \tag{2.8}$$

This will result in ND number of linear equations for the ND number of DOFs:

$$[\mathbf{K}_s]\begin{Bmatrix} u_1 \\ u_2 \\ \vdots \\ u_{ND} \end{Bmatrix} = \begin{Bmatrix} F_1 \\ F_2 \\ \vdots \\ F_{ND} \end{Bmatrix} \tag{2.9}$$

Or, in shorthand notation, $[\mathbf{K}_s]\{\mathbf{Q}_s\} = \{\mathbf{F}_s\}$, where $[\mathbf{K}_s]$ is the structural stiffness matrix, $\{\mathbf{Q}_s\}$ is the vector of displacements of all nodes in the model, and $\{\mathbf{F}_s\}$ is the vector of external forces, including the unknown reactions. The expanded form of Eq. (2.9) is given in Eq. (2.10) below:

$$
\begin{bmatrix}
k^1 + k^4 & -k^1 & -k^4 & 0 & 0 \\
-k^1 & k^1 + k^2 + k^3 & -k^3 & -k^2 & 0 \\
-k^4 & -k^3 & k^3 + k^4 + k^5 & -k^5 & 0 \\
0 & -k^2 & -k^5 & k^2 + k^5 + k^6 & -k^6 \\
0 & 0 & 0 & -k^6 & k^6
\end{bmatrix}
\begin{Bmatrix}
u_1 \\ u_2 \\ u_3 \\ u_4 \\ u_5
\end{Bmatrix}
=
\begin{Bmatrix}
F_1 \\ F_2 \\ F_3 \\ F_4 \\ F_5
\end{Bmatrix}
$$

or,

$$[\mathbf{K}_s]\{\mathbf{Q}_s\} = \{\mathbf{F}_s\} \tag{2.10}$$

Note that in Eq. (2.10) the parentheses ( ) for the superscript element numbers are omitted for the purpose of ease of typesetting. They should not be confused for exponents of k.

The properties of the structural stiffness matrix $[\mathbf{K}_s]$ are similar to those of the element stiffness matrix: square, symmetric, singular, and positive semi-definite. In addition, when nodes are numbered properly, $[\mathbf{K}_s]$ will be a banded matrix. It should be noted that when the boundary displacements in $\{\mathbf{Q}_s\}$ are known (usually equal to zero*), corresponding forces in $\{\mathbf{F}_s\}$ are unknown reactions. In the present illustration, $u_1 = u_5 = 0$, and corresponding forces (reactions) F_1 and F_5 are unknown.

We will impose the boundary conditions as follows. First, we ignore the equations for which the RHS forces are unknown and strike out the corresponding rows in $[\mathbf{K}_s]$. This is called striking-the-rows. Then we eliminate the columns in $[\mathbf{K}_s]$ that are multiplied by the zero values of displacements of the boundary nodes. This is called striking-the-columns. It may be noted that if n^{th} row is eliminated (struck), then the nth column will also be eliminated (struck). This process results in a system of equations given by $[\mathbf{K}]\{\mathbf{Q}\} = \{\mathbf{F}\}$, where $[\mathbf{K}]$ is the global stiffness matrix, $\{\mathbf{Q}\}$ is the vector of unknown DOFs, and $\{\mathbf{F}\}$ is the vector of known forces. The global stiffness matrix will be square, symmetric, and **positive definite** and hence non-singular. Usually $[\mathbf{K}]$ will also be banded. In large systems, that is, in models with large numbers of DOFs, $[\mathbf{K}]$ will be a sparse matrix with a small proportion of non-zero numbers in a diagonal band.

After striking the rows and columns corresponding to zero DOFs (u_1 and u_5) in Eq. (2.10), we obtain the global equations as follows:

$$
\begin{bmatrix}
k^1 + k^2 + k^3 & -k^3 & -k^2 \\
-k^3 & k^3 + k^4 + k^5 & -k^5 \\
-k^2 & -k^5 & k^2 + k^5 + k^6
\end{bmatrix}
\begin{Bmatrix}
u_2 \\ u_3 \\ u_4
\end{Bmatrix}
=
\begin{Bmatrix}
F_2 \\ F_3 \\ F_4
\end{Bmatrix}
$$

or,

$$[\mathbf{K}]\{\mathbf{Q}\} = \{\mathbf{F}\} \tag{2.11}$$

In principle, the solution can be obtained as $\{\mathbf{Q}\} = [\mathbf{K}]^{-1}\{\mathbf{F}\}$. Once the unknown DOFs are determined, the spring forces can be obtained using Eq. (2.1). The support reactions can be obtained from either the nodal equilibrium equations (2.7) or the structural equations (2.10).

---

*Non-zero or prescribed DOFs will be dealt in Chapter 5.

Rigid Body-Spring System

Find the displacements of the rigid bodies shown in Figure 2.1. Assume that the only non-zero force is $F_3 = 1,000\,\text{N}$. Determine the element forces (tensile/compressive) in the springs. What are the reactions at the walls? Assume the bodies can undergo only translation in the horizontal direction. The spring constants (N/mm) are $k^{(1)} = 500$, $k^{(2)} = 400$, $k^{(3)} = 600$, $k^{(4)} = 200$, $k^{(5)} = 400$, and $k^{(6)} = 300$.

SOLUTION   The element equilibrium equations are as follows:

$$\left\{ \begin{matrix} f_1^{(1)} \\ f_2^{(1)} \end{matrix} \right\} = 500 \begin{bmatrix} 1 & -1 \\ -1 & 1 \end{bmatrix} \left\{ \begin{matrix} u_1 \\ u_2 \end{matrix} \right\}; \quad \left\{ \begin{matrix} f_2^{(2)} \\ f_4^{(2)} \end{matrix} \right\} = 400 \begin{bmatrix} 1 & -1 \\ -1 & 1 \end{bmatrix} \left\{ \begin{matrix} u_2 \\ u_4 \end{matrix} \right\}$$

$$\left\{ \begin{matrix} f_2^{(3)} \\ f_3^{(3)} \end{matrix} \right\} = 600 \begin{bmatrix} 1 & -1 \\ -1 & 1 \end{bmatrix} \left\{ \begin{matrix} u_2 \\ u_3 \end{matrix} \right\}; \quad \left\{ \begin{matrix} f_1^{(4)} \\ f_3^{(4)} \end{matrix} \right\} = 200 \begin{bmatrix} 1 & -1 \\ -1 & 1 \end{bmatrix} \left\{ \begin{matrix} u_1 \\ u_3 \end{matrix} \right\} \qquad (2.12)$$

$$\left\{ \begin{matrix} f_3^{(5)} \\ f_4^{(5)} \end{matrix} \right\} = 400 \begin{bmatrix} 1 & -1 \\ -1 & 1 \end{bmatrix} \left\{ \begin{matrix} u_3 \\ u_4 \end{matrix} \right\}; \quad \left\{ \begin{matrix} f_4^{(6)} \\ f_5^{(6)} \end{matrix} \right\} = 300 \begin{bmatrix} 1 & -1 \\ -1 & 1 \end{bmatrix} \left\{ \begin{matrix} u_4 \\ u_5 \end{matrix} \right\}$$

The nodal equilibrium equations are:

$$f_1^{(1)} + f_1^{(4)} = F_1 = R_1$$

$$f_2^{(1)} + f_2^{(2)} + f_2^{(3)} = F_2 = 0$$

$$f_3^{(3)} + f_3^{(4)} + f_3^{(5)} = F_3 = 1000 \qquad (2.13)$$

$$f_4^{(2)} + f_4^{(5)} + f_4^{(6)} = F_4 = 0$$

$$f_5^{(6)} = F_5 = R_5$$

where R_1 and R_5 are unknown reaction forces at Nodes 1 and 5, respectively. In the above equation, F_2 and F_4 are equal to zero because no EXTERNAL FORCES act on those nodes. Combining Eqs. (2.12) and (2.13), we obtain the equation $[\mathbf{K}_s]\{\mathbf{Q}_s\} = \{\mathbf{F}_s\}$,

$$100 \begin{bmatrix} 7 & -5 & -2 & 0 & 0 \\ -5 & 15 & -6 & -4 & 0 \\ -2 & -6 & 12 & -4 & 0 \\ 0 & -4 & -4 & 11 & -3 \\ 0 & 0 & 0 & -3 & 3 \end{bmatrix} \left\{ \begin{matrix} u_1 \\ u_2 \\ u_3 \\ u_4 \\ u_5 \end{matrix} \right\} = \left\{ \begin{matrix} R_1 \\ 0 \\ 1000 \\ 0 \\ R_5 \end{matrix} \right\} \qquad (2.14)$$

After implementing the boundary conditions at Nodes 1 and 5 (striking the rows and columns corresponding to zero displacements), we obtain the global equations $[\mathbf{K}]\{\mathbf{Q}\} = \{\mathbf{F}\}$:

$$100 \begin{bmatrix} 15 & -6 & -4 \\ -6 & 12 & -4 \\ -4 & -4 & 11 \end{bmatrix} \left\{ \begin{matrix} u_2 \\ u_3 \\ u_4 \end{matrix} \right\} = \left\{ \begin{matrix} 0 \\ 1000 \\ 0 \end{matrix} \right\}$$

By inverting the global stiffness matrix, the unknown displacements can be obtained as $u_2 = 0.854\,\text{mm}$, $u_3 = 1.55\,\text{mm}$, and $u_4 = 0.875\,\text{mm}$.

The forces in the springs are computed using $P = k(u_j - u_i)$:

$$P^{(1)} = 427\,\text{N}; \quad P^{(2)} = 8.3\,\text{N}; \quad P^{(3)} = 419\,\text{N}$$
$$P^{(4)} = 310\,\text{N}; \quad P^{(5)} = -271\,\text{N}; \quad P^{(6)} = -263\,\text{N}$$

Wall reactions, R_1 and R_5, can be computed either from Eq. (2.14) after substituting for the displacements or from Eqs. (2.12) and (2.13) as $R_1 = -737$ N; $R_5 = -263$ N. Both reactions are negative, meaning that they act on the structure (the system) from right to left.

2.2   UNIAXIAL BAR ELEMENT

The FE analysis procedure for the spring-force system in the previous section can easily be extended to uniaxial bars. Plane and space trusses consist of uniaxial bars, and hence a detailed study of uniaxial bar element will provide the basis for analysis of trusses. Typical problems that can be solved using uniaxial bar elements are shown in Figure 2.4. A uniaxial bar is a slender two-force member where the length is much larger than the cross-sectional dimensions. The bar can have varying cross-sectional area, $A(x)$, and consists of different materials, i.e., varying Young's modulus, $E(x)$. Both concentrated forces F and distributed force $p(x)$ can be applied. The distributed forces can be applied over a portion of the bar. The forces F and $p(x)$ are considered positive if they act in the positive direction of the x-axis. Both ends of the bar can be fixed, making it a statically indeterminate problem. Solving this problem by solving the differential equation of equilibrium could be difficult, if not impossible. However, this problem can be readily solved using FE analysis.

2.2.1   FE Formulation for Uniaxial Bar

The FE analysis procedures for the uniaxial bar are as follows:

1. Discretize the bar into a number of elements. The criteria for determining the size of the elements will become obvious after learning the properties of the element. It is assumed that each element has a constant axial rigidity, *EA*, throughout its length. It may vary from element to element.

2. The elements are connected at nodes. Thus, more than one element can share a node. There will be nodes at points where the bar is supported.

3. External forces are applied only at the nodes, and they must be point forces (concentrated forces). If distributed forces are applied on the bar, they have to be approximated as point forces acting at nodes. At the bar boundary, if displacement is specified, then the reaction is unknown. The reaction will be the external force acting on the boundary node. If a specified external force acts on the boundary, then the corresponding displacement is unknown.

4. The deformation of the bar is determined by the axial displacements of the nodes. That is, the nodal displacements are the DOFs in the FEM. Thus, the DOFs are u_1, $u_2, u_3, \ldots, u_N$, where N is the total number of nodes.

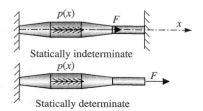

Figure 2.4 Typical one-dimensional bar problems

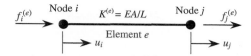

Figure 2.5 Uniaxial bar finite element

The objective of the FE analysis is to determine: (i) unknown DOFs (u_i); (ii) axial force resultant (P) in each element; and (iii) support reactions.

We will use the *direct stiffness method* to derive the element stiffness matrix. Consider the free-body diagram of a typical Element e, as illustrated in Figure 2.5. Forces and displacements are defined as positive when they are in the positive x-direction. The element has two nodes, namely, i and j. Node i will be the first node, and Node j will be called the second node. The convention is that the line i-j will be in the positive direction of the x-axis. The displacements of the nodes are u_i and u_j.

The forces acting at the two ends of the free-body are: $f_i^{(e)}$ and $f_j^{(e)}$. The superscript denotes the element number, and the subscripts denote the node numbers. The (lowercase) force f denotes the internal force as opposed to the (uppercase) external force F_i acting on the nodes. Since we do not know the direction of f, we will assume that all forces act in the positive direction. It should be noted that the nodal displacements do not need a superscript, as they are unique to the nodes. However, the internal force acting at a node may be different for different elements connected to the same node.

First, we will determine a relation between the f's and u's of the Element e. For equilibrium of the element, we have

$$f_i^{(e)} + f_j^{(e)} = 0 \tag{2.15}$$

From elementary mechanics of materials, the force is proportional to the elongation of the element. The elongation of the bar element is denoted by $\Delta^{(e)} = u_j - u_i$. Then, similar to the spring element, where $f = k\Delta$, the force equilibrium of the one-dimensional bar element can be written, as

$$f_j^{(e)} = \left(\frac{AE}{L}\right)^{(e)} (u_j - u_i)$$

$$f_i^{(e)} = -f_j^{(e)} = \left(\frac{AE}{L}\right)^{(e)} (u_i - u_j)$$

where A, E, and L, respectively, are the area of cross-section, Young's modulus, and length of the element. Using matrix notation, the above equations can be written as

$$\left\{ \begin{array}{c} f_i^{(e)} \\ f_j^{(e)} \end{array} \right\} = \left(\frac{AE}{L}\right)^{(e)} \begin{bmatrix} 1 & -1 \\ -1 & 1 \end{bmatrix} \left\{ \begin{array}{c} u_i \\ u_j \end{array} \right\} \tag{2.16}$$

Equation (2.16) is called the element equilibrium equation, which relates the nodal forces of Element e to the corresponding nodal displacements. One can notice that Eq. (2.16) is similar to Eq. (2.5) of the spring element. Equation (2.16) for each element can be written in a compact form as

$$\{\mathbf{f}^{(e)}\} = [\mathbf{k}^{(e)}]\{\mathbf{q}^{(e)}\}, \quad e = 1, 2, \ldots, N_e \tag{2.17}$$

where $[\mathbf{k}^{(e)}]$ is the element stiffness matrix of Element e, $\{\mathbf{q}^{(e)}\}$ is the vector of nodal displacements of the Element e, and N_e is the total number of elements in the model.

Note that the element stiffness matrix in Eq. (2.16) is singular. The fact that the element stiffness matrix does not have an inverse has a physical significance. If the nodal

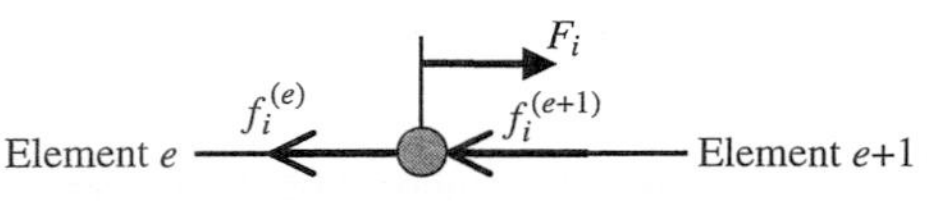

Figure 2.6 Force equilibrium at Node i

displacements of an element are specified, then the element forces can be uniquely determined by performing the matrix multiplication in Eq. (2.16). On the other hand, if the forces acting on the element are given, the nodal displacements cannot be uniquely determined because one can always translate the element by adding a rigid body displacement without affecting the forces acting on it. Thus, it is always necessary to remove the rigid body motion by fixing some displacements at nodes.

2.2.2 Nodal Equilibrium

Consider the free-body diagram of a typical node i. It is connected to, say, Elements e and $e + 1$. Then, the forces acting on the nodes are the external force F_i and reactions to the element forces as shown in Figure 2.6. The internal forces are applied in the negative x-direction because they are the reaction to the forces acting on the element. The sum of the forces acting on Node i must be equal to zero:

$$F_i - f_i^{(e)} - f_i^{(e+1)} = 0$$

or

$$f_i^{(e)} + f_i^{(e+1)} = F_i \tag{2.18}$$

In general, the external force acting on a node is equal to the sum of all of the internal forces acting on different elements connected to the node, and Eq. (2.18) can be generalized as

$$F_i = \sum_{e=1}^{i_e} f_i^{(e)} \tag{2.19}$$

where i_e is the number of elements connected to Node i, and the sum is carried out over all the elements connected to Node i.

2.2.3 Assembly

The next to step is to eliminate the internal forces from Eq. (2.18) using Eq. (2.17) in order to obtain a relation between the unknown displacements $\{\mathbf{Q}_s\}$ and known forces $\{\mathbf{F}_s\}$. This step results in a process called assembly of the element stiffness matrices. We substitute for f's from Eq. (2.17) into Eq. (2.19) in order to find a relation between the nodal displacements and forces F. The force equilibrium in Eq. (2.19) can be written for each DOF at each node, yielding a relation between the external forces and displacements as

$$[\mathbf{K}_s]\{\mathbf{Q}_s\} = \{\mathbf{F}_s\} \tag{2.20}$$

In the above equation, $[\mathbf{K}_s]$ is the structural stiffness matrix which characterize the load-deflection behavior of the entire structure; $\{\mathbf{Q}_s\}$ is the vector of all nodal displacements, known and unknown; and $\{\mathbf{F}_s\}$ is the vector of external forces acting at the nodes, including the unknown reactions.

There is a mechanical procedure by which the element stiffness matrices $[\mathbf{k}^{(e)}]$ can be assembled to obtain $[\mathbf{K}_s]$. We will assign a row-address and column-address for each entry in $[\mathbf{k}^{(e)}]$ and $[\mathbf{K}_s]$. The column address of a column is the DOF that the column multiplies with in the equilibrium equation. For example, the column addresses of the first and second column in $[\mathbf{k}^{(e)}]$ are u_i and u_j, respectively. The column addresses of columns 1, 2, 3,... in $[\mathbf{K}_s]$ are u_1, u_2, u_3 respectively. The row addresses and column addresses are always symmetric. That is, the row address of ith row is same as the column address of ith column. Having determined the row and column addresses of $[\mathbf{k}^{(e)}]$ and $[\mathbf{K}_s]$, assembly of the element stiffness matrices can be done in a mechanical way. Each of the four entries (boxes) of an element stiffness matrix is transferred to the box in $[\mathbf{K}_s]$ with corresponding row and column addresses.

It is important to discuss the properties of the structural stiffness matrix $[\mathbf{K}_s]$. After assembly, the matrix $[\mathbf{K}_s]$

1. is square;
2. is symmetric;
3. is positive semi-definite;
4. has a determinant equal to zero and does not have an inverse (singular); and
5. has diagonal entries of the matrix greater than or equal to zero.

For a given $\{\mathbf{Q}_s\}$, $\{\mathbf{F}_s\}$ can be determined uniquely; however, for a given $\{\mathbf{F}_s\}$, $\{\mathbf{Q}_s\}$ cannot be determined uniquely because an arbitrary rigid-body displacement can be added to $\{\mathbf{Q}_s\}$ without affecting $\{\mathbf{F}_s\}$.

2.2.4   Boundary Conditions

Before we solve Eq. (2.20), we need to impose the displacement boundary conditions, i.e., use the known nodal displacements in Eq. (2.20). Mathematically, this means to make the global stiffness matrix positive definite so that the unknown displacements can be uniquely determined. Let us assume that the total size of $[\mathbf{K}_s]$ is $m \times m$. From the m equations, we will discard the equations for which we do not know the right-hand side (unknown reaction forces). This is called "striking-the-rows". The structural stiffness matrix becomes rectangular, as the number of equations is less than m. Now we delete the columns that will multiply into prescribed zero displacements in $\{\mathbf{Q}_s\}$. Usually if the ith row is deleted, then i^{th} column will also be deleted. Thus, we will be deleting as many columns as we did rows. This procedure is called "striking-the-columns." Now the stiffness matrix becomes square with size $n \times n$, where n is the number of unknown displacements. The resulting equations can be written as

$$[\mathbf{K}]\{\mathbf{Q}\} = \{\mathbf{F}\} \tag{2.21}$$

where $[\mathbf{K}]$ is the global stiffness matrix, $\{\mathbf{Q}\}$ are the unknown displacements, and $\{\mathbf{F}\}$ are the known external forces applied to nodes. The global stiffness matrix is always a positive definite matrix that has an inverse. It is square symmetric, and diagonal elements of the matrix are positive, i.e., $K_{ii} > 0$. Thus, the displacements $\{\mathbf{Q}\}$ can be solved uniquely for a given set of nodal forces $\{\mathbf{F}\}$.

2.2.5   Calculation of Element Forces and Reaction Forces

Now that all the DOFs are known, the element force in Element e can be determined using Eq. (2.16). The axial force resultant $P^{(e)}$ in the Element e is given by

$$P^{(e)} = \left(\frac{AE}{L}\right)^{(e)} \Delta^{(e)} = \left(\frac{AE}{L}\right)^{(e)} (u_j - u_i) \tag{2.22}$$

Another method of determining the axial force resultant distribution along an element length is as follows. Consider the element in equation (2.16). At the first node or Node i, the axial force is given by $P_i = -f_i$. That is, if f_i acts in the positive direction, that end is under compression. If f_i is in the negative direction, the element is under tension. On the other hand, the opposite is true at the second node, Node j. In that case, $P_j = +f_j$. Then, we can modify Eq. (2.16) as

$$\left\{\begin{array}{c} -P_i^{(e)} \\ +P_j^{(e)} \end{array}\right\} = \left(\frac{AE}{L}\right)^{(e)} \begin{bmatrix} 1 & -1 \\ -1 & 1 \end{bmatrix} \left\{\begin{array}{c} u_i \\ u_j \end{array}\right\} \tag{2.23}$$

It happens that $P_i = P_j$, and hence we use a single variable $P^{(e)}$ to denote the axial force in an element as shown in Eq. (2.22).

It is important to realize that according to the convention used in structural mechanics, the reactions are forces acting on the structure exerted by the supports. There are two methods of determining the support reactions. The straightforward method is to use Eq. (2.20) to determine the unknown $\{\mathbf{F}_s\}$. However, in some FE programs, the structural stiffness matrix $[\mathbf{K}_s]$ is never assembled. The striking of rows and columns is performed at element level and the global stiffness matrix $[\mathbf{K}]$ is assembled directly. In such situations, Eq. (2.19) is used to compute the reactions. For example, reaction at the ith node is obtained by computing the internal forces in the elements connected to Node i and summing all the internal forces.

EXAMPLE 2.2   *Clamped-Clamped Uniaxial Bar*

Use FEM to determine the axial force P in each portion, AB and BC, of the uniaxial bar shown in Figure 2.7. What are the support reactions? Young's modulus is $E = 100\,\text{GPa}$; areas of cross-sections of the two portions AB and BC are, respectively, $1 \times 10^{-4}\,\text{m}^2$ and $2 \times 10^{-4}\,\text{m}^2$ and $F = 10,000\,\text{N}$. The force F is applied at the cross-section at B.

SOLUTION   Since the applied force is a concentrated or point force, it is sufficient to use two elements, AB and BC. The Nodes A, B, and C, respectively, will be Nodes 1, 2, and 3.

Using Eq. (2.16), the element stiffness matrices for two elements are first calculated by

$$[\mathbf{k}^{(1)}] = \frac{10^{11} \times 10^{-4}}{0.25} \begin{bmatrix} 1 & -1 \\ -1 & 1 \end{bmatrix} = 10^7 \begin{bmatrix} 4 & -4 \\ -4 & 4 \end{bmatrix} \begin{array}{l} u_1 \\ u_2 \end{array}$$

$$[\mathbf{k}^{(2)}] = \frac{10^{11} \times 2 \times 10^{-4}}{0.4} \begin{bmatrix} 1 & -1 \\ -1 & 1 \end{bmatrix} = 10^7 \begin{bmatrix} 5 & -5 \\ -5 & 5 \end{bmatrix} \begin{array}{l} u_2 \\ u_3 \end{array}$$

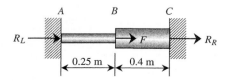

Figure 2.7 Two clamped uniaxial bar structures

Note that the row addresses are written against each row in the element stiffness matrices, and column addresses are shown above each column. Using Eqs. (2.19) and (2.20), the two elements are assembled to produce the following structural equilibrium equations:

$$10^7 \begin{bmatrix} \overset{u_1}{4} & \overset{u_2}{-4} & \overset{u_3}{0} \\ -4 & 4+5 & -5 \\ 0 & -5 & 5 \end{bmatrix} \begin{Bmatrix} u_1 \\ u_2 \\ u_3 \end{Bmatrix} = \begin{Bmatrix} F_1 \\ 10{,}000 \\ F_3 \end{Bmatrix} \tag{2.24}$$

Note that Nodes 1 and 3 are fixed and have unknown reaction forces. After deleting the rows and columns corresponding to zero DOFs (u_1 and u_3), we obtain: $[\mathbf{K}]\{\mathbf{Q}\} = \{\mathbf{F}\}$:

$$10^7 [9]\{u_2\} = \{10{,}000\} \quad \Rightarrow \quad u_2 = 1.111 \times 10^{-4}\,\text{m}$$

Note that the final equation turns out to be a scalar equation because there is only one DOF that is free. By collecting all DOFs, the vector of nodal displacements can be obtained as $\{\mathbf{Q}_s\}^T = \{u_1, u_2, u_3\} = \{0,\ 1.111 \times 10^{-4},\ 0\}$. After solving for the unknown nodal displacements, the axial forces of the elements can be computed from $P = (AE/L)(u_j - u_i)$, as

$$P^{(1)} = 4 \times 10^7 (u_2 - u_1) = 4{,}444\,\text{N}$$
$$P^{(2)} = 5 \times 10^7 (u_3 - u_2) = -5{,}556\,\text{N}$$

Note that the first element is under tension, while the second is under compressive force.

The reaction forces can be calculated from the first and third rows in Eq. (2.24) using the calculated nodal DOFs, as

$$R_L = F_1 = -4 \times 10^7 u_2 = -4{,}444\,\text{N}$$
$$R_R = F_3 = -5 \times 10^7 u_2 = -5{,}556\,\text{N}$$

Alternatively, from the equilibrium between internal and external forces [Eq. (2.19)], the two reaction forces can be calculated using the internal forces, as

$$R_L = -P^{(1)} = -4{,}444N$$
$$R_R = +P^{(2)} = -5{,}556N$$

Note that both reaction forces are in the negative x-direction.

EXAMPLE 2.3   *Three Uniaxial Bar Elements*

Consider an assembly of three two-force members, as shown in Figure 2.8. Motion is restricted to one-dimension along the x-axis. Determine the displacement of the rigid member, element forces, and reaction forces from the wall. Assume $K_1 = 50\,\text{N/cm}$, $K_2 = 30\,\text{N/cm}$, $K_3 = 70\,\text{N/cm}$, and $F_1 = 40\,\text{N}$.

SOLUTION   The assembly consists of three elements and four nodes. Figure 2.9 illustrates the free-body diagram of the system with node and element numbers.

We write down the stiffness matrix of each element and the row addresses. From now on, we will not show the column addresses over the stiffness matrices.

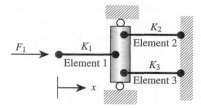

Figure 2.8 One-dimensional structure with three uniaxial bar elements

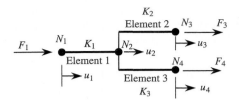

Figure 2.9 Finite element model

Element 1:

$$[\mathbf{k}^{(1)}] = \begin{bmatrix} K_1 & -K_1 \\ -K_1 & K_1 \end{bmatrix} \begin{matrix} u_1 \\ u_2 \end{matrix}$$

Element 2:

$$[\mathbf{k}^{(2)}] = \begin{bmatrix} K_2 & -K_2 \\ -K_2 & K_2 \end{bmatrix} \begin{matrix} u_2 \\ u_3 \end{matrix}$$

Element 3:

$$[\mathbf{k}^{(3)}] = \begin{bmatrix} K_3 & -K_3 \\ -K_3 & K_3 \end{bmatrix} \begin{matrix} u_2 \\ u_4 \end{matrix}$$

After assembling the element stiffness matrices, we obtain the following structural stiffness matrix:

$$[\mathbf{K}_s] = \underbrace{\begin{bmatrix} K_1 & -K_1 & 0 & 0 \\ -K_1 & (K_1 + K_2 + K_3) & -K_2 & -K_3 \\ 0 & -K_2 & K_2 & 0 \\ 0 & -K_3 & 0 & K_3 \end{bmatrix}}_{\text{Structural Stiffness Matrix}} \begin{matrix} u_1 \\ u_2 \\ u_3 \\ u_4 \end{matrix}$$

The equation $[\mathbf{K}_s]\{\mathbf{Q}_s\} = \{\mathbf{F}_s\}$ takes the following form:

$$\begin{bmatrix} K_1 & -K_1 & 0 & 0 \\ -K_1 & (K_1 + K_2 + K_3) & -K_2 & -K_3 \\ 0 & -K_2 & K_2 & 0 \\ 0 & -K_3 & 0 & K_3 \end{bmatrix} \begin{Bmatrix} u_1 \\ u_2 \\ u_3 \\ u_4 \end{Bmatrix} = \begin{Bmatrix} F_1 \\ F_2 \\ F_3 \\ F_4 \end{Bmatrix} \qquad (2.25)$$

The next step is to substitute boundary conditions and solve for unknown displacements. At the boundaries (Nodes 1, 3, and 4), either the externally applied load or the displacement is specified. Substituting for the stiffnesses K_1, K_2, and K_3, $F_1 = 40\,\text{N}$ and $F_2 = 0$, and $u_3 = u_4 = 0$ in Eq. (2.25), we obtain

$$\begin{Bmatrix} F_1 = 40 \\ F_2 = 0 \\ F_3 = R_3 \\ F_4 = R_4 \end{Bmatrix} = \begin{bmatrix} 50 & -50 & 0 & 0 \\ -50 & (50 + 30 + 70) & -30 & -70 \\ 0 & -30 & 30 & 0 \\ 0 & -70 & 0 & 70 \end{bmatrix} \begin{Bmatrix} u_1 \\ u_2 \\ u_3 = 0 \\ u_4 = 0 \end{Bmatrix} \qquad (2.26)$$

Next, we delete the rows and columns corresponding to zero displacements. In this example, the third and fourth rows and columns correspond to zero displacements. Deleting these rows and columns, we obtain the global equations in the form $[\mathbf{K}]\{\mathbf{Q}\} = \{\mathbf{F}\}$, where $[\mathbf{K}]$ is the global stiffness matrix:

$$\begin{bmatrix} 50 & -50 \\ -50 & 150 \end{bmatrix} \begin{Bmatrix} u_1 \\ u_2 \end{Bmatrix} = \begin{Bmatrix} 40 \\ 0 \end{Bmatrix} \qquad (2.27)$$

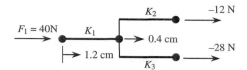

Figure 2.10 Free-body diagram of the structure

The unknown displacements u_1 and u_2 can be obtained by solving Eq. (2.27), as

$$u_1 = 1.2\,\text{cm} \quad \text{and} \quad u_2 = 0.4\,\text{cm} \tag{2.28}$$

By collecting all DOFs, the vector of nodal displacements can be obtained as

$$\{\mathbf{Q}_s\}^T = \{u_1, u_2, u_3, u_4\} = \{1.2,\, 0.4,\, 0,\, 0\}.$$

Next, we substitute u_1 and u_2 into rows 3 and 4 in Eq. (2.26) to calculate the reaction forces F_3 and F_4:

$$
\begin{aligned}
F_3 &= 0u_1 - 30u_2 + 30u_3 + 0u_4 = -12\,\text{N} \\
F_4 &= 0u_1 - 70u_2 + 0u_3 + 70u_4 = -28\,\text{N}
\end{aligned}
\tag{2.29}
$$

Based on the results obtained, we can now redraw the free-body diagram of the system, as shown in Figure 2.10. Both reaction forces are in the negative x-direction.

2.2.6   FE Program Organization

Since the FE analysis follows a standard procedure, as described in the preceding section, it is possible to make a general-purpose FE program. Commercial FE programs typically consist of three parts: preprocessor, FE solver, and postprocessor. A preprocessor allows the user to define the structure, divide it into number of elements, identify the nodes and their coordinates, define connectivity between various elements, and define material properties and the loads. Developments in computer graphics and CAD technology have resulted in sophisticated preprocessors that let the users create models and define various properties interactively on the computer terminal itself. A post-processor takes the FE analysis results and presents them in a user-friendly graphical form. Again, developments in software and graphics have resulted in very sophisticated animations to help the analysts better understand the results of a FE model. This book is mostly concerned with the principles involved in the development and operation of the core FE program, which computes the stiffness matrix and assembles and solves the final set of equations. This will be discussed further in Chapter 7. In addition, a brief introduction is provided to perform FE analysis using commercial programs in the Appendix. Also in the Appendices, various FE analysis programs are introduced, including Pro/Mechanica, NEiNASTRAN, ANSYS, and MATLAB Toolbox.

2.3   PLANE TRUSS ELEMENTS

This section presents the formulation of stiffness matrix and general procedures for solving the two-dimensional or plane truss using the direct stiffness method. A truss consisting of two elements is used to illustrate the solution procedures.

Consider the plane truss consisting of two elements or members, as shown in Figure 2.11. A horizontal force $F = 50\,\text{N}$ is applied at the top node. Although the elements of the truss are uniaxial bars, the methods described in the previous section cannot be readily applied to this problem for two reasons: the two elements are not in the same direction but are inclined at different angles, and the external forces at a node are applied in both x- and y-directions.

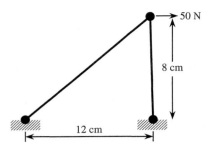

50 N

8 cm

12 cm

Figure 2.11 A plane truss consisting of two members

However, the element stiffness matrix of uniaxial bar elements will be applicable to individual elements of the truss if we consider a local coordinate system. For a plane truss element, the following two coordinate systems can be defined:

1. The global coordinate system, x-y for the entire structure.
2. A local coordinate system, $\bar{x}$-$\bar{y}$ for a particular element such that the $\bar{x}$-axis is along the length of the element.

Referring to Figure 2.12, the force-displacement relation of a truss element can be written in the local coordinate system as

$$\left\{ \begin{array}{c} f_{1\bar{x}} \\ f_{2\bar{x}} \end{array} \right\} = \frac{EA}{L} \begin{bmatrix} 1 & -1 \\ -1 & 1 \end{bmatrix} \left\{ \begin{array}{c} \bar{u}_1 \\ \bar{u}_2 \end{array} \right\} \tag{2.30}$$

where E, A, and L, respectively, are the Young's modulus, the area of cross-section, and the length of the element, and EA/L corresponds to the spring constant K in Eq. (2.16).

Note that the force and displacement are represented in the local coordinate system. In order to make the above equation more general, let us consider the transverse displacement $\bar{v}_1$ and $\bar{v}_2$ in the $\bar{y}$-direction. Corresponding transverse forces at each node can be defined as $f_{1\bar{y}}$ and $f_{2\bar{y}}$. However, in the truss element, these forces do not exist and hence they are equated to zero. Then the above stiffness matrix (system equations in matrix form) can be expanded to incorporate the force and displacements in the $\bar{y}$-direction as shown below.

$$\left\{ \begin{array}{c} f_{1\bar{x}} \\ f_{1\bar{y}} \\ f_{2\bar{x}} \\ f_{2\bar{y}} \end{array} \right\} = \frac{EA}{L} \begin{bmatrix} 1 & 0 & -1 & 0 \\ 0 & 0 & 0 & 0 \\ -1 & 0 & 1 & 0 \\ 0 & 0 & 0 & 0 \end{bmatrix} \left\{ \begin{array}{c} \bar{u}_1 \\ \bar{v}_1 \\ \bar{u}_2 \\ \bar{v}_2 \end{array} \right\} \tag{2.31}$$

The expanded local stiffness matrix in the above equation is

1. is a square matrix;
2. is symmetric; and
3. has diagonal elements of the matrix greater than or equal to zero.

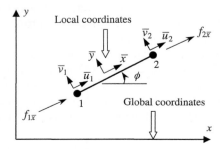

y

Local coordinates

$\bar{v}_2$ $\bar{u}_2$ $f_{2\bar{x}}$

$\bar{y}$ $\bar{x}$

$\bar{v}_1$

$\bar{u}_1$ ϕ

2

1

Global coordinates

$f_{1\bar{x}}$

x

Figure 2.12 Local and global coordinate systems

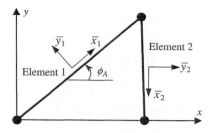

Figure 2.13 Local coordinate systems of the two-bar truss

The above stiffness matrix is valid only for the particular element, Element 1 in the above example. It cannot be applied to other elements because the local coordinates $\bar{x}$-$\bar{y}$ are different for different elements. The local coordinates for element Element 2 are shown in Figure 2.13.

In order to develop a system of equations that connect all elements in the truss, we need to transform the force-displacement relations, e.g., Eq. (2.31), to the global coordinates, which is common for all elements of the truss. This requires the use of vector coordinate transformation.

2.3.1 Coordinate Transformation

Since forces and displacements are vectors, we can use the vector transformation to find the relation between the displacements in local and global coordinates at a node. For example, the local displacements of Node 1 can be written as

$$\left\{ \begin{array}{c} \bar{u}_1 \\ \bar{v}_1 \end{array} \right\} = \left[\begin{array}{cc} \cos\phi & \sin\phi \\ -\sin\phi & \cos\phi \end{array} \right] \left\{ \begin{array}{c} u_1 \\ v_1 \end{array} \right\}$$

A similar relation for Node 2 will be

$$\left\{ \begin{array}{c} \bar{u}_2 \\ \bar{v}_2 \end{array} \right\} = \left[\begin{array}{cc} \cos\phi & \sin\phi \\ -\sin\phi & \cos\phi \end{array} \right] \left\{ \begin{array}{c} u_2 \\ v_2 \end{array} \right\}$$

Actually, we can combine the above relations to obtain

$$\underbrace{\left\{ \begin{array}{c} \bar{u}_1 \\ \bar{v}_1 \\ \bar{u}_2 \\ \bar{v}_2 \end{array} \right\}}_{\text{local}} = \left[\begin{array}{cccc} \cos\phi & \sin\phi & 0 & 0 \\ -\sin\phi & \cos\phi & 0 & 0 \\ 0 & 0 & \cos\phi & \sin\phi \\ 0 & 0 & -\sin\phi & \cos\phi \end{array} \right] \underbrace{\left\{ \begin{array}{c} u_1 \\ v_1 \\ u_2 \\ v_2 \end{array} \right\}}_{\text{global}}$$

The above relation between local and global displacements can be written using a shorthand notation such as

$$\{\bar{\mathbf{q}}\} = [\mathbf{T}]\{\mathbf{q}\} \tag{2.32}$$

where $\{\bar{\mathbf{q}}\}$ and $\{\mathbf{q}\}$ are the element DOFs in the local and global coordinates, respectively, and $[\mathbf{T}]$ is the *transformation matrix*. In some literature, $[\mathbf{T}]$ is called the rotation matrix. Since forces are also vectors, the forces $\{\bar{\mathbf{f}}\}$ in element coordinates are related to $\{\mathbf{f}\}$ in global coordinates as

$$
\underbrace{\begin{Bmatrix} f_{1\bar{x}} \\ f_{1\bar{y}} \\ f_{2\bar{x}} \\ f_{2\bar{y}} \end{Bmatrix}}_{\text{local}} = \begin{bmatrix} \cos\phi & \sin\phi & 0 & 0 \\ -\sin\phi & \cos\phi & 0 & 0 \\ 0 & 0 & \cos\phi & \sin\phi \\ 0 & 0 & -\sin\phi & \cos\phi \end{bmatrix} \underbrace{\begin{Bmatrix} f_{1x} \\ f_{1y} \\ f_{2x} \\ f_{2y} \end{Bmatrix}}_{\text{global}}
$$

or in shorthand notation

$$\{\bar{\mathbf{f}}\} = [\mathbf{T}]\{\mathbf{f}\} \tag{2.33}$$

In the following section, we will express the local element equation (2.31) to the global coordinate using transformation relations in Eqs. (2.32) and (2.33). Once all element equations are expressed in the global coordinate, they can be assembled using procedures similar to those of uniaxial bar elements.

2.3.2 Element Stiffness Matrix in Global Coordinates

For a single truss element, using the above coordinate transformation equation, we can proceed to transform the element stiffness matrix from local to global coordinates.

Consider the truss element arbitrarily positioned in two-dimensional space, as shown in Figure 2.14.

The force-displacement equations can be expressed in the local coordinates as:

$$
\begin{Bmatrix} f_{1\bar{x}} \\ f_{1\bar{y}} \\ f_{2\bar{x}} \\ f_{2\bar{y}} \end{Bmatrix} = \frac{EA}{L} \underbrace{\begin{bmatrix} 1 & 0 & -1 & 0 \\ 0 & 0 & 0 & 0 \\ -1 & 0 & 1 & 0 \\ 0 & 0 & 0 & 0 \end{bmatrix}}_{\text{element stiffness matrix}} \begin{Bmatrix} \bar{u}_1 \\ \bar{v}_1 \\ \bar{u}_2 \\ \bar{v}_2 \end{Bmatrix} \tag{2.34}
$$

In shorthand notation, Eq. (2.34) takes the following form:

$$\{\bar{\mathbf{f}}\} = [\bar{\mathbf{k}}]\{\bar{\mathbf{q}}\} \tag{2.35}$$

Substitution of Eqs. (2.32) and (2.33) into Eq. (2.34) yields

$$[\mathbf{T}]\{\mathbf{f}\} = [\bar{\mathbf{k}}][\mathbf{T}]\{\mathbf{q}\}.$$

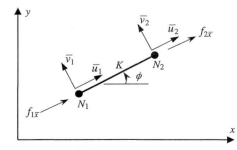

Figure 2.14 Definition of two-dimensional truss element

Multiplying both sides of the equation with $[\mathbf{T}]^{-1}$,

$$\{\mathbf{f}\} = [\mathbf{T}]^{-1}[\overline{\mathbf{k}}][\mathbf{T}]\{\mathbf{q}\}$$

$\underbrace{\qquad}_{\text{global}} \qquad \underbrace{\qquad}_{\text{global}}$

or,

$$\{\mathbf{f}\} = [\mathbf{k}]\{\mathbf{q}\} \tag{2.36}$$

The element stiffness matrix $[\mathbf{k}]$ in global coordinates is now be expressed in terms of $[\overline{\mathbf{k}}]$ as

$$[\mathbf{k}] = [\mathbf{T}]^{-1}[\overline{\mathbf{k}}][\mathbf{T}] \tag{2.37}$$

It can be shown that the inverse of the transformation matrix $[\mathbf{T}]$ is equal to its transpose, and hence $[\mathbf{k}]$ can be written as

$$[\mathbf{k}] = [\mathbf{T}]^{T}[\overline{\mathbf{k}}][\mathbf{T}] \tag{2.38}$$

Performing the matrix multiplication in Eq. (2.38), we obtain an explicit expression for $[\mathbf{k}]$ as

$$[\mathbf{k}] = \frac{EA}{L}\begin{bmatrix} \cos^2\phi & \cos\phi\sin\phi & -\cos^2\phi & -\cos\phi\sin\phi \\ \cos\phi\sin\phi & \sin^2\phi & -\cos\phi\sin\phi & -\sin^2\phi \\ -\cos^2\phi & -\cos\phi\sin\phi & \cos^2\phi & \cos\phi\sin\phi \\ -\cos\phi\sin\phi & -\sin^2\phi & \cos\phi\sin\phi & \sin^2\phi \end{bmatrix} \tag{2.39}$$

From Eq. (2.39), it is clear that the element stiffness matrix of a plane truss element depends on the length L, axial rigidity EA, and the angle of orientation ϕ. As mentioned earlier, element stiffness matrix is symmetric. Its determinant is equal to zero, and hence it does not have an inverse. Furthermore, the element stiffness matrix is positive semi-definite, and the diagonal elements of the matrix are either equal to zero or greater than zero.

EXAMPLE 2.4   *Two-Bar Truss*

The two-bar truss shown in Figure 2.15 has circular cross-sections with diameter of 0.25 cm and Young's modulus $E = 30 \times 10^6\,\text{N/cm}^2$. An external force $F = 50\,\text{N}$ is applied in the horizontal direction at Node 2. Calculate the displacement of each node and stress in each element.

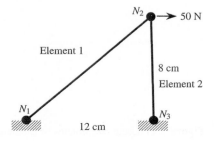

Figure 2.15 Two-bar truss structure

SOLUTION

Element 1:

In the local coordinate system shown in Figure 2.16, the force-displacement equations for Element 1 is given in Eq. (2.31), which can be transformed to the global coordinates similar to the one in Eq. (2.36), to yield

$$\{\mathbf{f}^{(1)}\} = [\mathbf{k}^{(1)}]\{\mathbf{q}^{(1)}\} \tag{2.40}$$

Since the orientation angle of the element is $\phi_1 = 33.7°$, the element equation in the global coordinates can be obtained using the stiffness matrix in Eq. (2.39), as

$$\begin{Bmatrix} f_{1x}^{(1)} \\ f_{1y}^{(1)} \\ f_{2x}^{(1)} \\ f_{2y}^{(1)} \end{Bmatrix} = 102150 \begin{bmatrix} 0.692 & 0.462 & -0.692 & -0.462 \\ 0.462 & 0.308 & -0.462 & -0.308 \\ -0.692 & -0.462 & 0.692 & 0.462 \\ -0.462 & -0.308 & 0.462 & 0.308 \end{bmatrix} \begin{Bmatrix} u_1 \\ v_1 \\ u_2 \\ v_2 \end{Bmatrix}$$

Element 2:

For Element 2, the same procedure can be applied with the orientation angle of the element being $\phi_2 = -90°$(see Figure 2.17). In the global coordinates, the element equations become

$$\begin{Bmatrix} f_{2x}^{(2)} \\ f_{2y}^{(2)} \\ f_{3x}^{(2)} \\ f_{3y}^{(2)} \end{Bmatrix} = 184125 \begin{bmatrix} 0 & 0 & 0 & 0 \\ 0 & 1 & 0 & -1 \\ 0 & 0 & 0 & 0 \\ 0 & -1 & 0 & 1 \end{bmatrix} \begin{Bmatrix} u_2 \\ v_2 \\ u_3 \\ v_3 \end{Bmatrix} \tag{2.41}$$

Note that the orientation is measured in the counterclockwise direction from the positive *x*-axis of the global coordinates.

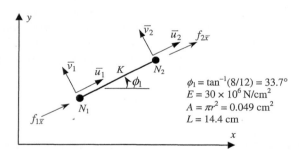

$\phi_1 = \tan^{-1}(8/12) = 33.7°$
$E = 30 \times 10^6 \text{ N/cm}^2$
$A = \pi r^2 = 0.049 \text{ cm}^2$
$L = 14.4 \text{ cm}$

Figure 2.16 Local coordinates of Element 1

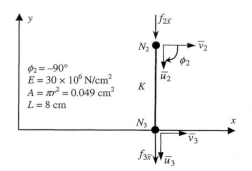

$\phi_2 = -90°$
$E = 30 \times 10^6 \text{ N/cm}^2$
$A = \pi r^2 = 0.049 \text{ cm}^2$
$L = 8 \text{ cm}$

Figure 2.17 Local coordinates of Element 2

Now, we are ready to assemble the global stiffness matrix of the structure. Summing the two sets of force-displacement equations in the global coordinates:

$$
\begin{Bmatrix} F_{1x} \\ F_{1y} \\ F_{2x} \\ F_{2y} \\ F_{3x} \\ F_{3y} \end{Bmatrix} =
\begin{bmatrix}
70687 & 47193 & -70687 & -47193 & 0 & 0 \\
47193 & 31462 & -47193 & -31462 & 0 & 0 \\
-70687 & -47193 & 70687 & 47193 & 0 & 0 \\
-47193 & -31462 & 47193 & 215587 & 0 & -184125 \\
0 & 0 & 0 & 0 & 0 & 0 \\
0 & 0 & 0 & -184125 & 0 & 184125
\end{bmatrix}
\begin{Bmatrix} u_1 \\ v_1 \\ u_2 \\ v_2 \\ u_3 \\ v_3 \end{Bmatrix}
$$

Element 1

Element 2

Note that two element stiffness matrices overlap at DOFs corresponding to u_2 and v_2 because the two elements are connected at Node 2. Next, apply the following known boundary conditions:

(a) Nodes 1 and 3 are fixed; therefore, displacement components of these two nodes are zero (u_1, v_1 and u_3, v_3).

(b) The only applied external forces are at Node 2: $F_{2x} = 50\,\text{N}$ and $F_{2y} = 0\,\text{N}$.

$$
\begin{Bmatrix} F_{1x} \\ F_{1y} \\ 50 \\ 0 \\ F_{3x} \\ F_{3y} \end{Bmatrix} =
\begin{bmatrix}
70687 & 47193 & -70687 & -47193 & 0 & 0 \\
47193 & 31462 & -47193 & -31462 & 0 & 0 \\
-70687 & -47193 & 70687 & 47193 & 0 & 0 \\
-47193 & -31462 & 47193 & 215587 & 0 & -184125 \\
0 & 0 & 0 & 0 & 0 & 0 \\
0 & 0 & 0 & -184125 & 0 & 184125
\end{bmatrix}
\begin{Bmatrix} 0 \\ 0 \\ u_2 \\ v_2 \\ 0 \\ 0 \end{Bmatrix}
$$

We first delete the columns corresponding to zero displacements. In this example, the third and fourth columns correspond to non-zero displacements. We keep these two columns and strike out all other columns, to obtain

$$
\begin{Bmatrix} F_{1x} \\ F_{1y} \\ 50 \\ 0 \\ F_{3x} \\ F_{3y} \end{Bmatrix} =
\begin{bmatrix}
-70687 & -47193 \\
-47193 & -31462 \\
70687 & 47193 \\
47193 & 215587 \\
0 & 0 \\
0 & -184125
\end{bmatrix}
\begin{Bmatrix} u_2 \\ v_2 \end{Bmatrix}
\tag{2.42}
$$

Because F_{1x}, F_{1y}, F_{3x}, and F_{3y} are unknown reaction forces, we delete those rows corresponding to the unknown reaction forces. Then, finally we have the following 2×2 matrix equation for the nodal displacements u_2 and v_2:

$$
\begin{Bmatrix} 50 \\ 0 \end{Bmatrix} =
\begin{bmatrix} 70687 & 47193 \\ 47193 & 215587 \end{bmatrix}
\begin{Bmatrix} u_2 \\ v_2 \end{Bmatrix}
$$

Since the global stiffness matrix in the above equation is positive-definite, it is possible to invert it to solve for the unknown nodal displacements:

$$
u_2 = 8.28 \times 10^{-4}\,\text{cm}
$$
$$
v_2 = -1.81 \times 10^{-4}\,\text{cm}
$$

-60.2 N 60.2 N   **Figure 2.18** Element force for Element 1 in
     1   2   local coordinates

Substituting the known u_2 and v_2 values into the matrix equation (2.42), we solve for the reaction forces:

$$\begin{Bmatrix} F_{1x} \\ F_{1y} \\ F_{3x} \\ F_{3y} \end{Bmatrix} = \begin{bmatrix} -70687 & -47193 \\ -47193 & -31462 \\ 0 & 0 \\ 0 & -184125 \end{bmatrix} \begin{Bmatrix} 8.28 \times 10^{-4} \\ -1.81 \times 10^{-4} \end{Bmatrix} = \begin{Bmatrix} -50 \\ -33.39 \\ 0 \\ 33.39 \end{Bmatrix} N$$

Since the truss element is a two-force member, it is clear that the reaction force at Node 3 is in the vertical direction, and the reaction force at Node 1 is parallel to the direction of the element.

To determine the normal stress in each truss member, one option is to use the displacement transformation equations to transform the results from the global coordinate system back to the local coordinate system. For example, the nodal displacements of Element 1 in the local coordinate system can be obtained from Eq. (2.32), as

$$\begin{Bmatrix} \bar{u}_1 \\ \bar{v}_1 \\ \bar{u}_2 \\ \bar{v}_2 \end{Bmatrix} = \begin{bmatrix} .832 & .555 & 0 & 0 \\ -.555 & .832 & 0 & 0 \\ 0 & 0 & .832 & .555 \\ 0 & 0 & -.555 & .832 \end{bmatrix} \begin{Bmatrix} 0 \\ 0 \\ u_2 \\ v_2 \end{Bmatrix} = \begin{Bmatrix} 0 \\ 0 \\ 5.89 \times 10^{-4} \\ -6.11 \times 10^{-4} \end{Bmatrix}$$

Then, the local force-displacement equations (2.35) can be used to calculate the element forces, as

$$\begin{Bmatrix} f_{1\bar{x}} \\ f_{1\bar{y}} \\ f_{2\bar{x}} \\ f_{2\bar{y}} \end{Bmatrix} = \frac{EA}{L} \begin{bmatrix} 1 & 0 & -1 & 0 \\ 0 & 0 & 0 & 0 \\ -1 & 0 & 1 & 0 \\ 0 & 0 & 0 & 0 \end{bmatrix} \begin{Bmatrix} 0 \\ 0 \\ 5.89 \times 10^{-4} \\ -6.11 \times 10^{-4} \end{Bmatrix} = \begin{Bmatrix} -60.2 \\ 0 \\ 60.2 \\ 0 \end{Bmatrix} N \quad (2.43)$$

Equation (2.43) represents the forces acting on the element in the local coordinate system. As expected, there is no force component in the $\bar{y}$-direction (local y-direction). In the $\bar{x}$-direction (local x-direction), the two nodes have the same magnitude of internal forces but in the opposite direction. As can be seen in Figure 2.18, two equal and opposite forces act on the truss element, which results in tensile stresses in the element. In general, the sign of the force f_{jx} at Node j (second node) will be the same as the sign of the force resultant P. In the present example, the element force of Element 1 is positive and has the magnitude of 60.2 N. Therefore, the normal stress in Element 1 is tensile and can be calculated as $(60.2/0.049) = 1228 \text{ N/cm}^2$.

Another method to calculate the axial force in a truss element is described below. This follows the method used in deriving Eq. (2.22). For a truss element at an arbitrary orientation, Eq. (2.22) is modified as

$$P^{(e)} = \left(\frac{AE}{L}\right)^{(e)} \Delta^{(e)} = \left(\frac{AE}{L}\right)^{(e)} (\bar{u}_j - \bar{u}_i) \quad (2.44)$$

Using the transformation relations in Eq. (2.32), one can express the displacements in global coordinates. Let $l = \cos\phi$ and $m = \sin\phi$. Then, the expression for P takes the following form:

$$P^{(e)} = \left(\frac{AE}{L}\right)^{(e)} ((lu_j + mv_j) - (lu_i + mv_i))$$

$$= \left(\frac{AE}{L}\right)^{(e)} (l(u_j - u_i) + m(v_j - v_i)) \quad (2.45)$$

In the above example, the basic principles of FE analysis were used: (1) derive the force displacement relations of each truss member, (2) assemble the equations to obtain the global equations, and (3) solve for unknown displacements. However, in practical problems with a large number of elements, one need not write the equations of equilibrium for each element. We would like to develop a systematic procedure that is suitable for a large number of elements. In this method, each element is assigned a first node and second node. These node numbers are denoted by i and j. The choice of the first and second nodes is arbitrary; however, it has to be consistent throughout the solution of the problem. The orientation of the element is defined by the angle the direction i-j makes with the positive x-axis, and it is denoted by ϕ. The direction cosines of the element are $l = \cos\phi$, $m = \sin\phi$. We assign the row and column addresses to the element stiffness matrix, as shown below. The element stiffness matrix in Eq. (2.39) can be written in terms of l and m with the row and column addresses as

$$[\mathbf{k}] = \left(\frac{EA}{L}\right)^{(e)} \begin{array}{c} \\ \begin{bmatrix} l^2 & lm & -l^2 & -lm \\ lm & m^2 & -lm & -m^2 \\ -l^2 & -lm & l^2 & lm \\ -lm & -m^2 & lm & m^2 \end{bmatrix} \end{array} \begin{array}{l} u_i \\ v_i \\ u_j \\ v_j \end{array} \qquad (2.46)$$

$$\begin{array}{cccc} u_i & v_i & u_j & v_j \end{array}$$

As illustrated in the following example, the row and column addresses are useful in assembling the element stiffness matrices into the global stiffness matrix. One can note that the row addresses are transpose of the column addresses.

EXAMPLE 2.5   *Plane Truss with Three Elements*

The plane truss shown in Figure 2.19 consists of three members connected to each other and to the walls by pin joints. The members make equal angles with each other, and Element 2 is vertical. The members are identical to each other with properties: Young's modulus $E = 206$ GPa, cross-sectional area $A = 1 \times 10^{-4}\,\mathrm{m}^2$, and length $L = 1$ m. An inclined force $F = 20,000$ N is applied at Node 1. Solve for the displacements at Node 1 and stresses in the three elements.

SOLUTION   Based on Figure 2.19 the element properties, node connectivity, and direction cosines can be calculated as shown in Table 2.1.

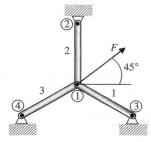

Figure 2.19 Plane structure with three truss elements

Table 2.1 Element Properties and Direction Cosines of Plane Truss in Example 2.5

Element	AE/L	$i \rightarrow j$	ϕ	$l = \cos\phi$	$m = \sin\phi$
1	206×10^5	$1 \rightarrow 3$	$-\pi/6$	0.866	-0.5
2	206×10^5	$1 \rightarrow 2$	$\pi/2$	0	1
3	206×10^5	$1 \rightarrow 4$	$-5\pi/6$	-0.866	-0.5

Then, using Eq. (2.46), the element stiffness matrices in the global coordinates system can be obtained, as

$$[\mathbf{k}^{(1)}] = 206 \times 10^5 \begin{bmatrix} 0.750 & -0.433 & -0.750 & 0.433 \\ -0.433 & 0.250 & 0.433 & -0.250 \\ -0.750 & 0.433 & 0.750 & -0.433 \\ 0.433 & -0.250 & -0.433 & 0.250 \end{bmatrix} \begin{matrix} u_1 \\ v_1 \\ u_3 \\ v_3 \end{matrix}$$

$$[\mathbf{k}^{(2)}] = 206 \times 10^5 \begin{bmatrix} 0 & 0 & 0 & 0 \\ 0 & 1 & 0 & -1 \\ 0 & 0 & 0 & 0 \\ 0 & -1 & 0 & 1 \end{bmatrix} \begin{matrix} u_1 \\ v_1 \\ u_2 \\ v_2 \end{matrix}$$

$$[\mathbf{k}^{(3)}] = 206 \times 10^5 \begin{bmatrix} 0.750 & 0.433 & -0.750 & -0.433 \\ 0.433 & 0.250 & -0.433 & -0.250 \\ -0.750 & -0.433 & 0.750 & 0.433 \\ -0.433 & -0.250 & 0.433 & 0.250 \end{bmatrix} \begin{matrix} u_1 \\ v_1 \\ u_4 \\ v_4 \end{matrix}$$

Note that row addresses are indicated on the RHS of the stiffness matrices, and they are useful in assembling the structural stiffness matrix. One can easily identify the column addresses, although they are not written above the stiffness matrices. The structural stiffness matrix and the FE equations are obtained by assembling the three element stiffness matrices:

$$206 \times 10^5 \begin{bmatrix} 1.5 & 0 & 0 & 0 & -0.750 & 0.433 & -0.750 & -0.433 \\ & 1.5 & 0 & -1 & 0.433 & -0.250 & -0.433 & -0.250 \\ & & 0 & 0 & 0 & 0 & 0 & 0 \\ & & & 1 & 0 & 0 & 0 & 0 \\ & & & & 0.750 & -0.433 & 0 & 0 \\ & \text{Symmetric} & & & & 0.250 & 0 & 0 \\ & & & & & & 0.750 & 0.433 \\ & & & & & & & 0.250 \end{bmatrix} \begin{Bmatrix} u_1 \\ v_1 \\ u_2 \\ v_2 \\ u_3 \\ v_3 \\ u_4 \\ v_4 \end{Bmatrix} = \begin{Bmatrix} F_{1x} \\ F_{1y} \\ F_{2x} \\ F_{2y} \\ F_{3x} \\ F_{3y} \\ F_{4x} \\ F_{4y} \end{Bmatrix}$$

Or,

$$[\mathbf{K}_s]\{\mathbf{Q}_s\} = \{\mathbf{F}_s\}$$

Now, the known external forces and displacements are applied to the above matrix equation. First, the inclined force at Node 1 is decomposed into x- and y-directions, as

$$F_{1x} = 20000 \cdot \cos(\pi/4) = 14,142$$
$$F_{1y} = 20000 \cdot \sin(\pi/4) = 14,142$$

In addition, since Nodes 2, 3, and 4 are fixed, their displacements are equal to zero:

$$u_2 = v_2 = u_3 = v_3 = u_4 = v_4 = 0.$$

We delete the rows and columns in matrix $[K_s]$ corresponding to those DOFs that have zero displacements. Then, the global FE equations are written in the form $[K]\{Q\} = \{F\}$:

$$206 \times 10^5 \begin{bmatrix} 1.5 & 0 \\ 0 & 1.5 \end{bmatrix} \begin{Bmatrix} u_1 \\ v_1 \end{Bmatrix} = \begin{Bmatrix} 14,142 \\ 14,142 \end{Bmatrix}$$

The global stiffness matrix is now positive-definite and invertible. After solving for unknown displacements, we have

$$u_1 = 0.458 \, \text{mm}$$
$$v_1 = 0.458 \, \text{mm}$$

By including all other zero displacements, the vector of nodal displacements can be written as $\{Q_s\}^T = \{0.458, 0.458, 0, 0, 0, 0, 0, 0\}$.

Force in each element can be obtained using Eq. (2.45) and element properties given in Table 2.1. For example, the force in Element 1 is

$$P^{(1)} = 206 \times 10^5 (0.866(u_3 - u_1) - 0.5(v_3 - v_1)) = -3,450 \, \text{N}$$

The same calculation can be repeated for other elements to obtain:

$$P^{(2)} = -9,440 \, \text{N}$$
$$P^{(3)} = 12,900 \, \text{N}$$

The negative values of $P^{(1)}$ and $P^{(2)}$ indicate compressive forces in those elements. The stresses of the elements can be obtained by dividing the force by the area of cross cross-section:

$$\sigma^{(1)} = -34.5 \, \text{MPa}$$
$$\sigma^{(2)} = -94.4 \, \text{MPa}$$
$$\sigma^{(3)} = 129 \, \text{MPa}$$

Once element properties, node connectivity, and direction cosines of all elements are listed as in Table 2.1, it is easy to make a computer program that can build the global stiffness matrix and solve for unknown displacements.

2.4   THREE-DIMENSIONAL TRUSS ELEMENTS (SPACE TRUSS)

This section presents the formulation of stiffness matrix and general procedures for solving the three-dimensional or space truss using the direct stiffness method. The procedure is similar to that of plane truss except that an additional DOF is added at each node.

2.4.1   Three-Dimensional Coordinate Transformation

The coordinate transformation used for the two-dimensional truss element can be generalized to the three-dimensional truss elements for space trusses.

A space-truss element has three DOFs, u, v, and w, at each node. Thus, the space-truss element is a 2-Node 6–DOF element. Corresponding to the three displacements at each node, there are three forces, f_x, f_y and f_z. The displacements and forces can also be expresses in local or elemental coordinate system $\bar{x}$-$\bar{y}$-$\bar{z}$, as shown in Figure 2.20. The

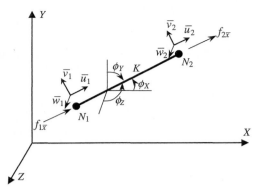

Figure 2.20 Three-dimensional coordinates transformation

forces and displacements in the local coordinate system are related by a simple equation similar to Eq. (2.30), as

$$\left\{ \begin{array}{c} f_{i\bar{x}} \\ f_{j\bar{x}} \end{array} \right\} = \frac{AE}{L} \begin{bmatrix} 1 & -1 \\ -1 & 1 \end{bmatrix} \left\{ \begin{array}{c} \bar{u}_i \\ \bar{u}_j \end{array} \right\}$$

or,

$$\{\bar{\mathbf{f}}\} = [\bar{\mathbf{k}}]\{\bar{\mathbf{q}}\} \tag{2.47}$$

To assemble truss elements, the element equation (2.47) must be transformed into the global coordinate system. The transformation of displacements at a node follows the three-dimensional vector transformations rule:

$$\left\{ \begin{array}{c} \bar{u}_i \\ \bar{u}_j \end{array} \right\} = \begin{bmatrix} l & m & n & 0 & 0 & 0 \\ 0 & 0 & 0 & l & m & n \end{bmatrix} \left\{ \begin{array}{c} u_i \\ v_i \\ w_i \\ u_j \\ v_j \\ w_j \end{array} \right\}$$

or,

$$\{\bar{\mathbf{q}}\} = [\mathbf{T}]\{\mathbf{q}\} \tag{2.48}$$

where l, m, and n are the directions cosines of the element connecting Nodes i and j, which can be calculated from the element length L and the nodal coordinates as shown below:

$$l = \frac{x_j - x_i}{L}, \ m = \frac{y_j - y_i}{L}, \ n = \frac{z_j - z_i}{L}$$

$$\tag{2.49}$$

$$L = \sqrt{(x_j - x_i)^2 + (y_j - y_i)^2 + (z_j - z_i)^2}$$

Similarly, the nodal forces in the local coordinate system can be transformed as

$$\left\{ \begin{array}{c} f_{ix} \\ f_{iy} \\ f_{iz} \\ f_{jx} \\ f_{jy} \\ f_{jz} \end{array} \right\} = \begin{bmatrix} l & 0 \\ m & 0 \\ n & 0 \\ 0 & l \\ 0 & m \\ 0 & n \end{bmatrix} \left\{ \begin{array}{c} f_{i\bar{x}} \\ f_{j\bar{x}} \end{array} \right\}$$

or,

$$\{\mathbf{f}\} = [\mathbf{T}]^T \{\bar{\mathbf{f}}\} \tag{2.50}$$

2.4.2 Stiffness Matrix

Substituting for $\bar{\mathbf{q}}$ from Eq. (2.48) into Eq. (2.47) and multiplying both sides of the equation with $[\mathbf{T}]^T$, we obtain

$$[\mathbf{T}]^T \{\bar{\mathbf{f}}\} = [\mathbf{T}]^T \{\bar{\mathbf{k}}\}[\mathbf{T}]\{\mathbf{q}\}$$

or,

$$\{\mathbf{f}\} = [\mathbf{k}]\{\mathbf{q}\} \tag{2.51}$$

where $[\mathbf{k}]$ is the element stiffness matrix that relates the nodal forces and displacements expressed in the global coordinates. From Eq. (2.51), it is clear that $[\mathbf{k}]$ is obtained as the product of $[\mathbf{T}]^T [\bar{\mathbf{k}}][\mathbf{T}]$. An explicit form of $[\mathbf{k}]$ is given below.

$$[\mathbf{k}] = \frac{EA}{L} \begin{bmatrix} l^2 & lm & ln & -l^2 & -lm & -ln \\ & m^2 & mn & -lm & -m^2 & -mn \\ & & n^2 & -ln & -mn & -n^2 \\ & & & l^2 & lm & ln \\ & \text{sym} & & & m^2 & mn \\ & & & & & n^2 \end{bmatrix} \begin{matrix} u_i \\ v_i \\ w_i \\ u_j \\ v_j \\ w_j \end{matrix} \tag{2.52}$$

Where the row DOFs are shown next to the matrix. Assembling $[\mathbf{k}]$ into the structural stiffness matrix $[\mathbf{K}_s]$ and deleting the rows and columns to obtain the global stiffness matrix $[\mathbf{K}]$ follow procedures similar to those of plane truss.

EXAMPLE 2.6   *Space Truss*

Use the FEM to determine the displacements and forces in the space truss, shown in Figure 2.21. The coordinates of the nodes in meter units are given in Table 2.2. Assume Young's modulus

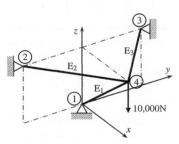

Figure 2.21 Three-bar space truss structure

Table 2.2 Nodal Coordinates of Space Truss Structure in Example 2.6

Node	x	y	z
1	0	0	0
2	0	−1	1
3	0	1	1
4	1	0	1

$E = 70$ GPa and area of cross cross-section $A = 1$ cm^2. The magnitude of the downward force (negative z-direction) at Node 4 is equal to 10,000 N.

SOLUTION   The first step is to determine the direction cosines of the elements. Their length and direction cosines are calculated using the formulas in Eq. (2.49).

Element	$i \rightarrow j$	L (m)	l	m	n
1	$1 \rightarrow 4$	$\sqrt{2}$	$\sqrt{2}/2$	0	$\sqrt{2}/2$
2	$2 \rightarrow 4$	$\sqrt{2}$	$\sqrt{2}/2$	$\sqrt{2}/2$	0
3	$3 \rightarrow 4$	$\sqrt{2}$	$\sqrt{2}/2$	$-\sqrt{2}/2$	0

Using Eq. (2.52), element stiffness matrices can be constructed. Since all nodes are fixed except for Node 4, the rows and columns corresponding to zero DOF can be deleted at this stage, and the element stiffness matrix only contains active DOFs. In the case of Element 1, for example, the 6 × 6 element stiffness matrix in Eq. (2.52) involves two nodes, $i = 1$ and $j = 4$. Since Node 1 is fixed, the three rows and columns that correspond to Node 1 can be deleted at the element level. Then, the 3 × 3 reduced element stiffness matrix can be obtained. By repeating the procedure for all three elements, we can obtain the following reduced element stiffness matrices:

$$[\mathbf{k}^{(1)}] = 35\sqrt{2} \times 10^5 \begin{bmatrix} 0.5 & 0 & 0.5 \\ 0 & 0 & 0 \\ 0.5 & 0 & 0.5 \end{bmatrix} \begin{matrix} u_4 \\ v_4 \\ w_4 \end{matrix}$$

$$[\mathbf{k}^{(2)}] = 35\sqrt{2} \times 10^5 \begin{bmatrix} 0.5 & 0.5 & 0 \\ 0.5 & 0.5 & 0 \\ 0 & 0 & 0 \end{bmatrix} \begin{matrix} u_4 \\ v_4 \\ w_4 \end{matrix}$$

$$[\mathbf{k}^{(3)}] = 35\sqrt{2} \times 10^5 \begin{bmatrix} 0.5 & -0.5 & 0 \\ -0.5 & 0.5 & 0 \\ 0 & 0 & 0 \end{bmatrix} \begin{matrix} u_4 \\ v_4 \\ w_4 \end{matrix}$$

After assembly, we obtain the global equations in the form $[\mathbf{K}]\{\mathbf{Q}\} = \{\mathbf{F}\}$:

$$35\sqrt{2} \times 10^5 \begin{bmatrix} 1.5 & 0 & 0.5 \\ 0 & 1.0 & 0 \\ 0.5 & 0 & 0.5 \end{bmatrix} \begin{Bmatrix} u_4 \\ v_4 \\ w_4 \end{Bmatrix} = \begin{Bmatrix} 0 \\ 0 \\ -10,000 \end{Bmatrix}$$

The above global stiffness matrix is positive definite, as the displacement boundary conditions have already been implemented. By solving the global matrix equation, the unknown nodal displacements are obtained as

$$u_4 = 2.020 \times 10^{-3} \text{ m}$$
$$v_4 = 0$$
$$w_4 = -6.061 \times 10^{-3} \text{ m}$$

To calculate the element forces, the displacements of nodes of each element have to be transformed to local coordinates using the relation $\{\bar{\mathbf{q}}\} = [\mathbf{T}]\{\mathbf{q}\}$ in Eq. (2.48). Then, the element forces are calculated using $\{\bar{\mathbf{f}}\} = [\bar{\mathbf{k}}]\{\bar{\mathbf{q}}\}$. The axial force resultant P can be obtained from the element force $f_{\bar{j}\bar{x}}$ for the corresponding element. For Element 1, the nodal displacements in the local coordinate system can be obtained as

$$\begin{Bmatrix} \bar{u}_1 \\ \bar{u}_4 \end{Bmatrix}^{(1)} = \frac{\sqrt{2}}{2} \begin{bmatrix} 1 & 0 & 1 & 0 & 0 & 0 \\ 0 & 0 & 0 & 1 & 0 & 1 \end{bmatrix} \begin{Bmatrix} u_1 = 0 \\ v_1 = 0 \\ w_1 = 0 \\ u_4 \\ v_4 \\ w_4 \end{Bmatrix} = \begin{Bmatrix} 0 \\ -2.857 \end{Bmatrix} \times 10^{-3} \text{m}$$

From the force-displacement relation, the element force can be obtained as

$$\begin{Bmatrix} f_{1\bar{x}} \\ f_{4\bar{x}} \end{Bmatrix}^{(1)} = \left(\frac{AE}{L}\right)^{(1)} \begin{bmatrix} 1 & -1 \\ -1 & 1 \end{bmatrix} \begin{Bmatrix} \bar{u}_1 \\ \bar{u}_4 \end{Bmatrix} = \begin{Bmatrix} 14,141 \\ -14,141 \end{Bmatrix}$$

Thus, the element force is

$$P^{(1)} = (f_{4\bar{x}})^{(1)} = -14,141\,N$$

The above calculations are repeated for Elements 2 and 3:

$$\begin{Bmatrix} \bar{u}_2 \\ \bar{u}_4 \end{Bmatrix}^{(2)} = \frac{\sqrt{2}}{2}\begin{bmatrix} 1 & 1 & 0 & 0 & 0 & 0 \\ 0 & 0 & 0 & 1 & 1 & 0 \end{bmatrix}\begin{Bmatrix} u_2 \\ v_2 \\ w_2 \\ u_4 \\ v_4 \\ w_4 \end{Bmatrix} = \begin{Bmatrix} 0 \\ 1.428 \end{Bmatrix} \times 10^{-3}m$$

$$\begin{Bmatrix} f_{2\bar{x}} \\ f_{4\bar{x}} \end{Bmatrix}^{(2)} = \left(\frac{AE}{L}\right)^{(2)} \begin{bmatrix} 1 & -1 \\ -1 & 1 \end{bmatrix} \begin{Bmatrix} \bar{u}_2 \\ \bar{u}_4 \end{Bmatrix} = \begin{Bmatrix} -7,070 \\ +7,070 \end{Bmatrix}$$

$$P^{(2)} = (f_{4x})^{(2)} = +7,070\,N$$

$$\begin{Bmatrix} \bar{u}_3 \\ \bar{u}_4 \end{Bmatrix}^{(3)} = \frac{\sqrt{2}}{2}\begin{bmatrix} 1 & -1 & 0 & 0 & 0 & 0 \\ 0 & 0 & 0 & 1 & -1 & 0 \end{bmatrix}\begin{Bmatrix} u_3 \\ v_3 \\ w_3 \\ u_4 \\ v_4 \\ w_4 \end{Bmatrix} = \begin{Bmatrix} 0 \\ 1.428 \end{Bmatrix} 10^{-3}m$$

$$\begin{Bmatrix} f_{3\bar{x}} \\ f_{4\bar{x}} \end{Bmatrix}^{(3)} = \left(\frac{AE}{L}\right)^{(3)} \begin{bmatrix} 1 & -1 \\ -1 & 1 \end{bmatrix} \begin{Bmatrix} \bar{u}_2 \\ \bar{u}_4 \end{Bmatrix} = \begin{Bmatrix} -7,070 \\ +7,070 \end{Bmatrix}$$
$$P^{(3)} = (f_{4x})^{(3)} = +7,070\,N$$

Alternately, we can calculate the axial forces in an element using an equation similar to Eq. (2.45). For three-dimensional elements, this equation takes the following form:

$$P^{(e)} = \left(\frac{AE}{L}\right)^{(e)}(l(u_j - u_i) + m(v_j - v_i) + n(w_j - w_i)) \tag{2.53}$$

Note that Element 1 is in compression, while Elements 2 and 3 are in tension.

2.5   THERMAL STRESSES

Thermal stresses in structural elements appear when they are subjected to a temperature change from the reference temperature. At the reference temperature, as shown in Figure 2.22(a), if there are no external loads acting on the structure, then there will be no

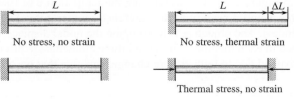

No stress, no strain

No stress, thermal strain

Thermal stress, no strain

(a) at $T = T_{ref}$

(b) at $T = T_{ref} + \Delta T$

Figure 2.22 Effects of temperature change to the structure

stresses; stresses and strains vanish simultaneously. When the temperature of one or more elements in a structure is changed as shown in Figure 2.22 (b), then the members tend to expand. However, if the expansion is partially constrained by the surrounding members, then stresses will develop due to the constraint. The constraining members will also experience a force as a reaction to this constraint. This reaction in turn produces thermal stresses in the members. The same idea can be extended to thermal stresses in a solid, if we imagine the solid to contain many small elements and each restraining others from expanding or contracting due to temperature change.

The linear relation between stresses and strains for linear elastic solids in Section 1.3 is valid only when the temperature remains constant and in the absence of residual stresses. In the presence of a temperature differential, i.e., when the temperature is different from the reference temperature, we need to use the thermo-elastic stress-strain relations. Such a relation in one dimension is

$$\sigma = E(\varepsilon - \alpha \Delta T) \tag{2.54}$$

where σ is the uniaxial stress, ε is the total strain, E is the Youngs modulus, α is the coefficient of thermal expansion (CTE), and ΔT is the difference between the operating temperature and the reference temperature. From Eq. (2.54), it is clear that the *reference temperature* is defined as the temperature at which both stress and strain vanish simultaneously when there is no external load. Equation (2.54) states that the stress is caused by and proportional to the mechanical strain, which is the difference between the total strain $\varepsilon = \Delta L / L$ and the *thermal strain* $\alpha \Delta T$. The strain-stress relation now takes the following form:

$$\varepsilon = \frac{\sigma}{E} + \alpha \Delta T \tag{2.55}$$

The total strain is the sum of the mechanical strain caused by the stresses and the thermal strain caused by the temperature rise. Similarly, thermo-elastic stress-strain relations can be developed in two and three dimensions.[3] The above stress-strain relation can be converted into the force-displacement relation by multiplying Eq. (2.54) by the area of cross-section of the uniaxial bar, A:

$$P = AE\left(\frac{\Delta L}{L} - \alpha \Delta T\right) = AE\frac{\Delta L}{L} - AE\alpha \Delta T \tag{2.56}$$

where the first term in the parentheses is the total strain or simply strain, and the second the thermal strain.

Before we introduce the formal method of solving a thermal stress problem using FE, it will be instructive to discuss the method of superposition for solving thermal stress problems.

2.5.1   Method of Superposition

Consider the truss shown in Figure 2.23 (a). Assume that the temperature of Element 2 is raised by ΔT, and Elements 1 and 3 remain at the reference temperature. There are no external forces acting at Node 4. The objective is to compute the nodal displacements and forces that will be developed in each member. First, if all three elements are disconnected, then only Element 2 will expand due to temperature change. Imagine that we apply a pair

---

[3] A.P. Boresi and R.J. Schmidt, *Advanced mechanics of Materials*, 6th ed, John Wiley & Sons, New York, 2003.

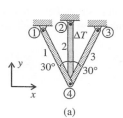
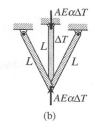
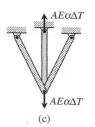

(a)   (b)   (c)

Figure 2.23 A three-element truss: (a) The middle element is subjected to a temperature rise. This is the given problem. (b) A pair of compressive forces is applied to Element 2 to prevent it from expanding. This is called Problem I. (c) The forces in Problem I are reversed. No thermal stresses are involved in this problem. This is called Problem II

of equal and opposite forces on the two nodes of Element 2, such that the forces restrain the thermal expansion of the element [see Figure 2.23 (b)]. This force can be determined by setting $\Delta L = 0$ in Eq. (2.56), and it is equal to $-AE\alpha\Delta T$. That is, a compressive force is required to prevent Element 2 from expanding. Hence the force resultant on Element 2, $P^{(2)}$, is compressive with magnitude equal to $AE\alpha\Delta T$. If several members are subjected to temperature changes, then a corresponding pair of forces is applied to each element.

The solution to this problem, which will be called Problem I, is obvious: the nodal displacements are all equal to zero because no element is allowed to expand or contract, and the force in Element 2 is equal to $-AE\alpha\Delta T$. There are no forces in Elements 1 and 3. However, this is not the problem we want to solve. The pair of forces applied to Element 2 was not there in the original problem. Hence, we have to remove these extraneous forces.

To remove the fictitious force in Figure 2.23(b), we superpose the results from Problem II, where there is no thermal effect, but the forces applied in Problem I are all reversed. This is depicted in Figure 2.23(c). Sometimes the forces acting in Problem II are called fictitious thermal forces, as they do not actually exist. Problem II is a standard truss problem, and hence the FEM we have already discussed can be used to determine the nodal displacements and the element forces. The solution of the problem in Figure 2.23(a) can be obtained by adding the solutions from both Problems I and II.

EXAMPLE 2.7 *Thermal Stresses in a Plane Truss*

Solve the nodal displacements and element forces of the plane truss problem in Figure 2.23. Use the following numerical data: $AE = 10^7$ N, $L = 1$ m, $\alpha = 10^{-5}/°C$, $\Delta T = 100°C$.

SOLUTION Solution to Problem I is as follows:

$$u_4 = v_4 = 0$$
$$P^{(1)} = 0, \quad P^{(2)} = -AE\alpha\Delta T = -10,000\,\text{N}, \quad P^{(3)} = 0 \tag{2.57}$$

The Problem II is depicted in Figure 2.23(c). The element properties and direction cosines are listed in Table 2.3.

Table 2.3 Element Properties and Direction Cosines for Truss Structure in Figure 2.23

Element	$i \rightarrow j$	AE/L(N/m)	$AE\alpha\Delta T$ (N)	ϕ (Degrees)	$l = \cos\phi$	$m = \sin\phi$
1	$1 \rightarrow 4$	10^7	0	-60	$1/2$	$-\sqrt{3}/2$
2	$2 \rightarrow 4$	10^7	10,000	-90	0	-1
3	$3 \rightarrow 4$	10^7	0	240	$-1/2$	$-\sqrt{3}/2$

The element stiffness matrices in the global coordinates are written below. For convenience, the rows and columns corresponding to zero DOFs are deleted, and only those corresponding to active DOFs are shown.

$$[\mathbf{k}^{(1)}] = \frac{10^7}{4} \begin{bmatrix} 1 & -\sqrt{3} \\ -\sqrt{3} & 3 \end{bmatrix} \begin{matrix} u_4 \\ v_4 \end{matrix}$$

$$[\mathbf{k}^{(2)}] = 10^7 \begin{bmatrix} 0 & 0 \\ 0 & 1 \end{bmatrix} \begin{matrix} u_4 \\ v_4 \end{matrix} \tag{2.58}$$

$$[\mathbf{k}^{(3)}] = \frac{10^7}{4} \begin{bmatrix} 1 & \sqrt{3} \\ \sqrt{3} & 3 \end{bmatrix} \begin{matrix} u_4 \\ v_4 \end{matrix}$$

Assembling the element stiffness matrices, we obtain the global stiffness matrix $[\mathbf{K}]$. The only external force for this problem is $F_{4y} = -10,000 \, \text{N}$.

$$\frac{10^7}{4} \begin{bmatrix} 2 & 0 \\ 0 & 10 \end{bmatrix} \begin{Bmatrix} u_4 \\ v_4 \end{Bmatrix} = \begin{Bmatrix} 0 \\ -10,000 \end{Bmatrix} \tag{2.59}$$

Since the global stiffness matrix is positive definite, the solution for displacements can be obtained as:

$$\begin{aligned} u_4 &= 0 \\ v_4 &= -0.4 \times 10^{-3} \, \text{m} \end{aligned} \tag{2.60}$$

The force resultants in the elements for Problem II can be obtained from Eq. (2.45). Substituting the element properties and displacements, we obtain:

$$\begin{aligned} P^{(1)} &= 3,464 \, \text{N} \\ P^{(2)} &= 4,000 \, \text{N} \\ P^{(3)} &= 3,464 \, \text{N} \end{aligned} \tag{2.61}$$

Then, the solution (displacements and forces) to the given problem is the sum of solutions to Problems I and II, as shown in Table 2.4. Note that Elements 1 and 3 are in tension, while Element 2 is in compression.

If there were external forces acting at Node 4 in the given problem, then they can be added to the fictitious forces in Problem II.

Table 2.4   Solution of Thermal Stresses in a Truss Using the Superposition Method

Variable	Problem I	Problem II	Final solution
u_4	0	0	0
v_4	0	-0.4×10^{-3} m	-0.4×10^{-3} m
$P^{(1)}$	$-AE\alpha\Delta T^{(1)} = 0$	3,464 N	3,464 N
$P^{(2)}$	$-AE\alpha\Delta T^{(2)} = -10,000$ N	4,000 N	$-6,000$ N
$P^{(3)}$	$-AE\alpha\Delta T^{(3)} = 0$	3,464 N	3,464 N

2.5.2   Thermal Stresses Using FEA

In using the FEM for thermal stress problems, we combine the two problems in the previous subsection as one problem and solve for displacements and forces simultaneously. This procedure is similar to the superposition method. Consider the element equilibrium equation in Eq. (2.30) for the uniaxial bar element. It states that the forces acting on an element are the product of element stiffness matrix and the vector of nodal displacements, i.e., $\{\bar{\mathbf{f}}\} = [\bar{\mathbf{k}}]\{\bar{\mathbf{q}}\}$. This is similar to the linear elastic stress-strain relation at the element level, where stresses are linear combination of strains at a point, $\{\sigma\} = [E]\{\varepsilon\}$. However, we notice that in the presence of temperature differential, the stresses are not linear combination of strains, [see Eq. (2.54)]. A similar adjustment also has to be made at the element level equation. In the presence of thermal stresses, Eq. (2.35) can be modified as

$$\left\{\bar{\mathbf{f}}^{(e)}\right\} = \left[\bar{\mathbf{k}}^{(e)}\right]\left\{\bar{\mathbf{q}}^{(e)}\right\} - \left\{\bar{\mathbf{f}}_T^{(e)}\right\} \qquad (2.62)$$

where the element thermal force vector $\{\bar{\mathbf{f}}_T^{(e)}\}$ in the local coordinate system is given by

$$\left\{\bar{\mathbf{f}}_T^{(e)}\right\} = AE\alpha\Delta T \begin{Bmatrix} -1 \\ 0 \\ +1 \\ 0 \end{Bmatrix} \begin{matrix} \bar{u}_i \\ \bar{v}_i \\ \bar{u}_j \\ \bar{v}_j \end{matrix} \qquad (2.63)$$

Note that the row addresses or the DOF corresponding to each force is indicated next to the force vector. Multiplying both sides of Eq. (2.62) by the transpose of the transformation matrix, $[\mathbf{T}]^T$, and also using $\{\bar{\mathbf{q}}\} = [\mathbf{T}]\{\mathbf{q}\}$, we obtain

$$\{\mathbf{f}\} = [\mathbf{k}]\{\mathbf{q}\} - \{\mathbf{f}_T\} \qquad (2.64)$$

where $[\mathbf{k}]$ is the element stiffness matrix defined in Eq. (2.39) and $\{\mathbf{f}_T\}$ is the thermal force vector in the global coordinates given by

$$\{\mathbf{f}_T\} = AE\alpha\Delta T \begin{Bmatrix} -l \\ -m \\ +l \\ +m \end{Bmatrix} \begin{matrix} u_i \\ v_i \\ u_j \\ v_j \end{matrix} \qquad (2.65)$$

The vector $\{\mathbf{f}_T\}$ has four rows, and its row addresses are in the same order as those of $[\mathbf{k}]$.

If Eq. (2.65) is substituted in the nodal equilibrium equations, we will obtain the global equations at the structural level as

$$[\mathbf{K}_s]\{\mathbf{Q}_s\} = \{\mathbf{F}_s\} + \{\mathbf{F}_{Ts}\} \qquad (2.66)$$

where $\{\mathbf{F}_{Ts}\}$ is the thermal load vector, which is obtained by assembling $\{\mathbf{f}_T\}$ of various elements. It is clear from Eq. (2.66) that the increase in temperature is equivalent to adding an additional force to the member. After striking out the rows and columns corresponding to zero DOFs, we obtain the global equations as

$$[\mathbf{K}]\{\mathbf{Q}\} = \{\mathbf{F}\} + \{\mathbf{F}_T\} \qquad (2.67)$$

The assembly of $\{\mathbf{F}_T\}$ is similar to that of stiffness matrix. Equation (2.67) is solved to obtain the unknown displacements $\{\mathbf{Q}\}$. To find the forces in elements, we must use Eq. (2.56).

EXAMPLE 2.8 *Thermal Stresses in a Plane Truss Using FEA*

Solve the thermal stress problem in Example 2.7 using finite element method.

SOLUTION The element stiffness matrices are already given in Eq. (2.58). The thermal force vectors are written below. For convenience, the rows and columns corresponding to zero DOF are deleted and only those corresponding to active DOFs are shown.

$$\left\{\mathbf{f}_T^{(1)}\right\} = AE\alpha\Delta T^{(1)} \begin{Bmatrix} 1/2 \\ -\sqrt{3}/2 \end{Bmatrix} = \begin{Bmatrix} 0 \\ 0 \end{Bmatrix} \begin{matrix} u_4 \\ v_4 \end{matrix} \qquad (2.68)$$

$$\{\mathbf{f}_T^{(2)}\} = AE\alpha\Delta T^{(2)} \begin{Bmatrix} 0 \\ -1 \end{Bmatrix} = \begin{Bmatrix} 0 \\ -10,000 \end{Bmatrix} \begin{matrix} u_4 \\ v_4 \end{matrix}$$

$$\{\mathbf{f}_T^{(3)}\} = AE\alpha\Delta T^{(3)} \begin{Bmatrix} 1/2 \\ -\sqrt{3}/2 \end{Bmatrix} = \begin{Bmatrix} 0 \\ 0 \end{Bmatrix} \begin{matrix} u_4 \\ v_4 \end{matrix}$$

Note that there is no thermal force vector for Elements 1 and 3 because they are at the reference temperature. The row addresses are shown next to the thermal force vector in Eq. (2.68).

Assembling the element stiffness matrices, we obtain the global stiffness matrix $[\mathbf{K}]$, and assembling the element thermal force vectors $\{\mathbf{f}_T\}$, we obtain the global thermal force vectors $\{\mathbf{F}_T\}$ as

$$\{\mathbf{F}_T\} = \begin{Bmatrix} 0 \\ -10,000 \end{Bmatrix} \tag{2.69}$$

The solution for displacements is obtained using the global equations

$$[\mathbf{K}]\{\mathbf{Q}\} = \{\mathbf{F}\} + \{\mathbf{F}_T\} \tag{2.70}$$

Since there are no external forces in the present problem, $\{\mathbf{F}\} = \{0\}$. Hence the global equations are:

$$\frac{10^7}{4} \begin{bmatrix} 2 & 0 \\ 0 & 10 \end{bmatrix} \begin{Bmatrix} u_4 \\ v_4 \end{Bmatrix} = \begin{Bmatrix} 0 \\ -10,000 \end{Bmatrix} \tag{2.71}$$

The solution to the above equations is obtained as

$$\boxed{\begin{matrix} u_4 = 0 \\ v_4 = -0.4 \times 10^{-3}\,\text{m} \end{matrix}}$$

The force resultants in the elements are obtained from Eq. (2.56):

$$\begin{aligned} P &= AE\left(\frac{\Delta L}{L} - \alpha\Delta T\right) \\ &= \frac{AE}{L}\left[l(u_j - u_i) + m(v_j - v_i)\right] - AE\alpha\Delta T \end{aligned} \tag{2.72}$$

Substituting the element properties and displacements, we obtain

$$\boxed{\begin{matrix} P^{(1)} = 3,464\,\text{N} \\ P^{(2)} = -6,000\,\text{N} \\ P^{(3)} = 3,464\,\text{N} \end{matrix}}$$

The FE solution for displacements and forces above can be compared with those obtained from the superposition method presented in Table 2.4.

One can check the force equilibrium at Node 4. The three forces acting on Node 4 are shown in Figure 2.24. Summing the forces in the x- and y-directions,

$$\begin{aligned} \sum F_x &= -P^{(1)}\sin 30 + P^{(2)}\sin 30 = 0 \\ \sum F_y &= P^{(1)}\cos 30 - P^{(2)} + P^{(3)}\cos 30 \\ &= 3464 \times \frac{\sqrt{3}}{2} - 6000 + 3464 \times \frac{\sqrt{3}}{2} \\ &= 0 \end{aligned} \tag{2.73}$$

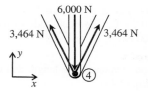

Figure 2.24 Force equilibrium at Node 4

Thermal stress analysis of space trusses follows the same procedures. The thermal force vector $\{\mathbf{f}_T\}$ is a 6×1 matrix and is given by

$$\{\mathbf{f}_T\}^T = AE\alpha\Delta T \begin{bmatrix} \overset{u_i}{-l} & \overset{v_i}{-m} & \overset{w_i}{-n} & \overset{u_j}{+l} & \overset{v_j}{+m} & \overset{w_j}{+n} \end{bmatrix} \tag{2.74}$$

In the above equation, $\{\mathbf{f}_T\}^T$ is given as a row matrix with addresses shown above the elements of the matrix. Another difference between two- and three-dimensional thermal stress problems is in the calculation of force in an element. An equation similar to Eq. (2.72) for the three-dimensional truss element can be derived as

$$\begin{aligned} P &= AE\left(\frac{\Delta L}{L} - \alpha\Delta T\right) \\ &= \frac{AE}{L}[l(u_j - u_i) + m(v_j - v_i) + n(w_j - w_i)] - AE\alpha\Delta T \end{aligned} \tag{2.75}$$

where l, m, and n are the direction cosines of the element.

EXAMPLE 2.9   *Thermal Stresses in a Space Truss*

Use the FEM to determine the displacements and forces in the space truss, shown in Figure 2.25. The coordinates of the nodes in meter units are given in Table 2.5. The temperature of Element 1

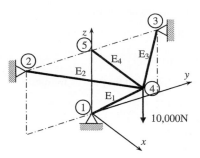

Figure 2.25 Three-bar space truss structure

Table 2.5   Nodal Coordinates of Space Truss Structure in Example 2.9

Node	x	y	z
1	0	0	0
2	0	−1	1
3	0	1	1
4	1	0	1
5	0	0	1

is raised by 100°C above the reference temperature. Assume Young's modulus $E = 70$ GPa and area of cross-section $A = 1$ cm^2. Assume CTE $\alpha = 20 \times 10^{-6}/°C$. The magnitude of the downward force (negative z-direction) at Node 4 is equal to 10,000 N.

SOLUTION   The first step is to determine the direction cosines of the elements. The length and direction cosines of each element are calculated using the formulas in Eq. (2.49).

Element	$i \rightarrow j$	L (m)	l	m	n
1	$1 \rightarrow 4$	$\sqrt{2}$	$\sqrt{2}/2$	0	$\sqrt{2}/2$
2	$2 \rightarrow 4$	$\sqrt{2}$	$\sqrt{2}/2$	$\sqrt{2}/2$	0
3	$3 \rightarrow 4$	$\sqrt{2}$	$\sqrt{2}/2$	$-\sqrt{2}/2$	0
4	$5 \rightarrow 4$	1	1	0	0

Stiffness matrices of Elements 1 through 3 are the same as in the Example 2.6. The stiffness matrix of Element 4 is as follows:

$$[\mathbf{k}^{(4)}] = 70 \times 10^5 \begin{bmatrix} 1 & 0 & 0 \\ 0 & 0 & 0 \\ 0 & 0 & 0 \end{bmatrix} \begin{matrix} u_4 \\ v_4 \\ w_4 \end{matrix}$$

We need to calculate the thermal force vector for Element 1 only as its temperature is different from the reference temperature. Using the formula in Eq. (2.74), we obtain

$$\{\mathbf{f}_\mathbf{T}^{(1)}\}^T = 7000\sqrt{2} \begin{matrix} u_1 & v_1 & w_1 & u_4 & v_4 & w_4 \\ [-1 & 0 & -1 & 1 & 0 & 1] \end{matrix} \tag{2.76}$$

After assembly, we obtain the global equations in the form $[\mathbf{K}]\{\mathbf{Q}\} = \{\mathbf{F}\} + \{\mathbf{F}_\mathbf{T}\}$:

$$35\sqrt{2} \times 10^5 \begin{bmatrix} 1.5 + \sqrt{2} & 0 & 0.5 \\ 0 & 1.0 & 0 \\ 0.5 & 0 & 0.5 \end{bmatrix} \begin{Bmatrix} u_4 \\ v_4 \\ w_4 \end{Bmatrix} = \begin{Bmatrix} 0 \\ 0 \\ -10,000 \end{Bmatrix} + \begin{Bmatrix} 9900 \\ 0 \\ 9900 \end{Bmatrix}$$

Solving the above equation, the unknown nodal displacements are obtained as

$$u_4 = 0.8368 \times 10^{-3} \text{ m}$$
$$v_4 = 0$$
$$w_4 = -0.8772 \times 10^{-3} \text{ m}$$

The forces in the elements can be calculated using Eq. (2.75), and they are as follows:

$$P^{(1)} = -14,141 \text{ N}$$
$$P^{(2)} = +2,929 \text{ N}$$
$$P^{(3)} = +2,929 \text{ N}$$
$$P^{(4)} = +5,858 \text{ N}$$

One can verify that the force equilibrium is satisfied at Node 4.

2.6   PROJECTS

Project 2.1.   Analysis and Design of a Space Truss

A space truss, as shown in Figure 2.26, consists of 25 truss members. Initially, all members have the same circular cross-sections with diameter $d = 2.0$ in. At nodes 1 and 2, a constant force $F = 60,000$ lb is applied in the y-direction. Four nodes (7, 8, 9, and 10) are fixed on the ground. The frame structure is made of a steel material whose properties are

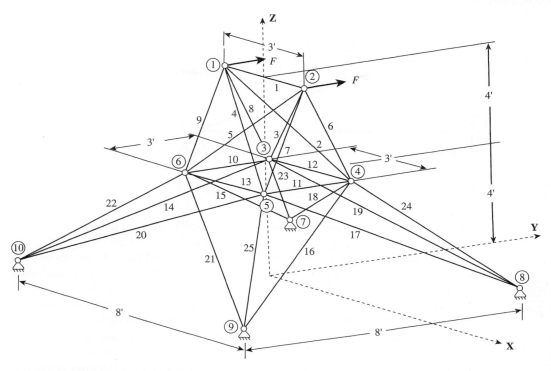

Figure 2.26 25-member space truss

Young's modulus $E = 3 \times 10^7$ psi, Poisson's ratio $\nu = 0.3$, yield stress $\sigma_Y = 37,000$ psi, and mass density $\rho = 7.3 \times 10^{-4}$ lb $\cdot$ sec^2/in^4. The safety factor $N = 1.5$ is used. Due to the manufacturing constraints, the diameter of the truss members should vary between 0.1 in and 2.5 in.

1. Solve the initial truss structure using truss FEs. Provide a plot that shows labels for elements and nodes along with boundary conditions. Provide deformed geometry of the structure and a table of stress in each element.

2. Minimize the structural weight by changing the cross-sectional diameter of each truss element, while all members are safe under the given yield stress and the safety factor. You can use the symmetric geometry of the structure. Identify zero-force members. For zero-force members, use the lower bound of the cross-sectional diameter. Provide deformed geometry at the optimum design along with a table of stress at each element. Provide structural weights of initial and optimum designs.

Project 2.2.   Analysis and Design of a Plane Truss 1

The truss shown in Figure 2.27 has two elements. The members are made of aluminum hollow square cross-section. The outer dimension of the square is 12 mm, and the inner dimension is 9 mm. (The wall thickness is 1.5 mm in all four sides). Assume Young's modulus $E = 70$ GPa and yield strength $\sigma_Y = 70$ MPa. The magnitude of the force at Node 1 (F) is equal to 1,000 N.

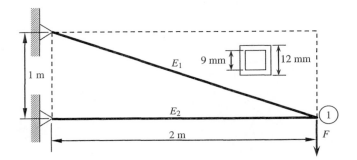

Figure 2.27 Plane truss and design domain for Project 2.2

1. Use FEM to determine the displacements at Node 1 and axial forces in Elements 1 and 2. Use von Mises' yield theory to determine if the elements will yield or not. Use Euler buckling load ($P_{cr} = \pi^2 EI/L^2$) to determine if the elements under compressive loads will buckle. In the above expression, P_{cr} is the axial compressive force, E is the Young's modulus, I is the moment of inertia of the cross-section given by $I = (a_o^4 - a_i^4)/12$ where a_o and a_i, respectively, are the outer and inner dimensions of the hollow square cross section, and L is the length of the element.

2. Redesign the truss so that both the stress and buckling constraints are satisfied with a Safety Factor of N not less than 2 for stresses and N not less than 1.2 for buckling. Your design goal should be to reduce the weight of the truss as much as possible. The truss should be contained within the virtual rectangle shown by the dashed lines. Node 1 must be present to take the downward load $F = 1,000$ N. The nodes at the left wall have to be fixed completely. Nodes not attached to the wall have to be completely free to move in the x and y directions. Use the same cross-section for all elements. Calculate the mass of the truss you have designed. Assume density of aluminum as 2,800 kg/m^3.

   Draw the truss you have designed and provide the nodal coordinates and element connectivity in the form of a table. Results should also include the nodal displacements, forces in each member, and safety factors for stresses and buckling for each element in the form of tables.

Projects 2.3.   Analysis and Design of a Plane Truss 2

Consider a plane truss in Figure 2.28. The horizontal and vertical members have length L, while inclined members have length $\sqrt{2}L$. Assume the Young's modulus $E = 100$ GPa, cross-sectional area $A = 1.0$ cm^2, and $L = 0.3$ m.

1. Use a FE program to determine the deflections and element forces for the following three load cases. Present your results in the form of a table.

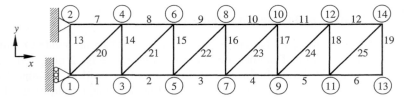

Figure 2.28 Plane truss and design domain for Project 2.3

Load Case A: $F_{x13} = F_{x14} = 10{,}000\,\text{N}$

Load Case B: $F_{y13} = F_{y14} = 10{,}000\,\text{N}$

Load Case C: $F_{x13} = 10{,}000\,\text{N}$ and $F_{x14} = -10{,}000\,\text{N}$

2. Assuming that the truss behaves like a cantilever beam, one can determine the equivalent cross-sectional properties of the beam from the results for Cases A through C above. The three beam properties are axial rigidity $(EA)_{eq}$ (this is different from the AE of the truss member), flexural rigidity $(EI)_{eq}$, and shear rigidity $(GA)_{eq}$. Let the beam length be equal to $l(l = 6 \times 0.3 = 1.8\,\text{m})$.

   The axial deflection of a beam due to an axial force F is given by

$$u_{tip} = \frac{Fl}{(EA)_{eq}} \tag{2.77}$$

   The transverse deflection due to a transverse force F at the tip is

$$v_{tip} = \frac{Fl^3}{3(EI)_{eq}} + \frac{Fl}{(GA)_{eq}} \tag{2.78}$$

   In Eq.(2.78), the first term on the RHS represents the deflection due to flexure and the second term due to shear deformation. In the elementary beam theory (Euler-Bernoulli beam theory), we neglect the shear deformation, as it is usually much smaller than the flexural deflection.

   The transverse deflection due to an end couple C is given by

$$v_{tip} = \frac{Cl^2}{2(EI)_{eq}} \tag{2.79}$$

   Substitute the average tip deflections obtained in Part 1 in Eqs. (2.77)–(2.79) to compute the equivalent section properties: $(EA)_{eq}$, $(EI)_{eq}$, and $(GA)_{eq}$.

   You may use the average of deflections at Nodes 13 and 14 to determine the equivalent beam deflections.

3. Verify the beam model by adding two more bays to the truss $(l = 8 \times 0.3 = 2.4\,\text{m})$. Compute the tip deflections of the extended truss for the three load cases A–C using the FE program. Compare the FE results with deflections obtained from the equivalent beam model [Eqs. (2.77)–(2.79)].

Project 2.4.   Fully Stressed Design of a Ten-Bar Truss

Fully stressed design is often used for truss structures. The idea is that we should remove material from members that are not fully stressed unless prevented by minimum cross-sectional area constraint. Practically, at every design cycle, the new cross-sectional area can be found using the following relation:

$$A_{new}^{(e)} = \frac{\sigma_{old}^{(e)}}{\sigma_{allowable}^{(e)}} A_{old}^{(e)}$$

   A 10-bar truss structure shown in Figure 2.29 is under two loads, P_1 and P_2. The design goal is to minimize the weight, W, by varying the cross-sectional areas, A_i, of the truss members. The stress of the member should be less than the allowable stress with the safety factor. For manufacturing reasons, the cross-sectional areas should be greater than the minimum value. Input data are summarized in the table. Find the optimum design using fully stressed design.

Parameters	Values
Dimension, b	360 inches
Safety factor, S_F	1.5
Load, P_1	66.67 kips
Load, P_2	66.67 kips
Density, ρ	0.1 lb/in^3
Modulus of elasticity, E	10^4 ksi
Allowable stress, $\sigma_{\text{allowable}}$	25 ksi*
Initial area A_i	1.0 in^2
Minimum cross-sectional area	0.1 in^2

*for Element 9, allowable stress is 75 ksi

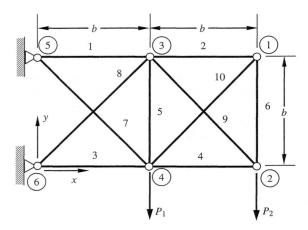

Figure 2.29 Ten-bar truss structure for Project 2.4

2.7   EXERCISE

1. Three rigid bodies, 2, 3, and 4, are connected by four springs, as shown in the figure. A horizontal force of 1,000 N is applied on Body 4, as shown in the figure. Find the displacements of the three bodies and the forces (tensile/compressive) in the springs. What is the reaction at the wall? Assume the bodies can undergo only translation in the horizontal direction. The spring constants (N/mm) are $k_1 = 400$, $k_2 = 500$, $k_3 = 500$, $k_4 = 300$.

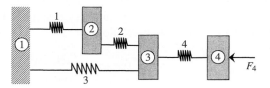

2. Three rigid bodies, 2, 3, and 4, are connected by six springs as shown in the figure. The rigid walls are represented by 1 and 5. A horizontal force $F_3 = 1000$ N is applied on Body 3 in the direction shown in the figure. Find the displacements of the three bodies and the forces (tensile/compressive) in the springs. What are the reactions at the walls? Assume the bodies can undergo only translation in the horizontal direction. The spring constants (N/mm) are $k_1 = 500$, $k_2 = 400$, $k_3 = 600$, $k_4 = 200$, $k_5 = 400$, $k_6 = 300$.

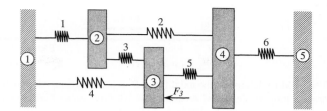

3. Consider the spring-rigid body system described in Problem 2. What force F_2 should be applied on Body 2 to keep it from moving? How will this affect the support reactions?
   Hint: Impose the boundary condition $u_2 = 0$ in the FEM and solve for displacements u_3 and u_4. Then, the force F_2 will be the reaction at Node 2.

4. Four rigid bodies, 1, 2, 3, and 4, are connected by four springs as shown in the figure. A horizontal force of 1,000 N is applied on Body 1 as shown in the figure. Using FE analysis, (a) find the displacements of the two bodies (1 and 3), (2) find the element force (tensile/compressive) of spring 1, and (3) the reaction force at the right wall (Body 2). Assume the bodies can undergo only translation in the horizontal direction. The spring constants (N/mm) are $k_1 = 400$, $k_2 = 500$, $k_3 = 500$, and $k_4 = 300$. Do not change node and element numbers.

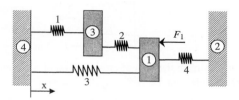

5. Determine the nodal displacements and reaction forces using the direct stiffness method. Calculate the nodal displacements and element forces using the FE program.

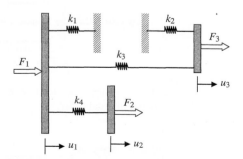

6. In the structure shown below, rigid blocks are connected by linear springs. Imagine that only horizontal displacements are allowed. Write the global equilibrium equations $[\mathbf{K}]\{\mathbf{Q}\} = \{\mathbf{F}\}$ after applying displacement boundary conditions in terms of spring stiffness k_i, displacement DOFs u_i, and applied loads F_i.

7. A structure is composed of two one-dimensional bar elements. When 10 N force is applied to node 2, calculate displacement vector $\{\mathbf{Q}\}^T = \{u_1, u_2, u_3\}$ using the finite element method.

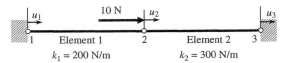

8. Use FEM to determine the axial force P in each portion, AB and BC, of the uniaxial bar. What are the support reactions? Assume $E = 100$ GPa, area of cross sections of the two portions AB and BC are, respectively, 10^{-4}m^2 and $2 \times 10^{-4} \text{m}^2$, and $F = 10,000$ N. The force F is applied at the cross section at B.

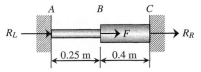

9. Consider a tapered bar of circular cross-section. The length of the bar is 1 m, and the radius varies as $r(x) = 0.050 - 0.040x$, where r and x are in meters. Assume Young's modulus $= 100$ MPa. Both ends of the bar are fixed, and $F = 10,000$ N is applied at the center. Determine the displacements, axial force distribution, and wall reactions using four elements of equal length.

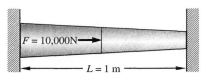

Hint: To approximate the area of cross-section of a bar element, use the geometric mean of the end areas of the element, i.e. $A^{(e)} = \sqrt{A_i A_j} = \pi r_i r_j$.

10. The stepped bar shown in the figure is subjected to a force at the center. Use FEM to determine the displacement at the center and reactions R_L and R_R.

   Assume: $E = 100$ GPa, area of cross sections of the three portions shown are, respectively, 10^{-4}m^2, $2 \times 10^{-4} \text{m}^2$, and 10^{-4}m^2, and $F = 10,000$ N.

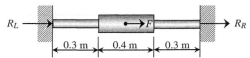

11. Using the direct stiffness matrix method, find the nodal displacements and the forces in each element and the reactions.

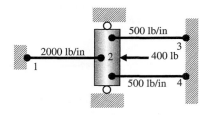

12. A stepped bar is clamped at one end and subjected to concentrated forces as shown.
   Note: The node numbers are not in usual order!

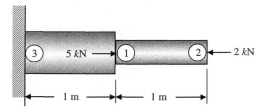

Assume: $E = 100$ GPa, small area of cross-section $= 1\,\text{cm}^2$, and large area of cross-section $= 2\,\text{cm}^2$.

(a) Write the element stiffness matrices of Elements 1 and 2 showing the row addresses.

(b) Assemble the above element stiffness matrices to obtain the structural level equations in the form $[\mathbf{K}_s]\{\mathbf{Q}_s\} = \{\mathbf{F}_s\}$.

(c) Delete the rows and columns corresponding to zero DOF to obtain the global equations in the form of $[\mathbf{K}]\{\mathbf{Q}\} = \{\mathbf{F}\}$.

(d) Determine the displacements and element forces.

13. The uniaxial bar FE equation can be used for other types of engineering problems, if the proper analogy is applied. For example, consider the piping network shown in the figure. Each section of the network can be modeled using a FE. If the flow is laminar and steady, we can write the equations for a single pipe element as

$$q_i = K(P_i - P_j)$$
$$q_j = K(P_j - P_i)$$

where q_i and q_j are fluid flow at nodes i and j, respectively; P_i and P_j are fluid pressure at nodes i and j, respectively; and K is

$$K = \frac{\pi D^4}{128 \mu L}$$

where D is the diameter of the piper, μ is the viscosity, and L is the length of the pipe. The fluid flow going into the system is considered positive. The viscosity of the fluid is $9 \times 10^{-4}\,\text{Pa} \cdot \text{s}$.

(a) Write the element matrix equation for the flow in the pipe element.

(b) The net flow rates into nodes 1 and 2 are 10 and 15 m^3/s, respectively. The pressures at the nodes 6, 7, and 8 are all zero. The net flow rate into the nodes 3, 4, and 5 are all zero. What is the outflow rate for elements 4, 6, and 7?

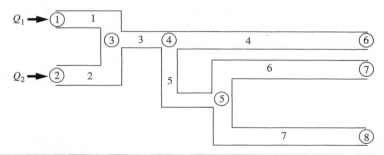

Elem	1	2	3	4	5	6	7
D(mm)	40	40	50	25	40	25	25
L(m)	1	1	1	4	2	3	3

14. For a two-dimensional truss structure, as shown in the figure, determine displacements of the nodes and normal stresses developed in the members using the direct stiffness method. Use $E = 30 \times 10^6\,\text{N/cm}^2$ and a diameter of the circular cross-section of 0.25 cm.

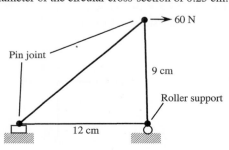

15. For a two-dimensional truss structure, as shown in the figure, determine displacements of the nodes and normal stresses developed in the members using a FE program in Appendices. Use $E = 30 \times 10^6 \, \text{N/cm}^2$ and a diameter of the circular cross section of 0.25 cm.

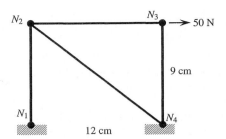

16. The truss structure shown in the figure supports force F at Node 2. FEM is used to analyze this structure using two truss elements as shown.

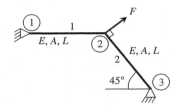

(a) Compute the transformation matrix for Elements 1 and 2.

(b) Compute the element stiffness matrices for both elements in the global coordinate system.

(c) Assemble the element stiffness matrices and force vectors to structural matrix equation $[\mathbf{K}_s]\{\mathbf{Q}_s\} = \{\mathbf{F}_s\}$ before applying boundary conditions.

(d) Solve the FE equation after applying the boundary conditions. Write nodal displacements in the global coordinates.

(e) Compute the stress in the Element 1. Is it tensile or compressive?

17. The truss structure shown in the figure supports the force F. FEM is used to analyze this structure using two truss elements as shown. Area of cross-section (for all elements) $= A = 2 \, \text{in}^2$, Young's modulus $= E = 30 \times 10^6$ psi. Both the elements are of equal length $L = 10$ ft.

(a) Compute the transformation matrix for Elements 1 and 2 to transform between the global coordinate system and the local coordinate system for each element.

(b) Compute the stiffness matrix for the Elements 1 and 2.

(c) Assemble the structural matrix equation $[\mathbf{K}_s]\{\mathbf{Q}_s\} = \{\mathbf{F}_s\}$ (without applying the boundary conditions).

(d) It is determined after solving the final equations that the displacement components of the node 1 are $u_1 = 1.5 \times 10^{-2}$ in, $v_1 = -0.5 \times 10^{-2}$ in. Compute the applied load F.

(e) Compute stress and strain in Element 1.

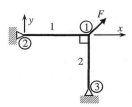

18. Use FEM to solve the plane truss shown below. Assume $AE = 10^6$N, $L = 1$ m. Determine the nodal displacements, forces in each element, and the support reactions.

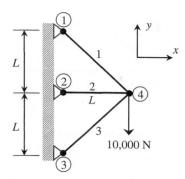

19. The plane truss shown in the figure has two elements and three nodes. Calculate the 4×4 element stiffness matrices. Show the row addresses clearly. Derive the final equations (after applying boundary conditions) for the truss in the form of $[\mathbf{K}]\{\mathbf{Q}\} = \{\mathbf{F}\}$. What are nodal displacements and the element forces? Assume $E = 10^{11}$ Pa, $A = 10^{-4}$ m^2, $L = 1$ m, $F = 14,142$ N.

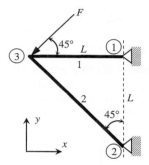

20. Use FEM to solve the two plane truss problems shown in the figure below. Assume $AE = 10^6$ N, $L = 1$ m. Before solving the global equations $[\mathbf{K}]\{\mathbf{Q}\} = \{\mathbf{F}\}$, find the determinant of $[\mathbf{K}]$. Does $[\mathbf{K}]$ have an inverse? Explain your answer.

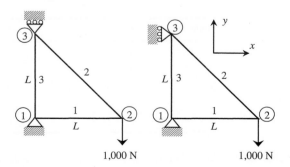

21. Determine the member force and axial stress in each member of the truss shown in the figure using one of FE analysis programs in the Appendix. Assume that Young's modulus is 10^4 psi and all cross-sections are circular with a diameter of 2 in. Compare the results with the exact solutions that are obtained from free-body diagram.

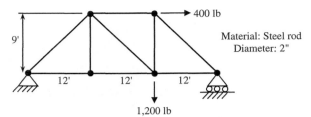

22. Determine the normal stress in each member of the truss structure. All joints are ball-joint and the material is steel, whose Young's modulus is $E = 210\,\text{GPa}$.

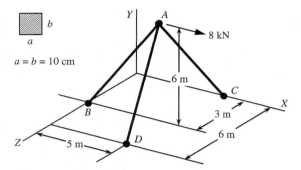

23. The space truss shown has four members. Determine the displacement components of Node 5 and the force in each member. The node numbers are numbers in the circle in the figure. The dimensions of the imaginary box that encloses the truss are: $1 \times 1 \times 2\,\text{m}$. Assume $AE = 10^6\,\text{N}$. The coordinates of the nodes are given in the table below:

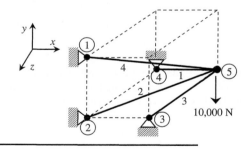

Node	x	y	z
1	0	1	2
2	0	0	2
3	1	0	2
4	0	0	0
5	1	0	0

24. The uniaxial bar shown below can be modeled as a one-dimensional truss. The bar has the following properties: $L = 1\,\text{m}$, $A = 10^{-4}\,\text{m}^2$, $E = 100\,\text{GPa}$, and $\alpha = 10^{-4}/°\text{C}$. From the stress-free initial state, a force of 5,000 N is applied at Node 2 and the temperature is lowered by 100°C below the reference temperature.

(a) Calculate the global matrix equation after applying boundary conditions.
(b) Solve for the displacement u_2.
(c) What is the element force P in the bar?

25. In the structure shown below, the temperature of **Element 2** is **100°C above** the reference temperature. An external force of 20,000 N is applied in the x-direction (horizontal direction) at Node 2. Assume $E = 10^{11}$ Pa, $A = 10^{-4}$ m^2, and $\alpha = 10^{-5}/$°C.

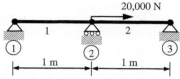

(a) Write down the stiffness matrices and thermal force vectors for each element.
(b) Write down the global matrix equations.
(c) Solve the global equations to determine the displacement at Node 2.
(d) Determine the forces in each element. State whether it is tension or compression.
(e) Show that force equilibrium is satisfied at Node 2.

26. Use FEA to determine the nodal displacements in the plane truss shown in Figure (a). The temperature of Element 2 is 100°C above the reference temperature, i.e., $\Delta T^{(2)} = 100$°C. Compute the force in each element. Show that the force equilibrium is satisfied at Node 3. Assume $L = 1$ m, $AE = 10^7$ N, $\alpha = 5 \times 10^{-6}/$°C.

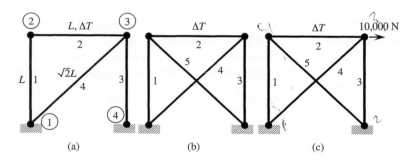

(a) (b) (c)

27. Repeat Problem 26 for the new configuration with Element 5 added, as shown in Figure (b).
28. Repeat Problem 26 with an external force added at Node 3, as shown in Figure (c).
29. The properties of the members of the truss in the left side of the figure are given in the table. Calculate the nodal displacement and element forces. Show that force equilibrium is satisfied as Node 3.

Elem	L (m)	A (cm^2)	E (GPa)	$\alpha(/$°C)	$\Delta T($°C)
1	1	1	100	20×10^{-6}	0
2	1	1	100	20×10^{-6}	0
3	1	1	100	20×10^{-6}	0
4	1	1	100	20×10^{-6}	0
5	$\sqrt{2}$	1	100	20×10^{-6}	-200

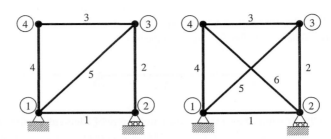

30. Repeat Problem 29 for the truss on the right side of the figure. Properties of Element 6 are same as those of Element 5, but $\Delta T = 0\,^{\circ}\text{C}$.

31. The truss shown in the figure supports the force $F = 2{,}000\,\text{N}$. Both elements have the same axial rigidity of $AE = 10^7\,\text{N}$, thermal expansion coefficient of $\alpha = 10^{-6}/^{\circ}\text{C}$, and length $L = 1\,\text{m}$. While the temperature at Element 1 remains constant, that of Element 2 is dropped by 100°C.

 (a) Write the 4×4 element stiffness matrices $[\mathbf{k}]$ and the 4×1 thermal force vectors $\{\mathbf{f}_T\}$ for Elements 1 and 2. Show the row addresses clearly.

 (b) Assemble two elements and apply the boundary conditions to obtain the global matrix equation in the form of $[\mathbf{K}]\{\mathbf{Q}\} = \{\mathbf{F}\}+\{\mathbf{F}_T\}$.

 (c) Solve for the nodal displacements.

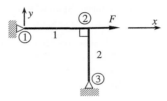

32. FEA was used to solve the truss problem shown below. The solution for displacements was obtained as $u_2 = 1\,\text{mm}$, $v_2 = -1\,\text{mm}$, $u_3 = 2\,\text{mm}$, and $v_3 = -1\,\text{mm}$.

 (a) Determine the axial forces P in Elements 2 and 4.

 (b) The forces in Elements 3 and 5 are found to be as follows: $P^{(3)} = -2{,}000\,\text{N}$, $P^{(5)} = 7{,}070\,\text{N}$. Determine the support reactions R_{y4} at Node 4 using the nodal equilibrium equations.

The element properties are listed in the following table.

Elem	$i \rightarrow j$	AE [N]	L [m]	$\Delta T[^{\circ}\text{C}]$	$\alpha[I/^{\circ}\text{C}]$	ϕ
1	1, 2	10^7	1	-100	10^{-6}	90°
2	2, 3	10^7	1	0	10^{-6}	0°
3	3, 4	10^7	1	$+100$	10^{-6}	-90°
4	1, 3	10^7	1.414	$+200$	10^{-6}	45°
5	2, 4	10^7	1.414	$+200$	10^{-6}	-45°

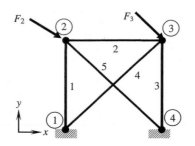

33. Use FEM to solve the plane truss shown below. Assume $AE = 10^6$ N, $L = 1$ m, $\alpha = 20 \times 10^{-6}$/C. The temperature of Element 1 is 100°C below the reference temperature, while Elements 2 and 3 are in the reference temperature. Determine the nodal displacements, forces in each element, and support reactions. Show that the nodal equilibrium is satisfied at Node 4.

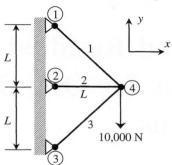

Chapter 3

Weighted Residual and Energy Methods for One-Dimensional Problems

In the previous chapter, we developed the finite element equations for the truss element using the direct stiffness method. However, this method becomes impractical when element formulations are complicated or multidimensional problems are considered. Thus, we need to develop a more systematic approach to construct the finite element equations for general engineering problems. In fact, the finite element method can be applied to any engineering problem that is governed by a differential equation. There are two other methods of deriving the finite element equations: weighted residual method and energy method. In the first part of this chapter, we will consider ordinary differential equations that occur commonly in engineering problems and derive the corresponding finite element equations through the weighted residual method, in particular using the Galerkin method. Energy methods are alternative methods that are very powerful and amenable for approximate solutions when solving structures that are more realistic. In the second part of this chapter, we will use the principle of minimum potential energy to derive finite element equations of discrete systems and uniaxial bar.

3.1 EXACT VS. APPROXIMATE SOLUTION

3.1.1 Exact Solution

Many engineering problems, such as the deformation of a beam and heat conduction in a solid, can be described using a differential equation. The differential equation along with boundary conditions is called the *boundary value problem*. A simple, one-dimensional example of a boundary value problem is

$$\frac{d^2u}{dx^2} + p(x) = 0, \quad 0 \le x \le 1$$

$$\left.\begin{array}{l} u(0) = 0 \\[2mm] \dfrac{du}{dx}(1) = 1 \end{array}\right\} \quad \text{Boundary conditions} \tag{3.1}$$

The above differential equation describes the displacements in a uniaxial bar subjected to a distributed force $p(x)$ along its axis. The first boundary condition prescribes the value of the solution at a given point and is called the *essential boundary condition*. The term *displacement boundary condition* or *kinematic boundary condition* is also used in the context of solid mechanics. On the other hand, the second boundary condition prescribes the value of derivative, du/dx, at $x = 1$ and is called the *natural boundary condition*. In solid mechanics, the term *force boundary condition* or *stress boundary condition* is also used. The solution $u(x)$ of Eq. (3.1) satisfies the differential equation at every point in the domain, and it needs to be, at least, a twice-differentiable function because the differential equation includes the second-order derivative of $u(x)$. In this text, the solution $u(x)$ of Eq. (3.1) will be referred to as the *exact solution*.

When the geometry is complicated, it is not trivial to solve for $u(x)$ analytically. Since the solution that satisfies the differential equation and boundary conditions can have a complicated expression, an infinite series solution is often employed.

3.1.2   Approximate Solution

In using the weighted residual method, we seek an approximate solution $\tilde{u}(x) \approx u(x)$ for Eq. (3.1). *We choose $\tilde{u}(x)$ such that it satisfies the essential boundary conditions of the problem, but not necessarily the natural boundary conditions.* In the present example, $\tilde{u}(x)$ must satisfy the condition $\tilde{u}(0) = 0$. Since $\tilde{u}(x)$ is an approximation of the exact solution $u(x)$, it will not satisfy the differential equation in Eq. (3.1) and hence it will not be identically equal to zero. We will call the resulting function the *residual* and denote it by $R(x)$:

$$\frac{d^2\tilde{u}}{dx^2} + p(x) = R(x) \tag{3.2}$$

Our goal now is to minimize the error or the residual as much as possible. Instead of trying to make $R(x)$ vanish everywhere, we will make $R(x)$ equal to zero in some average sense. Consider the following integral:

$$\int_0^1 R(x)W(x)dx = 0 \tag{3.3}$$

where $W(x)$ is an arbitrary function called the *weight function*. The above equation requires the integral of the weighted residual equal to zero. If Eq. (3.3) is satisfied for several different linearly independent weight functions, then $R(x)$ will approach zero, and the approximate solution $\tilde{u}(x)$ will approach the exact solution $u(x)$. The choice of the weight functions leads to different weighted residual methods such as least square error method, collocation method, Petrov-Galerkin method, and Galerkin method.[1] In this book, we will discuss only the Galerkin method, which is popular in deriving the finite element equations for many engineering problems.

3.2   GALERKIN METHOD

The approximate solution $\tilde{u}(x)$ is expressed as a sum of a number of functions called *trial functions*:

$$\tilde{u}(x) = \sum_{i=1}^{N} c_i \phi_i(x) \tag{3.4}$$

where N is the number of terms used, $\phi_i(x)$ are known *trial functions*, and c_i are coefficients to be determined using the weighted residual method. Since the approximate solution is a linear combination of the trial functions, the accuracy of approximation depends on them. The trial functions and coefficients are chosen such that $\tilde{u}(x)$ must satisfy the

---

[1] R. D. Cook, Concepts and Applications of Finite Element Analysis 4th ed., John Wiley & Sons, New York, 2007

essential boundary conditions of the problem. In the present example in Eq. (3.1), $\tilde{u}(x)$ must satisfy $\tilde{u}(0) = 0$.

The *Galerkin method* differs from other weighted residual methods in that the N weight functions are same as the N number of trial functions, $\phi_i(x)$. Thus, we obtain the following N number of weighted residual equations:

$$\int_0^1 R(x)\phi_i(x)dx = 0, \quad i = 1, \ldots, N \tag{3.5}$$

Substituting for $R(x)$ from Eq. (3.2) we obtain

$$\int_0^1 \left(\frac{d^2\tilde{u}}{dx^2} + p(x)\right)\phi_i(x)dx = 0, \quad i = 1, \ldots, N \tag{3.6}$$

Since the function $p(x)$ is known, we will take the term containing it to the right-hand side (RHS) to obtain

$$\int_0^1 \frac{d^2\tilde{u}}{dx^2}\phi_i(x)dx = -\int_0^1 p(x)\phi_i(x)dx, \quad i = 1, \ldots, N \tag{3.7}$$

In the view of approximation in Eq. (3.4), the above N equations can be used to solve for the N unknown coefficients c_i. However, we will use integration by parts as shown below to reduce the order of differentiation of $\tilde{u}$:

$$\frac{d\tilde{u}}{dx}\phi_i\Big|_0^1 - \int_0^1 \frac{d\tilde{u}}{dx}\frac{d\phi_i}{dx}dx = -\int_0^1 p(x)\phi_i(x)dx, \quad i = 1, \ldots, N \tag{3.8}$$

Note that the boundary term on the left-hand side (LHS) of the above equation has the term $d\tilde{u}/dx$. For this, we will not use the approximation but will use the actual boundary condition given in Eq. (3.1). After rearrangement, Eq. (3.8) takes the following form:

$$\int_0^1 \frac{d\phi_i}{dx}\frac{d\tilde{u}}{dx}dx = \int_0^1 p(x)\phi_i(x)dx + \frac{du}{dx}(1)\phi_i(1) - \frac{du}{dx}(0)\phi_i(0), \quad i = 1, \ldots, N \tag{3.9}$$

Note that the orders of differentiation for both the approximate solution and the trial functions are identical. To rewrite the above equation explicitly in terms of unknown coefficients c_i, the approximation in Eq. (3.4) is substituted to obtain

$$\int_0^1 \frac{d\phi_i}{dx}\sum_{j=1}^N c_j\frac{d\phi_j}{dx}dx = \int_0^1 p(x)\phi_i(x)dx + \frac{du}{dx}(1)\phi_i(1) - \frac{du}{dx}(0)\phi_i(0), \quad i = 1, \ldots, N \tag{3.10}$$

Note that the approximation is applied only to the LHS, not to the boundary terms on the RHS. The above N equations can be written in a compact form as

$$\sum_{j=1}^N K_{ij}c_j = F_i, \quad i = 1, \ldots, N \tag{3.11}$$

or in matrix form as

$$\underset{(N\times N)}{[\mathbf{K}]}\ \underset{(N\times 1)}{\{\mathbf{c}\}} = \underset{(N\times 1)}{\{\mathbf{F}\}} \tag{3.12}$$

where the matrices $[\mathbf{K}]$ and $\{\mathbf{F}\}$ are defined as

$$K_{ij} = \int_0^1 \frac{d\phi_i}{dx}\frac{d\phi_j}{dx}dx \tag{3.13}$$

and

$$F_i = \int_0^1 p(x)\phi_i(x)dx + \frac{du}{dx}(1)\phi_i(1) - \frac{du}{dx}(0)\phi_i(0) \qquad (3.14)$$

In Eq. (3.12), the numbers in parentheses below the matrices refer to the size of the respective matrices. Note that $[\mathbf{K}]$ is symmetric as $K_{ij} = K_{ji}$. Equations in Eq. (3.11) are solved to determine the N unknown coefficients c_i in the approximate solution. We will demonstrate the method for the problem defined by Eq. (3.1) in the following example.

EXAMPLE 3.1   *Galerkin Solution of a Second-Order Differential Equation, Case 1*

Solve the differential equation of Eq. (3.1) for $p(x) = 1$. Use two trial functions, $\phi_1(x) = x$ and $\phi_2(x) = x^2$.

SOLUTION   Since $N = 2$, the approximate solution takes the following form:

$$\tilde{u}(x) = \sum_{i=1}^{2} c_i\phi_i(x) = c_1 x + c_2 x^2 \qquad (3.15)$$

We will discuss the method of choosing the trial functions later. Note that the approximation above satisfies the essential boundary condition $u(0) = 0$. The derivatives of the trial functions are $\phi_1'(x) = 1$ and $\phi_2'(x) = 2x$. Substituting for ϕ_i and its derivatives in Eq. (3.13), the components of matrix $[\mathbf{K}]$ can be obtained as

$$K_{11} = \int_0^1 (\phi_1')^2 dx = 1$$

$$K_{12} = K_{21} = \int_0^1 (\phi_1'\phi_2')dx = 1$$

$$K_{22} = \int_0^1 (\phi_2')^2 dx = \frac{4}{3}$$

In a similar way, the components of vector $\{\mathbf{F}\}$ can be calculated by substituting for ϕ_i and the boundary condition $du(1)/dx = 1$ in Eq. (3.14) as

$$F_1 = \int_0^1 \phi_1(x)dx + \phi_1(1) - \frac{du}{dx}(0)\phi_1(0) = \frac{3}{2}$$

$$F_2 = \int_0^1 \phi_2(x)dx + \phi_2(1) - \frac{du}{dx}(0)\phi_2(0) = \frac{4}{3}$$

Note that the value of $du(0)/dx$ is not required because the trial functions are zero at $x = 0$. Thus, the following matrix $[\mathbf{K}]$ and vector $\{\mathbf{F}\}$, respectively, can be obtained:

$$[\mathbf{K}] = \frac{1}{3}\begin{bmatrix} 3 & 3 \\ 3 & 4 \end{bmatrix}$$

and

$$\{\mathbf{F}\} = \frac{1}{6}\begin{Bmatrix} 9 \\ 8 \end{Bmatrix}$$

Solving for $\{\mathbf{c}\} = [\mathbf{K}]^{-1}\{\mathbf{F}\}$, we obtain $c_1 = 2$ and $c_2 = -\frac{1}{2}$. Thus, the approximate solution becomes

$$\tilde{u}(x) = 2x - \frac{x^2}{2} \qquad (3.16)$$

The exact solution for this case can be obtained by integrating Eq. (3.1) twice and using the boundary conditions to evaluate the arbitrary constants of integration. One can note that the solution in Eq. (3.16) is also the exact solution to the problem. This happened because the approximate solution contains the two terms, x and x^2, which are also in the exact solution.

EXAMPLE 3.2   *Galerkin Solution of a Second-Order Differential Equation, Case 2*

Solve the differential equation of Eq. (3.1) for $p(x) = x$. Use the same trial functions as in Example 3.1.

SOLUTION   We will use the same form of approximation as given in Eq. (3.15). This will change only the vector $\{\mathbf{F}\}$, and the matrix $[\mathbf{K}]$ will remain the same. The vector $\{\mathbf{F}\}$ is calculated as

$$\{\mathbf{F}\} = \frac{1}{12} \left\{ \begin{matrix} 16 \\ 15 \end{matrix} \right\}$$

Solving for $\{\mathbf{c}\}$ we obtain $c_1 = \frac{19}{12}$ and $c_2 = -\frac{1}{4}$. Then the approximate solution becomes

$$\tilde{u}(x) = \frac{19}{12}x - \frac{x^2}{4}$$

The exact solution for this case can be obtained by integrating Eq. (3.1) twice and using the boundary conditions to evaluate the arbitrary constants of integration. The exact solution is

$$u(x) = \frac{3}{2}x - \frac{x^3}{6}$$

The exact and approximate solutions and their derivatives are compared in Figure 3.1. From Figure 3.1, one can note that the results for $u(x)$ and $\tilde{u}(x)$ agree quite well in the entire domain of the problem. Nevertheless, there is some difference in the results for the derivatives of $u(x)$ and $\tilde{u}(x)$, the maximum error being approximately 8%.

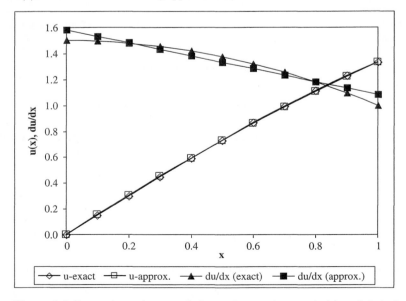

Figure 3.1 Comparison of exact solution and approximate solution and their derivatives for Example 3.2

EXAMPLE 3.3   *Galerkin Solution of a Second-Order Differential Equation, Case 3*

Solve the differential equation given below for $p(x) = x^2$.

$$\frac{d^2u}{dx^2} + p(x) = 0, \quad 0 \le x \le 1$$

$$\left.\begin{array}{l} u(0) = 0 \\ u(1) = 0 \end{array}\right\} \text{ Boundary conditions}$$

SOLUTION   Note that both the boundary conditions are essential boundary conditions and the assumed solution must satisfy both boundary conditions. Let the approximate solution be

$$\tilde{u}(x) = \sum_{i=1}^{2} c_i \phi_i(x) = c_1 x(x-1) + c_2 x^2(x-1) \tag{3.17}$$

From Eq. (3.4), one can recognize that the trial functions are

$$\phi_1(x) = x(x-1), \quad \phi_2(x) = x^2(x-1)$$

The derivatives of the trial functions are

$$\phi_1'(x) = 2x - 1, \quad \phi_2'(x) = 3x^2 - 2x$$

Substituting for ϕ_i' in Eq. (3.13), we derive the components of matrix $[\mathbf{K}]$ as

$$K_{11} = \int_0^1 (\phi_1')^2 dx = \frac{1}{3}$$

$$K_{12} = K_{21} = \int_0^1 (\phi_1'\phi_2') dx = \frac{1}{6}$$

$$K_{22} = \int_0^1 (\phi_2')^2 dx = \frac{2}{15}$$

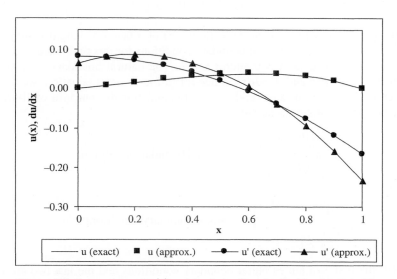

Figure 3.2 Comparison of $u(x)$ and its derivative obtained by the Galerkin method for Example 3.3

Substituting $p(x) = x^2$ in Eq. (3.14), we derive the vector $\{\mathbf{F}\}$ as

$$F_1 = \int_0^1 x^2 \phi_1(x)dx + \frac{du}{dx}(1)\cancel{\phi_1(1)} - \frac{du}{dx}(0)\cancel{\phi_1(0)} = -\frac{1}{20}$$

$$F_2 = \int_0^1 x^2 \phi_2(x)dx + \frac{du}{dx}(1)\cancel{\phi_2(1)} - \frac{du}{dx}(0)\cancel{\phi_2(0)} = -\frac{1}{30}$$

Note that we do not know the boundary values of du/dx. Still we are able to evaluate F_1 and F_2 because the values of ϕ_1 and ϕ_2 at the boundaries are equal to zero.

Solving $[\mathbf{K}]\{\mathbf{c}\} = \{\mathbf{F}\}$, we obtain $c_1 = -\frac{1}{15}$ and $c_2 = -\frac{1}{6}$. Substituting for c_i in Eq. (3.17), we obtain the expression for the approximate solution as

$$\tilde{u}(x) = \frac{x}{15} + \frac{x^2}{10} - \frac{x^3}{6}$$

The exact solutions are as follows:

$$u(x) = \frac{1}{12}x(1 - x^3)$$

It may be seen that the exact solution is a quartic polynomial, whereas the approximate solution is cubic in x. The two solutions and their respective derivatives are compared in Figure 3.2.

The comparison between $u(x)$ and $\tilde{u}(x)$ is excellent. However, the derivatives deviate from one another. This is typical of approximate methods such as Galerkin method and, in general, the finite element method. If we assume a three-term solution that includes x^4 term in the approximation, then the approximate solution will be the same as the exact solution.

3.3   HIGHER-ORDER DIFFERENTIAL EQUATIONS

In the second-order differential equation in Eq. (3.1), we perform integration by parts once so that the orders of differentiation for both the approximate solution and the trial functions are identical. We will now illustrate the Galerkin method for a fourth-order ordinary differential equation. Consider the following equation:

$$\frac{d^4w}{dx^4} - p(x) = 0, \quad 0 \le x \le L \tag{3.18}$$

which is the governing equation for deflection $w(x)$ of a beam of length L subjected to transverse loading $p(x)$. The essential boundary conditions for this problem are specifying w or the derivative dw/dx at $x = 0$ and/or $x = L$. The natural boundary conditions are known values of d^2w/dx^2 and d^3w/dx^3. Since it is a fourth-order equation, we need four boundary conditions, two at each end of the domain. For the purpose of illustration, assume the boundary conditions as follows:

$$\left.\begin{array}{l} w(0) = 0 \\ \dfrac{dw}{dx}(0) = 0 \end{array}\right\} \text{ Essential boundary conditions}$$

$$\left.\begin{array}{l} \dfrac{d^2w}{dx^2}(L) = M \\ \dfrac{d^3w}{dx^3}(L) = -V \end{array}\right\} \text{ Natural boundary conditions} \tag{3.19}$$

In Eq. (3.19) the first two are essential boundary conditions and the last two are natural boundary conditions. Physically, the beam is clamped at $x = 0$, and moment M and shear force V are applied at $x = L$.

As before, we assume the approximate solution as $w(x) = \tilde{w}(x)$. Substituting in the governing equation, we obtain the residual as

$$\frac{d^4\tilde{w}}{dx^4} - p(x) = R(x) \tag{3.20}$$

We assume the approximate solution in a series form as

$$\tilde{w}(x) = \sum_{i=1}^{N} c_i \phi_i(x) \tag{3.21}$$

such that $\tilde{w}(x)$ satisfies the essential boundary conditions. The integral of the weighted residual takes the form

$$\int_0^L \left(\frac{d^4\tilde{w}}{dx^4} - p(x) \right) \phi_i(x)\,dx = 0, \quad i = 1, \ldots, N \tag{3.22}$$

where ϕ_i are used as the weight functions. We will take the term containing $p(x)$ to the RHS and use integration by parts to reduce the order of differentiation of $\tilde{w}$:

$$\left. \frac{d^3\tilde{w}}{dx^3} \phi_i \right|_0^L - \left. \frac{d^2\tilde{w}}{dx^2} \frac{d\phi_i}{dx} \right|_0^L + \int_0^L \frac{d^2\tilde{w}}{dx^2} \frac{d^2\phi_i}{dx^2}\,dx = \int_0^L p(x)\phi_i(x)\,dx, \quad i = 1, \ldots, N \tag{3.23}$$

It should be noted that we have used integration by parts twice to reduce the order of differentiation from four to two. Equation (3.23) can be rewritten as

$$\int_0^L \frac{d^2\tilde{w}}{dx^2} \frac{d^2\phi_i}{dx^2}\,dx = \int_0^L p(x)\phi_i(x)\,dx - \left. \frac{d^3\tilde{w}}{dx^3} \phi_i \right|_0^L + \left. \frac{d^2\tilde{w}}{dx^2} \frac{d\phi_i}{dx} \right|_0^L, \quad i = 1, \ldots, N \tag{3.24}$$

Substituting for $\tilde{w}(x)$ from Eq. (3.21) into Eq. (3.24) we obtain

$$\int_0^L \sum_{j=1}^{N} c_j \frac{d^2\phi_j}{dx^2} \frac{d^2\phi_i}{dx^2}\,dx = \int_0^L p(x)\phi_i(x)\,dx - \left. \frac{d^3\tilde{w}}{dx^3} \phi_i \right|_0^L + \left. \frac{d^2\tilde{w}}{dx^2} \frac{d\phi_i}{dx} \right|_0^L, \quad i = 1, \ldots, N \tag{3.25}$$

Note that the assumed solution is substituted only on the LHS of the above equations. For the boundary terms involving $\tilde{w}$, we will use the actual boundary conditions. Equations (3.25) can be written in a compact form as

$$\underset{N \times N}{[\mathbf{K}]} \underset{N \times 1}{\{\mathbf{c}\}} = \underset{N \times 1}{\{\mathbf{F}\}} \tag{3.26}$$

where $\{\mathbf{c}\}$ is the column vector of unknown coefficients c_j, and the elements of matrices $[\mathbf{K}]$ and $\{\mathbf{F}\}$, respectively, are defined as

$$K_{ij} = \int_0^L \frac{d^2\phi_i}{dx^2} \frac{d^2\phi_j}{dx^2}\,dx \tag{3.27}$$

and

$$F_i = \int_0^L p(x)\phi_i(x)\,dx - \left. \frac{d^3 w}{dx^3} \phi_i \right|_0^L + \left. \frac{d^2 w}{dx^2} \frac{d\phi_i}{dx} \right|_0^L \tag{3.28}$$

EXAMPLE 3.4 *Galerkin Solution of a Fourth-Order Differential Equation*

Solve the fourth-order differential equation in Eq. (3.18) with the boundary conditions in Eq. (3.19). Assume that $L = 1$, $p(x) = 1$, $V = 1$, and $M = 2$.

SOLUTION To proceed further with the illustration, we assume that $N = 2$ and assume the trial functions ϕ_i as follows:

$$\phi_1 = x^2, \quad \phi_2 = x^3$$

Note that the functions ϕ_i are such that $\tilde{w}(x)$ satisfies the essential boundary conditions given in Eq. (3.19). The second derivatives of the trial functions become $\phi_1'' = 2$ and $\phi_2'' = 6x$. Substituting these two derivatives in Eq. (3.27), we obtain the components of the matrix $[\mathbf{K}]$ as

$$K_{11} = \int_0^1 (\phi_1'')^2 dx = 4$$

$$K_{12} = K_{21} = \int_0^1 (\phi_1'' \phi_2'') dx = 6$$

$$K_{22} = \int_0^1 (\phi_2'')^2 dx = 12$$

Thus, the matrix $[\mathbf{K}]$ becomes

$$[\mathbf{K}] = \begin{bmatrix} 4 & 6 \\ 6 & 12 \end{bmatrix}$$

Similarly, substituting for ϕ_i and the natural boundary conditions in Eq. (3.28), we obtain the components of the vector $\{\mathbf{F}\}$ as

$$F_1 = \int_0^1 x^2 dx + V\phi_1(1) + \frac{d^3w(0)}{dx^3}\phi_1\cancel{(0)} + M\phi_1'(1) - \frac{d^2w(0)}{dx^2}\phi_1'\cancel{(0)} = \frac{16}{3}$$

$$F_2 = \int_0^1 x^3 dx + V\phi_2(1) + \frac{d^3w(0)}{dx^3}\phi_2\cancel{(0)} + M\phi_2'(1) - \frac{d^2w(0)}{dx^2}\phi_2'\cancel{(0)} = \frac{29}{4}$$

Thus, the vector $\{\mathbf{F}\}$ becomes

$$\left\{ \begin{array}{c} F_1 \\ F_2 \end{array} \right\} = \left\{ \begin{array}{c} \dfrac{16}{3} \\ \dfrac{29}{4} \end{array} \right\}$$

Solving for c_j using Eq. (3.26) and substituting in Eq. (3.21), we obtain the approximate solution as

$$\tilde{w}(x) = \frac{41}{24}x^2 - \frac{1}{4}x^3 \tag{3.29}$$

The exact solution is obtained by integrating Eq. (3.18) four times and applying the boundary conditions to evaluate the arbitrary constants of integration:

$$w(x) = \frac{1}{24}x^4 - \frac{1}{3}x^3 + \frac{7}{4}x^2 \tag{3.30}$$

The approximate and exact solutions are compared in Table 3.1 by calculating the functions and its derivatives. One can note that the solution obtained using the Galerkin method compares very well to the exact solution up to the second derivative. The error in the third derivative is significant, as high as 50%. In fact the third derivative is a constant for the approximate solution whereas for the exact solution it is a linear function in x. The second and third derivatives are compared in Figure 3.3.

Table 3.1 Comparison of Approximate and Exact Solutions

	$w(x)$			$w'(x)$			$w''(x)$			$w'''(x)$		
x	Exact	Approx.	% Error	Exact	Approx.	% Error	Exact	Approx.	% Error	Exact	Approx.	% Error
0.0	0.000	0.000	0.00	0.000	0.000	0.00	3.500	3.42	−2.38	−2.0	−1.5	−25.0
0.1	0.017	0.017	−1.97	0.340	0.334	−1.76	3.305	3.27	−1.16	−1.9	−1.5	−21.1
0.2	0.067	0.066	−1.58	0.661	0.653	−1.21	3.120	3.12	−0.11	−1.8	−1.5	−16.7
0.3	0.149	0.147	−1.23	0.965	0.958	−0.73	2.945	2.97	0.74	−1.7	−1.5	−11.8
0.4	0.260	0.257	−0.92	1.251	1.247	−0.32	2.780	2.82	1.32	−1.6	−1.5	−6.3
0.5	0.398	0.396	−0.65	1.521	1.521	0.00	2.625	2.67	1.59	−1.5	−1.5	0.0
0.6	0.563	0.561	−0.43	1.776	1.780	0.23	2.480	2.52	1.48	−1.4	−1.5	7.1
0.7	0.753	0.751	−0.24	2.017	2.024	0.35	2.345	2.37	0.92	−1.3	−1.5	15.4
0.8	0.966	0.965	−0.11	2.245	2.253	0.36	2.220	2.22	−0.15	−1.2	−1.5	25.0
0.9	1.202	1.202	−0.03	2.462	2.468	0.24	2.105	2.07	−1.82	−1.1	−1.5	36.4
1.0	1.458	1.458	0.00	2.667	2.667	0.00	2.000	1.92	−4.17	−1.0	−1.5	50.0

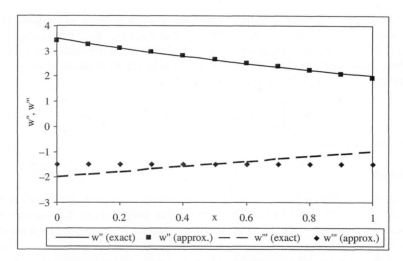

Figure 3.3 Comparison of w'' and w''' for the beam problem in Example 3.4

3.4 FINITE ELEMENT APPROXIMATION

3.4.1 Domain Discretization

In the previous sections, we approximated the solution to the differential equation by a series of functions in the entire domain of the problem. In such a case, it is difficult to obtain the trial functions that satisfy the essential boundary conditions. An important idea of the finite element method is to divide the entire domain into a set of simple sub-domains or *finite elements*. The finite elements are connected with adjacent elements by sharing their nodes. Then within each finite element, the solution is approximated in a simple polynomial form. For example, let us assume that the domain is one-dimensional and the exact solution $u(x)$ is given as a dashed curve in Figure 3.4. When the entire domain is divided into sub-domains (finite elements), it is possible to approximate the solution using piecewise

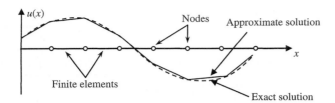

Figure 3.4 Piecewise linear approximation of the solution for one-dimensional problem

continuous linear polynomials, as shown in Figure 3.4. Within each element, the approximate solution is linear. Two adjacent elements have the same solution value at the shared node. As can be seen in the figure, when more elements are used, the approximate piecewise linear solution will converge to the exact solution. In addition, the approximation can be more accurate if higher-order polynomials are used in each element.

Various types of finite elements can be used depending on the domain that needs to be discretized and the order of polynomials used to approximate the solution. Table 3.2 illustrates several types of finite elements that are often used in one-, two-, and three-dimensional problems.

After dividing the domain into finite elements, the integrations to compute $[\mathbf{K}]$ and $\{\mathbf{F}\}$, e.g., Eqs. (3.27) and (3.28), are performed over each element. For example, let us assume that the one-dimensional domain $(0, 1)$ is divided into 10 equal-sized finite elements. Then, the integral in Eq. (3.27) can be written as the summation of integrals over each element:

$$\int_0^1 \square\,dx = \int_0^{0.1} \square\,dx + \int_{0.1}^{0.2} \square\,dx + \cdots + \int_{0.9}^1 \square\,dx \tag{3.31}$$

where $\square$ is the integrand.

3.4.2 Trial Solution

After the domain is divided into a set of simple-shaped elements, the solution within an element is approximated in the form of simple polynomials. Let us consider a one-dimensional problem in Eq. (3.1) in which the domain is discretized by n number of

Table 3.2 Different types of finite elements

Element	Name
○——○	1D linear element
△	2D triangular element
▭	2D rectangular element
△ (tetrahedron)	3D tetrahedron element
▭ (hexahedron)	3D hexahedron element

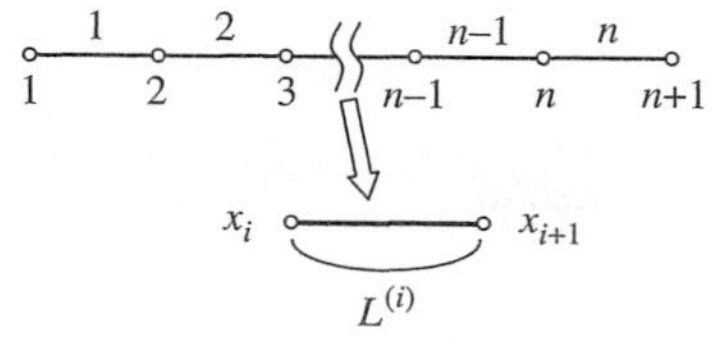

Figure 3.5 Domain discretization of one-dimensional problem

elements, as shown in Figure 3.5. For this specific example, each element is composed of two end nodes. The trial solution is constructed in the element using the solution values at these nodes.

For example, the i-th element connects two nodes at $x = x_i$ and $x = x_{i+1}$. If we want to interpolate the solution using two nodal values, then the linear polynomial is the appropriate choice because it has two unknowns. Thus, the solution is approximated by

$$\tilde{u}(x) = a_0 + a_1 x, \quad x_i \leq x \leq x_{i+1} \tag{3.32}$$

Note that the trial solution in the above equation is only defined in the i-th element. Although we can determine two coefficients a_0 and a_1, they do not have a physical meaning. Instead, the unknown coefficients a_0 and a_1 in Eq. (3.32) will be expressed in terms of the nodal solutions $\tilde{u}(x_i)$ and $\tilde{u}(x_{i+1})$. By substituting these two nodal values, we have

$$\begin{cases} \tilde{u}(x_i) & = u_i = a_0 + a_1 x_i \\ \tilde{u}(x_{i+1}) & = u_{i+1} = a_0 + a_1 x_{i+1} \end{cases} \tag{3.33}$$

where u_i and u_{i+1} are the solution values at the two end nodes. Then, by solving Eq. (3.33), two unknown coefficients, a_0 and a_1, can be represented by the nodal solution u_i and u_{i+1}. After substituting the two coefficients into Eq. (3.32), the approximate solution can be expressed in terms of the nodal solutions as

$$\tilde{u}(x) = \underbrace{\frac{x_{i+1} - x}{L^{(i)}}}_{N_i(x)} u_i + \underbrace{\frac{x - x_i}{L^{(i)}}}_{N_{i+1}(x)} u_{i+1} \tag{3.34}$$

where $L^{(i)} = x_{i+1} - x_i$ is the length of the i-th element. Now the approximate solution for $u(x)$ in Eq. (3.32) can be rewritten as

$$\tilde{u}(x) = N_i(x) u_i + N_{i+1}(x) u_{i+1}, \quad x_i \leq x \leq x_{i+1} \tag{3.35}$$

where the functions $N_i(x)$ and $N_{i+1}(x)$ are called *interpolation functions* for obvious reasons. The expression in Eq. (3.35) shows that the solution $\tilde{u}(x)$ is interpolated using its nodal values u_i and u_{i+1}. $N_i(x) = 1$ at $x = x_i$ and $N_i(x) = 0$ at $x = x_{i+1}$, while $N_{i+1}(x) = 1$ at $x = x_{i+1}$ and $N_{i+1}(x) = 0$ at $x = x_i$. Interpolation functions $N_i(x)$ and $N_{i+1}(x)$ are also called *shape functions*, a term used in solid mechanics, as the functions describe the deformed shape of a solid or structure.

Note that the approximate solution in Eq. (3.35) is similar to that of the weighted residual method in Eq. (3.5). In this case, the interpolation function corresponds to the trial function. The difference is that the approximation in Eq. (3.35) is written in terms of solution values at nodes, whereas the coefficients c_i in the approximation in Eq. (3.4) do not have any physical meanings. Detailed relation between interpolation and trial functions will be discussed in the following section.

To explain the accuracy of approximation, the interpolated solution and its gradients for two continuous elements are illustrated in Figure 3.6. Note that in this particular interpolation, the solution is approximated by piecewise linear function and its gradient is

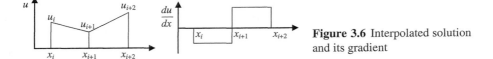

Figure 3.6 Interpolated solution and its gradient

constant within an element. Accordingly, the gradients are not continuous at the element interface. In structural problems, the solution $u(x)$ often represents displacement of the structure and its gradient is stress or strain. Thus, the approximation yields a continuous displacement but discontinuous stress and strain between elements. Many commercial finite element programs provide the stress values at nodes and display a smooth change of stresses across elements. However, users must be careful because these nodal stress values are the average of values for different elements connected to a node.

3.4.3   Galerkin Method

Before we apply the Galerkin method, in the finite element scheme we have to identify the relation between the interpolation functions $N_i(x)$ and the trial functions $\phi_i(x)$. Remember we expressed the approximate solution in terms of trial functions $\phi_i(x)$ with unknown coefficients c_i. The finite element approximation in Eq. (3.35) can also be written as

$$\tilde{u}(x) = \sum_{i=1}^{N_D} u_i\phi_i(x) \tag{3.36}$$

where u_i are the nodal values of the functions and N_D is the total number of nodes in the finite element model. In the case of one-dimensional problems with linear interpolation between nodes, $N_D = N_E + 1$, where N_E is the number of elements. Comparing Eq. (3.35) and (3.36), the expression for $\phi_i(x)$ can be written as

$$\phi_i(x) = \begin{cases} 0, & 0 \le x \le x_{i-1} \\ N_i^{(i-1)}(x) = \dfrac{x - x_{i-1}}{L^{(i-1)}}, & x_{i-1} < x \le x_i \\ N_i^{(i)}(x) = \dfrac{x_{i+1} - x}{L^{(i)}}, & x_i < x \le x_{i+1} \\ 0, & x_{i+1} < x \le x_{N_D} \end{cases} \tag{3.37}$$

In the above equations, the superscripts $(i-1)$ and (i) denote the element numbers. Thus the major difference between the weighted residual method described in the previous section and the finite element method is that the function $\phi_i(x)$ does not exist in the entire domain, but it exists only in elements connected to Node i.

The derivative of $\phi_i(x)$ is also derived as

$$\frac{d\phi_i(x)}{dx} = \begin{cases} 0, & 0 \le x \le x_{i-1} \\ \dfrac{1}{L^{(i-1)}}, & x_{i-1} < x \le x_i \\ -\dfrac{1}{L^{(i)}}, & x_i < x \le x_{i+1} \\ 0, & x_{i+1} < x \le x_{N_D} \end{cases} \tag{3.38}$$

Function $\phi_i(x)$ and its derivative are plotted in Figure 3.7.

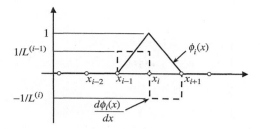

Figure 3.7 Function $\phi_i(x)$ and its derivative

EXAMPLE 3.5   *Finite Element Solution of a Differential Equation*

Solve the differential equation given in Eq. (3.1) for $p(x) = 1$ using the Galerkin finite element method. Assume that the domain $0 \leq x \leq 1$ is divided by two equal-length elements.

SOLUTION   Since the domain is divided by two elements of equal length, there exist three nodes located at $x = 0, 0.5$, and 1.0. The unknown coefficients—analogous to c_i in the previous examples—are the values of u at the nodes u_1, u_2, and u_3. Then the approximate solution is written as

$$\tilde{u}(x) = u_1\phi_1(x) + u_2\phi_2(x) + u_3\phi_3(x) \tag{3.39}$$

and the functions ϕ_i are defined as follows using Eq. (3.37):

$$\phi_1(x) = \begin{cases} 1 - 2x, & 0 \leq x \leq 0.5 \\ 0, & 0.5 < x \leq 1 \end{cases}$$

$$\phi_2(x) = \begin{cases} 2x, & 0 \leq x \leq 0.5 \\ 2 - 2x, & 0.5 < x \leq 1 \end{cases}$$

$$\phi_3(x) = \begin{cases} 0, & 0 \leq x \leq 0.5 \\ -1 + 2x, & 0.5 < x \leq 1 \end{cases}$$

The three functions are plotted in Figure 3.8. It may be noted that ϕ_i are such that the variation of $\tilde{u}(x)$ is linear between nodes.

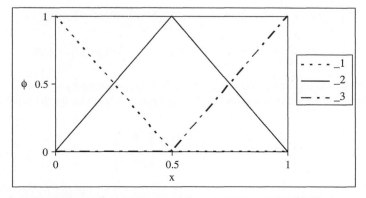

Figure 3.8 Trial function $\phi_i(x)$ for two equal-length finite elements

The derivatives of ϕ_i are as follows:

$$\frac{d\phi_1(x)}{dx} = \begin{cases} -2, & 0 \leq x \leq 0.5 \\ 0, & 0.5 < x \leq 1 \end{cases}$$

$$\frac{d\phi_2(x)}{dx} = \begin{cases} 2, & 0 \leq x \leq 0.5 \\ -2, & 0.5 < x \leq 1 \end{cases}$$

$$\frac{d\phi_3(x)}{dx} = \begin{cases} 0, & 0 \leq x \leq 0.5 \\ 2, & 0.5 < x \leq 1 \end{cases}$$

We will evaluate $[\mathbf{K}]$ and $\{\mathbf{F}\}$ using Eqs. (3.13) and (3.14) for the case $p(x) = 1$. Note that $N = 3$ for this case. We will show the procedures for K_{12} and K_{22}. Evaluation of other terms of $[\mathbf{K}]$ is left as an exercise for the reader. Since two elements are involved, all integrations are divided into two parts.

$$K_{12} = \int_0^1 \frac{d\phi_1}{dx} \frac{d\phi_2}{dx} dx$$

$$= \int_0^{0.5} (-2)(2)dx + \int_{0.5}^1 (0)(-2)dx$$

$$= -2$$

$$K_{22} = \int_0^1 \frac{d\phi_2}{dx} \frac{d\phi_2}{dx} dx$$

$$= \int_0^{0.5} 4dx + \int_{0.5}^1 4dx$$

$$= 4$$

The first component of $\{\mathbf{F}\}$ can be calculated from Eq. (3.14):

$$F_1 = \int_0^{0.5} 1 \times (1 - 2x)dx + \int_{0.5}^1 1 \times (0)dx + \frac{du}{dx}(1)\phi_1(1) - \frac{du}{dx}(0)\phi_1(0)$$

$$= 0.25 - \frac{du}{dx}(0)$$

It should be noted that F_1 could not be computed as we do not know the boundary condition $du(0)/dx$. We will use F_1 as an unknown term. Computation of F_2 and F_3 is straightforward, as shown below:

$$F_2 = \int_0^{0.5} 2xdx + \int_{0.5}^1 (2 - 2x)dx + \frac{du}{dx}(1)\phi_2(1) - \frac{du}{dx}(0)\phi_2(0) = 0.5$$

$$F_3 = \int_0^{0.5} 0dx + \int_{0.5}^1 (-1 + 2x)dx + \frac{du}{dx}(1)\phi_3(1) - \frac{du}{dx}(0)\phi_3(0) = 1.25$$

Note that we have used the boundary condition $du(1)/dx = 1$ in computing F_3.

After calculating all necessary components, the final matrix equations for the finite element analysis become

$$\begin{bmatrix} 2 & -2 & 0 \\ -2 & 4 & -2 \\ 0 & -2 & 2 \end{bmatrix} \begin{Bmatrix} u_1 \\ u_2 \\ u_3 \end{Bmatrix} = \begin{Bmatrix} F_1 \\ 0.5 \\ 1.25 \end{Bmatrix} \qquad (3.40)$$

where the nodal solutions u_1, u_2, and u_3 are required. We will discard the first of the three equations, as we do not know the RHS of that equation. This is called striking-the-row. Refer to the

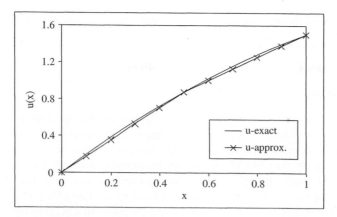

Figure 3.9 Exact solution $u(x)$ and finite element solution $\tilde{u}(x)$

explanation for Eq. (2.11) in Chapter 2. Furthermore, since the essential boundary condition $u_1 = 0$ is given, we can eliminate the first column (striking-the-column). Then the last two equations take the following form:

$$\begin{bmatrix} 4 & -2 \\ -2 & 2 \end{bmatrix} \begin{Bmatrix} u_2 \\ u_3 \end{Bmatrix} = \begin{Bmatrix} 0.5 \\ 1.25 \end{Bmatrix} \tag{3.41}$$

Note that the matrix in Eq. (3.40) is not positive definite, but the one in Eq. (3.41) is. Solving Eq. (3.41), we obtain the nodal solutions as $u_2 = 0.875$ and $u_3 = 1.5$. Note that the nodal solution at Node 1 is given as the essential boundary condition; i.e., $u_1 = 0.0$. Once the nodal solutions are available, the approximate solution can be obtained from Eq. (3.39) as

$$\tilde{u}(x) = \begin{cases} 1.75x, & 0 \le x \le 0.5 \\ 0.25 + 1.25x, & 0.5 \le x \le 1 \end{cases} \tag{3.42}$$

Due to the linear trial functions, the approximate finite element solution $\tilde{u}(x)$ is composed of piecewise linear polynomials. It is coincidental that the exact solution and the FE solution agree at the nodes ($x = 0.5$ and $x = 1$). The approximate solution from Eq. (3.42) and the exact solution from Eq. (3.16) are compared in Figure 3.9. The maximum error is about 8%.

The derivative du/dx is plotted in Figure 3.10, and one can note the large discrepancy in the derivatives of the two solutions. Since the approximate solution is a piecewise linear polynomial, its

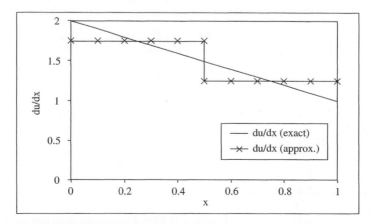

Figure 3.10 Derivatives of the exact and finite element solutions

derivative is constant in each element. Accordingly, the derivative is not continuous at the element boundary.

3.5   FORMAL PROCEDURE

Although Example 3.5 illustrated the implementation of the Galerkin method in the finite element scheme, it is not still general enough to implement in computer codes. Hence, we will present a slightly modified form of Galerkin finite element formulation.

In this method, we apply the Galerkin method to one element at a time. Let us consider a general element, say Element e in Figure 3.11. It has two nodes, say, i and j. The choice of Nodes i and j is such that $x_j > x_i$. We also introduce a local coordinate ξ such that $\xi = 0$ at Node i and $\xi = 1$ at Node j. The relation between x and ξ for Element e is given by

$$x = x_i(1 - \xi) + x_j \xi \quad \text{or} \quad \xi = \frac{x - x_i}{x_j - x_i} = \frac{x - x_i}{L^{(e)}} \tag{3.43}$$

where $L^{(e)}$ denotes the length of the element. As will be seen later, the use of local coordinate ξ is a matter of convenience as it helps in expressing the interpolation functions and their integrations in an elegant manner. The approximate solution within the Element e is given by

$$\tilde{u}(x) = u_i N_1(x) + u_j N_2(x) \tag{3.44}$$

where the interpolations functions N_1 and N_2 can be conveniently expressed as a function of the variable ξ

$$N_1(\xi) = (1 - \xi)$$
$$N_2(\xi) = \xi \tag{3.45}$$

In terms of the variable x, the interpolation functions take the form

$$N_1(x) = \left(1 - \frac{x - x_i}{L^{(e)}}\right)$$
$$N_2(x) = \frac{x - x_i}{L^{(e)}} \tag{3.46}$$

We will use the interpolation functions in Eq. (3.45). Using the chain rule of differentiation, the derivatives of N_i are

$$\frac{dN_1}{dx} = \frac{dN_1}{d\xi}\frac{d\xi}{dx} = -\frac{1}{L^{(e)}}$$
$$\frac{dN_2}{dx} = \frac{dN_2}{d\xi}\frac{d\xi}{dx} = +\frac{1}{L^{(e)}} \tag{3.47}$$

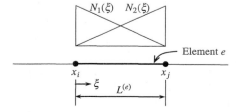

Figure 3.11 One-dimensional finite element with interpolation functions

One can easily verify that the interpolation functions satisfy the following relation:

$$N_1(x_i) = 1, \quad N_1(x_j) = 0$$
$$N_2(x_i) = 0, \quad N_2(x_j) = 1$$

(3.48)

In addition, $N_1(x)$ linearly decreases from x_i to x_j, while $N_2(x)$ linearly increases. One can also verify that the above functions yield

$$\tilde{u}(x_i) = u_i$$
$$\tilde{u}(x_j) = u_j$$

(3.49)

where u_i and u_j are nodal solution at Nodes i and j, respectively. Equation (3.49) is an important property of interpolation.

From Eq. (3.44), the derivative of $\tilde{u}(x)$ is obtained as

$$\frac{d\tilde{u}}{dx} = u_i \frac{dN_1}{dx} + u_j \frac{dN_2}{dx}$$

(3.50)

The above equation can be written in a matrix form, using the chain rule of differentiation in Eq. (3.47), as

$$\frac{d\tilde{u}}{dx} = \left\lfloor \frac{dN_1}{dx} \quad \frac{dN_2}{dx} \right\rfloor \left\{ \begin{array}{c} u_1 \\ u_2 \end{array} \right\}$$
$$= \frac{1}{L^{(e)}} \left\lfloor \frac{dN_1}{d\xi} \quad \frac{dN_2}{d\xi} \right\rfloor \left\{ \begin{array}{c} u_1 \\ u_2 \end{array} \right\}$$

(3.51)

We apply the Galerkin method described in Section 3.2 at the element level. Then, N_1 and N_2 will be the two trial functions. A set of two equations similar to Eq. (3.10) can be written for the Element e:

$$\int_{x_i}^{x_j} \frac{dN_i}{dx} \frac{d\tilde{u}}{dx} dx = \int_{x_i}^{x_j} p(x)N_i(x)dx + \frac{du}{dx}(x_j)N_i(x_j) - \frac{du}{dx}(x_i)N_i(x_i), \quad i = 1, 2$$

(3.52)

Now the variable x can be changed to ξ by substituting for $d\tilde{u}/dx$ from Eq. (3.51) and dN_i/dx from Eq. (3.47). The integration domain can also be changed by using the relation: $dx = L^{(e)}d\xi$. After changing the variable, we obtain

$$\frac{1}{L^{(e)}} \int_0^1 \frac{dN_i}{d\xi} \left\lfloor \frac{dN_1}{d\xi} \quad \frac{dN_2}{d\xi} \right\rfloor d\xi \cdot \left\{ \begin{array}{c} u_1 \\ u_2 \end{array} \right\} = L^{(e)} \int_0^1 p(x)N_i(\xi)d\xi + \frac{du}{dx}(x_j)N_i(1)$$
$$- \frac{du}{dx}(x_i)N_i(0), \quad i = 1, 2$$

(3.53)

Note that the variable x still remains in $p(x)$. When a specific form of $p(x)$ is given, it will be converted into a function of ξ using the relation in Eq. (3.43). We do not need to convert $du(x_j)/dx$ and $du(x_i)/dx$ because the boundary conditions do not use the approximation scheme. The two equations in Eq. (3.53) can be written in a matrix form as

$$[\mathbf{k}^{(e)}]\{\mathbf{u}^{(e)}\} = \{\mathbf{f}^{(e)}\} + \left\{ \begin{array}{c} -\dfrac{du}{dx}(x_i) \\[2mm] +\dfrac{du}{dx}(x_j) \end{array} \right\}$$

(3.54)

where

$$\underset{2\times2}{[\mathbf{k}^{(e)}]} = \frac{1}{L^{(e)}} \int_0^1 \begin{bmatrix} \left(\dfrac{dN_1}{d\xi}\right)^2 & \dfrac{dN_1}{d\xi}\dfrac{dN_2}{d\xi} \\[2ex] \dfrac{dN_2}{d\xi}\dfrac{dN_1}{d\xi} & \left(\dfrac{dN_2}{d\xi}\right)^2 \end{bmatrix} d\xi \tag{3.55}$$

$$= \frac{1}{L^{(e)}} \begin{bmatrix} 1 & -1 \\ -1 & 1 \end{bmatrix}$$

$$\{\mathbf{f}^{(e)}\} = L^{(e)} \int_0^1 p(x) \begin{Bmatrix} N_1(\xi) \\ N_2(\xi) \end{Bmatrix} d\xi \tag{3.56}$$

and

$$\{\mathbf{u}^{(e)}\} = \begin{Bmatrix} u_i \\ u_j \end{Bmatrix} \tag{3.57}$$

In arriving at Eq. (3.54) from Eq. (3.53), we have used the boundary values of the inter-polation functions given in Eq. (3.48). Equation (3.54) is the element-level equivalent of the global equation derived in Eq. (3.12). One can derive an equation similar to Eq. (3.54) for each element $e = 1, 2, \ldots, N_E$, where N_E is the number of elements.

The RHS of these equations contain terms that are derivatives at the nodes $du(x_i)/dx$ and $du(x_j)/dx$, which are not generally known. However, the second equation for Element e can be added to the first equation of Element $e + 1$ to eliminate the derivative term. To illustrate this point consider the equations for Elements 1 and 2. Two element matrix equations are

$$\begin{bmatrix} k_{11} & k_{12} \\ k_{21} & k_{22} \end{bmatrix}^{(1)} \begin{Bmatrix} u_1 \\ u_2 \end{Bmatrix} = \begin{Bmatrix} f_1 \\ f_2 \end{Bmatrix}^{(1)} + \begin{Bmatrix} -\dfrac{du}{dx}(x_1) \\ +\dfrac{du}{dx}(x_2) \end{Bmatrix} \tag{3.58}$$

$$\begin{bmatrix} k_{11} & k_{12} \\ k_{21} & k_{22} \end{bmatrix}^{(2)} \begin{Bmatrix} u_2 \\ u_3 \end{Bmatrix} = \begin{Bmatrix} f_2 \\ f_3 \end{Bmatrix}^{(2)} + \begin{Bmatrix} -\dfrac{du}{dx}(x_2) \\ +\dfrac{du}{dx}(x_3) \end{Bmatrix} \tag{3.59}$$

We want to combine these two matrix equations into one, which is called the *assembly process*. The assembled matrix equation will have three unknowns: $u_1, u_2,$ and u_3. Equation (3.58) will be added to the first two rows, while Eq. (3.59) will be added to the last two rows. When the second equation in Eq. (3.58) and the first equation in Eq. (3.59) are added together, the boundary term, $du(x_2)/dx$, is canceled. Thus, the assembled matrix equation becomes

$$\begin{bmatrix} k_{11}^{(1)} & k_{12}^{(1)} & 0 \\ k_{21}^{(1)} & k_{22}^{(1)} + k_{11}^{(2)} & k_{12}^{(2)} \\ 0 & k_{21}^{(2)} & k_{22}^{(2)} \end{bmatrix} \begin{Bmatrix} u_1 \\ u_2 \\ u_3 \end{Bmatrix} = \begin{Bmatrix} f_1^{(1)} \\ f_2^{(1)} + f_2^{(2)} \\ f_3^{(2)} \end{Bmatrix} + \begin{Bmatrix} -\dfrac{du}{dx}(x_1) \\ 0 \\ \dfrac{du}{dx}(x_3) \end{Bmatrix} \tag{3.60}$$

This process can be continued for successive elements, and the $2 \times N_E$ equations for the N_E elements will reduce to $N_E + 1$ number of equations. In fact, $N_E + 1 = N_D$, which is equal to the number of nodes. The N_D equations will take the form

$$
\begin{bmatrix}
k_{11}^{(1)} & k_{12}^{(1)} & 0 & \cdots & 0 \\
k_{21}^{(1)} & k_{22}^{(1)} + k_{11}^{(2)} & k_{12}^{(2)} & \cdots & 0 \\
0 & k_{221}^{(2)} & k_{22}^{(2)} + k_{11}^{(2)} & \cdots & 0 \\
\vdots & \vdots & \vdots & \ddots & \vdots \\
0 & 0 & 0 & k_{21}^{(N_E)} & k_{22}^{(N_E)}
\end{bmatrix}_{(N_D \times N_D)}
\begin{Bmatrix}
u_1 \\ u_2 \\ u_3 \\ \vdots \\ u_N
\end{Bmatrix}_{(N_D \times 1)}
=
\begin{Bmatrix}
f_1^{(1)} \\ f_2^{(1)} + f_2^{(2)} \\ f_3^{(2)} + f_3^{(3)} \\ \vdots \\ f_N^{(N_E)}
\end{Bmatrix}_{(N_D \times 1)}
+
\begin{Bmatrix}
-\dfrac{du}{dx}(x_1) \\ 0 \\ 0 \\ \vdots \\ +\dfrac{du}{dx}(x_N)
\end{Bmatrix}_{(N_D \times 1)}
\tag{3.61}
$$

In compact form, the above equation is written as

$$[\mathbf{K}]\{\mathbf{u}\} = \{\mathbf{F}\} \tag{3.62}$$

In general, the global matrix $[\mathbf{K}]$ will be singular and hence the equations cannot be solved directly. However, the matrix will be nonsingular after implementing the boundary conditions. It may be noted that there are N_D unknowns in the N_D equations. At the boundaries $(x = 0$ and $x = 1)$, either u (the essential boundary condition) or du/dx (the natural boundary condition) will be specified. We will illustrate the method in the following example.

EXAMPLE 3.6   *Three-Element Solution of a Differential Equation*

Using three elements of equal length, solve the differential equation given below for $p(x) = x$.

$$\frac{d^2 u}{dx^2} + p(x) = 0, \qquad 0 \le x \le 1$$

$$\left. \begin{aligned} u(0) &= 0 \\ u(1) &= 0 \end{aligned} \right\} \qquad \text{Boundary conditions}$$

SOLUTION   Since the elements are of equal length, each element has a length of $L^{(e)} = \frac{1}{3}$. Substituting in Eq. (3.55), the element stiffness matrices for the three elements can be derived as

$$[\mathbf{k}^{(e)}]_{2 \times 2} = \frac{1}{L^{(e)}} \begin{bmatrix} 1 & -1 \\ -1 & 1 \end{bmatrix} = \begin{bmatrix} 3 & -3 \\ -3 & 3 \end{bmatrix}, \qquad (e = 1, 2, 3)$$

Note that the element stiffness matrices for the three elements are identical. Now the variable in $p(x) = x$ can be changed to ξ using Eq. (3.43), as

$$p(\xi) = x_i(1 - \xi) + x_j \xi$$

Substituting this expression in Eq. (3.56), the element force vectors for the three elements can be derived as

$$
\begin{aligned}
\{\mathbf{f}^{(e)}\} &= L^{(e)} \int_0^1 p(x) \begin{Bmatrix} N_1(\xi) \\ N_2(\xi) \end{Bmatrix} d\xi \\
&= L^{(e)} \int_0^1 [x_i(1 - \xi) + x_j \xi] \begin{Bmatrix} 1 - \xi \\ \xi \end{Bmatrix} d\xi \\
&= L^{(e)} \begin{Bmatrix} \dfrac{x_i}{3} + \dfrac{x_j}{6} \\ \dfrac{x_i}{6} + \dfrac{x_j}{3} \end{Bmatrix}, \qquad (e = 1, 2, 3)
\end{aligned}
$$

Substituting for the element lengths and nodal coordinates

$$\left\{ \begin{matrix} f_1^{(1)} \\ f_2^{(1)} \end{matrix} \right\} = \frac{1}{54} \left\{ \begin{matrix} 1 \\ 2 \end{matrix} \right\}, \quad \left\{ \begin{matrix} f_2^{(2)} \\ f_3^{(2)} \end{matrix} \right\} = \frac{1}{54} \left\{ \begin{matrix} 4 \\ 5 \end{matrix} \right\}, \quad \left\{ \begin{matrix} f_3^{(3)} \\ f_4^{(3)} \end{matrix} \right\} = \frac{1}{54} \left\{ \begin{matrix} 7 \\ 8 \end{matrix} \right\}$$

Now the global matrix $[\mathbf{K}]$ and vector $\{\mathbf{F}\}$ can be assembled using Eq. (3.61), as

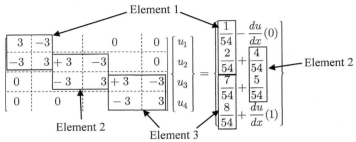

We discard the first and last rows, as we do not know the RHS of these equations (striking-the-rows). Furthermore, we note that $u_1 = u_4 = 0$. Thus, these two variables are removed, and the first and last columns of matrix $[\mathbf{K}]$ are deleted (striking-the-columns). Then, the four global equations reduce to two equations:

$$\begin{bmatrix} 6 & -3 \\ -3 & 6 \end{bmatrix} \left\{ \begin{matrix} u_2 \\ u_3 \end{matrix} \right\} = \frac{1}{9} \left\{ \begin{matrix} 1 \\ 2 \end{matrix} \right\}$$

Solving the above matrix equation, we obtain $u_2 = 4/81$ and $u_3 = 5/81$. Then, using the interpolation functions in Eq. (3.44), the approximate solution at each element can be expressed as

$$\tilde{u}(x) = \begin{cases} \dfrac{4}{27}x, & 0 \le x \le \dfrac{1}{3} \\[2mm] \dfrac{4}{81} + \dfrac{1}{27}\left(x - \dfrac{1}{3}\right), & \dfrac{1}{3} \le x \le \dfrac{2}{3} \\[2mm] \dfrac{5}{81} - \dfrac{5}{27}\left(x - \dfrac{2}{3}\right), & \dfrac{2}{3} \le x \le 1 \end{cases} \tag{3.63}$$

The exact solution can be obtained by integrating the governing differential equation twice and applying the two essential boundary conditions to solve for the constants:

$$u(x) = \frac{1}{6}x(1 - x^2) \tag{3.64}$$

The exact and approximate solutions are plotted in Figure 3.12. The value of the approximate solution at Nodes 2 and 3 coincide with that of the exact solution, and it is actually a coincidence.

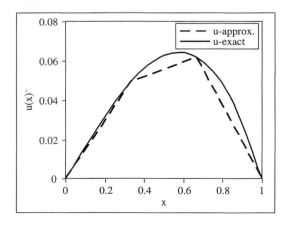

Figure 3.12 Comparison of exact and approximate solution for Example 3.6

Otherwise, one can note that the three-element solution is a poor approximation of the exact solution, and more elements are needed to obtain a more accurate solution. This is because the finite element solution is a linear function between nodes, whereas the exact solution is a cubic polynomial in x.

Sections 3.4 and 3.5 describe two different methods of implementing the Galerkin method in the finite element scheme. Although they appear slightly different, they result in the same matrix equations. The first method described in Section 3.4 mimics the analytical method described in the preceding Section 3.2. In the global finite element method (Section 3.2), the global matrix $[\mathbf{K}]$ is directly calculated using Eq. (3.13). However, in the local finite element method, we first compute the element matrix $[\mathbf{k}]$ for each element, and they are assembled to form the global matrix. Furthermore, the methods described in Section 3.5 are amenable to easy coding in computer programs. The element matrix $[\mathbf{k}]$ is calculated for each element using the element properties, such as the length of the element [Eq. (3.55)]. This procedure could be automated easily as the interpolation functions expressed in the variable ξ have the same form for all elements. The RHS matrix $\{\mathbf{f}\}$ can also be calculated easily for each element. Then, the assembling of the element matrices $[\mathbf{k}]$ and $\{\mathbf{f}\}$ to obtain the global matrices $[\mathbf{K}]$ and $\{\mathbf{F}\}$, respectively, can be performed in a more mechanical way as described in Chapter 2. Such considerations are important in using the finite element method for large problems involving tens of thousands of elements.

3.6   ENERGY METHODS

Energy methods are alternative methods that are very powerful and amenable for approximate solutions when solving structures that are more realistic. Finite element equations can also be derived using energy methods. Some of the energy methods are Castigliano's theorems, principle of minimum potential energy, principle of minimum complementary potential energy, unit load method, and Rayleigh-Ritz method. The principle of virtual work is the fundamental principle from which all of the aforementioned methods are derived.

3.6.1   Principle of Virtual Work for a Particle

The principle of virtual work for a particle states that for a particle in equilibrium, the virtual work is identically equal to zero. *Virtual displacement is any small arbitrary (imaginary, not real) displacement that is consistent with the kinematic constraints of the particle. Virtual work is the work done by the (real) external forces through the virtual displacements.*

For a particle in equilibrium, the sum of the forces acting on it in each coordinate direction must be equal to zero:

$$\sum F_x = 0, \quad \sum F_y = 0, \quad \sum F_z = 0 \tag{3.65}$$

Let the *virtual displacements* in the x-, y-, and z-direction be δu, δv, and δw, respectively. Then it is obvious that

$$\delta u \sum F_x = 0, \quad \delta v \sum F_y = 0, \quad \delta w \sum F_z = 0 \tag{3.66}$$

or, in other words, the *virtual work* is equal to zero:

$$\delta W = \delta u \sum F_x + \delta v \sum F_y + \delta w \sum F_z = 0 \tag{3.67}$$

Conversely, if the virtual work of a particle is equal to zero for arbitrary virtual displacements, then the particle is in equilibrium under the applied forces. In fact, the principle is another statement of equilibrium equations given in Eq. (3.65). This seemingly trivial principle leads to some important results when applied to deformable bodies.

3.6.2 Principle of Virtual Work for Deformable Bodies

As an example, we will consider the one-dimensional case, uniaxial bar, as shown in Figure (3.13). The bar is fixed at $x = 0$ and under the body force B_x in the x-direction as well as a concentrated force at F at $x = L$. In general the cross-sectional area $A(x)$ can vary along the length of the bar L.

From Section 1.4, the stress equilibrium equation can be written as

$$\frac{d\sigma_x}{dx} + B_x = 0 \tag{3.68}$$

where B_x is the body force expressed in force per unit volume. In the uniaxial bar, the stress is constant over the cross-section. Consider a virtual displacement field $\delta u(x)$ along the length of the bar. Since Eq. (3.68) is true at every point in every cross-section of the bar, the following must be true for any $\delta u(x)$:

$$\int_0^L \int_A \left(\frac{d\sigma_x}{dx} + B_x \right) \delta u(x) dA dx = 0 \tag{3.69}$$

By using the definition of axial force resultant P, i.e., $P = A\sigma_x$, Eq. (3.69) can be written as

$$\int_0^L \left(\frac{dP}{dx} + b_x \right) \delta u(x) dx = 0 \tag{3.70}$$

where the body force b_x is force per unit length of the bar. Using integration by parts in Eq. (3.70), we obtain

$$P\delta u \Big|_0^L - \int_0^L P \frac{d(\delta u)}{dx} dx + \int_0^L b_x \delta u(x) dx = 0 \tag{3.71}$$

Since the bar is fixed at $x = 0$, $u(0) = 0$. That means, $\delta u(0)$ is also equal to zero because *the virtual displacement should be consistent with the displacement constraints of the body*. Since F is the force applied at $x = L$, $P(L) = F$. We also define virtual strains $\delta\varepsilon(x)$ that are created by the virtual displacement field $\delta u(x)$:

$$\delta\varepsilon(x) = \frac{d(\delta u)}{dx} \tag{3.72}$$

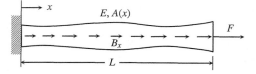

Figure 3.13 Uniaxial bar under body force B_x and concentrated force F

Then, using Eq. (3.72), Eq. (3.71) can be written as

$$F\delta u(L) + \int_0^L b_x \delta u(x)dx = \int_0^L P\delta\varepsilon(x)dx \tag{3.73}$$

Note that the LHS is the work done by external forces, F and b_x, while the RHS is the work done by internal force resultant, $P(x)$. We define the external virtual work δW_e and internal virtual work δW_i as follows:

$$\delta W_e = F\delta u(L) + \int_0^L b_x \delta u(x)dx \tag{3.74}$$

$$\delta W_i = -\int_0^L \int_A \sigma_x \delta\varepsilon(x)dAdx = -\int_0^L P\delta\varepsilon(x)dx \tag{3.75}$$

The negative sign is added in internal virtual work because it is positive when work is done to the system. Then, Eq. (3.73) can be written as

$$\delta W_e + \delta W_i = 0 \tag{3.76}$$

Equation (3.76) states that for a bar in equilibrium, the sum of external and internal virtual work is zero for every virtual displacement field, and it constitutes the principle of virtual work for a one-dimensional deformable body. This principle can be derived for three-dimensional bodies also, except the definition of δW_e and δW_i take different forms. The external virtual work will be the sum of work done by all (real) external forces through the corresponding virtual displacements. Considering both distributed forces (surface tractions t_x, t_y and t_z) and concentrated forces (F_x, F_y and F_z) we obtain

$$\delta W_e = \int_S (t_x\delta u + t_y\delta v + t_z\delta w)dS + \sum_i (F_{xi}\delta u_i + F_{yi}\delta v_i + F_{zi}\delta w_i) \tag{3.77}$$

In Eq. (3.77), the integration is performed over the surface of the three-dimensional solid. The internal virtual work is defined by

$$\delta W_i = -\int_V (\sigma_x\delta\varepsilon_x + \sigma_y\delta\varepsilon_y + \cdots + \tau_{xy}\delta\gamma_{xy}) \, dV \tag{3.78}$$

where the integration is performed over the volume of the body, V.

3.6.3 Variation of a Function

Virtual displacements in the previous section can be considered as a variation of real displacements. To explain it further, consider x-directional displacement $u(x)$ and those neighboring displacements that are described by arbitrary virtual displacement $\delta u(x)$ and small parameter $\tau > 0$, as

$$u_\tau(x) = u(x) + \tau\delta u(x) \tag{3.79}$$

Thus, for given virtual displacement, these neighboring displacements are controlled by one parameter τ. $u_\tau(x)$ is the perturbed displacement. Similar to the first variation in calculus, the variation of displacement can be obtained by

$$\left.\frac{du_\tau(x)}{d\tau}\right|_{\tau=0} = \delta u(x) \tag{3.80}$$

It is obvious that the virtual displacement $\delta u(x)$ is the *displacement variation*.

An important requirement of the virtual displacement is that it must satisfy kinematic constraints. It means that when the value of displacement is prescribed at a point, the perturbed displacement must have the same value. To satisfy this requirement, the displacement variation must be equal to zero at that point. For example, let displacement boundary condition be given as $u(0) = 1$. Then, the displacement variation must satisfy $\delta u(0) = 0$.

The variation of a complex function can be obtained using the chain rule of differentiation. For example, let a function $f(u)$ depends on displacement. The variation of $f(u)$ can be obtained using

$$\delta f = \left.\frac{df(u_\tau)}{d\tau}\right|_{\tau=0} = \frac{df}{du}\delta u \tag{3.81}$$

An important property of the variation is that it is independent of differentiation with respect to space coordinates. For example, consider the variation of strain:

$$\delta \varepsilon_x = \delta\left(\frac{du}{dx}\right) = \frac{d(\delta u)}{dx}$$

The concept of variation plays an important role in understanding structural equilibrium, which will be derived in the following section.

3.6.4 Principle of Minimum Potential Energy

Consider the strain energy density in a one-dimensional body in Figure (3.13), which is given by

$$U_0 = \frac{1}{2}\sigma_x\varepsilon_x = \frac{1}{2}E\varepsilon_x^2 \tag{3.82}$$

Suppose the displacement field is perturbed slightly by $\delta u(x)$, then it will cause a slight change in the strain field also. Then the variation in the strain energy density is given by

$$\delta U_0 = E\varepsilon_x\delta\varepsilon_x = \sigma_x\delta\varepsilon_x \tag{3.83}$$

The change in strain energy of the bar is expressed as

$$\delta U = \int_0^L \int_A \delta U_0 \, dA \, dx = \int_0^L \int_A \sigma_x\delta\varepsilon_x \, dA \, dx = \int_0^L P\delta\varepsilon_x \, dx \tag{3.84}$$

Comparing Eq. (3.75) and Eq. (3.84), we obtain

$$\delta U = -\delta W_i \tag{3.85}$$

Next, we define the potential energy of external forces. Consider a force F at $x = L$ and the corresponding displacement $u(L)$. If there were an additional displacement of $\delta u(L)$, then the force would have done additional work of $F\delta u(L)$. Then we can claim that the force F has lost some potential to do work or its potential has been reduced by $F\delta u(L)$. If we denote the potential of the force by V, then the change in potential due to the variation $\delta u(L)$ is given by

$$\delta V = -F\delta u(L) \tag{3.86}$$

Since the force F does not vary through this change in displacement, we can write δV as

$$\delta V = -\delta(Fu(L)) \tag{3.87}$$

Thus, the potential energy of external forces is negative of the external forces multiplied by displacements at that point. Since the body force is distributed in $0 \leq x \leq L$, the potential energy must be integrated over the domain. Then the potential energy takes the following form:

$$V = -Fu(L) - \int_0^L b_x u(x) dx \tag{3.88}$$

The negative sign in front of the work term in the above equation is sometimes confusing to students. It can be easily understood, if one considers the gravitational potential energy on the surface of the earth. From elementary physics we know that the potential energy of a body of mass m at a height h from the datum is given by $V = mgh$. Since gravity acts downwards and height is positive upwards, the potential energy should be written as $V = -Fh$, where F is the force of gravity or the weight of the body, and the magnitude of F is equal to mg. If we replace the gravitational force by an external force, then one obtains Eq. (3.88).

Comparing Eqs. (3.74) and (3.88) one can note that the change in the potential of external forces is equal to the negative of the external virtual work:

$$\delta V = -\delta W_e \tag{3.89}$$

Substituting for the external and internal virtual work terms from Eq. (3.85) and (3.89) into Eq. (3.76), the principle of virtual work takes the form

$$\delta U + \delta V = 0 \quad \text{or} \quad \delta(U + V) = 0 \tag{3.90}$$

We define the total potential energy Π as the sum of the strain energy and the potential of external forces, i.e., $\Pi = U + V$. Then Eq. (3.90) takes the form

$$\delta \Pi = 0 \tag{3.91}$$

Equation (3.91) is the *Principle of Minimum Potential Energy*, and it can be stated as follows:

Of all displacement configurations of a solid consistent with its displacement (kinematic) constraints, the actual one that satisfies the equilibrium equations is given by the minimum value of total potential energy.

The variation of total potential energy is expressed in terms of differentiation with respect to displacements. Using chain rule of differentiation,

$$\delta \Pi = \frac{d\Pi}{du} \delta u = 0 \tag{3.92}$$

Since the above equation must be satisfied for all displacement variations that are consistent with kinematic constraints, the equilibrium can be found by setting the derivative of total potential energy equal to zero. In practice, the displacement is often approximated using discrete parameters, such as coefficients c_i in Eq. (3.4) or nodal solution u_i in Eq. (3.55). In such a case, the differentiation in the above equation can be performed with respect to the unknown coefficients. This process will be explained in the following section.

3.6.5 Application of Principle of Minimum Potential Energy to Discrete Systems

Deformation of discrete systems can be defined by a finite number of variables. Consider the assemblage of springs shown in Figure 3.14. In the context of finite element method, the springs are called elements and the masses that are connected to the elements are

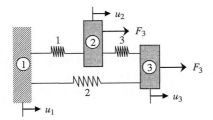

Figure 3.14 Example of a discrete system with finite number of degrees of freedom

called the nodes. External forces are applied only at the nodes. At least one node has to be fixed to prevent rigid body displacement and thus obtain a unique solution for displacements. The strain energy of the system and the potential energy of the forces are expressed in terms of the displacements. Then the total potential energy is minimized with respect to the displacements, which yields as many linear equations as the number of unknown displacements. The procedures are illustrated in the flowing example.

EXAMPLE 3.7 *Energy Method for a Discrete System*

Consider a system of rigid bodies connected by springs, as shown in Figure 3.14. The bodies are assumed to move only in the horizontal direction. Further, we consider only the static problem, and hence the actual mass effects will not be considered. External forces are applied on the rigid bodies 2 and 3 as shown. The objectives are to determine the displacement of each body, forces in the springs, and support reaction. Use the following values: $k^{(1)} = 100\,\text{N/mm}$, $k^{(2)} = 200\,\text{N/mm}$, $k^{(3)} = 150$, N/mm, $F_2 = 1,000\,\text{N}$, and $F_3 = 500\,\text{N}$.

SOLUTION The discrete degrees of freedom (DOFs) in the above problem are the displacements u_1, u_2 and u_3. However, the displacement at Node 1 is known, and hence the number of DOFs in this problem is equal to two. It should be noted that the force F_1 corresponding to the known displacement u_1 is the reactions exerted by the support, and it is unknown. The total potential energy of the system consists of two parts: strain energy of the system U, and potential energy V of the external forces F_2 and F_3.

The strain energy stored in any of the springs, say Spring 1, can be expressed in terms of the displacements of its nodes, u_1 and u_2. The strain energy $U^{(1)}$ of the spring is then

$$U^{(1)} = \frac{1}{2}k^{(1)}(u_2 - u_1)^2$$

The above expression can be written in a matrix form:

$$U^{(1)} = \frac{1}{2} \underset{(1\times 2)}{\lfloor u_1 \quad u_2 \rfloor} \underset{(2\times 2)}{\begin{bmatrix} k^{(1)} & -k^{(1)} \\ -k^{(1)} & k^{(1)} \end{bmatrix}} \underset{(2\times 1)}{\begin{Bmatrix} u_1 \\ u_2 \end{Bmatrix}}$$

In the above equation, the numbers in parentheses indicate the size of the matrix. Similarly, the strain energies in Springs 2 and 3 can be written as

$$U^{(2)} = \frac{1}{2} \lfloor u_1 \quad u_3 \rfloor \begin{bmatrix} k^{(2)} & -k^{(2)} \\ -k^2 & k^2 \end{bmatrix} \begin{Bmatrix} u_1 \\ u_3 \end{Bmatrix}$$

$$U^{(3)} = \frac{1}{2} \lfloor u_2 \quad u_3 \rfloor \begin{bmatrix} k^{(3)} & -k^{(3)} \\ -k^{(3)} & k^{(3)} \end{bmatrix} \begin{Bmatrix} u_2 \\ u_3 \end{Bmatrix}$$

Since the strain energy is a scalar, we can add energy in all springs to obtain the strain energy of the system:

$$U = \sum_{e=1}^{3} U^{(e)}$$

It can be shown that the sum of the strain energy terms can be written again in a matrix form that includes all the DOFs in the system:

$$U = \frac{1}{2} \lfloor u_1 \quad u_2 \quad u_3 \rfloor \begin{bmatrix} k^{(1)} + k^{(2)} & -k^{(1)} & -k^{(2)} \\ -k^{(1)} & k^{(1)} + k^{(3)} & -k^{(3)} \\ -k^{(2)} & -k^{(3)} & k^{(2)} + k^{(3)} \end{bmatrix} \begin{Bmatrix} u_1 \\ u_2 \\ u_3 \end{Bmatrix} \tag{3.93}$$

or symbolically

$$U = \frac{1}{2} \{Q\}^T [K] \{Q\} \tag{3.94}$$

where the column vector $\{Q\}$ represents the three DOFs and $[K]$ is the square matrix in Eq. (3.93). The potential energy of the external forces is given by

$$\begin{aligned} V &= -(F_1 u_1 + F_2 u_2 + F_3 u_3) \\ &= - \lfloor u_1 \quad u_2 \quad u_3 \rfloor \begin{Bmatrix} F_1 \\ F_2 \\ F_3 \end{Bmatrix} \\ &= -\{Q\}^T \{F\} \end{aligned} \tag{3.95}$$

where $\{F\}$ is the column vector of external forces. The total potential energy is the sum of the strain energy and the potential of external forces:

$$\Pi = U + V = \frac{1}{2} \{Q\}^T [K] \{Q\} - \{Q\}^T \{F\} \tag{3.96}$$

In the above equation, the total potential energy is written in terms of the vector of nodal displacements $\{Q\} = \{u_1, u_2, u_3\}^T$. Thus, the differentiation in Eq. (3.92) can be applied to $\{Q\}$. According to the principle of minimum potential energy, of all possible u_i the displacements that will satisfy the equilibrium equations minimize the total potential energy. The total potential energy is minimized with respect to the DOFs:

$$\frac{\partial \Pi}{\partial u_1} = 0, \quad \frac{\partial \Pi}{\partial u_2} = 0, \quad \frac{\partial \Pi}{\partial u_3} = 0$$

or, $$\frac{\partial \Pi}{\partial \{Q\}} = 0 \tag{3.97}$$

The above minimization procedure results in the following equations (refer to Ch 0):

$$[K] \begin{Bmatrix} u_1 \\ u_2 \\ u_3 \end{Bmatrix} = \begin{Bmatrix} F_1 \\ F_2 \\ F_3 \end{Bmatrix} \tag{3.98}$$

In the above system of equations, the three unknowns are u_2, u_3 and F_1. Substituting numerical values in Eq. (3.98), we obtain

$$\begin{bmatrix} 300 & -100 & -200 \\ -100 & 250 & -150 \\ -200 & -150 & 350 \end{bmatrix} \begin{Bmatrix} 0 \\ u_2 \\ u_3 \end{Bmatrix} = \begin{Bmatrix} F_1 \\ 1,000 \\ 500 \end{Bmatrix} \tag{3.99}$$

The solution to the above system of equations is given by $u_2 = 6.538 \, \text{mm}$, $u_3 = 4.231 \, \text{mm}$, and $F_1 = -1,500 \, \text{N}$. The forces in the springs are calculated using

$$P^{(e)} = k^{(e)} (u_j - u_i) \tag{3.100}$$

Using the results for displacements, the forces in the springs are obtained as

$$P^{(1)} = k^{(1)}(u_2 - u_1) = 654\text{N}$$
$$P^{(2)} = k^{(2)}(u_3 - u_1) = 846\text{N} \tag{3.101}$$
$$P^{(3)} = k^{(3)}(u_3 - u_2) = -346\text{N}$$

Note that Springs 1 and 2 are in tension, while Spring 3 is in compression.

3.6.6   Rayleigh–Ritz Method

The principle of minimum potential energy is easily applied to discrete systems (e.g., a number of springs connected together), wherein the unknown DOFs are finite. In that case, the method yields exact solution. For continuous systems, e.g., beam, uniaxial bar, where the DOFs are infinite, an approximate method has to be used. Rayleigh-Ritz method is one such method for continuous systems.

In the Rayleigh–Ritz method, a continuous system is approximated as a discrete system with finite number of DOFs. This is accomplished by approximating the displacements by a function containing finite number of coefficients to be determined. The total potential energy is then evaluated in terms of the unknown coefficients. Then the principle of minimum potential energy is applied to determine the best set of coefficients by minimizing the total potential energy with respect to the coefficients. The solution thus obtained may not be exact. It is the best solution from among the family of solutions that can be obtained from the assumed displacement functions. In the following, we demonstrate the method to a uniaxial bar problem.

The steps involved in solving the uniaxial bar problem using the Rayleigh Ritz method are as follows.

1. Assume displacement of the bar in the form $u(x) = c_1 f_1(x) + \cdots + c_n f_n(x)$, where the c_i are coefficients to be determined. Assume $u(x)$ **must** satisfy the displacement boundary conditions, e.g., $u(0) = 0$, or $u(L) = 0$, where L is the length of the bar.

2. Determine the strain energy U in the bar using the formula in Eq. (3.105) in terms of c_i.

3. Find the potential of external forces V using formulas of types given in Eq. (3.106).

4. The total potential energy is obtained as $\Pi(c_1, c_2, \cdots c_n) = U + V$

5. Apply the principle of minimum potential energy to determine the coefficients $c_1, c_2, \cdots, c_n$.

6. After solving for the constants in the assumed displacement field, find the axial force resultant using $P(x) = AEdu/dx$.

EXAMPLE 3.8   *Uniaxial Bar Using the Rayleigh–Ritz Method*

The uniaxial bar shown in Figure 3.15 is of uniform cross-sectional area A and length L. It is clamped at the left end and subjected to a concentrated force F at the right end as shown. In addition, a uniformly distributed load $b_x(x)$ acts along the length of the bar. Use the Rayleigh–Ritz method to determine the displacements $u(x)$, the axial force resultant $P(x)$, and the support reaction. The Young's modulus of the material of the bar is E. Provide numerical results for the case $L = 1\text{m}$, $A = 100\text{mm}^2$, $E = 100\text{GPa}$, $F = 10\text{kN}$, $b_x = 10\text{kN/m}$.

b_x

F

Figure 3.15 Uniaxial bar subject to distributed and concentrated forces

SOLUTION   In applying the Rayleigh-Ritz method, we approximate the displacement field using a simple but physically reasonable function. The assumed displacement must satisfy the displacement boundary conditions; in the present case $u(0) = 0$. Actually, one needs some experience to make reasonable approximations. Let us assume a quadratic polynomial in x as the approximate displacement field:

$$u(x) = c_0 + c_1 x + c_2 x^2 \tag{3.102}$$

where c_i are coefficients to be determined. However, we have to make sure that the assumed displacement field satisfies the displacement boundary conditions *a priori*. In the present case, $u(0) = 0$, and this is accomplished by setting $c_0 = 0$. Then the assumed displacements take the form

$$u(x) = c_1 x + c_2 x^2 \tag{3.103}$$

Actually, we have reduced the continuous system to a discrete system with two DOFs. The coefficients c_1 and c_2 are the two DOFs as they determine the deformed configurations of the bar. The next step is to express the strain energy in the bar in terms of the unknown coefficients. The stain energy of the bar is obtained by integrating the strain energy per unit length U_L over the length of the bar:

$$U = \int_0^L U_L(x)dx = \int_0^L \frac{1}{2}AE\varepsilon_x^2 dx = \int_0^L \frac{1}{2}AE\left(\frac{du}{dx}\right)^2 dx \tag{3.104}$$

In Eq. (3.104), ε_x is the strain in the bar as a function of x. The strain in the uniaxial bar is given by $\varepsilon_x = du/dx$. Using the assumed displacements in Eq. (3.103), we express the strains in terms of c_i as $\varepsilon_x = c_1 + 2c_2 x$. Substituting in Eq. (3.104), the strain energy in the bar is obtained as

$$\begin{aligned} U(c_1, c_2) &= \frac{1}{2}AE\int_0^L (c_1 + 2c_2 x)^2 dx \\ &= \frac{1}{2}AE\left(Lc_1^2 + 2L^2 c_1 c_2 + \frac{4}{3}L^3 c_2^2\right) \end{aligned} \tag{3.105}$$

The potential energy of the forces acting on the bar can be derived as follows:

$$\begin{aligned} V(c_1, c_2) &= -\int_0^L b_x(x)u(x)dx - (-F)u(L) \\ &= -\int_0^L b_x\left(c_1 x + c_2 x^2\right)dx + F\left(c_1 L + c_2 L^2\right) \\ &= c_1\left(FL - b_x\frac{L^2}{2}\right) + c_2\left(FL^2 - b_x\frac{L^3}{3}\right) \end{aligned} \tag{3.106}$$

The total potential energy is then $\Pi(c_1, c_2) = U + V$. According to the principle of minimum potential energy, the best set of coefficients c_i is obtained by minimizing Π. We take the partial derivatives of Π with respect to c_i and equate them to zero.

$$\begin{aligned} \frac{\partial \Pi}{\partial c_1} &= AELc_1 + AEL^2 c_2 + FL - b_x\frac{L^2}{2} = 0 \\ \frac{\partial \Pi}{\partial c_2} &= AEL^2 c_1 + \frac{4}{3}AEL^3 c_2 + FL^2 - b_x\frac{L^3}{3} = 0 \end{aligned} \tag{3.107}$$

The c_i are obtained by solving the two equations in Eq. (3.107). Using the numerical values for the various bar properties and the forces, we obtain

$$10^7 c_1 + 10^7 c_2 = -5,000$$
$$10^7 c_1 + \frac{4 \times 10^7}{3} c_2 = -6,667$$

(3.108)

Solving the above equations, we obtain $c_1 = 0$ and $c_2 = -0.5 \times 10^{-3}$. Substituting for c_i in Eq. (3.103), the displacement field is obtained as $u(x) = -0.5 \times 10^{-3} x^2$. The axial force in the bar is obtained using $P(x) = AEdu/dx = -10,000x$. The support reaction is given by $R = -P(0) = 0$. It should be noted that the above solution obtained using the Rayleigh–Ritz method is indeed the exact solution to the problem. The reason is that the exact solution is a quadratic polynomial in x, and we have assumed the same form of solution in the Rayleigh-Ritz method also.

3.7   EXERCISE

1. Use the Galerkin method to solve the following boundary value problem using (a) one-term approximation and (b) two-term approximation. Compare your results with the exact solution by plotting them on the same graph.

$$\frac{d^2u}{dx^2} + x^2 = 0, \quad 0 < x < 1$$

$$\left. \begin{array}{l} u(0) = 1 \\ u(1) = 0 \end{array} \right\} \quad \text{Boundary conditions}$$

   Hint: Use the following one- and two-term approximations:

   One-term approximation:

$$\begin{aligned} \tilde{u}(x) &= (1-x) + c_1\phi_1(x) \\ &= (1-x) + c_1 x(1-x) \end{aligned}$$

   Two-term approximation:

$$\begin{aligned} \tilde{u}(x) &= (1-x) + c_1\phi_1(x) + c_2\phi_2(x) \\ &= (1-x) + c_1 x(1-x) + c_2 x^2(1-x) \end{aligned}$$

   The exact solution is $u(x) = 1 - x(x^3 + 11)/12$

   The approximate solution is split into two parts. The first term satisfies the given essential boundary conditions exactly; i.e., $u(0) = 1$ and $u(1) = 0$. The rest of the solution containing the unknown coefficients vanishes at the boundaries.

2. Solve the differential equation in Problem 1 using (a) two and (b) three finite elements. Use the local Galerkin method described in section 3.4. Plot the exact solution and two- and three-element solutions on the same graph. Similarly plot the derivative du/dx. **Note:** The boundary conditions are not homogeneous. The boundary condition $u(0) = 1$ has to be used in solving the final equations.

3. Using the Galerkin method, solve the following differential equation with the approximate solution in the form of $\tilde{u}(x) = c_1 x + c_2 x^2$. Compare the approximate solution with the exact one by plotting them on a graph. Also compare the derivatives du/dx and $d\tilde{u}/dx$.

$$\frac{d^2u}{dx^2} + x^2 = 0, \quad 0 \le x \le 1$$

$$\left. \begin{array}{l} u(0) = 0 \\ \frac{du}{dx}(1) = 1 \end{array} \right\} \quad \text{Boundary conditions}$$

4. The one-dimensional heat conduction problem can be expressed by the following differential equation:

$$k\frac{d^2T}{dx^2} + Q = 0, \quad 0 < x < L$$

where k is the thermal conductivity, $T(x)$ is the temperature, and Q is heat generated per unit length. Q, the heat generated per unit length, is assumed constant. Two essential boundary conditions are given at both ends: $T(0) = T(L) = 0$. Calculate the approximate temperature $T(x)$ using Galerkin method. Compare the approximate solution with the exact one.
   Hint: Start with assumed solution in the following form: $\tilde{T}(x) = c_0 + c_1 x + c_2 x^2$, and then make it to satisfy the two essential boundary conditions.

5. Solve the one-dimensional heat conduction Problem 4 using the Rayleigh-Ritz method. For the heat conduction problem, the total potential can be defined as

$$\Pi = \int_0^L \left[\frac{1}{2} k \left(\frac{dT}{dx} \right)^2 - QT \right] dx$$

Use the approximate solution $\tilde{T}(x) = T_1 \phi_1(x) + T_2 \phi_2(x) + T_3 \phi_3(x)$ where the trial functions are given in Eq. (3.37) with $N_D = 3$ and $x_1 = 0$, $x_2 = L/2$, and $x_3 = L$. Compare the approximate temperature with the exact one by plotting them on a graph.

6. Consider the following differential equation:

$$\frac{d^2u}{dx^2} + u + x = 0, \quad 0 < x < 1$$

$$u(0) = 0$$

$$\left. \frac{du}{dx} \right|_{x=1} = 1$$

   Assume a solution of the form:

$$\tilde{u}(x) = c_1 x + c_2 x^2$$

   Calculate the unknown coefficients using Galerkin method. Compare $u(x)$ and $du(x)/dx$ with the exact solution $u(x) = 3.7 \sin x - x$ by plotting the solution.

7. Solve the differential equation in Problem 6 for the following boundary conditions using Galerkin method:

$$u(0) = 1, \quad u(1) = 2$$

   Assume the approximate solution as

$$\tilde{u}(x) = \phi_0(x) + c_1 \phi_1(x)$$

   where $\phi_0(x)$ is a function that satisfies the essential boundary conditions and $\phi_1(x)$ is the weight function that satisfies the homogeneous part of the essential boundary conditions; i.e., $\phi_1(0) = \phi_1(1) = 0$. Hence, assume the functions as follows:

$$\phi_0(x) = 1 + x, \quad \phi_1(x) = x(1 - x)$$

   Compare the approximate solution with the exact solution by plotting their graphs. The exact solution can be derived as

$$u(x) = 2.9231 \sin x + \cos x - x$$

8. Consider the following boundary value problem:

$$2\frac{d^2u}{dx^2} + 3u = 0, \quad 0 < x < 2$$

$$u(0) = 1 \quad \text{and} \quad \frac{du}{dx}(2) = 1$$

Using equal-length two finite elements, calculate unknown $u(x)$ and its derivative. Compare the finite element solution with the exact solution.

9. Consider the following boundary value problem:

$$\frac{d}{dx}\left(x\frac{du}{dx}\right) = \frac{2}{x^2} \quad 1 < x < 2$$

$$u(1) = 2$$

$$\frac{du}{dx}(2) = -\frac{1}{4}$$

(a) When two equal-length finite elements are used to approximate the problem, write interpolation functions and their derivatives.

(b) Calculate the approximate solution using the Galerkin method.

10. The boundary value problem for a clamped-clamped beam can be written as

$$\frac{d^4w}{dx^4} - p(x) = 0, \quad 0 < x < 1$$

$$w(0) = w(1) = \frac{dw}{dx}(0) = \frac{dw}{dx}(1) = 0 : \text{boundary conditions}$$

When a uniformly distributed load is applied, i.e., $p(x) = p_0$, calculate the approximate beam deflection $\tilde{w}(x)$, using the Galerkin method.

Hint: Assume the approximate deflection as $\tilde{w}(x) = c\phi(x) = cx^2(1-x)^2$.

11. The boundary value problem for a cantilevered beam can be written as

$$\frac{d^4w}{dx^4} - p(x) = 0, \quad 0 < x < 1$$

$$w(0) = \frac{dw}{dx}(0) = 0, \frac{d^2w}{dx^2}(1) = 1, \frac{d^3w}{dx^3}(1) = -1 : \text{boundary conditions}$$

Assume $p(x) = x$. Assuming the approximate deflection in the form $\tilde{w}(x) = c_1\phi_1(x) + c_2\phi_2(x) = c_1x^2 + c_2x^3$. Solve for the boundary value problem using the Galerkin method. Compare the approximate solution to the exact solution by plotting the solutions on a graph.

12. Repeat Problem 11 by assuming $\tilde{w}(x) = \sum_{i=1}^{3} c_i\phi_i(x) = c_1x^2 + c_2x^3 + c_3x^4$

13. Consider a finite element with three nodes, as shown in the figure. When the solution is approximated using $u(x) = N_1(x)u_1 + N_2(x)u_2 + N_3(x)u_3$, calculate the interpolation functions $N_1(x)$, $N_2(x)$, and $N_3(x)$.

Hint: Start with assumed solution in the following form: $u(x) = c_0 + c_1x + c_2x^2$.

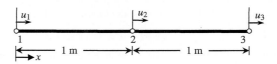

14. A vertical rod of elastic material is fixed at both ends with constant cross-sectional area A, Young's modulus E, and height of L under the distributed load f per unit length. The vertical deflection $u(x)$ of the rod is governed by the following differential equation:

$$AE\frac{d^2u}{dx^2} + f = 0$$

Using three elements of equal length, solve for $u(x)$ and compare it with the exact solution. Use the following numerical values: $A = 10^{-4}\text{m}^2$, $E = 10\,\text{GPa}$, $L = 0.3\text{m}$, $f = 10^6\text{N/m}$.

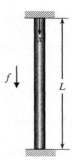

15. A bar component in the figure is under the uniformly distributed load q due to gravity. For linear elastic material with Young's modulus E and uniform cross-sectional area A, the governing differential equation can be written as

$$AE\frac{\partial^2 u}{\partial x^2} + q = 0$$

where $u(x)$ is the downward displacement. The bar is fixed at the top and free at the bottom. Using Galerkin method and two equal-length finite element, answer the following questions.

(a) Starting from the above differential equation, derive an integral equation using the Galerkin method.

(b) Write the expression of boundary conditions at $x = 0$ and $x = L$. Identify whether they are essential or natural boundary conditions.

(c) Derive the assembled finite element matrix equation, and solve it after applying boundary conditions.

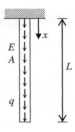

16. Consider a tapered bar of circular cross-section. The length of the bar is 1 m, and the radius varies as $r(x) = 0.050 - 0.040x$, where r and x are in meters. Assume Young's modulus $= 100$ MPa. Both ends of the bar are fixed, and a uniformly distributed load of 10,000 N/m is applied along the entire length of the bar. Determine the displacements, axial force distribution, and wall reactions using

(a) three elements of equal length;

(b) four elements of equal length.

Compare your results with the exact solution by plotting $u(x)$ and $P(x)$ curves for each case (a) and (b) and the exact solution. How do the finite element results for the left and right wall reactions, R_L and R_R, compare with the exact solution?

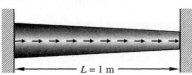

$L = 1$ m

Hint: To approximate the area of cross-section of a bar element, use the geometric mean of the end areas of the element, i.e., $A^{(e)} = \sqrt{A_i A_j} = \pi r_i r_j$. The exact solution is obtained by solving the following differential equation with the boundary conditions $u(0) = 0$ and $u(1) = 0$:

$$\frac{d}{dx}\left(A(x)E\frac{du}{dx}\right) = -p(x) = -10,000$$

The axial force distribution is found from $P(x) = A(x)Edu/dx$. The wall reactions are $R_L = -P(0)$ and $R_R = P(1)$.

17. A tapered bar with circular cross-section is fixed at $x = 0$, and an axial force of 0.3×10^6 N is applied at the other end. The length of the bar (L) is 0.3 m, and the radius varies as $r(x) = 0.03 - 0.07x$, where r and x are in meters. Use three equal-length finite elements to determine the displacements, axial force resultants and support reactions. Compare your FE solutions with the exact solution by plotting u vs. x and P(element force) vs. x. Use $E = 10^{10}$ Pa.

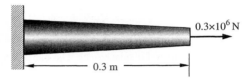

0.3×10⁶ N

0.3 m

18. The stepped bar shown in the figure is subjected to a force at the center. Use FEM to determine the displacement field $u(x)$, axial force distribution $P(x)$, and reactions R_L and R_R.
Assume $E = 100$ GPa areas of cross-sections of the three portions shown are, respectively, 10^{-4}, 2×10^{-4} and 10^{-4} m², and $F = 10,000$ N.

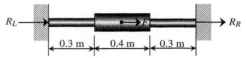

R_L F R_R

0.3 m   0.4 m   0.3 m

19. A bar shown in the figure is modeled using three equal-length bar elements. The total length of the bar is $L_T = 1.5$ m, and the radius of the circular cross-section is $r = 0.1$ m. When Young's modulus $E = 207$ GPa and distributed load $q = 1,000$ N/m, calculate displacement and stress using one of finite element analysis programs in Appendix. Compare your finite element solution with the exact solution. Provide XY-graphs of displacement and stress with respect to bar length. Explain why the finite element solutions are different from the exact solutions.

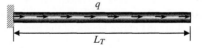

q

L_T

20. Consider the tapered bar in Problem 17. Use the Rayleigh-Ritz method to solve the same problem. Assume the displacement in the form of $u(x) = a_0 + a_1 x + a_2 x^2$. Compare the solutions for $u(x)$ and $P(x)$ with the exact solution given below by plotting them.

$$u(x) = \frac{Fx}{\pi Er(0)r(x)}, \quad P(x) = EA(x)\frac{du}{dx} = F$$

21. Consider the tapered bar in Problem 16. Use the Rayleigh-Ritz method to solve the same problem. Assume the displacement in the form of $u(x) = (x - 1)(c_1 x + c_2 x^2)$.

Chapter 4

Finite Element Analysis of Beams and Frames

In Chapter 2, the finite element equations of a truss were obtained using the direct stiffness method. Similar direct methods for beams are quite complicated, and such methods are impossible for plates and two-dimensional and three-dimensional solids. In Chapter 3 we introduced the Galerkin method and the principle of minimum potential energy for beam problems. In this chapter we will formally derive the finite element equations for beams using the energy method. The same finite element equation can be obtained using the principle of virtual work, which is addressed in some advanced textbooks.[1]

In the previous chapter we learned that in the finite element method, the displacements in an element are interpolated using $u = \lfloor \mathbf{N} \rfloor \{\mathbf{q}\}$ type expression in which $\lfloor \mathbf{N} \rfloor$ is the row vector of shape functions and $\{\mathbf{q}\}$ is the vector of nodal displacements or in general nodal degrees of freedom (DOFs). In the case of beam finite element, the nodal DOFs include the vertical (or, transverse) deflection as well as the rotation. Using this interpolation scheme, the stiffness matrix and applied load vector are derived and solved for the nodal DOFs.

After a review of the elementary beam theory in Section 4.1, we will first present the Rayleigh-Ritz method in Section 4.2. The formal development of the interpolation functions for the beam finite elements is presented in Section 4.3. The principle of minimum potential energy for beam elements is presented in Section 4.4. Once the nodal DOFs are obtained, the bending moment and shear force can be obtained by differentiating the interpolation relation for deflection, which is explained in Section 4.5. In Section 4.6, we present the frame finite element, which combines the action of a uniaxial bar and a beam.

4.1 REVIEW OF ELEMENTARY BEAM THEORY

Unlike the uniaxial bar, a beam can carry a transverse load, and the slope of the beam can change along the span. Let us consider a beam with its longitudinal axis parallel to the x-axis. We will consider beam cross-sections that are symmetric about the plane of loading (xy plane), and all applied loads will reside in this plane. The y-axis passes through the centroid of the cross-section, which is called the *neutral axis*. In the elementary beam theory, which is also called the *Euler-Bernoulli beam* theory, we assume that the transverse deflection is independent of y and is a function of x only. That is, the deflection of the beam is represented by $v(x)$, which is also called the *deflection curve*. The displacement in the x direction is represented by $u(x, y)$. The Euler-Bernoulli beam theory is based

---

[1] T. J. R. Hughes, *The Finite Element Method*, Prentice-Hall, Englewood Cliffs, NJ, 1987

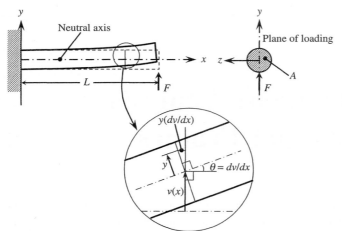

Figure 4.1 Deflection of a plane Euler-Bernoulli beam

on the assumption that plane sections normal to the beam axis remain plane and normal to the axis after deformation. Then, the displacement field $u(x, y)$ can be written as

$$u(x, y) = u_0(x) - y\frac{dv}{dx} \tag{4.1}$$

where u_0 is the x-directional displacement of the beam along the neutral axis and $\theta = dv/dx$ is the slope of the beam (see Figure 4.1).

From Eq. (4.1), the normal strain in the beam is derived as

$$\varepsilon_{xx} = \frac{\partial u}{\partial x} = \frac{du_0}{dx} - y\frac{d^2 v}{dx^2} \tag{4.2}$$

The term du_0/dx or ε_0 represents the strain along the beam axis (x-axis), where $y = 0$. One may note that the strain varies linearly in y at a given cross-section of the beam. The term $-d^2v/dx^2$ is an approximation for the *curvature* of the deflection curve. The normal strain ε_{yy} vanishes everywhere, as we have assumed v is independent of y. We assume a state of plane stress normal to the z-axis. Then, the stress-strain relation is the same as uniaxial problem. The normal stress σ_{xx} in the beam cross-section is given by

$$\sigma_{xx} = E\varepsilon_{xx} = E\varepsilon_0 - Ey\frac{d^2 v}{dx^2} \tag{4.3}$$

The axial force resultant P and the bending moment M at a cross-section can be obtained by integrating the axial stress, as

$$P = \int_A \sigma_{xx}dA, \quad M = -\int_A y\sigma_{xx}dA \tag{4.4}$$

where integration is performed over the cross-sectional area, A. Substituting for σ_{xx} from Eq. (4.3), the axial force and bending moment in the above equation take the following forms:

$$P = E\varepsilon_0 \int_A dA - E\frac{d^2 v}{dx^2}\int_A ydA$$

$$M = -E\varepsilon_0 \int_A ydA + E\frac{d^2 v}{dx^2}\int_A y^2 dA \tag{4.5}$$

Figure 4.2 Positive directions for axial force, shear force, and bending moment

In the above equation, ε_0 and the curvature terms are outside the integral because they are a function of the x-coordinate only. Since the choice of the beam axis (x-axis) is such that it passes through the centroid of the cross-section, the first moment of the area, $\int y\, dA$, vanishes and we can recognize the second moment of inertia of the cross-section, $I = \int y^2\, dA$, in the expression for bending moment M in the above equation. Now the expressions for P and M take the following forms:

$$P = EA\varepsilon_0$$

$$M = EI\frac{d^2 v}{dx^2} \tag{4.6}$$

where A and I are, respectively, the area and moment of inertia of the cross-section. It should be noted that the moment of inertia I is about the z-axis passing through the centroid of the cross-section, which is usually denoted by I_{zz} or I_z. The terms EA and EI are called, respectively, the *axial rigidity* and *flexural rigidity* of the beam cross-section. We will assume the beam does not have any net axial force, i.e., $P = 0$. Then, from Eq. (4.6), $\varepsilon_0 = 0$. Thus, the only equation we need is $M = EI(d^2 v/dx^2)$. The second part of Eq. (4.6) will be called the *beam constitutive relation*, or the *moment-curvature relation*. The relationship between the bending moment and the curvature is linear with the proportionality constant of flexural rigidity.

Since P and M are derived from the stress σ_{xx}, their sign conventions are similar to those of stresses rather than those of forces and couples. Positive P, V_y, and M are illustrated in Figure 4.2. Note that V_y is the transverse shear force acting on the beam cross-section.

A beam can be subjected to concentrated forces and couples, F_i and C_i, and distributed transverse force $p(x)$, as shown in Figure 4.3. Note that F and p are considered positive when they act in the positive y direction, whereas a counterclockwise couple is considered positive. The common units of the distributed force p are N/m and lb/in.

Another set of equations that will complement the moment-curvature relation of Eq. (4.6) are the beam equilibrium equations. Consider the free-body diagram of the infinitesimal beam shown in Figure 4.3. The shear force acting at a cross-section is denoted by V_y. Force equilibrium in the y direction requires

$$\sum f_y = 0 \Rightarrow p(x)dx + \left(V_y + \frac{dV_y}{dx}dx\right) - V_y = 0$$

or

$$\boxed{\frac{dV_y}{dx} = -p(x)} \tag{4.7}$$

Similarly, taking moments about the z-axis passing through the right face of the element yields

$$-M + \left(M + \frac{dM}{dx}dx\right) - (pdx)\frac{dx}{2} + V_y dx = 0$$

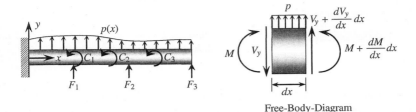

Free-Body-Diagram

Figure 4.3 Equilibrium of infinitesimal beam section under various loadings

or

$$V_y = -\frac{dM}{dx} \tag{4.8}$$

In Eq. (4.8), $(pdx)dx/2$ term is ignored because it is a higher-order term. The equations in the boxes in Eqs. (4.7) and (4.8) are the equilibrium equations of the beam. Combining these two equations with the beam constitutive relation $M = EI(d^2v/dx^2)$, we obtain the governing differential equation of the beam

$$EI\frac{d^4v}{dx^4} = p(x) \tag{4.9}$$

The above equation is a fourth-order differential equation, and as discussed in the previous chapter, it requires four boundary conditions [see Eq. (3.19)].

The stresses in the beam cross-section can be determined from the bending moment and shear force resultants. The expression for stress σ_{xx} in the absence of axial force P can be obtained from Eq. (4.3) as

$$\sigma_{xx} = -Ey\frac{d^2v}{dx^2} \tag{4.10}$$

Substituting for the curvature from Eq. (4.6), we obtain

$$\sigma_{xx}(x, y) = -\frac{M(x)y}{I} \tag{4.11}$$

Because of the Euler-Bernoulli beam theory assumptions that $v(x)$ is independent of y and the assumed form of $u(x, y)$ in Eq. (4.1), we obtain the shear strain γ_{xy} as

$$\gamma_{xy} = \frac{\partial u}{\partial y} + \frac{\partial v}{\partial x} = -\frac{\partial v}{\partial x} + \frac{\partial v}{\partial x} = 0 \tag{4.12}$$

That is, the Euler-Bernoulli beam theory predicts zero shear strain. However, we know that a beam cross-section is subjected to shear stresses, which results in the transverse shear force resultant V_y, as derived in Eq. (4.8). According to the theory of elasticity, there are non-zero shear strains in a beam, but they are small compared to the normal strains. The average shear stress in a cross-section is given by V_y/A, although the maximum shear stress depends on the cross-sectional geometry. For example, in a rectangular cross-section, the shear stress has a parabolic variation through the thickness with maximum value at the center equal to 1.5 times the average shear stress.

We will use the principle of minimum potential energy, Rayleigh-Ritz method to be specific, to derive the finite element equations. Referring to Chapter 3, the potential energy is defined as

$$\Pi = U + V \tag{4.13}$$

where U is the strain energy and V is the potential energy of external forces. In the Rayleigh-Ritz method, the deflection of the beam is expressed in terms of unknown coefficients. The objective is then to represent the potential energy in terms of these coefficients and then to differentiate it with respect to the coefficients in order to minimize the potential energy.

The energy method requires an expression for strain energy in the beam, which is derived as follows. The strain energy density at a point in the beam is given by

$$U_0 = \frac{1}{2}\sigma_{xx}\varepsilon_{xx} = \frac{1}{2}E(\varepsilon_{xx})^2 = \frac{1}{2}E\left(-y\frac{d^2v}{dx^2}\right)^2 = \frac{1}{2}Ey^2\left(\frac{d^2v}{dx^2}\right)^2 \tag{4.14}$$

where we have substituted for strain from Eq. (4.2) with $\varepsilon_0 = 0$. In the context of beams, we define another strain energy density term, which is called the strain energy per unit length of the beam, U_L. It is derived by integrating U_0 over the entire cross-section at a given x:

$$U_L(x) = \int_A U_0(x, y, z)\, dA \tag{4.15}$$

Substituting for U_0 from Eq. (4.14) in Eq. (4.15) yields an expression for strain energy per unit length of the beam:

$$U_L(x) = \int_A \frac{1}{2}Ey^2\left(\frac{d^2v}{dx^2}\right)^2 dA = \frac{1}{2}E\left(\frac{d^2v}{dx^2}\right)^2\int_A y^2 dA$$

or

$$\boxed{U_L(x) = \frac{1}{2}EI\left(\frac{d^2v}{dx^2}\right)^2} \tag{4.16}$$

The units for U_L are J/m or N or lb · in/in. By substituting the moment-curvature relation of Eq. (4.6), we obtain an expression for U_L in terms of M as $U_L = M^2/2EI$.

The strain energy U in the beam can be derived as

$$U = \int_0^L U_L(x)\, dx = \frac{1}{2}\int_0^L EI\left(\frac{d^2v}{dx^2}\right)^2 dx \tag{4.17}$$

Figure 4.3 shows positive directions of concentrated forces, couples, and distributed loads. Using these notations, the potential energy of external forces and moments can be represented by

$$V = -\int_0^L p(x)v(x)\, dx - \sum_{i=1}^{N_F} F_i v(x_i) - \sum_{i=1}^{N_C} C_i \frac{dv(x_i)}{dx} \tag{4.18}$$

where N_F and N_C are, respectively, the number of concentrated forces and couples applied to the beam.

Thus, the total potential energy in Eq. (4.13) can be represented using the transverse deflection and slope (derivative of the deflection), as

$$\Pi = U + V$$
$$= \frac{1}{2}\int_0^L EI\left(\frac{d^2v}{dx^2}\right)^2 dx - \int_0^L p(x)v(x)\,dx - \sum_{i=1}^{N_F} F_i v(x_i) - \sum_{i=1}^{N_C} C_i \frac{dv(x_i)}{dx} \qquad (4.19)$$

The principle of minimum potential energy in Chapter 3 states that the beam is in equilibrium when the potential energy has its minimum value. We will present the Rayleigh-Ritz method, followed by the finite element method.

4.2   RAYLEIGH-RITZ METHOD

As discussed in Chapter 3, the Rayleigh-Ritz method can be used for continuous systems. In the Rayleigh-Ritz method, a continuous system is approximated as a discrete system with a finite number of DOFs. This is accomplished by approximating the displacements by a function containing a finite number of coefficients to be determined. The total potential energy is then evaluated in terms of the unknown coefficients. Then the principle of minimum potential energy is applied to determine the best set of coefficients by minimizing the total potential energy with respect to the coefficients. The solution thus obtained may not be exact. It is the best solution from among the family of solutions that can be obtained from the assumed displacement functions. In the following, we demonstrate the method for beam problems. The steps involved in solving the beam problem using the Rayleigh-Ritz method are as follows.

1. Assume a deflection shape for the beam in the following form: $v(x) = c_1 f_1(x) + c_2 f_2(x) \ldots \ldots + c_n f_n(x)$, where c_i are coefficients to be determined and $f_i(x)$ are known basis functions. The deflection curve $v(x)$ **must** satisfy the displacement boundary conditions, e.g., deflection $v = 0$ or slope $dv/dx = 0$.

2. Determine the strain energy U in the beam using the formula in Eq. (4.17) in terms of c_i.

3. Find the potential energy of external forces V using formulas of types given in Eq. (4.18).

4. The total potential energy is obtained as $\Pi(c_1, c_2, \ldots c_n) = U + V$.

5. Apply the principle of minimum potential energy to determine the coefficients $c_1, c_2, \ldots c_n$.

EXAMPLE 4.1   *Rayleigh-Ritz Method for a Simply Supported Beam*

Consider a simply supported beam of length L subjected to a uniformly distributed transverse load $p(x) = p_0$, as shown in Fig. 4.4. Use the Rayleigh-Ritz method to determine the transverse deflection $v(x)$ of the beam. Assume the flexural rigidity of beam cross-section is EI.

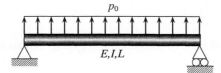

Figure 4.4 Simply supported beam under uniformly distributed load

SOLUTION   As in the case of uniaxial bar, we assume the deflection of the beam using some convenient functions that satisfy the displacement boundary conditions. In the case of the simply supported beam, the deflection must be zero at both ends of the beam. We will deviate from typical polynomial forms and assume a sinusoidal function, as follows:

$$v(x) = C \sin \frac{\pi x}{L} \tag{4.20}$$

where C is the coefficient to be determined by minimizing the total potential energy. Here we have approximated the beam as a one DOF system. The strain energy of the beam is given in Eq. (4.17). Substituting for the deflection from Eq. (4.20) into Eq. (4.17), we obtain

$$U = \frac{C^2 EI \pi^4}{4L^3} \tag{4.21}$$

The potential energy of the external forces is derived as

$$V = -\int_0^L p(x) v(x) dx = -\int_0^L p_0 C \sin \frac{\pi x}{L} dx = -\frac{2 p_0 L}{\pi} C \tag{4.22}$$

Then the total potential energy is derived as

$$\Pi = U + V = \frac{EI \pi^4}{4L^3} C^2 - \frac{2 p_0 L}{\pi} C \tag{4.23}$$

The principle of minimum potential energy requires $d\Pi/dC = 0$. That is,

$$\frac{d\Pi}{dC} = \frac{EI \pi^4}{2L^3} C - \frac{2 p_0 L}{\pi} = 0 \Rightarrow C = \frac{4 p_0 L^4}{EI \pi^5} \tag{4.24}$$

The maximum deflection at the center is equal to the value of the coefficient C, which can be written as $C = P_0 L^4 / 76.5 EI$. The exact deflection is given by $C = P_0 L^4 / 76.8 EI$. The bending moment and the shear force are derived from the approximate deflection using the following relations:

$$
\begin{aligned}
M(x) &= EI \frac{d^2 v}{dx^2} = -EIC \frac{\pi^2}{L^2} \sin \frac{\pi x}{L} = -\frac{4 p_0 L^2}{\pi^3} \sin \frac{\pi x}{L} \\
V_y(x) &= -EI \frac{d^3 v}{dx^3} = -EIC \frac{\pi^3}{L^3} \cos \frac{\pi x}{L} = -\frac{4 p_0 L}{\pi^2} \cos \frac{\pi x}{L}
\end{aligned}
\tag{4.25}
$$

The exact solution for this problem can be found in any mechanics of materials book.[2] The exact expressions for deflection, bending moment, and shear force are as follows:

$$
\begin{aligned}
v(x) &= \frac{1}{EI} \left(\frac{p_0 L^3}{24} x - \frac{p_0 L}{12} x^3 + \frac{p_0}{24} x^4 \right) \\
M(x) &= -\frac{p_0 L}{2} x + \frac{p_0}{2} x^2 \\
V_y(x) &= \frac{p_0 L}{2} - p_0 x
\end{aligned}
\tag{4.26}
$$

Let us define non-dimensional deflection, bending moment, and shear force as

$$\bar{x} = \frac{x}{L}, \quad \bar{v} = \frac{384 EI}{5 p_0 L^4} v, \quad \overline{M} = \frac{1}{p_0 L^2} M, \quad \overline{V} = \frac{1}{p_0 L} V_y \tag{4.27}$$

The approximate solution is compared with the exact solution by plotting the non-dimensional values of deflection and force and moment resultants over the length of the beam in Figure 4.5(a)–(c). It turns out that the approximate deflection is close to the exact one. However, the

---

[2] A. P. Boresi and R. J. Schmidt, *Advanced Mechanics of Materials*, 6th ed., John Wiley & Sons, Hoboken, NJ, 2003

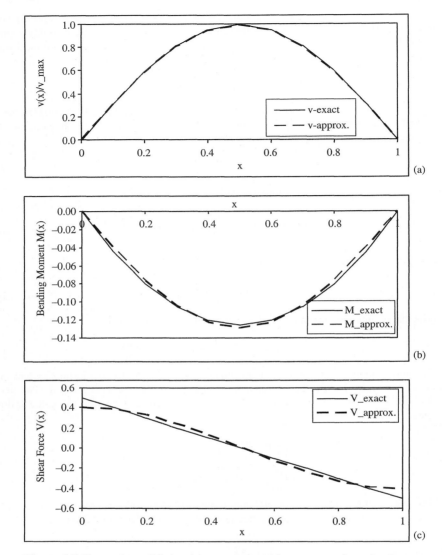

Figure 4.5 Comparison of finite element results with exact ones for a simply supported beam; (a) deflection, (b) bending moment, and (c) shear force

error increases for the bending moment as well as the shear force. Note that the exact shear force is linear but the approximate shear force has cosine term because the deflection is approximated by a sine function.

EXAMPLE 4.2   *Rayleigh-Ritz Method for a Cantilevered Beam*

Consider a cantilevered beam subjected to a distributed load $p(x) = -p_0$, a transverse force F, and a couple C at the right end. Use the Rayleigh-Ritz method to obtain the deflection, shear force, and bending moment distribution along the length of the beam. Assume the following numerical values: $E = 100\text{GPa}, I = 10^{-7}\text{m}^4, L = 1\text{m}, p_0 = 300\text{N/m}, F = 500\text{N}, C = 100\text{N} \cdot \text{m}$

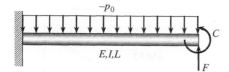

Figure 4.6 Cantilevered beam under uniformly distributed load

SOLUTION We will assume a polynomial in x to approximate the deflection $v(x)$. We will use a cubic polynomial, $v(x) = a + bx + c_1x^2 + c_2x^3$, to describe the beam deflection. The essential boundary conditions of the beam are as follows: at $x = 0$, $v(0) = 0$, $dv(0)/dx = 0$. These conditions must be satisfied by the assumed deflections. This requires that $a = b = 0$. Hence, the assumed deflection takes the form

$$v(x) = c_1x^2 + c_2x^3 \tag{4.28}$$

Using the expression for strain energy in a beam given in Eq. (4.17), we obtain

$$U = \frac{EI}{2}\int_0^L (2c_1 + 6c_2x)^2 dx \tag{4.29}$$

It is not necessary to perform the integration in Eq. (4.29) at this stage. Actually, we require the derivatives of U with respect to c_i, and it can be obtained by performing the differentiation under the integral as shown below:

$$\frac{\partial U}{\partial c_1} = 2EI\int_0^L (2c_1 + 6c_2x)dx = EI(4Lc_1 + 6L^2c_2)$$
$$\frac{\partial U}{\partial c_2} = 6EI\int_0^L (2c_1 + 6c_2x)xdx = EI(6L^2c_1 + 12L^3c_2) \tag{4.30}$$

The potential energy of the external forces can be derived as follows:

$$V(c_1, c_2) = -\int_0^L (-p_0)v(x)dx - Fv(L) - C\frac{dv}{dx}(L)$$
$$= -\int_0^L (-p_0)(c_1x^2 + c_2x^3)dx - F(c_1L^2 + c_2L^3) - C(2c_1L + 3c_2L^2) \tag{4.31}$$
$$= c_1\left(\frac{p_0L^3}{3} - FL^2 - 2CL\right) + c_2\left(\frac{p_0L^4}{4} - FL^3 - 3CL^2\right)$$

The principle of minimum potential energy requires Π be minimum with respect to the coefficient c_i. Thus, the potential energy is differentiated with respect to the two unknown coefficients to obtain

$$\frac{\partial\Pi}{\partial c_1} = \frac{\partial U}{\partial c_1} + \frac{\partial V}{\partial c_1} = 0$$
$$\frac{\partial\Pi}{\partial c_2} = \frac{\partial U}{\partial c_2} + \frac{\partial V}{\partial c_2} = 0 \tag{4.32}$$

Substituting from Eqs. (4.30) and (4.31) into Eq. (4.32), we obtain two equations for c_i:

$$EI(4Lc_1 + 6L^2c_2) = -\frac{p_0L^3}{3} + FL^2 + 2CL$$
$$EI(6L^2c_1 + 12L^3c_2) = -\frac{p_0L^4}{4} + FL^3 + 3CL^2 \tag{4.33}$$

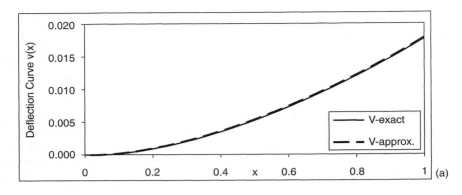

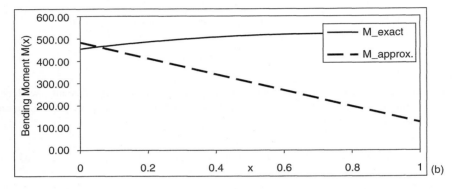

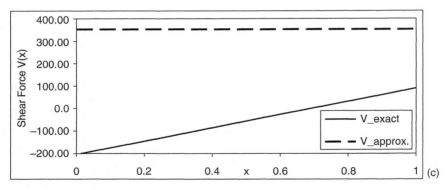

Figure 4.7 Comparison of finite element results with exact ones for a cantilevered beam; (a) deflection, (b) bending moment, and (c) shear force

Substituting the numerical values for the beam properties and loads, we obtain

$$10^4(4c_1 + 6c_2) = 600$$
$$10^4(6c_1 + 12c_2) = 725 \tag{4.34}$$

The solution is: $c_1 = 23.75 \times 10^{-3}$, $c_2 = -5.833 \times 10^{-3}$. Substituting for c_i in Eq. (4.28), we obtain the beam deflection $v(x)$ as

$$v(x) = 10^{-3}(23.75x^2 - 5.833x^3) \tag{4.35}$$

The exact solution for deflection is given by

$$v(x) = \frac{1}{24EI}(5400x^2 - 800x^3 - 300x^4)$$

The deflection, bending moment, and shear force resultants are compared with the exact solution in Figure 4.7. It turns out that the exact solution is a quartic polynomial, while the approximate solution is cubic. This explains the error in the deflection curve in Figure 4.7(a). If a fourth-order polynomial is used in Eq. (4.28), the approximate solution will be the same as the exact solution. The errors usually increase for the bending moment and the shear force, as they are derivatives of the deflection curve.

In the following sections, we develop the finite element version of the Rayleigh-Ritz method.

4.3   FINITE ELEMENT INTERPOLATION

The finite element method is different from the Rayleigh-Ritz method in the aspect that the approximation is performed in the element, rather than the entire structure. By using the property of interpolation functions, it is trivial to satisfy the essential boundary conditions in this approach. In this section, we will present the interpolation method similar to that in Chapter 3, but specialized for the beam element. This interpolation will be used in approximating the potential energy in the following section.

Consider a beam shown in Figure 4.3. It is subjected to a distributed force and several concentrated or point forces and couples. Our goal is to determine the deflection curve $v(x)$ of the beam, the bending moment distribution $M(x)$, and shear force resultant $V_y(x)$ along the length of the beam. The first step in finite element analysis is to divide the beam into a number of elements. An element is connected to the other element at nodes. Concentrated forces and couples can be applied only at nodes; that means that there should be nodes at points where concentrated forces and couples are applied. In this text, we will consider a beam element that consists of two end nodes. In general, however, it is possible to have more than two nodes in an element. The positive directions of applied forces and couples are shown in Figure 4.8. The distributed load $p(x)$ can vary along the x-axis. However, we will only consider the cases of either constant or linear distribution.

The DOF in beam elements are the transverse deflection v and the rotation θ of the cross-section that is also equal to the slope dv/dx. The transverse deflection is positive in the positive y direction, whereas a counter-clockwise rotation of the cross-section is considered positive. Consider a typical element shown in Figure 4.9. Our goal is to interpolate the deflection at any point on the element in terms of nodal DOFs $v_1, \theta_1, v_2,$ and θ_2. We first define a vector of nodal DOFs, as

$$\{\mathbf{q}\} = \{\, v_1 \quad \theta_1 \quad v_2 \quad \theta_2 \,\}^T \tag{4.36}$$

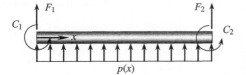

Figure 4.8 Positive directions for forces and couples in a beam element

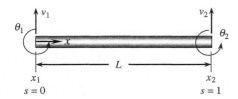

Figure 4.9 Nodal displacements and rotations for the beam element

It is convenient to define a parameter s, which varies from 0 to 1 within the element so that a unified derivation is possible for an element of any length. The parameter s can be defined as (see Figure 4.9)

$$s = \frac{x - x_1}{L}, \quad ds = \frac{1}{L}dx, \quad dx = Lds, \quad \frac{ds}{dx} = \frac{1}{L} \tag{4.37}$$

where x_1 is the x-coordinate of Node 1 (first node of the element). Thus, the deflection curve can be written in terms of parameter s; i.e., $v(s)$.

Our goal is to interpolate the deflection $v(s)$ in terms of the nodal DOFs. Since the beam element has four nodal values, it is appropriate to use a cubic function to approximate the deflection:

$$v(s) = a_0 + a_1 s + a_2 s^2 + a_3 s^3 \tag{4.38}$$

where a_0, a_1, a_2, and a_3 are the constants to be determined in terms of the nodal DOFs. In addition to the deflection, the slope or rotation θ is also included as the nodal values. Thus, it is necessary to differentiate $v(s)$ with respect to x. However, $v(s)$ in Eq. (4.38) is expressed in terms of parameter s. The expression for the slope can be obtained using the chain rule of differentiation, as

$$\frac{dv}{dx} = \frac{dv}{ds}\frac{ds}{dx} = \frac{1}{L}\left(a_1 + 2a_2 s + 3a_3 s^2\right) \tag{4.39}$$

Note that the relation $ds/dx = 1/L$ is from Eq. (4.37).

The expression of the strain energy in Eq. (4.16) contains the second derivative of the transverse displacement, whose expression can be obtained by differentiating the above equation as

$$\frac{d^2 v}{dx^2} = \frac{d}{ds}\left(\frac{dv}{dx}\right)\frac{ds}{dx} = \frac{1}{L^2}\left(2a_2 + 6a_3 s\right) \tag{4.40}$$

Now the four coefficients $(a_0, a_1, a_2, \text{ and } a_3)$ need to be expressed in terms of nodal values. By definition, the vertical displacement at the left end of the element ($s = 0$) is v_1 and the slope is θ_1. By evaluating Eqs. (4.38) and (4.39) at $s = 0$, we can calculate a_0 and a_1 as

$$v(0) = v_1 = a_0$$

$$\frac{dv}{dx}(0) = \theta_1 = \frac{a_1}{L}, \quad a_1 = L\theta_1$$

In the same way, we can evaluate (4.38) and (4.39) at $s = 1$ to obtain the following simultaneous system equations:

$$v(1) = v_2 = v_1 + L\theta_1 + a_2 + a_3$$

$$\frac{dv}{dx}(1) = \theta_2 = \frac{1}{L}(L\theta_1 + 2a_2 + 3a_3)$$

By solving the above two equations for a_2 and a_3, we have

$$a_2 = -3v_1 - 2L\theta_1 + 3v_2 - L\theta_2$$
$$a_3 = 2v_1 + L\theta_1 - 2v_2 + L\theta_2$$

Thus, all unknown coefficients are expressed in terms of nodal values. By substituting these coefficients into Eq. (4.38), we have

$$
\begin{aligned}
v(s) = {} & v_1 + L\theta_1 s + (-3v_1 - 2L\theta_1 + 3v_2 - L\theta_2)s^2 \\
& + (2v_1 + L\theta_1 - 2v_2 + L\theta_2)s^3
\end{aligned}
\tag{4.41}
$$

Since our goal is to express $v(s)$ in terms of nodal values, Eq. (4.41) can be rearranged to obtain

$$
\begin{aligned}
v(s) = {} & (1 - 3s^2 + 2s^3)v_1 \\
& + L(s - 2s^2 + s^3)\theta_1 \\
& + (3s^2 - 2s^3)v_2 \\
& + L(-s^2 + s^3)\theta_2
\end{aligned}
\tag{4.42}
$$

An important concept in finite element approximation is the definition of the **shape functions**, which are the coefficients of the nodal values. The deflection $v(s)$ in (4.42) can be written in the form $v(s) = N_1 v_1 + N_2 \theta_1 + N_3 v_2 + N_4 \theta_2$. The coefficients of the nodal DOFs in Eq. (4.42) are called the *shape functions* of the beam element:

$$
\boxed{
\begin{aligned}
N_1(s) &= 1 - 3s^2 + 2s^3 \\
N_2(s) &= L(s - 2s^2 + s^3) \\
N_3(s) &= 3s^2 - 2s^3 \\
N_4(s) &= L(-s^2 + s^3)
\end{aligned}
}
\tag{4.43}
$$

Equation (4.41) can then be written in matrix form as

$$
v(s) = [\,N_1(s) \quad N_2(s) \quad N_3(s) \quad N_4(s)\,]
\begin{Bmatrix} v_1 \\ \theta_1 \\ v_2 \\ \theta_2 \end{Bmatrix}
$$

or,

$$
\boxed{v(s) = \lfloor\,\mathbf{N}\,\rfloor\{\mathbf{q}\}}\,,
\tag{4.44}
$$

where $\lfloor\,\mathbf{N}\,\rfloor$ is a row vector of shape functions of the beam element. Equation (4.44) approximates the deflection of the beam using the values of deflections and slopes at the two nodes. For example, evaluation of Eq. (4.44) at $s = 1/2$ will provide the beam deflection at the center of the element.

Figure 4.10 shows the plot of the shape functions. These shape functions are also called the *Hermite polynomials*. Note that when $s = 0$, all shape functions are zero except for N_1, which is equal to unity. Thus, Eq. (4.44) yields $v(0) = v_1$, which is the desired result. When $s = 1$, the only non-zero shape function is N_3. Thus, the approximation in Eq. (4.44) yields $v(1) = v_2$.

Note that the interpolation of the beam deflection in Eq. (4.44) is valid within an element. If the beam consists of more than one element, the interpolation must be

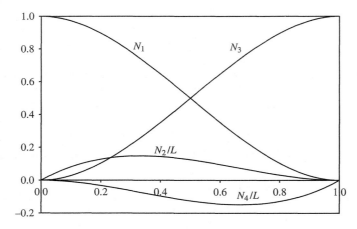

Figure 4.10 Shape functions of the beam element

performed in each element. Two adjacent elements will have continuous deflection and slope, as they share the nodal values.

The second derivative of the deflection in Eq. (4.40) can be derived as

$$\frac{d^2 v}{dx^2} = \frac{1}{L^2}\frac{d^2 v}{ds^2} = \frac{1}{L^2}[-6+12s, \; L(-4+6s), \; 6-12s, \; L(-2+6s)]\begin{Bmatrix} v_1 \\ \theta_1 \\ v_2 \\ \theta_2 \end{Bmatrix}$$

or

$$\boxed{\frac{d^2 v}{dx^2} = \frac{1}{L^2}\underset{1\times 4}{\lfloor \mathbf{B} \rfloor}\underset{4\times 1}{\{\mathbf{q}\}}}\tag{4.45}$$

where the row vector $\lfloor \mathbf{B} \rfloor$ is, in general, called the *strain-displacement vector*. In the context of beams, it relates the curvature of the beam to the nodal displacements. Note that the vector $\lfloor \mathbf{B} \rfloor$ is a linear function of the parameter s. Thus, the curvature varies in a linear fashion within an element. The reader can verify that the curvature term in (4.45) can also be written as

$$\frac{d^2 v}{dx^2} = \frac{1}{L^2}\underset{1\times 4}{\lfloor \mathbf{q}^T \rfloor}\underset{4\times 1}{\{\mathbf{B}^T\}}\tag{4.46}$$

Equation (4.45) provides some information about the quality of interpolation. If the loads acting on a beam result in a constant or linear variation of curvature or bending moment, then the interpolation in Eq. (4.45) will represent it accurately. However, when a problem requires a higher order variation of curvature (bending moment) with respect to x, then the interpolation is only an approximation. In such a case, several beam elements will be required to approximate the higher order variation of bending moment with a reasonable accuracy.

Using Eq. (4.45), the bending moment and shear force in the beam element can also be calculated in terms of nodal DOFs, as

$$M(s) = EI\frac{d^2 v}{dx^2} = \frac{EI}{L^2}\lfloor \mathbf{B} \rfloor\{\mathbf{q}\}\tag{4.47}$$

$$V_y = -\frac{dM}{dx} = -EI\frac{d^3 v}{dx^3} = \frac{EI}{L^3}[-12 \quad -6L \quad 12 \quad -6L]\{\mathbf{q}\}\tag{4.48}$$

It is interesting to note that the bending moment is a linear function of s, or x, while the shear force is constant throughout the element. This is because the assumed deflection is a cubic function as in Eq. (4.38). Since the stress is proportional to the bending moment as shown in Eq. (4.11), the maximum stress occurs at the location of the maximum bending moment. Since the bending moment varies linearly, the maximum stress always occurs at one of the nodes of the beam element. However, caution is required because the bending moments between two adjacent elements are discontinuous at the node.

Equation (4.45) can be used to approximate the strain energy in the following section.

EXAMPLE 4.3   *Interpolation in a Beam Element*

Consider a cantilevered beam as shown in Figure 4.11. The beam is approximated using one beam finite element. The nodal values of the beam element are given as $\{\mathbf{q}\} = \{0,\ 0,\ -0.1,\ -0.2\}^T$. Calculate the deflection and slope at the midpoint of the beam. Assume $L = 1$ m.

SOLUTION   The value of the parameter s in the middle of the element is $1/2$. Thus, the shape functions in Eq. (4.43) are evaluated with $s = 1/2$:

$$N_1\left(\tfrac{1}{2}\right) = \frac{1}{2}, \quad N_2\left(\tfrac{1}{2}\right) = \frac{L}{8}, \quad N_3\left(\tfrac{1}{2}\right) = \frac{1}{2}, \quad N_4\left(\tfrac{1}{2}\right) = -\frac{L}{8}$$

Thus, from Eq. (4.44), the vertical deflection at the midpoint of the element is

$$
\begin{aligned}
v\left(\tfrac{1}{2}\right) &= N_1\left(\tfrac{1}{2}\right)v_1 + N_2\left(\tfrac{1}{2}\right)\theta_1 + N_3\left(\tfrac{1}{2}\right)v_2 + N_4\left(\tfrac{1}{2}\right)\theta_2 \\
&= \frac{1}{2} \times 0 + \frac{L}{8} \times 0 + \frac{1}{2} \times v_2 - \frac{L}{8} \times \theta_2 \\
&= \frac{v_2}{2} - \frac{L\theta_2}{8} \\
&= -0.025
\end{aligned}
$$

Next, the slope of the element is defined as $\theta = dv/dx$. Using (4.39), the slope can be expressed as

$$
\begin{aligned}
\frac{dv}{dx} = \frac{1}{L}\frac{dv}{ds} &= \frac{1}{L}\left(v_1\frac{dN_1}{ds} + \theta_1\frac{dN_2}{ds} + v_2\frac{dN_3}{ds} + \theta_2\frac{dN_4}{ds}\right) \\
&= v_1\frac{1}{L}(-6s + 6s^2) + \theta_1\left(1 - 4s + 3s^2\right) \\
&\quad + v_2\frac{1}{L}(6s - 6s^2) + \theta_2(-2s + 3s^2)
\end{aligned}
$$

The slope in the middle of the beam can be obtained by substituting $s = 1/2$ in the above equation. Then

$$
\begin{aligned}
\theta\left(\tfrac{1}{2}\right) &= -\frac{3}{2L}v_1 - \frac{1}{4}\theta_1 + \frac{3}{2L}v_2 - \frac{1}{4}\theta_2 \\
&= -0.1
\end{aligned}
$$

Figure 4.11 Cantilevered beam element with nodal displacements

4.4   FINITE ELEMENT EQUATION FOR THE BEAM ELEMENT

As you may recall, one of the steps in finite element analysis is to express the strain energy of the solid in terms of nodal DOFs. In this section, we will derive the finite element equation using the principle of minimum potential energy. Let us consider a beam with the total length of L_T. The beam is divided into *NEL* number of beam elements with equal length L. The elements do not have to be of the same length, but the assumption makes the explanation simple. We further assume that the cross-sectional area remains constant within an element. The beam is under concentrated forces and couples at the nodes and distributed load $p(x)$.

The strain energy in a beam can be formally written in terms of strain energy per unit length as

$$U = \int_0^{L_T} U_L(x)dx = \sum_{e=1}^{NEL} \int_{x_1^{(e)}}^{x_2^{(e)}} U_L(x)dx = \sum_{e=1}^{NEL} U^{(e)} \tag{4.49}$$

where $U^{(e)}$ is the strain energy of Element e. The integration is performed over each beam element and summed over *NEL* number of elements. Note that $x_1^{(e)}$ and $x_2^{(e)}$, respectively, are the x-coordinates of the first and second nodes of Element e. Substituting for U_L from Eq. (4.16), we obtain

$$U^{(e)} = EI \int_{x_1^{(e)}}^{x_2^{(e)}} \frac{1}{2} \left(\frac{d^2 v}{dx^2}\right)^2 dx = \frac{EI}{L^3} \int_0^1 \frac{1}{2} \left(\frac{d^2 v}{ds^2}\right)^2 ds \tag{4.50}$$

In the above equation, we have used the relation $dx = Lds$; see Eq. (4.37). The expression of the strain energy in Eq. (4.50) contains the second-order derivative of deflection $v(s)$. Using the interpolation in Eqs. (4.45) and (4.46), we can write the second-order derivative term as

$$\left(\frac{d^2 v}{ds^2}\right)^2 = \left(\frac{d^2 v}{ds^2}\right)\left(\frac{d^2 v}{ds^2}\right) = \{\,\mathbf{q}^{(e)}\,\}^T \underset{1\times 4}{\lfloor \mathbf{B} \rfloor}^T \underset{1\times 4}{\lfloor \mathbf{B} \rfloor} \underset{1\times 4}{\{\,\mathbf{q}^{(e)}\,\}}$$

Note that for the first curvature term we have used $\{\mathbf{q}\}^T \lfloor \mathbf{B} \rfloor^T$ and for the second $\lfloor \mathbf{B} \rfloor \{\mathbf{q}\}$. Substituting the above relation in Eq. (4.50), the strain energy of Element e is derived as

$$U^{(e)} = \frac{1}{2}\{\,\mathbf{q}^{(e)}\,\}^T \left[\frac{EI}{L^3} \int_0^1 \lfloor \mathbf{B} \rfloor^T \lfloor \mathbf{B} \rfloor \, ds\right]^{(e)} \{\,\mathbf{q}^{(e)}\,\}$$

$$= \frac{1}{2}\{\,\mathbf{q}^{(e)}\,\}^T [\mathbf{k}^{(e)}]\{\,\mathbf{q}^{(e)}\,\} \tag{4.51}$$

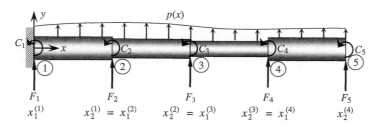

Figure 4.12 Finite element models using four beam elements

where $[\mathbf{k}^{(e)}]$ is the element stiffness matrix of the beam finite element. After integrating, the stiffness matrix can be derived as

$$[\mathbf{k}^{(e)}] = \frac{EI}{L^3} \int_0^1 \begin{bmatrix} -6 + 12s \\ L(-4 + 6s) \\ 6 - 12s \\ L(-2 + 6s) \end{bmatrix} [\, -6 + 12s \quad L(-4 + 6s) \quad 6 - 12s \quad L(-2 + 6s) \,] \, ds$$

or

$$[\mathbf{k}^{(e)}] = \frac{EI}{L^3} \begin{bmatrix} 12 & 6L & -12 & 6L \\ 6L & 4L^2 & -6L & 2L^2 \\ -12 & -6L & 12 & -6L \\ 6L & 2L^2 & -6L & 4L^2 \end{bmatrix} \tag{4.52}$$

which is a symmetric 4×4 matrix. Note that the element stiffness matrix is proportional to the flexural rigidity EI and inversely proportional to L^3. The stiffness matrices of all the elements can then be calculated from the element properties. The strain energy in the beam can then be obtained by summing strain energies of the individual elements as in Eq. (4.49):

$$U = \sum_{e=1}^{NEL} U^{(e)} = \frac{1}{2} \sum_{e=1}^{NEL} \{\mathbf{q}^{(e)}\}^T [\mathbf{k}^{(e)}] \{\mathbf{q}^{(e)}\}$$

or,

$$U = \frac{1}{2} \{\mathbf{Q}_s\}^T [\mathbf{K}_s] \{\mathbf{Q}_s\} \tag{4.53}$$

where $\{\mathbf{Q}_s\}$ is the column matrix of all DOFs in the beam and $[\mathbf{K}_s]$ is the *structural stiffness matrix* obtained by assembling the element stiffness matrices. The assembly procedure, which is similar to that of uniaxial bar and truss elements, will be illustrated in the examples to follow.

EXAMPLE 4.4   *Assembly of Two Beam Elements*

Consider a stepped beam structure modeled using two beam elements. The flexural rigidity of Elements 1 and 2 are, respectively, $2EI$ and EI. Construct element stiffness matrices and assemble them to build the structural stiffness matrix.

SOLUTION   Since each node has two DOFs, the vector of total DOFs are defined by

$$\{\mathbf{Q}_s\}^T = \{\, v_1 \quad \theta_1 \quad v_2 \quad \theta_2 \quad v_3 \quad \theta_3 \,\}$$

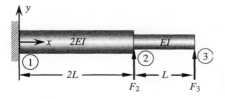

Figure 4.13 Finite element model of a stepped cantilevered beam

From Eq. (4.52), the element stiffness matrices of two beam elements can be obtained as

$$
[\mathbf{k}^{(1)}] = \frac{EI}{L^3}
\begin{array}{c}
\begin{array}{cccc} v_1 & \theta_1 & v_2 & \theta_2 \end{array} \\
\begin{bmatrix}
3 & 3L & -3 & 3L \\
3L & 4L^2 & -3L & 2L^2 \\
-3 & -3L & 3 & -3L \\
3L & 2L^2 & -3L & 4L^2
\end{bmatrix}
\begin{array}{c} v_1 \\ \theta_1 \\ v_2 \\ \theta_2 \end{array}
\end{array}
$$

$$
[\mathbf{k}^{(2)}] = \frac{EI}{L^3}
\begin{array}{c}
\begin{array}{cccc} v_2 & \theta_2 & v_3 & \theta_3 \end{array} \\
\begin{bmatrix}
12 & 6L & -12 & 6L \\
6L & 4L^2 & -6L & 2L^2 \\
-12 & -6L & 12 & -6L \\
6L & 2L^2 & -6L & 4L^2
\end{bmatrix}
\begin{array}{c} v_2 \\ \theta_2 \\ v_3 \\ \theta_3 \end{array}
\end{array}
$$

Corresponding row and column locations at the global DOFs are denoted in the above equation.

$$
[\mathbf{K}_s] = \frac{EI}{L^3}
\begin{bmatrix}
3 & 3L & -3 & 3L & 0 & 0 \\
3L & 4L^2 & -3L & 2L^2 & 0 & 0 \\
-3 & -3L & 15 & 3L & -12 & 6L \\
3L & 2L^2 & 3L & 8L^2 & -6L & 2L^2 \\
0 & 0 & -12 & -6L & 12 & -6L \\
0 & 0 & 6L & 2L^2 & -6L & 4L^2
\end{bmatrix}
$$

Note that the structural stiffness matrix $[\mathbf{K}_s]$ is singular. Applying the displacement boundary conditions is equivalent to deleting the first two rows and columns, to yield the global stiffness matrix $[\mathbf{K}]$.

The next step is to derive an expression for the potential energy of external forces. If there are only concentrated forces and couples acting on the beam, the potential energy V can be written as

$$
V = -\sum_{i=1}^{ND}(F_i v_i + C_i \theta_i) \tag{4.54}
$$

where F_i and C_i, respectively, are the transverse force and couple acting at Node i and the total number of nodes in the beam model is denoted by ND. The above expression for V can be written in a matrix form as

$$
V = -\lfloor v_1\, \theta_1\, v_2 \ldots \ldots \theta_{ND} \rfloor
\begin{Bmatrix}
F_1 \\ C_1 \\ F_2 \\ \vdots \\ C_{ND}
\end{Bmatrix}
= -\{\mathbf{Q}_s\}^T\{\mathbf{F}_s\} \tag{4.55}
$$

where $\{\mathbf{F}_s\}$ is the vector of nodal forces.

If there are distributed forces acting on the beam, then they have to be converted into equivalent nodal forces. Let us assume that the distributed load acting on the beam is given by $p(x)$. Then the potential energy of this load is given by

$$
V = -\int_0^{L_T} p(x)v(x)\, dx \tag{4.56}
$$

where L_T is the total length of the beam. The integral can be broken down to integrals over each element as

$$V = -\sum_{e=1}^{NEL} \int_{x_1^{(e)}}^{x_2^{(e)}} p(x)v(x)\,dx = \sum_{e=1}^{NEL} V^{(e)} \tag{4.57}$$

In the above equation, $V^{(e)}$ is the contribution to V by Element e, which can be derived as

$$-V^{(e)} = \int_{x_1^{(e)}}^{x_2^{(e)}} p(x)v(x)\,dx = L\int_0^1 p(s)v(s)\,ds \tag{4.58}$$

In the above, $p(s)$ should be expressed as a function of s using the change of variables given by Eq. (4.37). Now we will use the shape functions to express $v(s)$ within an element in terms of nodal displacements, using Eq. (4.44) to obtain

$$
\begin{aligned}
-V^{(e)} &= L^{(e)}\int_0^1 p(s)(v_1 N_1 + \theta_1 N_2 + v_2 N_3 + \theta_2 N_4)\,ds \\[2mm]
&= v_1\left(L^{(e)}\int_0^1 p(s)N_1\,ds\right) + \theta_1\left(L^{(e)}\int_0^1 p(s)N_2\,ds\right) \\[2mm]
&\quad + v_2\left(L^{(e)}\int_0^1 p(s)N_3\,ds\right) + \theta_2\left(L^{(e)}\int_0^1 p(s)N_4\,ds\right) \\[2mm]
&= v_1 F_1^{(e)} + \theta_1 C_1^{(e)} + v_2 F_2^{(e)} + \theta_2 C_2^{(e)}
\end{aligned} \tag{4.59}
$$

The terms inside the parentheses in the above equation are called *work equivalent loads* contributed by Element e and are denoted by $F_1^{(e)}$, $C_1^{(e)}$, $F_2^{(e)}$ and $C_2^{(e)}$. These loads can be calculated from the distributed load p, the shape functions, and element length L. One can note that a transverse load p results not only in two concentrated forces at the nodes but also results in couples acting at the nodes. If a node belongs to more than one element, then the contributions from all elements at that node must be added together along with any applied concentrated forces and couples.

EXAMPLE 4.5   ***Work-Equivalent Loads for a Uniformly Distributed Load***

Calculate the work-equivalent loads for a beam element of length L under uniformly distributed load $p(x) = p$.

SOLUTION   In the formula for the potential energy of the distributed load in Eq. (4.59), $p(x) = p$ can be moved out from the integral because it is a constant. Thus, calculation of the work-equivalent nodal forces involves integration of the shape functions.

$$F_1 = pL\int_0^1 N_1(s)\,ds = pL\int_0^1 (1 - 3s^2 + 2s^3)\,ds = \frac{pL}{2}$$

$$C_1 = pL\int_0^1 N_2(s)\,ds = pL^2\int_0^1 (s - 2s^2 + s^3)\,ds = \frac{pL^2}{12}$$

$$F_2 = pL\int_0^1 N_3(s)\,ds = pL\int_0^1 (3s^2 - 2s^3)\,ds = \frac{pL}{2}$$

$$C_2 = pL\int_0^1 N_4(s)\,ds = pL^2\int_0^1 (-s^2 + s^3)\,ds = -\frac{pL^2}{12}$$

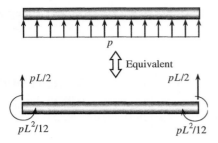

Figure 4.14 Work-equivalent nodal forces for the distributed load

Thus, the work-equivalent nodal forces for the uniformly distributed load become

$$\{\mathbf{F}\}^T = \left\{ \begin{array}{cccc} \dfrac{pL}{2} & \dfrac{pL^2}{12} & \dfrac{pL}{2} & -\dfrac{pL^2}{12} \end{array} \right\}$$

Figure 4.14 illustrates the equivalent nodal forces. The equal transverse nodal force $pL/2$ is applied to the two nodes, and the couple of $qL^2/12$ is applied at the two nodes with the opposite signs (see Figure 4.14).

Using the strain energy in Eq. (4.53) and the potential energy of the applied loads in Eq. (4.55), the total potential energy of the beam can be written as

$$\Pi = U + V = \frac{1}{2}\{\mathbf{Q}_s\}^T[\mathbf{K}_s]\{\mathbf{Q}_s\} - \{\mathbf{Q}_s\}^T\{\mathbf{F}_s\}. \qquad (4.60)$$

The principle of minimum potential energy can be applied to determine the unknown DOFs $\{\mathbf{Q}_s\}$. One can recognize Π above as a quadratic function in $\{\mathbf{Q}_s\}$. Using the minimization principle derived in Chapter 0 [Section 0.6] for a quadratic form, the stationary condition of Π with respect to $\{\mathbf{Q}_s\}$ yields

$$[\mathbf{K}_s]\{\mathbf{Q}_s\} = \{\mathbf{F}_s\} \qquad (4.61)$$

The above equations are the structural equations for the beam. In general the structural stiffness matrix $[\mathbf{K}_s]$ is singular. We can apply the boundary conditions by deleting the rows and columns corresponding to zero DOFs to obtain the global matrix equation as

$$[\mathbf{K}]\{\mathbf{Q}\} = \{\mathbf{F}\}$$

This process is the same as that of the truss elements in Chapter 2. The only difference is that the boundary conditions include not only the displacement but also the slope or rotation of the beam.

EXAMPLE 4.6   *Clamped-Clamped Beam Element*

A beam of length 2 m is clamped at both ends and subjected to a transverse force of 240 N at the center. Assume $EI = 1000\,\mathrm{Nm}^2$. Use beam finite elements to determine the deflections and slopes at $x = 0.5$, 1.0, and 1.5 m.

SOLUTION   It is sufficient to use two elements of equal length for this problem. The element stiffness matrices of the two elements will be the same, although the row and column addresses will be

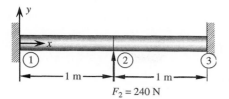

Figure 4.15 Finite element model of a clamped-clamped beam

different. (See Chapter 2 for row and column addresses.) Using the formula in Eq. (4.52), we can write the element stiffness matrices as

$$
\left[\mathbf{k}^{(1)}\right] = 1000 \begin{array}{c} \\ \begin{array}{cccc} v_1 & \theta_1 & v_2 & \theta_2 \end{array} \\ \left[\begin{array}{cccc} 12 & 6 & -12 & 6 \\ 6 & 4 & -6 & 2 \\ -12 & -6 & 12 & -6 \\ 6 & 2 & -6 & 4 \end{array}\right] \begin{array}{c} v_1 \\ \theta_1 \\ v_2 \\ \theta_2 \end{array} \end{array}
$$

$$
\left[\mathbf{k}^{(2)}\right] = 1000 \begin{array}{c} \\ \begin{array}{cccc} v_2 & \theta_2 & v_3 & \theta_3 \end{array} \\ \left[\begin{array}{cccc} 12 & 6 & -12 & 6 \\ 6 & 4 & -6 & 2 \\ -12 & -6 & 12 & -6 \\ 6 & 2 & -6 & 4 \end{array}\right] \begin{array}{c} v_2 \\ \theta_2 \\ v_3 \\ \theta_3 \end{array} \end{array}
$$

Assembling the element stiffness matrices, we obtain the 6×6 structural stiffness matrix $[\mathbf{K}_s]$. Also, the vector of applied nodal forces includes reactions at the clamped walls, as $\{\mathbf{F}_s\}^T = \{ F_1 \ C_1 \ 240 \ 0 \ F_3 \ C_3 \}$. Note that F_1 and F_3 are reaction forces at the walls, whereas C_1 and C_2 are the reaction couples. The structural matrix equation is then obtained as

$$
1000 \begin{bmatrix} 12 & 6 & -12 & 6 & 0 & 0 \\ 6 & 4 & -6 & 2 & 0 & 0 \\ -12 & -6 & 24 & 0 & -12 & 6 \\ 6 & 2 & 0 & 8 & -6 & 2 \\ 0 & 0 & -12 & -6 & 12 & -6 \\ 0 & 0 & 6 & 2 & -6 & 4 \end{bmatrix} \begin{Bmatrix} v_1 \\ \theta_1 \\ v_2 \\ \theta_2 \\ v_3 \\ \theta_3 \end{Bmatrix} = \begin{Bmatrix} F_1 \\ C_1 \\ 240 \\ 0 \\ F_3 \\ C_3 \end{Bmatrix} \tag{4.62}
$$

As discussed before, the structural stiffness matrix above is singular. Since both ends are clamped, the vertical deflection and rotation at these locations are zero. These boundary conditions can be applied by deleting the rows and columns corresponding to v_1, θ_1, v_3, and θ_3 (first, second, fifth, and sixth rows and columns) to obtain the global matrix equation as

$$
1000 \begin{bmatrix} 24 & 0 \\ 0 & 8 \end{bmatrix} \begin{Bmatrix} v_2 \\ \theta_2 \end{Bmatrix} = \begin{Bmatrix} 240 \\ 0 \end{Bmatrix} \tag{4.63}
$$

Now the global stiffness matrix above is positive definite, and, thus, a unique solution can be obtained by multiplying inverse of the global stiffness matrix, to yield

$$
v_2 = 0.01
$$
$$
\theta_2 = 0.0
$$

As another approach, it is possible to assemble the global matrix equation (4.63) directly. This can be done by deleting the rows and columns corresponding to zero DOF at the element level and only assembling those rows and columns corresponding to non-zero DOF.

The deflection and slope at points in between nodes can be interpolated using the shape functions. The point $x = 0.5\,\text{m}$ is in the first element. Hence the deflection and slope at this point is interpolated using the DOFs belonging to Element 1. The local coordinate of this point is $s = 1/2$

[see Eq. (4.37)]. Then the deflection and slope are obtained using the method similar to that in Example 4.3, as

$$v\left(\tfrac{1}{2}\right) = v_1 N_1 \left(\tfrac{1}{2}\right) + \theta_1 N_2 \left(\tfrac{1}{2}\right) + v_2 N_3 \left(\tfrac{1}{2}\right) + \theta_2 N_4 \left(\tfrac{1}{2}\right) = 0.01 \times N_3 \left(\tfrac{1}{2}\right) = 0.005 \,\text{m}$$

$$\theta\left(\tfrac{1}{2}\right) = \frac{1}{L^{(1)}} v_2 \left. \frac{dN_3}{ds} \right|_{s=\frac{1}{2}} = 0.015 \,\text{rad}$$

Deflection and slope at $x = 1$ m are v_2 and θ_2. In this case, either Element 1 or 2 can be used. If Element 1 is used, the location $x = 1$ m corresponds to $s = 1$. Thus,

$$v(1) = v_1 N_1(1) + \theta_1 N_2(1) + v_2 N_3(1) + \theta_2 N_4(1) = 0.01 \times N_3(1) = 0.01 \,\text{m}$$

$$\theta(1) = \frac{1}{L^{(1)}} v_2 \left. \frac{dN_3}{ds} \right|_{s=1} = 0.0 \,\text{rad}$$

On the other hand, if Element 2 is used, the location $x = 1$ m corresponds to $s = 0$. Thus,

$$v(1) = v_2 N_1(0) + \theta_2 N_2(0) + v_3 N_3(0) + \theta_3 N_4(0) = 0.01 \times N_1(0) = 0.01 \,\text{m}$$

$$\theta(1) = \frac{1}{L^{(2)}} v_2 \left. \frac{dN_1}{ds} \right|_{s=0} = 0.0 \,\text{rad}$$

Note that the two results are identical. This is true because in the beam element, the vertical deflection, and the rotation are continuous at the connecting node.

To determine the deflection and slope at $x = 1.5$ m, we use the DOFs of Element 2. Since the location is the center of the element, $s = 1/2$. Substituting in the shape functions we obtain

$$v(1.5) = v_2 N_1 \left(\tfrac{1}{2}\right) + \theta_2 N_2 \left(\tfrac{1}{2}\right) + v_3 N_3 \left(\tfrac{1}{2}\right) + \theta_3 N_4 \left(\tfrac{1}{2}\right) = 0.01 \times N_1 \left(\tfrac{1}{2}\right) = 0.005 \,\text{m}$$

$$\theta(1.5) = \frac{1}{L^{(2)}} v_2 \left. \frac{dN_1}{ds} \right|_{s=\frac{1}{2}} = -0.015 \,\text{rad}$$

As expected, the deflection is symmetric with respect to $x = 1$. Thus, the deflections at $x = 0.5$ and $x = 1.5$ are the same, whereas the rotations are equal but opposite.

EXAMPLE 4.7   *Finite Element Analysis of a Cantilevered Beam*

A cantilever beam of length 1 m is subjected to a uniformly distributed load $p(x) = p_0 = 120 \,\text{N/m}$ and a clockwise couple 50 N-m at the tip. Use one element to determine the deflection curve $v(x)$ of the beam. What are the support reactions? Assume $EI = 1,000 \,\text{N-m}^2$.

SOLUTION   Since there is only one element, the structural stiffness matrix is the same as the element stiffness matrix:

$$[\mathbf{K}_s] = 1000 \begin{bmatrix} 12 & 6 & -12 & 6 \\ 6 & 4 & -6 & 2 \\ -12 & -6 & 12 & -6 \\ 6 & 2 & -6 & 4 \end{bmatrix} \begin{matrix} v_1 \\ \theta_1 \\ v_2 \\ \theta_2 \end{matrix}$$

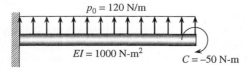

$p_0 = 120$ N/m

$EI = 1000$ N-m^2

$C = -50$ N-m

Figure 4.16 Cantilevered beam under uniformly distributed load and couple

The next step is to calculate the work equivalent loads for the distributed force. Substituting $p(x) = p_0$ in Eq. (4.59), we obtain

$$
\begin{Bmatrix} F_{1e} \\ C_{1e} \\ F_{2e} \\ C_{2e} \end{Bmatrix} = p_0 L \int_0^1 \begin{Bmatrix} 1 - 3s^2 + 2s^3 \\ (s - 2s^2 + s^3)L \\ 3s^2 - 2s^3 \\ (-s^2 + s^3)L \end{Bmatrix} ds = p_0 L \begin{Bmatrix} 1/2 \\ L/12 \\ 1/2 \\ -L/12 \end{Bmatrix} = \begin{Bmatrix} 60 \\ 10 \\ 60 \\ -10 \end{Bmatrix}
\tag{4.64}
$$

where the subscript e denotes the loads are equivalent loads. Note that the total force $p_0 L$ is equally divided between the two nodes. In addition, two equal and opposite couples of magnitude $p_0 L^2/12$ are also applied at the two nodes. These equivalent forces should be added to any concentrated forces and couples acting at the nodes. Thus, the force vector is written as

$$
\{\mathbf{F}_s\} = \begin{Bmatrix} F_1 + 60 \\ C_1 + 10 \\ 60 \\ -10 - 50 \end{Bmatrix}
$$

where F_1 and C_1 are the unknown reactions at the clamped end. The structural matrix equations become

$$
1000 \begin{bmatrix} 12 & 6 & -12 & 6 \\ 6 & 4 & -6 & 2 \\ -12 & -6 & 12 & -6 \\ 6 & 2 & -6 & 4 \end{bmatrix} \begin{Bmatrix} v_1 \\ \theta_1 \\ v_2 \\ \theta_2 \end{Bmatrix} = \begin{Bmatrix} F_1 + 60 \\ C_1 + 10 \\ 60 \\ -10 - 50 \end{Bmatrix}
$$

Now the boundary conditions for the cantilevered beam is $v_1 = \theta_1 = 0$. Since the first two columns are going to be multiplied by zeros and the first two rows contain unknown reactions on the right-hand side, we delete these two rows and columns to obtain the global matrix equation, as

$$
1000 \begin{bmatrix} 12 & -6 \\ -6 & 4 \end{bmatrix} \begin{Bmatrix} v_2 \\ \theta_2 \end{Bmatrix} = \begin{Bmatrix} 60 \\ -60 \end{Bmatrix}.
$$

The global stiffness matrix in the above equation is positive definite and can be inverted to obtain the unknown DOFs, as

$$
v_2 = -0.01 \, \text{m}
$$
$$
\theta_2 = -0.03 \, \text{rad}
$$

The deflection curve is computed using the shape functions as

$$
v(s) = -0.01 N_3(s) - 0.03 N_4(s) = -0.01 s^3
$$

where $s = x/L = x$. It is interesting to note that the tip deflection and rotations are exact. However, the deflection curve is approximate. The exact deflection curve is a quartic polynomial in x, whereas the approximate deflection is cubic. For comparison, the exact deflection curve is $v(x) = 0.005(x^4 - 4x^3 + x^2)$. Figure 4.17 compares the deflection and slope from the finite element and exact solutions.

The support reactions are determined from the first two of the four global equations:

$$
1000(-12v_2 + 6\theta_2) = F_1 + 60
$$
$$
1000(-6v_2 + 2\theta_2) = C_1 + 10
$$

Solving the above pair of equations, we obtain, $F_1 = -120$N, and $C_1 = -10$ as the reactions. The reader can verify that the reactions satisfy the force and moment equilibrium equations for the cantilever beam.

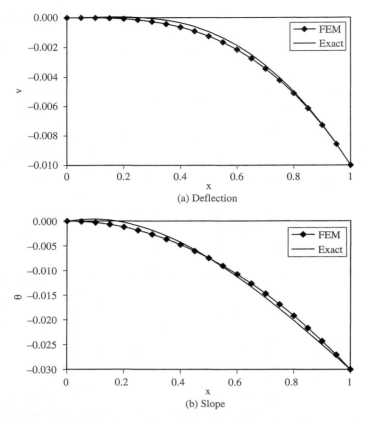

Figure 4.17 Comparison of beam deflection and rotation with exact solutions

4.5 BENDING MOMENT AND SHEAR FORCE DISTRIBUTION

After the nodal DOFs are determined, the bending moment $M(s)$ and shear force $V_y(s)$ distribution along the element can be calculated by substituting for the deflection curve into Eqs. (4.6) and (4.8).

Bending Moment:

$$M(s) = EI\frac{d^2 v}{dx^2} = \frac{EI}{L^2}\frac{d^2 v}{ds^2} = \frac{EI}{L^2}\lfloor \mathbf{B} \rfloor \{\mathbf{q}\}. \tag{4.65}$$

Shear Force:

$$V_y(s) = -\frac{dM}{dx} = -EI\frac{d^3 v}{dx^3} = -\frac{EI}{L^3}\frac{d^3 v}{ds^3}$$

$$= \frac{EI}{L^3}[-12 \quad -6L \quad 12 \quad -6L]\begin{Bmatrix} v_1 \\ \theta_1 \\ v_2 \\ \theta_2 \end{Bmatrix} \tag{4.66}$$

Note that the moment is a linear function of s, while the shear force is constant in an element. This is a limitation of the current beam element. If the loads on the beam are such that the shear force has a linear distribution, then several elements must be used so that the linear variation of the shear force can be better approximated by the piecewise constant shear force distribution.

We can use Eq. (4.65) to calculate the bending moment at the two nodes of the beam element. We will denote the bending moments at Nodes 1 and 2 by M_1 and M_2, respectively. They can be calculated as

$$M_1 = M(0) = \frac{EI}{L^2}[-6 \quad -4L \quad 6 \quad -2L)] \begin{Bmatrix} v_1 \\ \theta_1 \\ v_2 \\ \theta_2 \end{Bmatrix} \tag{4.67}$$

$$M_2 = M(1) = \frac{EI}{L^2}[6 \quad 2L \quad -6 \quad 4L] \begin{Bmatrix} v_1 \\ \theta_1 \\ v_2 \\ \theta_2 \end{Bmatrix} \tag{4.68}$$

From Eqs. (4.66) through (4.68), we can write the shear force and bending moments at the two nodes in a matrix form as shown below:

$$\begin{Bmatrix} -V_{y1} \\ -M_1 \\ +V_{y2} \\ M_2 \end{Bmatrix} = \frac{EI}{L^3} \begin{bmatrix} 12 & 6L & -12 & 6L \\ 6L & 4L^2 & -6L & 2L^2 \\ -12 & -6L & 12 & -6L \\ 6L & 2L^2 & -6L & 4L^2 \end{bmatrix} \begin{Bmatrix} v_1 \\ \theta_1 \\ v_2 \\ \theta_2 \end{Bmatrix} \tag{4.69}$$

One can note that the square matrix in the above equation is identical to the stiffness matrix of the beam element [see Eq. (4.52)]

After calculating the bending moment, the axial stress in the beam at a cross-section is calculated using

$$\sigma_x = -\frac{My}{I}. \tag{4.70}$$

It is clear that the maximum stress appears either at the top or bottom of the beam element, which corresponds to $y = \pm h/2$, where the beam height is h.

The computation of the shear stress is more complicated and depends on the shape of the cross-section. For a rectangular section of width b and height h, the shear stress distribution is given by

$$\tau_{xy}(y) = \frac{1.5V_y}{bh}\left(1 - \frac{4y^2}{h^2}\right) \tag{4.71}$$

The maximum shear stress occurs at the neutral axis of the beam $(y = 0)$, and it is zero at the top and bottom of the cross-section. The maximum shear stress is 1.5 times the average shear stress, which is equal to V_y/bh.

EXAMPLE 4.8   *Bending Moment and Shear Force for a Cantilevered Beam*

Calculate the bending moment and the shear force along the cantilevered beam in Example 4.7. Compare the results with exact solutions.

SOLUTION From Eq. (4.47), the bending moment can be written as

$$M(s) = \frac{EI}{L^2} \lfloor \mathbf{B} \rfloor \{\mathbf{q}\}$$

$$= \frac{EI}{L^2}[(-6 + 12s)v_1 + L(-4 + 6s)\theta_1 + (6 - 12s)v_2 + L(-2 + 6s)\theta_2]$$

$$= 1000[-0.01(6 - 12s) - 0.03(-2 + 6s)]$$

$$= -60s \, \text{N} \cdot \text{m}$$

From Eq. (4.48), the shear force can be written as

$$V_y = -\frac{EI}{L^3}[12v_1 + 6L\theta_1 - 12v_2 + 6L\theta_2]$$

$$= -1000[-12 \times (-0.01) + 6(-0.03)]$$

$$= 60 \, \text{N}$$

Since the deflection $v(s)$ is a cubic polynomial of s, the bending moment is linear and the shear force is constant. Figure 4.18 shows the bending moment and shear force along with the exact values. The errors are in general larger than those of deflection and rotation. Thus, in the case of distributed load, more than one element is required to reduce the errors.

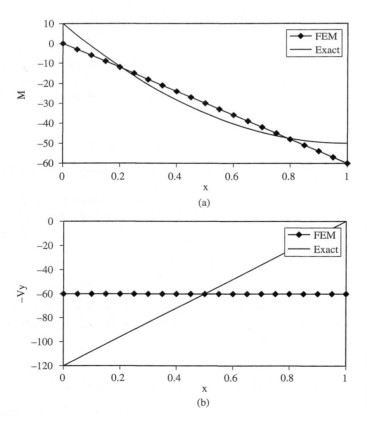

(a)

(b)

Figure 4.18 Comparison of bending moment and shear force with exact solutions

EXAMPLE 4.9 *Finite Element Analysis of a Simply Supported Beam*

Consider a simply supported beam under uniformly distributed load p, as shown in Figure 4.19. Using one beam element, calculate deflection curve, bending moment, and shear force. Compare the finite element solutions with exact solutions.

SOLUTION The whole beam is approximated using one beam finite element. The finite element equation can be written as

$$\frac{EI}{L^3} \begin{bmatrix} 12 & 6L & -12 & 6L \\ 6L & 4L^2 & -6L & 2L^2 \\ -12 & -6L & 12 & -6L \\ 6L & 2L^2 & -6L & 4L^2 \end{bmatrix} \begin{Bmatrix} v_1 \\ \theta_1 \\ v_2 \\ \theta_2 \end{Bmatrix} = \begin{Bmatrix} pL/2 \\ pL^2/12 \\ pL/2 \\ -pL^2/12 \end{Bmatrix} + \begin{Bmatrix} F_1 \\ 0 \\ F_2 \\ 0 \end{Bmatrix}$$

where F_1 and F_2 are unknown reactions at the both nodes. No external couples are applied at the two ends of the beam. However, the work-equivalent nodal forces yield non-zero couples at the both nodes. Since there is only one element, the structural equations are the same as the local element equations. The displacement boundary conditions are $v_1 = v_2 = 0$ for the simply supported beam. These boundary conditions can be imposed using the procedure that is explained in Section 2.1 (striking rows and columns that correspond to the zero displacement boundary conditions). Since v_1 and v_2 are fixed, the first and third rows and columns are removed. Then, the circled components in the equation below remain:

$$\frac{EI}{L^3} \begin{bmatrix} 12 & 6L & -12 & 6L \\ 6L & \boxed{4L^2} & -6L & \boxed{2L^2} \\ -12 & -6L & 12 & -6L \\ 6L & \boxed{2L^2} & -6L & \boxed{4L^2} \end{bmatrix} \begin{Bmatrix} 0 \\ \theta_1 \\ 0 \\ \theta_2 \end{Bmatrix} = \begin{Bmatrix} pL/2 + F_1 \\ pL^2/12 \\ pL/2 + F_2 \\ -pL^2/12 \end{Bmatrix}$$

The global equations corresponding to unknown DOFs are

$$\frac{EI}{L^3} \begin{bmatrix} 4L^2 & 2L^2 \\ 2L^2 & 4L^2 \end{bmatrix} \begin{Bmatrix} \theta_1 \\ \theta_2 \end{Bmatrix} = p \begin{Bmatrix} L^2/12 \\ -L^2/12 \end{Bmatrix}.$$

Solving this matrix equation yields the solution:

$$\theta_1 = \frac{pL^3}{24EI}, \quad \theta_2 = -\frac{pL^3}{24EI}$$

Thus, the two ends of the beam rotate and the slopes are equal to θ_1 and θ_2. Using Eq. (4.44), the displacement along the beam element can be approximated by

$$v(s) = \begin{bmatrix} N_1 & N_2 & N_3 & N_4 \end{bmatrix} \begin{Bmatrix} 0 \\ \dfrac{pL^3}{24EI} \\ 0 \\ -\dfrac{pL^3}{24EI} \end{Bmatrix} = \frac{pL^4(s - s^2)}{24EI} \tag{4.72}$$

Displacement $v(s)$ is a quadratic function of parameter s. It may be noted that the exact deflection is a quartic function of x. The maximum deflection occurs at the center ($s = 0.5$). Substituting for s in

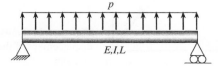

Figure 4.19 One element model with distributed force q

Eq. (4.72), we obtain the maximum deflection as $v_{max} = pL^4/96EI$. The exact solution for the maximum deflection is $v_{max,exact} = 5 pL^4/384EI$. In this example, the error in the maximum deflection is approximately 25%. It may be noted that the deflection from finite element analysis is smaller than the exact deflection; i.e., the beam looks stiffer than it is. This is typical of approximate methods such as finite element analysis. As the number of elements is increased, the solution will approach the exact solution.

The support reactions can be calculated from the first and third of the four global equations as $F_1 = F_2 = -pL/2$. The element-bending moment and shear force can be calculated as follows:

$$
M(s) = \frac{EI}{L^2}[-6 + 12s \quad L(-4 + 6s) \quad 6 - 12s \quad L(-2 + 6s)]
\begin{Bmatrix} 0 \\ \dfrac{pL^3}{24EI} \\ 0 \\ -\dfrac{pL^3}{24EI} \end{Bmatrix}
= -\frac{pL^2}{12}
$$

$$
V(s) = \frac{EI}{L^3}[-12 \quad -6L \quad 12 \quad -6L]
\begin{Bmatrix} 0 \\ \dfrac{pL^3}{24EI} \\ 0 \\ -\dfrac{pL^3}{24EI} \end{Bmatrix}
= 0
$$

Since no shear force appears in the element, this loading condition produces a pure (constant) bending moment. The exact solution for bending moment is a quadratic in x, and the shear force has a linear variation along the length of the beam. Thus, the one-element solution is not enough to estimate the bending moment and shear force.

One of the biggest dangers in using finite element analysis without understanding the basic principles is to believe the accuracy of the solution without verification. Many people simply believe the output results from the computer have to be accurate. In the case of trusses, we have shown that the finite element solution is exactly the same as the analytical solution. That is not true for beam elements. The FE solution for beams will be exact only when concentrated forces and couples act on the beam and the cross-section is uniform between the nodes. In the following, we compare the finite element and analytical solutions for the problem shown in Figure 4.19.

The analytical solution of the transverse displacement is given by

$$
v(s)_{analytical} = \frac{pL^4}{24EI}(s - 2s^3 + s^4) \tag{4.73}
$$

which is fourth-order function of s, while the finite element solution in Eq. (4.72) is quadratic in s. Figure 4.20 compares the analytical and finite element solutions for transverse displacements. The center deflection from the finite element analysis is only 80% of the analytical solution.

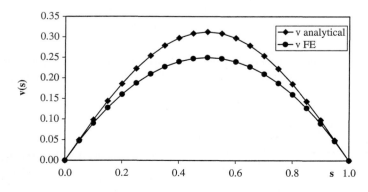

Figure 4.20 Transverse displacement of the beam element

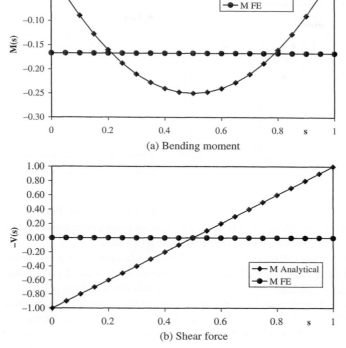

(a) Bending moment

(b) Shear force

Figure 4.21 Comparison of FE and analytical solutions for the beam shown in Figure 4.19 The beam is modeled by a single element.

The deviation of the finite element solution is more significant if the bending moment and shear force of the beam are compared. From the analytical solution, the bending moment and shear force of the beam can be calculated as

$$M(s)_{\text{analytical}} = \frac{qL^2}{2}(s^2 - s) \qquad (4.74)$$

$$V(s)_{\text{analytical}} = \frac{qL}{2}(2s - 1) \qquad (4.75)$$

Notice that the bending moment of the beam finite element was a constant function and the shear force was zero. Figure 4.21 compares the bending moment and shear force from the analytical and finite element solutions. One can note large differences between these solutions. However, the accuracy of the finite element solution can be improved by using more elements.

4.6 PLANE FRAME

A plane frame is a structure similar to a truss, except the members, in addition to axial force, can carry transverse shear force and bending moment. Thus, a frame member combines the action of a uniaxial bar and a beam. This is accomplished by connecting the members by a rigid joint such as a gusset plate, which transmits the shear force and bending moment. Welding the ends of the members will also be sufficient in some cases. The cross-sections of members connected to a joint (node) undergo the same rotation when the frame deforms. The nodes in a plane frame that is in the x-y plane have three DOFs, u, v,

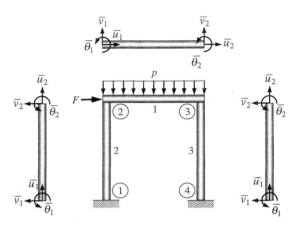

Figure 4.22 Frame structure and finite elements

and θ, displacements in the x and y directions, and rotation about the z-axis. Consequently, one can apply two forces and one couple at each node corresponding to the three DOFs. An example of common frame is depicted in Figure 4.22.

Consider the free-body diagram of a typical frame element shown in Figure 4.23. It has two nodes and three DOFs at each node. Each element has a local coordinate system. The local or element coordinate system $\bar{x} - \bar{y}$ is such that the $\bar{x}$-axis is parallel to the element. The positive $\bar{x}$ direction is from the first node to the second node of the element. The $\bar{y}$-axis is such that the $\bar{z}$-axis is in the same direction as the z-axis. In the local coordinate system, the displacements in $\bar{x}$ and $\bar{y}$ directions are, respectively, $\bar{u}$ and $\bar{v}$, and the rotation in the $\bar{z}$ direction is $\bar{\theta}$. Each node has these three DOFs. The forces acting on the element, in local coordinates, are $f_{\bar{x}1}$, $f_{\bar{y}1}$, and $\bar{c}_1$ at Node 1 and $f_{\bar{x}2}$, $f_{\bar{y}2}$, and $\bar{c}_2$ at Node 2. Our goal is to derive a relation between the six element forces and the six DOFs. It will be convenient to use the local coordinate system to derive the force-displacement relation as the axial effects and bending effects are uncoupled in the local coordinates.

The element forces and nodal displacements are vectors, and they can be transformed to the $\bar{x} - \bar{y}$ coordinate system as follows:

$$
\begin{Bmatrix} f_{\bar{x}1} \\ f_{\bar{y}1} \\ \bar{c}_1 \\ f_{\bar{x}2} \\ f_{\bar{y}2} \\ \bar{c}_2 \end{Bmatrix} =
\begin{bmatrix}
\cos\phi & \sin\phi & 0 & 0 & 0 & 0 \\
-\sin\phi & \cos\phi & 0 & 0 & 0 & 0 \\
0 & 0 & 1 & 0 & 0 & 0 \\
0 & 0 & 0 & \cos\phi & \sin\phi & 0 \\
0 & 0 & 0 & -\sin\phi & \cos\phi & 0 \\
0 & 0 & 0 & 0 & 0 & 1
\end{bmatrix}
\begin{Bmatrix} f_{x1} \\ f_{y1} \\ c_1 \\ f_{x2} \\ f_{y2} \\ c_2 \end{Bmatrix}
$$

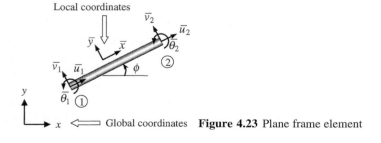

Local coordinates

Global coordinates   **Figure 4.23** Plane frame element

or,

$$\{\bar{\mathbf{f}}\} = [\mathbf{T}]\{\mathbf{f}\} \tag{4.76}$$

where the transformation matrix $[\mathbf{T}]$ is a function of the direction cosines of the element. Note that the size of the transformation matrix $[\mathbf{T}]$ in Chapter 2 was 4×4, as a two-dimensional truss element has four DOFs. The transformation matrix in Eq. (4.76) is basically the same, but the size is increased to 6×6 because of additional couples. However, the couple $\bar{c}$ is the same as the couple c, as local $\bar{z}$-axis is parallel to the global z-axis. For the same reason, $\bar{\theta} = \theta$. Since the displacements are also vectors, a similar relation can be used to connect the DOFs in the local and global coordinates:

$$
\begin{Bmatrix} \bar{u}_1 \\ \bar{v}_1 \\ \bar{\theta}_1 \\ \bar{u}_2 \\ \bar{v}_2 \\ \bar{\theta}_2 \end{Bmatrix} =
\begin{bmatrix}
\cos\phi & \sin\phi & 0 & 0 & 0 & 0 \\
-\sin\phi & \cos\phi & 0 & 0 & 0 & 0 \\
0 & 0 & 1 & 0 & 0 & 0 \\
0 & 0 & 0 & \cos\phi & \sin\phi & 0 \\
0 & 0 & 0 & -\sin\phi & \cos\phi & 0 \\
0 & 0 & 0 & 0 & 0 & 1
\end{bmatrix}
\begin{Bmatrix} u_1 \\ v_1 \\ \theta_1 \\ u_2 \\ v_2 \\ \theta_2 \end{Bmatrix}
$$

or,

$$\{\bar{\mathbf{q}}\} = [\mathbf{T}]\{\mathbf{q}\} \tag{4.77}$$

In the local coordinate system, the axial deformation (uniaxial bar) and bending deformations (beam) are uncoupled. The axial forces and axial displacements are related by the uniaxial bar stiffness matrix, as

$$\frac{EA}{L} \begin{bmatrix} 1 & -1 \\ -1 & 1 \end{bmatrix} \begin{Bmatrix} \bar{u}_1 \\ \bar{u}_2 \end{Bmatrix} = \begin{Bmatrix} f_{\bar{x}1} \\ f_{\bar{x}2} \end{Bmatrix} \tag{4.78}$$

On the other hand, the transverse force and couple are related to the transverse displacement and rotation by the bending stiffness matrix in Eq. (4.52), as

$$\frac{EI}{L^3}
\begin{bmatrix}
12 & 6L & -12 & 6L \\
6L & 4L^2 & -6L & 2L^2 \\
-12 & -6L & 12 & -6L \\
6L & 2L^2 & -6L & 4L^2
\end{bmatrix}
\begin{Bmatrix} \bar{v}_1 \\ \bar{\theta}_1 \\ \bar{v}_2 \\ \bar{\theta}_2 \end{Bmatrix} =
\begin{Bmatrix} f_{\bar{y}1} \\ \bar{c}_1 \\ f_{\bar{y}2} \\ \bar{c}_2 \end{Bmatrix} \tag{4.79}$$

In a sense, the plane frame element is a combination of two-dimensional truss and beam elements. Combining Eqs. (4.78) and (4.79), we obtain a relation between the element DOFs and forces in the local coordinate system:

$$
\begin{bmatrix}
a_1 & 0 & 0 & -a_1 & 0 & 0 \\
0 & 12a_2 & 6La_2 & 0 & -12a_2 & 6La_2 \\
0 & 6La_2 & 4L^2a_2 & 0 & -6La_2 & 2L^2a_2 \\
-a_1 & 0 & 0 & a_1 & 0 & 0 \\
0 & -12a_2 & -6La_2 & 0 & 12a_2 & -6La_2 \\
0 & 6La_2 & 2L^2a_2 & 0 & -6La_2 & 4L^2a_2
\end{bmatrix}
\begin{Bmatrix} \bar{u}_1 \\ \bar{v}_1 \\ \bar{\theta}_1 \\ \bar{u}_2 \\ \bar{v}_2 \\ \bar{\theta}_2 \end{Bmatrix} =
\begin{Bmatrix} f_{\bar{x}1} \\ f_{\bar{y}1} \\ \bar{c}_1 \\ f_{\bar{x}2} \\ f_{\bar{y}2} \\ \bar{c}_2 \end{Bmatrix} \tag{4.80}
$$

where

$$a_1 = \frac{EA}{L}, \quad a_2 = \frac{EI}{L^3}$$

It is clear that the equations at the first and fourth rows are the same as the uniaxial bar relation in Eq. (4.78), whereas the remaining four equations are the beam relation. Equation (4.80) can be written in symbolic notation as

$$\boxed{[\overline{\mathbf{k}}]\{\overline{\mathbf{q}}\} = \{\overline{\mathbf{f}}\}}$$ (4.81)

where $[\overline{\mathbf{k}}]$ is the element stiffness matrix in the local coordinate system.

As in the case of two-dimensional truss elements, the element matrix equation (4.81) cannot be used for assembly because different elements have different local coordinate systems. Thus, the element matrix equation needs to be transformed to the global coordinate system. Substituting for $\{\overline{\mathbf{f}}\}$ and $\{\overline{\mathbf{q}}\}$ from Eqs. (4.76) and (4.77), we obtain

$$[\overline{\mathbf{k}}][\mathbf{T}]\{\mathbf{q}\} = [\mathbf{T}]\{\mathbf{f}\}$$ (4.82)

Multiplying both sides of the above equation by $[\mathbf{T}]^{-1}$, we obtain

$$[\mathbf{T}]^{-1}[\overline{\mathbf{k}}][\mathbf{T}]\{\mathbf{q}\} = \{\mathbf{f}\}$$ (4.83)

It can be shown that for the transformation matrix $[\mathbf{T}]^T = [\mathbf{T}]^{-1}$. Hence, Eq. (4.83) can be written as

$$[\mathbf{T}]^T[\overline{\mathbf{k}}][\mathbf{T}]\{\mathbf{q}\} = \{\mathbf{f}\}$$

or

$$[\mathbf{k}]\{\mathbf{q}\} = \{\mathbf{f}\}$$ (4.84)

where $[\mathbf{T}]^T[\overline{\mathbf{k}}][\mathbf{T}] = [\mathbf{k}]$ is the element stiffness matrix of a plane frame element in the global coordinate system. One can verify that $[\mathbf{k}]$ is symmetric and positive semi-definite.

Assembly of the element stiffness matrix to form the structural stiffness matrix $[\mathbf{K}_s]$ follows the same steps as for earlier elements. After deleting the rows and columns corresponding to zero DOFs, we obtain the global equations in the form $[\mathbf{K}]\{\mathbf{Q}\} = \{\mathbf{F}\}$.

4.6.1 Calculation of Element Forces

After solving the global equations, the DOFs of all elements are known in the frame structure. Let us represent the DOF of a typical element by $\{\mathbf{q}\}$, which is a 6×1 column matrix. We will transform the DOFs into local coordinates using Eq. (4.77) to obtain $\{\overline{\mathbf{q}}\}$. Then the axial force in the element can be obtained using

$$P = \frac{AE}{L}(\overline{u}_2 - \overline{u}_1)$$ (4.85)

This relation is exactly the same as that of a uniaxial bar element. We note that the transverse shear force is constant throughout the element and the bending moment varies linearly. Thus, it is enough to find the nodal values. From Eq. (4.69), the nodal values of bending moments and shear forces can be found by

$$\begin{Bmatrix} -V_{\overline{y}1} \\ -\overline{M}_1 \\ +V_{\overline{y}2} \\ \overline{M}_2 \end{Bmatrix} = \frac{EI}{L^3} \begin{bmatrix} 12 & 6L & -12 & 6L \\ 6L & 4L^2 & -6L & 2L^2 \\ -12 & -6L & 12 & -6L \\ 6L & 2L^2 & -6L & 4L^2 \end{bmatrix} \begin{Bmatrix} \overline{v}_1 \\ \overline{\theta}_1 \\ \overline{v}_2 \\ \overline{\theta}_2 \end{Bmatrix}$$

4.7 PROJECT

I. This project is concerned with design of a bicycle frame using aluminum tubes. The schematic dimensions of the bicycle are shown in Figure 4.24. The following two load cases should be considered.

  (a) Vertical loads: When an adult rides the bike, the nominal load is estimated as a downward load of 900N at the seat position and a load of 300N at the pedal crank location. When a dynamic environment is simulated using the static analysis, the static loads are often multiplied by a dynamic load factor G. In this design project, use $G = 2$. Use ball-joint boundary conditions for the front dropout (location 1) and sliding boundary conditions for the rear dropouts (locations 5 and 6).

  (b) Horizontal impact: The frame should be able to withstand a horizontal load of 1,000N applied to the front dropout with rear dropouts constrained from any translational motion. For this load case, assume the front dropout can only move in the horizontal direction. Use $G = 2$.

Choose aluminum tubes of various diameters for the various members of the frame shown in Figure 4.24 such that the bicycle is as light as possible. The minimum outside diameter is 12mm and the wall thickness is 2mm. Approximate the frame as a plane frame by giving the same (x, y) coordinates for Nodes 5 and 6. Thus, all the nodes will be on the x-y plane. In addition to the dynamic load factor, use a safety factor of 1.5. Use von Mises failure stress criterion for yielding. For compression members, include buckling as additional criterion. The buckling load of a member is approximated as $P_{cr} = 2\pi^2 EI/L^2$ where L is the member length. Use a safety factor of 1.5 for buckling also.

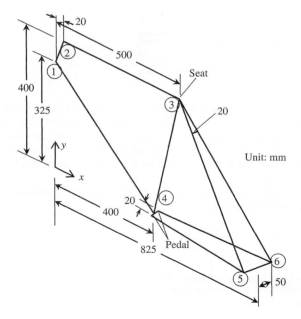

Figure 4.24 Bicycle frame structure

Properties of Aluminum

Material property	Value
Young's Modulus (E)	70 GPa
Poisson's Ratio (ν)	0.33
Density (ρ)	$2,580\,\text{kg/m}^3$
Yield Strength (σ_y)	210 MPa

The report should be easily readable and complete by itself, e.g., including introduction, approach, assumptions, results, conclusion, discussion, and references. In your report, you must include the following for each load case:

1. For each element the maximum normal stress, shear stress and maximum von Mises stress and safety factor at each node.

2. For each element under compress ($P < 0$), buckling load, actual axial force P and safety factor in buckling.

3. Nodal deflections at each node should be given. Calculate the weight of your frame.

4.8 EXERCISE

1. Repeat Example 4.2 with the approximate deflection in the following form: $v(x) = c_1 x^2 + c_2 x^3 + c_3 x^4$. Compare the deflection curve with the exact solution.

2. The deflection of the simply supported beam shown in the figure is assumed as $v(x) = cx(x - 1)$, where c is a constant. A force is applied at the center of the beam. Use the following properties: $EI = 1000\,\text{N-m}^2$ and $L = 1\,\text{m}$. First, (a) show that the above approximate solution satisfies displacement boundary conditions, and (b) use Rayleigh-Ritz method to determine c.

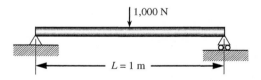

3. Use the Rayleigh-Ritz method to determine the deflection $v(x)$, bending moment $M(x)$, and shear force $V_y(x)$ for the beam shown in the figure. The bending moment and shear force are calculated from the deflection as $M(x) = EId^2v/dx^2$ and $V_y(x) = -EId^3v/dx^3$. Assume the displacement as $v(x) = c_0 + c_1 x + c_2 x^2 + c_3 x^3$, and $EI = 1,000\,\text{N-m}^2$, $L = 1\,\text{m}$, and $p_0 = 100\,\text{N/m}$, and $C = 100\,\text{N-m}$. Make sure that the displacement boundary conditions are satisfied a priori.

Hint: Potential energy of a couple is calculated as $V = -Cdv/dx$, where the rotation is calculated at the point of application of the couple.

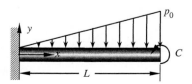

4. The right end of a cantilevered beam is resting on an elastic foundation that can be represented by a spring with spring constant $k = 1,000\,\text{N/m}$. A force of $1,000$ N acts at the center of the beam as shown. Use the Rayleigh-Ritz method to determine the deflection $v(x)$ and the force in the spring. Assume $EI = 1,000\,\text{N-m}^2$ and $v(x) = c_0 + c_1 x + c_2 x^2 + c_3 x^3$.

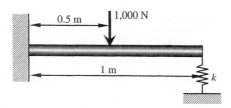

5. A cantilevered beam is modeled using one finite element. The nodal values of the beam element are given as

$$\{\mathbf{q}\} = \left\{0, 0, \frac{FL^3}{3EI}, \frac{FL^2}{2EI}\right\}^T.$$

Plot the deflection curve, bending moment, and shear force.

6. A simply supported beam with length L is under a concentrated vertical force $-F$ at the center. When two equal-length beam elements are used, the finite element analysis yields the following nodal DOFs:

$$\{\mathbf{Q_s}\}^T = \{v_1, \theta_1, v_2, \theta_2, v_3, \theta_3\} = \left\{0, -\frac{FL^2}{16EI}, -\frac{FL^3}{48EI}, 0, 0, \frac{FL^2}{16EI}\right\}.$$

Find the deflection curve $v(x)$ and compare it with the exact solution in a graph. Note: the exact deflection curve of the beam is, for $x \le \frac{1}{2}L, v_{\text{exact}}(x) = Fx(3L^2 - 4x^2)/48EI$ and symmetric for $x \ge \frac{1}{2}L$.

7. A simply supported beam with length L is under a uniformly distributed load $-p$. When two equal-length beam elements are used, the finite element analysis yields the following nodal DOFs:

$$\{\mathbf{Q_s}\}^T = \{v_1, \theta_1, v_2, \theta_2, v_3, \theta_3\} = \left\{0, -\frac{pL^3}{24EI}, -\frac{5\,pL^4}{384EI}, 0, 0, \frac{pL^3}{24EI}\right\}.$$

Find the deflection curve $v(x)$ and compare graphically with the exact solution. Note: the exact deflection curve of the beam is given as $v_{\text{exact}}(x) = -p(x^4 - 2Lx^3 + L^3x)/24EI$.

8. Consider a cantilevered beam with a Young's modulus E, moment of inertia I, height $2h$, and length L. A couple M_0 is applied at the tip of the beam. One beam finite element is used to approximate the structure.

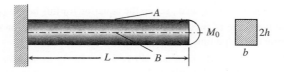

   (a) Calculate the tip displacement v and tip slope θ using the finite element equation.
   (b) Calculate the bending moment and shear force at the wall using the finite element equation.
   (c) Calculate the stress σ_{xx} at points A and B, which are at $L/2$.

9. The cantilever beam shown is modeled using one finite element. If the deflection at the nodes of a beam element are $\theta_1 = \theta_1 = 0$, and $v_2 = 0.01$ m and slope $\theta_2 = 0$, write the equation of the

deformed beam $v(s)$. In addition, compute the forces F_2 and M_2 acting on the beam to produce the above deformation in terms of E, I, and L.

Hint: Use the equations $[\mathbf{k}]\{\mathbf{q}\} = \{\mathbf{f}\}$ for the beam element.

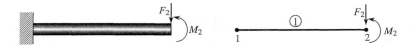

10. Let a uniform cantilevered beam of length L be supported at the loaded end so that this end cannot rotate, as shown in the figure. For the given moment of inertia I, Young's modulus E, and applied tip load P, calculate the deflection curve $v(x)$ using one beam element.

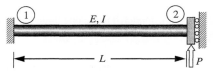

11. A cantilevered beam structure shown in the figure is under the distributed load. When $q = 1,000\,\text{N/m}$, $L_T = 1.5\,\text{m}$, $E = 207\,\text{GPa}$, and the radius of the circular cross-section $r = 0.1\,\text{m}$, solve the displacement of the neutral axis and stress on the top surface. Use three equal-length beam finite elements and the MATLAB program in Appendix. Compare the finite element solution with the exact solution. Provide the bending moment and shear force diagram from the finite element method and compare them with the exact solution. Explain why the finite element solutions are different from the exact solutions.

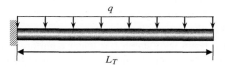

12. Model the beam shown in the figure using a two-node beam finite element.

$L = 1\,\text{m}$
$EI = 0.15\,\text{N·m}^2$
$f = 1\,\text{N/m}$

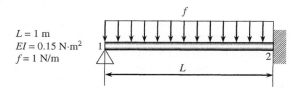

(a) Using a beam stiffness matrix, set up the equation for this beam: $([\mathbf{K}]\{\mathbf{Q}\} = \{\mathbf{F}\})$.

(b) Compute the angle of rotation at node 1, and write the equation of deformed shape of the beam using the shape functions.

(c) Explain why the answer obtained above is not likely to be very accurate. If you want to obtain a better answer for this problem using the finite element method, what would you do?

13. In this chapter, we derived the beam finite element equation using the principle of minimum potential energy. However, the same finite element equation can be derived from the Galerkin method, as in Section 3.3. The governing differential equation of the beam is

$$EI\frac{d^4 v}{dx^4} = f(x), \qquad x \in [0, L]$$

where $f(x)$ is the distributed load. In the case of a clamped beam, the boundary conditions are given by

$$v(0) = v(L) = \frac{dv}{dx}(0) = \frac{dv}{dx}(L) = 0$$

Using the Galerkin method and the interpolation scheme in Eq. (4.44), derive the finite element matrix equation when a constant distributed load $f(x) = q$ is applied along the beam.

14. Repeat the above derivation for the case of a cantilevered beam whose boundary condition is given by

$$v(0) = \frac{dv}{dx}(0) = \frac{d^2v}{dx^2}(L) = \frac{d^3v}{dx^3}(L) = 0$$

15. Solve the simply supported beam problem in Example 4.9 using the MATLAB program in Appendix. You can use either distributed load capability in the program or equivalent nodal load. Plot the vertical displacement and rotation along the span of the beam. Compare them to the analytical solutions. Also, plot the bending moment and shear force along the span of the beam. Compare them to the analytical solutions.

16. Consider a cantilevered beam with spring support at the end, as shown in the figure. Assume $E = 100$ ksi, $I = 1.0$ in^4, $L = 10$ in, $k = 200$ lb/in, beam height $h = 10$ in, and no gravity. The beam is subjected to a concentrated force $F = 100$ lb at the tip.

   (a) Using one beam element and one spring element, construct the structural matrix equation **before** applying boundary conditions. Clearly identify elements and nodes. Identify positive directions of all DOFs.

   (b) Construct the global matrix equation after applying all of the boundary conditions.

   (c) Solve the matrix equation and calculate the tip deflection.

   (d) Calculate the bending moment and shear force at the wall.

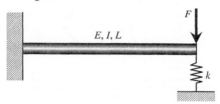

17. A beam is clamped at the left end and on a spring at the right end. The right support is such that the beam cannot rotate at that end. Thus, the only active DOF is v_2. A force $F = 3,000$ N acts downward at the right end as shown. The structure is modeled using two elements: one beam element and one spring element. The spring stiffness $k = 3,000$ N/m. The beam properties are $L = 1$ m, $EI = 1,000$ Nm^2.

   (a) Write the element stiffness matrices of both elements. Clearly show the DOFs.

   (b) Assemble the two element matrix equations and apply boundary conditions to obtain the global matrix equation.

   (c) Solve for unknown displacement v_2.

   (d) Calculate deflection v and bending moment M at $x = 0.8$ m.

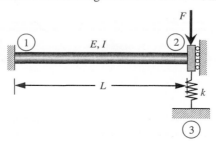

18. A linearly varying distributed load is applied to the beam finite element of length L. The maximum value of the load at the right side is q_0. Calculate "work equivalent" nodal forces and moment.

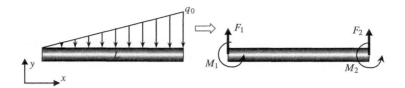

19. In general, a concentrated force can only be applied to the node. However, if we use the concept of "work-equivalent" load, we can convert the concentrated load within an element to corresponding nodal forces. A concentrated force P is applied at the center of one beam element of length L. Calculate "work equivalent" nodal forces and moments $\{F_1, M_1, F_2, M_2\}$ in terms of P and L. *Hint*: the work done by a concentrated force can be obtained by multiplying the force by displacement at that point.

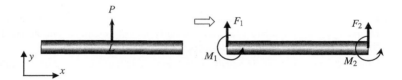

20. Use two equal-length beam elements to determine the deflection of the beam shown below. Estimate the deflection at Point B, which is at 0.5 m from the left support. $EI = 1000$ N-m^2.

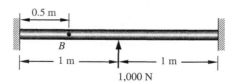

21. An external couple C_2 is applied at Node 2 in the beam shown below. When $EI = 10^5$ N-m^2, the rotations in radians at the three nodes are determined to be $\theta_1 = -0.025$, $\theta_2 = +0.05$, $\theta_3 = -0.025$

(a) Draw the shear force and bending moment diagrams for the entire beam.

(b) What is the magnitude of the couple C_2 applied at Node 2?

(c) What is the support reaction F_{y1} at Node 1?

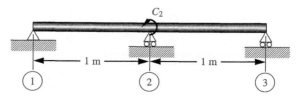

22. Two beam elements are used to model the structure shown in the figure. The beam is clamped to the wall at the left end (Node 1), supports a load $P = 100$ N at the center (Node 2), and is simply supported at the right end (Node 3). Elements 1 and 2 are two-node beam element each of length 0.05m and the flexural rigidity $EI = 0.15$ N-m^2.

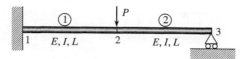

(a) Using the beam stiffness matrix, assemble the equations for the above model, apply boundary conditions and solve for the deflection/slope at Nodes 2 and 3.

(b) Write the equations and plot the deformed shape of each beam element showing defection and angles at all the nodes.

(c) What is the deflection v and slope θ at the midpoints of Element 1 and Element 2?

(d) What is the bending moments at the point of application of the load P?

(e) What are the reactions (bending moment and shear force) at the wall?

23. Consider the clamped-clamped beam shown below. Assume that there are no axial forces acting on the beam. Use two elements to solve the problem. (a) Determine the deflection and slope at $x = 0.5$, 1, and 1.5 m; (b) Draw the bending moment and shear force diagrams for the entire beam; (c) What are the support reactions? (d) Use the beam element shape functions to plot the deflected shape of the beam. Use $EI = 1,000$ N-m, $L = 1$ m, and $F = 1,000$ N.

24. The frame shown in the figure is clamped at the left end and supported on a hinged roller at the right end. The radius of circular cross-section $r = 0.05$ m. An axial force P and a couple C act at the right end. Assume the following numerical values: $L = 1$ m, $E = 80$ GPa, $P = 15,000$ N, $C = 1,000$ Nm.

(a) Use one element to determine the rotation θ at the right support.

(b) What is the deflection of the beam at $x = L/2$?

(c) What is the maximum tensile stress? Where does it occur?

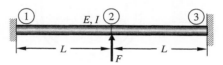

25. A frame is clamped on the left side and inclined roller supported on the right side, as shown in the figure. A uniformly distributed force q is applied and the roller contact surface is assumed to be non-frictional. When one frame finite element is used to approximate the structure, the nodal displacement vector can be defined as $\{d\} = \{u_1,\ v_1,\ \theta_1,\ u_2,\ v_2,\ q_2\}^T$, where u and v are x- and y-directional displacements, respectively, and θ is the rotation with respect to z-direction.

(a) Construct the 6×6 FE matrix equation before applying boundary conditions.

(b) Reduce the dimension of FE matrix equation to 2×2 by applying boundary conditions. You may need to use appropriate transformation.

(c) Solve the FE matrix equation and calculate nodal displacement vector $\{d\}$.

(d) Write the expression of FE vertical displacement $v(x)$, $0 < x < L$, and sketch the vertical displacement $v(x)$ of the frame.

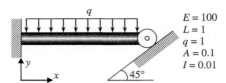

$$E = 100$$
$$L = 1$$
$$q = 1$$
$$A = 0.1$$
$$I = 0.01$$

26. A circular ring of square cross-section is subjected to a pair of forces $F = 10,000\,\text{N}$, as shown in the figure. Use a finite element analysis program to determine the compression of the ring, i.e., relative displacements of the points where forces are applied. Assume $E = 70\,\text{GPa}$. Determine the maximum values of the axial force P, bending moment M, and shear force V_r, and their respective locations.

   Hints: Divide the ring into 40 elements. Use $x = R\cos\theta$, and $y = R\sin\theta$ to determine the coordinates. Fix all DOFs at the bottom-most point to constrain rigid body translation and rotation. Otherwise, the global stiffness matrix will be singular.

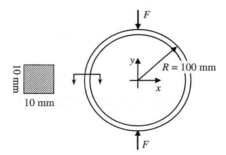

27. The ring in Problem 26 can be solved using a smaller model considering the symmetry. Use the ¼ model to determine the deflection and maximum force resultants. What are the appropriate boundary conditions for this model? Show that both models yield the same results

28. The frame shown in the figure is subjected to some forces at Nodes 2 and 3. The resulting displacements are given in the table below. Sketch the axial force, shear force, and bending moment diagrams for Element 3. What are the support reactions at Node 4? Assume $EI = 1000\,\text{N-m}^2$ and $EA = 10^7\,\text{m}^2$. Length of all elements $= 1\,\text{m}$

Node	u (mm)	v (mm)	θ (radian)
2	10	1	−0.1
3	11	−1	+0.1

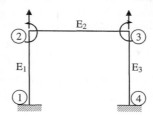

29. Solve the following frame structure using a finite element program in the Appendix. The frame is under a uniformly distributed load of $q = 1000\,\text{N/m}$ and has a circular cross-section with radius $r = 0.1\,\text{m}$. For material property, Young's modulus $E = 207\,\text{GPa}$. Plot the deformed geometry with an appropriate magnification factor, and draw a bending moment and shear force diagrams.

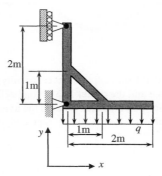

30. A cantilevered frame (Element 1) and a uniaxial bar (Element 2) are joined at Node 2 using a bolted joint as shown in the figure. Assume there is no friction at the joint. The temperature of Element 2 is raised by 200°C above the reference temperature. Both of the elements have the same length, $L = 1\,\text{m}$, Young's modulus $E = 1011\,\text{Pa}$, cross-sectional area $A = 10^{-4}\,\text{m}^2$. The frame has moment of inertia $I = 10^{-9}\,\text{m}^4$, while the bar has the coefficient of thermal expansion $\alpha = 20 \times 10^{-6}/\text{C}$. Using the finite element method, (a) determine displacements and rotation at Node 2; (b) determine the axial force in both elements; (c) determine the shear forces and bending moments in Element 1 at Nodes 1 and 2; and (d) draw the free-body diagram of Node 2 and show the force equilibrium is satisfied.

    Hint: Treat Element 1 as a plane frame element.

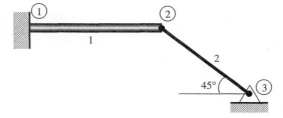

31. The figure shown below depicts a load cell made of aluminum. The ring and the stem both have square cross-section: $0.1 \times 0.1\,\text{m}^2$. Assume the Young's modulus is 72 GPa. The mean radius of the ring is 0.05 m. In a load cell, the axial load is measured from the average of strains at points P, Q, R, and S, as shown in the figure. Points P and S are on the outside surface, and Q and R are on the inside surface of the ring. Model the load cell using plane frame elements. The step portions may be modeled using one element each. Use about 20 elements to model the

entire ring. Compute the axial strain ε_{xx} at locations P, Q, R, and S for a load of 1,000 N. Draw the axial force, shear force, and bending moment diagrams for one quarter of the ring. The strain can be computed using the beam formula:

$$\sigma_{xx} = E\varepsilon_{xx} = \frac{P}{A} \pm \frac{Mc}{I}$$

where P is the axial force, M the bending moment, A the cross-sectional area, I the moment of inertia, and c the distance from the midplane.

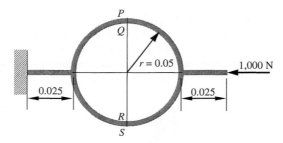

32. In Problem 31, assume that load is applied eccentrically; i.e., the distance between the line of action of the applied force and the center line of load cell, e, is not equal to zero. Calculate the strain at P, Q, R, and S for e = 0.002 m. What is the average of these strains? Comment on the results. Note: The eccentric load can be replaced by a central load and a couple of $1000 \times e$ N·m.

Chapter 5

Finite Elements for Heat Transfer Problems

5.1 INTRODUCTION

In this chapter, we will demonstrate the use of finite element analysis (FEA) in heat transfer problems, especially heat conduction in a solid. This is different from the thermal stress problem discussed in Chapter 2. For simplicity, we will derive the equations for one-dimensional heat transfer. Although, on the surface, the heat transfer problem looks different from the structural mechanics problem, there are a number of similarities between the two. The thermal conductivity is the material property that plays the role of Young's modulus, and the temperature gradient is analogous to strain. Similarly, heat flow across the boundary of the solid is analogous to the surface traction in structural analysis, and internally generated heat is similar to the body force. In heat transfer problems, we solve for the temperature field instead of the displacement field. Table 5.1 compares the terms that are used in structural mechanics and their counterparts in heat conduction. Thus, as far as the finite element method is concerned, the two problems are identical, if these terms are interpreted appropriately.

It is possible to couple the structural and heat transfer problems together. This is more practical because the structure deforms due to thermal strains caused by temperature changes. In addition to conduction and convection, heat is also transferred through radiation, which makes the problem nonlinear. However, such coupled problems are addressed in advanced finite element courses.

The matrix equation of the heat transfer problem is very similar to that of structural mechanics problems. In fact, the same finite element scheme can be used for the both problems. Starting from conservation of energy, we will obtain a matrix equation similar to the structural finite element equations, as shown below:

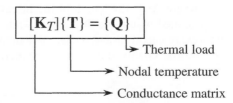

After applying the boundary conditions, the solution to the matrix equation will yield the nodal temperatures from which the temperature distribution within the solid can be calculated using the interpolation functions. Heat flow is calculated using the derivative of the temperature field.

Table 5.1   Analogy Between Structural and Heat Conduction Problems

Structural mechanics	Heat transfer
Displacement (vector)	Temperature (scalar)
Stress (tensor)	Heat flux (vector)
Displacement boundary conditions	Temperature boundary conditions
Traction boundary conditions	Surface heat input boundary conditions
Body force	Internal heat generation

5.2   FOURIER HEAT CONDUCTION EQUATION

When there is a temperature gradient in a solid, heat flows from the high-temperature region to the low-temperature region. The *Fourier law* of heat conduction states that the magnitude of the *heat flux* (heat flow per unit time) is proportional to the temperature gradient:

$$q_x = -kA \frac{dT}{dx} \tag{5.1}$$

where k, the *thermal conductivity*, is a material property, A is the area of cross section normal to the x-axis, and q_x is the *heat flux* in the x-direction. The unit of heat flux is Watts, and that of thermal conductivity is W/m/°C. The negative sign in Eq. (5.1) indicates that the direction of the heat flux is opposite to that of temperature gradient.

In Eq. (5.1), we have assumed that the temperature varies only along the x-axis, and is independent of y- and z-coordinates. This can happen in various situations. For example, in the case of heat transfer in a long wire, as shown in Figure 5.1(a), the temperature variation in the lateral directions can be ignored because their dimensions are small compared to that in the axial direction. In this case, there could be heat transfer in the lateral directions, but there is no temperature gradient. On the other hand, say in the case of a furnace wall shown in Figure 5.1(b), the temperature varies through the wall thickness, and hence there is heat flux in the thickness direction. However, there will not be any significant temperature gradient in the y- or z-directions, and hence the heat flow in those directions could be ignored. If the x-axis is parallel to the thickness direction, then the wall can also be modeled as a one-dimensional heat transfer problem.

We will derive the governing differential equation for one-dimensional heat conduction problems using the principle of *conservation of energy*. Consider an infinitesimal element (control volume) of the one-dimensional solid, as shown in Figure 5.2. The heat flux through a cross-section at x is given by q_x. The heat flux at the other end is then given by $q_x + (dq_x/dx)\Delta x$, where Δx is the length of the infinitesimal element. Let us assume that heat energy is generated within the element at a rate Q_g per unit volume. Examples of such heat generation are chemical and nuclear reactions and electrical resistance heating. If the system absorbs energy due to an endothermic reaction, then Q_g will be

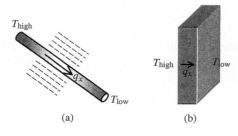

(a)   (b)

Figure 5.1 Examples of one-dimensional heat conduction problems; (a) Heat conduction in a thin long rod; (b) A furnace wall with dimensions in the y- and z-directions much greater than the thickness in the x direction

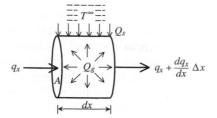

Figure 5.2 Energy balance in an infinitesimal volume

negative. Let us also assume heat enters the control volume through the lateral surfaces. We will consider two modes of heat transfer at the lateral surface. In the first type, a surface heat flow given by Q_s per unit area enters the control volume. The second mode will be convective heat transfer given by the following equation:

$$Q_h = h(T^\infty - T) \tag{5.2}$$

where h is the *convection coefficient*, T^∞ is the temperature of the surrounding fluid, and T is the surface temperature of the solid. The unit of the convection coefficient is $W/m^2/^\circ C$. Convection can also occur at the end faces of the one-dimensional body.

Consider an infinitesimal element shown in Figure 5.2. The principle of conservation of energy states that during a given time interval the sum of energy entering the element and the energy generated within the element should be equal to the sum of the energy leaving the element and the change in the internal energy. This relation can be written as

$$E_{\text{in}} + E_{\text{gen}} = \Delta U + E_{\text{out}} \tag{5.3}$$

where E_{in} is the energy entering the system, E_{out} the energy leaving the system, E_{gen} the energy generated within the system, and ΔU the increase of *internal energy*. In this text, we will derive the finite element equations only for steady-state problems wherein the temperature at a given cross-section remains constant. That is, T is only a function of x and is independent of time. In that case, the internal energy also remains constant and, hence, $\Delta U = 0$. Referring to Figure 5.2, Eq. (5.3) can be written as

$$\underbrace{q_x + Q_s P \Delta x + h(T^\infty - T)P\Delta x}_{E_{\text{in}}} + \underbrace{Q_g A \Delta x}_{E_{\text{gen}}} = \underbrace{\left(q_x + \frac{dq_x}{dx}\Delta x \right)}_{E_{\text{out}}} \tag{5.4}$$

In the above equation, A is the area of cross-section normal to the x-axis and P is the perimeter of the one-dimensional solid. The heat flux q_x can be cancelled as it appears on both sides of the above equation. Dividing by Δx throughout and letting Δx to approach zero, we obtain the following differential equation:

$$\frac{dq_x}{dx} = Q_g A + hP(T^\infty - T) + Q_s P, \quad 0 \le x \le L \tag{5.5}$$

The left-hand side (LHS) of the above equation represents the rate of change of heat flux along the length. The terms on the right-hand side (RHS) represent the heat generated and heat transferred into the system through the lateral surfaces. Our interest is in determining the temperature field $T(x)$. We use the Fourier law of heat conduction $q_x = -kAdT/dx$ in Eq. (5.1) in the above equation to obtain

$$\boxed{\frac{d}{dx}\left(kA\frac{dT}{dx} \right) + Q_g A + hP(T^\infty - T) + Q_s P = 0,} \quad 0 \le x \le L \tag{5.6}$$

Equation (5.6) is the governing differential equation for the steady-state one-dimensional heat conduction problem. The above differential equation can be uniquely solved when appropriate boundary conditions are provided. As has been discussed in Chapter 3, there are two types of boundary conditions: the essential and the natural boundary conditions. The temperature at the boundary is prescribed in the former, while the derivative of the temperature or the heat flux is prescribed in the latter. Let $x = 0$ and $x = L$ represent the boundaries where the essential and natural boundary conditions are prescribed. Then, the boundary condition can be written as

$$\begin{cases} T(0) = T_0 \\ kA \dfrac{dT}{dx}\Big|_{x=L} = q_L \end{cases} \tag{5.7}$$

where T_0 is the prescribed temperature at $x = 0$ and q_L is the prescribed heat flux at $x = L$. In Eq. (5.7), the first one is the essential boundary condition, while the second is the natural boundary condition. In deriving the natural boundary conditions we have used the sign convention that heat entering the body is considered positive and that leaving is negative. Equations (5.6) and (5.7) together constitute the boundary value problem.

5.3 FINITE ELEMENT ANALYSIS – DIRECT METHOD

We will first use an engineering approach to derive the finite element equations for the heat conduction problem. This is similar to the direct stiffness method we used for uniaxial bar elements. In this case, we do not use the differential equation but the principle of conservation of energy. Although the direct method is useful for discrete systems, it has limitations when complex heat transfer modes, such as heat generation and convection are present. Such cases will be treated using the Galerkin method.

The one-dimensional problem we are interested in is depicted in Figure 5.3. Consider a bar of varying cross-section, which we refer to as a thermodynamic system or simply a system. Heat enters (and also leaves) the system by various means. Heat can also be generated within the material volume. Examples of such situations are core in a nuclear reactor, chemical reaction, electric resistance heating or some other form of excitation using electromagnetic radiation. Heat can also enter through the lateral surface by convection caused by surrounding fluid. Another mode is radiation of heat into the system wherein the system is heated by exposure to sun or a hot flame. In the following derivation, we will ignore the heat transfer by radiation.

5.3.1 Element Conduction Equation

We divide the one-dimensional system into a number of elements. The elements are connected at nodes. In this idealized model, we assume that the heat can enter the system only though the nodes. Thus, all of the aforementioned modes of heat input should be

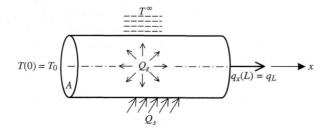

Figure 5.3 One-dimensional heat conduction of a long wire

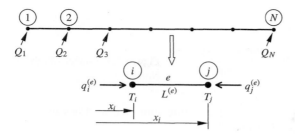

Figure 5.4 Finite elements for one-dimensional heat conduction problem

converted into equivalent nodal heat inputs. For that purpose, we can use a similar approach as the ''work-equivalent'' nodal forces in the structural elements. Let us assume that the heat input into the system at Node j is given by Q_j. The units of this heat input are Watts or Btu/s. The objective of finite element analysis is to determine the temperature distribution $T(x)$ along the length of the bar. Consider a typical element e of length $L^{(e)}$, as shown in Figure 5.4. It has two nodes, which we refer to as the first and second nodes, and the temperatures at these nodes are denoted by T_i and T_j, respectively. The heat going into the system is considered positive and that leaving negative.

The heat flow through the cross-section of the bar at Nodes i and j are denoted by $q_i^{(e)}$ and $q_j^{(e)}$. Using Fourier law of heat conduction we can relate the heat flow to the temperature as

$$q_i^{(e)} = -kA\frac{dT}{dx} = -kA\frac{(T_j - T_i)}{L^{(e)}} \tag{5.8}$$

In the above equation, the temperature gradient is approximated by $dT/dx = (T_j - T_i)/L^{(e)}$, which means that the temperature varies linearly along the x-axis.

Consider a typical element such as Element e shown in Figure 5.4. If the entire lateral surface is insulated, then from conservation of energy we have

$$q_i^{(e)} + q_j^{(e)} = 0 \tag{5.9}$$

Then, we can obtain the heat flow at Node j from Eqs. (5.8) and (5.9), as

$$q_j^{(e)} = +kA\frac{(T_j - T_i)}{L^{(e)}} \tag{5.10}$$

Equations (5.8) and (5.10) can be combined to obtain

$$\left\{\begin{matrix} q_i^{(e)} \\ q_j^{(e)} \end{matrix}\right\} = \frac{kA}{L^{(e)}}\begin{bmatrix} 1 & -1 \\ -1 & 1 \end{bmatrix}\left\{\begin{matrix} T_i \\ T_j \end{matrix}\right\} \tag{5.11}$$

where the square matrix on the RHS of the above equation is called the *element conductance matrix*. One can note the similarity between the above equation and the uniaxial bar element equation (2.16) in Chapter 2. If we interpret the thermal conductivity as Young's modulus, nodal temperatures as nodal displacements, and heat flows as element forces, then the two equations are analogous to each other. An equation similar to Eq. (5.11) can be derived for each element in the model.

5.3.2   Assembling Element Conduction Equations

Now consider the heat flow into Node 2, which is connected to Elements 1 and 2 (see Figure 5.5). The heat input Q_2 at Node 2 should be equal to the sum of the heat flow into Elements 1 and 2 through Node 2.

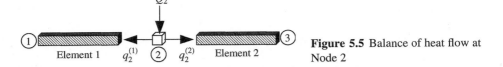

Figure 5.5 Balance of heat flow at Node 2

In general, heat input at Node i is equal to the sum of heat flows into all elements connected to Node i. This can be written as

$$Q_i = \sum_{e=1}^{N_i} q_i^{(e)} \tag{5.12}$$

where N_i is the number of elements connected to Node i. Substituting for $q_i^{(e)}$ from element conduction equation (5.11) that includes Node i, we obtain the global equations in the form

$$[\mathbf{K}_T]_{(N \times N)} \begin{Bmatrix} T_1 \\ T_2 \\ \vdots \\ T_N \end{Bmatrix} = \begin{Bmatrix} Q_1 \\ Q_2 \\ \vdots \\ Q_N \end{Bmatrix} \tag{5.13}$$

where $[\mathbf{K}_T]$ is the global conductance matrix obtained by assembling the element conductance matrices and N is the number of nodes in the model. The assembly procedure of $[\mathbf{K}_T]$ is similar to that of uniaxial bar elements described in Chapter 2. It must be mentioned that at each node either the temperature or the heat input should be known. Hence, the number of unknowns in Eq. (5.13) is always equal to the number of equations N. Unlike the structural mechanics problems, the rows and columns corresponding to prescribed temperature cannot be deleted. This is because in structural problems oftentimes the prescribed displacement at a node is equal to zero. However, in heat conduction problems, the prescribed temperature usually has a nonzero value. Thus, the action of ''striking-the-rows'' can still be used, but ''striking-the-columns'' cannot be used for heat conduction problems. Instead, the known temperature boundary condition has to be moved to the RHS of the global matrix equation. We will explain this operation in the following example.

EXAMPLE 5.1 *One-Dimensional Heat Conduction Using Finite Elements*

A heat conduction problem is modeled using four one-dimensional heat conduction elements, as shown in Figure 5.6. All elements are of the same length, $L = 1\,\text{m}$, cross-sectional area of $A = 1\,\text{m}^2$, and thermal conductivity of $k = 10\,\text{W/m/}°\text{C}$. The boundary conditions are given such that the temperature at Node 1 is prescribed as $T_1 = 200°\text{C}$, while there is no heat flux at Node 5; i.e., $Q_5 = 0$. In addition, heat enters at Nodes 2 with $Q_2 = 500$ Watts, and heat leaves from Node 4 with $Q_4 = -200$ Watts. Determine all nodal temperatures and the heat input at Node 1 in order to maintain the temperature distribution.

SOLUTION Since all elements are identical, the same element conduction equation can be used. The element matrix equations can be written as

Element 1: $10 \begin{bmatrix} 1 & -1 \\ -1 & 1 \end{bmatrix} \begin{Bmatrix} T_1 \\ T_2 \end{Bmatrix} = \begin{Bmatrix} q_1^{(1)} \\ q_2^{(1)} \end{Bmatrix}$

Element 2: $10 \begin{bmatrix} 1 & -1 \\ -1 & 1 \end{bmatrix} \begin{Bmatrix} T_2 \\ T_3 \end{Bmatrix} = \begin{Bmatrix} q_2^{(2)} \\ q_3^{(2)} \end{Bmatrix}$

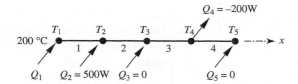

Figure 5.6 Finite elements for one-dimensional heat conduction problem

Element 3:
$$10\begin{bmatrix} 1 & -1 \\ -1 & 1 \end{bmatrix}\begin{Bmatrix} T_3 \\ T_4 \end{Bmatrix} = \begin{Bmatrix} q_3^{(3)} \\ q_4^{(3)} \end{Bmatrix}$$

Element 4:
$$10\begin{bmatrix} 1 & -1 \\ -1 & 1 \end{bmatrix}\begin{Bmatrix} T_4 \\ T_5 \end{Bmatrix} = \begin{Bmatrix} q_4^{(4)} \\ q_5^{(4)} \end{Bmatrix}$$

Now the conservation of energy can be applied at each node using Eq. (5.12). After replacing all element heat flows, q, by the element conduction equations, we can obtain the assembled global equations as

$$\begin{Bmatrix} Q_1 \\ Q_2 \\ Q_3 \\ Q_4 \\ Q_5 \end{Bmatrix} = \begin{Bmatrix} q_1^{(1)} \\ q_2^{(1)} + q_2^{(2)} \\ q_3^{(2)} + q_3^{(3)} \\ q_4^{(3)} + q_4^{(4)} \\ q_5^{(4)} \end{Bmatrix} = 10\begin{bmatrix} 1 & -1 & 0 & 0 & 0 \\ -1 & 2 & -1 & 0 & 0 \\ 0 & -1 & 2 & -1 & 0 \\ 0 & 0 & -1 & 2 & -1 \\ 0 & 0 & 0 & -1 & 1 \end{bmatrix}\begin{Bmatrix} T_1 \\ T_2 \\ T_3 \\ T_4 \\ T_5 \end{Bmatrix}$$

Since the temperature at Node 1 is prescribed ($T_1 = 200°C$), the heat flow at Node 1, Q_1, is unknown. However, external heat inputs at all other nodes are given. Thus, after substituting the known quantities in the above equations, we obtain

$$10\begin{bmatrix} 1 & -1 & 0 & 0 & 0 \\ -1 & 2 & -1 & 0 & 0 \\ 0 & -1 & 2 & -1 & 0 \\ 0 & 0 & -1 & 2 & -1 \\ 0 & 0 & 0 & -1 & 1 \end{bmatrix}\begin{Bmatrix} 200 \\ T_2 \\ T_3 \\ T_4 \\ T_5 \end{Bmatrix} = \begin{Bmatrix} Q_1 \\ 500 \\ 0 \\ -200 \\ 0 \end{Bmatrix}.$$

Note that each row has only one unknown, either the nodal heat input or temperature. Since Q_1 is unknown, we want to remove the first row so that we only have temperatures as unknowns. In Chapter 2, this was called the ''striking-the-row'' procedure. After that we have

$$10\begin{bmatrix} -1 & 2 & -1 & 0 & 0 \\ 0 & -1 & 2 & -1 & 0 \\ 0 & 0 & -1 & 2 & -1 \\ 0 & 0 & 0 & -1 & 1 \end{bmatrix}\begin{Bmatrix} 200 \\ T_2 \\ T_3 \\ T_4 \\ T_5 \end{Bmatrix} = \begin{Bmatrix} 500 \\ 0 \\ -200 \\ 0 \end{Bmatrix}$$

Unlike in structural problems, we cannot perform the "striking-the-column" procedure because the prescribed temperature T_1 is not equal to zero. Observing that the first column will be multiplied by the prescribed temperature, we move this column, after multiplying by T_1, to the RHS to yield

$$10 \begin{bmatrix} 2 & -1 & 0 & 0 \\ -1 & 2 & -1 & 0 \\ 0 & -1 & 2 & -1 \\ 0 & 0 & -1 & 1 \end{bmatrix} \begin{Bmatrix} T_2 \\ T_3 \\ T_4 \\ T_5 \end{Bmatrix} = \begin{Bmatrix} 500 \\ 0 \\ -200 \\ 0 \end{Bmatrix} + \begin{Bmatrix} 2000 \\ 0 \\ 0 \\ 0 \end{Bmatrix}$$

Note the square matrix is not singular anymore and we can solve for nodal temperatures:

$$\{\mathbf{T}\}^T = \{\, 200 \quad 230 \quad 210 \quad 190 \quad 190 \,\}°C$$

The unknown heat input Q_1 at Node 1 can be found from the deleted first equation of the global conduction equation as

$$Q_1 = 10T_1 - 10T_2 + 0T_3 + 0T_4 + 0T_5 = -300 \, \text{W}$$

Thus, to maintain a constant temperature of 200°C at Node 1, 300 W of heat must be removed from Node 1. Note that the sum of all nodal heat inputs is equal to zero, which is consistent with the steady-state heat conduction problem.

EXAMPLE 5.2   *Network of Heat Conduction Elements*

Consider a network of one-dimensional heat conduction elements, as shown in Figure 5.7. The temperature at Node 1 is fixed and heat inputs at other nodes are prescribed. Properties of the elements are listed in the table. Using the direct method, calculate the temperature at each node.

Element	$k(\text{W/m}°\text{C})$	A (m^2)	L (m)	Node i	Node j
1	10	1	1	1	2
2	15	1	$\sqrt{2}$	2	3
3	20	1	2	2	4
4	15	1	$\sqrt{2}$	3	4
5	10	1	1	4	5

SOLUTION   Using the element properties in the table, element conduction equations can be written as

Element 1:
$$10 \begin{bmatrix} 1 & -1 \\ -1 & 1 \end{bmatrix} \begin{Bmatrix} T_1 \\ T_2 \end{Bmatrix} = \begin{Bmatrix} q_1^{(1)} \\ q_2^{(1)} \end{Bmatrix}$$

Element 2:
$$\frac{15}{\sqrt{2}} \begin{bmatrix} 1 & -1 \\ -1 & 1 \end{bmatrix} \begin{Bmatrix} T_2 \\ T_3 \end{Bmatrix} = \begin{Bmatrix} q_2^{(2)} \\ q_3^{(2)} \end{Bmatrix}$$

Element 3:
$$10 \begin{bmatrix} 1 & -1 \\ -1 & 1 \end{bmatrix} \begin{Bmatrix} T_2 \\ T_4 \end{Bmatrix} = \begin{Bmatrix} q_2^{(3)} \\ q_4^{(3)} \end{Bmatrix}$$

Element 4:
$$\frac{15}{\sqrt{2}} \begin{bmatrix} 1 & -1 \\ -1 & 1 \end{bmatrix} \begin{Bmatrix} T_3 \\ T_4 \end{Bmatrix} = \begin{Bmatrix} q_3^{(4)} \\ q_4^{(4)} \end{Bmatrix}$$

Element 5:
$$10 \begin{bmatrix} 1 & -1 \\ -1 & 1 \end{bmatrix} \begin{Bmatrix} T_4 \\ T_5 \end{Bmatrix} = \begin{Bmatrix} q_4^{(5)} \\ q_5^{(5)} \end{Bmatrix}$$

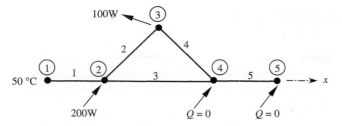

Figure 5.7 Network of heat conduction elements

Equation (5.12) can be applied for the balance of heat flow. Note that three elements are connected at Nodes 2 and 4. After replacing all heat flows by the element conduction equation, we can obtain the assembled matrix equations as

$$
\begin{Bmatrix} Q_1 \\ Q_2 \\ Q_3 \\ Q_4 \\ Q_5 \end{Bmatrix} = \begin{Bmatrix} q_1^{(1)} \\ q_2^{(1)} + q_2^{(2)} + q_2^{(3)} \\ q_3^{(2)} + q_3^{(4)} \\ q_4^{(3)} + q_4^{(4)} + q_4^{(5)} \\ q_5^{(5)} \end{Bmatrix} = \begin{bmatrix} 10 & -10 & 0 & 0 & 0 \\ -10 & 30.6 & -10.6 & -10 & 0 \\ 0 & -10.6 & 21.2 & -10.6 & 0 \\ 0 & -10 & -10.6 & 30.6 & -10 \\ 0 & 0 & 0 & -10 & 10 \end{bmatrix} \begin{Bmatrix} T_1 \\ T_2 \\ T_3 \\ T_4 \\ T_5 \end{Bmatrix}
$$

At each node, either temperature or heat flow, not both, should be known. For example, the temperature 50°C is prescribed at Node 1, while heat flows of 200, −100, 0, and 0 are prescribed at Nodes 2 through 5. The following matrix equation shows all known boundary conditions:

$$
\begin{bmatrix} 10 & -10 & 0 & 0 & 0 \\ -10 & 30.6 & -10.6 & -10 & 0 \\ 0 & -10.6 & 21.2 & -10.6 & 0 \\ 0 & -10 & -10.6 & 30.6 & -10 \\ 0 & 0 & 0 & -10 & 10 \end{bmatrix} \begin{Bmatrix} 50 \\ T_2 \\ T_3 \\ T_4 \\ T_5 \end{Bmatrix} = \begin{Bmatrix} Q_1 \\ 200 \\ -100 \\ 0 \\ 0 \end{Bmatrix}
$$

As with Example 5.1, the first row is deleted as the heat flow is unknown, and the first column is moved to the RHS after it is multiplied with $T_1 = 50°C$. Then the global conduction equation can be obtained as

$$
\begin{bmatrix} 30.6 & -10.6 & -10 & 0 \\ -10.6 & 21.2 & -10.6 & 0 \\ -10 & -10.6 & 30.6 & -10 \\ 0 & 0 & -10 & 10 \end{bmatrix} \begin{Bmatrix} T_2 \\ T_3 \\ T_4 \\ T_5 \end{Bmatrix} = \begin{Bmatrix} 700 \\ -100 \\ 0 \\ 0 \end{Bmatrix}
$$

Now, the global conductance matrix is positive definite and hence unique nodal temperatures can be obtained by solving the above set of equations. The vector of nodal temperature becomes

$$
\{\mathbf{T}\}^T = \{50 \quad 60 \quad 53.6 \quad 56.7 \quad 56.7\}°C
$$

After solving for the nodal temperature, the heat inputs at Node 1 can be obtained from the first row of the global equation as

$$
Q_1 = 10T_1 - 10T_2 + 0T_3 + 0T_4 + 0T_5 = -100\,\text{W}
$$

Thus, to maintain constant temperature of 70°C at Node 1, 100 W of heat must be removed. Note that the sum of all heat inputs Q_i is equal to zero as this is a steady-state problem.

5.4 GALERKIN METHOD FOR HEAT CONDUCTION PROBLEMS

The direct method described in the previous section used the Fourier law and the balance of heat flow. It does not require the differential equation and finite element interpolation. However, the direct method is not amenable when other modes of heat transfer such as heat generation and convection are present.

In this section, we use the Galerkin method discussed in Chapter 3 to derive the finite element equations for the one-dimensional heat conduction problem. Consider a typical element of length $L^{(e)}$, as shown in Figure 5.4. The element is composed of two nodes: the first Node i and the second Node j. In the two-node element, the temperature within the element is interpolated in terms of nodal temperatures as

$$\tilde{T}(x) = T_i N_i(x) + T_j N_j(x) \tag{5.14}$$

where the T_i and T_j are the temperatures of the first and second node of the element, respectively, and the tilde above the variable $\tilde{T}$ indicates that we are seeking an approximate solution. The interpolation functions are given by

$$N_i(x) = \left(1 - \frac{x - x_i}{L^{(e)}}\right), \quad N_j(x) = \frac{x - x_i}{L^{(e)}} \tag{5.15}$$

Note that $x_j - x_i = L^{(e)}$. In the vector notation, the above interpolation can be written as

$$\tilde{T}(x) = \lfloor \mathbf{N} \rfloor \{\mathbf{T}\} \tag{5.16}$$

where $\lfloor \mathbf{N} \rfloor = \lfloor N_i(x) \quad N_j(x) \rfloor$ is a row vector of shape functions and $\{\mathbf{T}\}$ is the column vector of nodal temperatures. The above interpolation is valid within the element; i.e., $x_i \leq x \leq x_j$. The above interpolation provides a linearly varying temperature field. In addition, since the heat flux is proportional to the temperature gradient, it is constant within an element.

$$\frac{d\tilde{T}}{dx} = \lfloor -\frac{1}{L^{(e)}} \quad \frac{1}{L^{(e)}} \rfloor \{\mathbf{T}\} = \lfloor \mathbf{B} \rfloor \{\mathbf{T}\} \tag{5.17}$$

where $\lfloor \mathbf{B} \rfloor$ is a constant row vector (1×2). Note that the vectors $\lfloor \mathbf{N} \rfloor$ and $\lfloor \mathbf{B} \rfloor$ are identical to those of the uniaxial bar element.

In the following, we will derive the element conduction equations when a heat source, Q_g, exists within the element. Other heat transfer mechanisms will be discussed later. In the presence of heat source the governing equation of the heat conduction problem in Eq. (5.6) is modified as

$$\frac{d}{dx}\left(kA\frac{dT}{dx}\right) + Q_g A = 0, \quad 0 \leq x \leq L \tag{5.18}$$

The approximation in Eq. (5.14) is substituted in the above governing equation to obtain

$$\frac{d}{dx}\left(k\frac{d\tilde{T}}{dx}\right) + AQ_g = R(x) \tag{5.19}$$

where $R(x)$ is the error or the *residual* as the approximation $\tilde{T}(x)$ will not satisfy the governing equation exactly. In the Galerkin method, the error is minimized in an average sense (see Chapter 3):

$$\int_{x_i}^{x_j} \left(\frac{d}{dx}\left(kA\frac{d\tilde{T}}{dx}\right) + AQ_g\right) N_i(x)dx = 0 \tag{5.20}$$

We use integration by parts to obtain

$$kA\frac{d\tilde{T}}{dx}N_i(x)\Big|_{x_i}^{x_j} - \int_{x_i}^{x_j} kA\frac{d\tilde{T}}{dx}\frac{dN_i}{dx}dx = -\int_{x_i}^{x_j} AQ_gN_i(x)dx \qquad (5.21)$$

Substituting for $\tilde{T}$ from Eq. (5.14) in the integrals on the LHS of Eq. (5.21), and rearranging the terms, we obtain

$$\int_{x_i}^{x_j} kA\left(T_i\frac{dN_i}{dx}+T_j\frac{dN_j}{dx}\right)\frac{dN_i}{dx}dx = \int_{x_i}^{x_j} AQ_gN_i(x)dx \qquad (5.22)$$

$$- q(x_j)N_i(x_j) + q(x_i)N_i(x_i)$$

One may note that we have substituted the Fourier's Law $q = -kAdT/dx$ in the boundary terms in the above equation. Note that the last term on the RHS becomes $q_i^{(e)} = q(x_i)$ because $N_i(x_i) = 1$ and $N_i(x_j) = 0$. On the other hand, for Node j, $q_j^{(e)} = -q(x_j)$ because the heat flow is positive when the heat enters the element (see Figure 5.4). Substituting for N_i and N_j and performing the integration, the above equation takes the following form:

$$\frac{kA}{L^{(e)}}(T_i - T_j) = Q_i^{(e)} + q_i^{(e)} \qquad (5.23)$$

where the equivalent heat input at Node i due to the heat generation term Q_g is given by

$$Q_i^{(e)} = \int_{x_i}^{x_j} AQ_gN_i(x)dx \qquad (5.24)$$

which is the *thermal load* at Node i corresponding to the distributed heat source, and $q_i^{(e)}$ is the heat flow across the cross-section at Node i into Element e.

We can repeat the procedure of minimizing the error with the weight function N_j to obtain

$$\frac{kA}{L^{(e)}}(T_j - T_i) = Q_j^{(e)} + q_j^{(e)} \qquad (5.25)$$

where

$$Q_j^{(e)} = \int_{x_i}^{x_j} AQ_gN_j(x)dx \qquad (5.26)$$

and $q_j^{(e)}$ is the heat flow across the cross-section at Node j. Note that the notation of $q_j^{(e)} = -q(x_j)$ is used. Equations (5.23) and (5.25) can be combined as

$$\frac{kA}{L^{(e)}}\begin{bmatrix} 1 & -1 \\ -1 & 1 \end{bmatrix}\begin{Bmatrix} T_i \\ T_j \end{Bmatrix} = \begin{Bmatrix} Q_i^{(e)} + q_i^{(e)} \\ Q_j^{(e)} + q_j^{(e)} \end{Bmatrix}$$

or

$$\boxed{[k_T^{(e)}]\{T\} = \{Q^{(e)}\} + \{q^{(e)}\}} \qquad (5.27)$$

where $[k_T^{(e)}]$ is the conductance matrix of Element e, $\{Q^{(e)}\}$ is the vector of thermal loads corresponding to the heat source, and $\{q^{(e)}\}$ is the vector of nodal heat flows across the cross-section.

When a uniform heat source exists in the element, Eqs. (5.24) and (5.26) yield

$$\left\{ Q^{(e)} \right\} = \int_{x_i}^{x_j} AQ_g \begin{bmatrix} N_i(x) \\ N_j(x) \end{bmatrix} dx = \frac{AQ_g L^{(e)}}{2} \left\{ \begin{matrix} 1 \\ 1 \end{matrix} \right\} \tag{5.28}$$

Note that $AQ_g L^{(e)}$ is the total heat generated in the element, and it is equally divided between the two nodes.

As mentioned before, the temperature is linear within an element. If the differential equation (5.18) is directly integrated with uniform heat source, the temperature will be a quadratic polynomial. Thus, the finite element solution is approximate, and the error will be reduced as more elements are used.

EXAMPLE 5.3   *Temperature Distribution in a Heat Chamber Wall*

Consider a heat chamber in which the temperature inside is maintained at a constant value of 200°C (see Figure 5.8). The chamber is covered by a 1.0-m-thick metal wall and outside is insulated so that no heat flows to the outer surface of the wall. There is a uniform heat source inside the wall generating $Q_g = 400 \, \text{W/m}^3$. The thermal conductivity of the wall is $k = 25 \, \text{W/m°C}$. Assuming that the temperature only varies in the x-direction, find the temperature distribution in the wall.

SOLUTION   From the assumption that the temperature only varies in the thickness direction, it is possible to model a unit slice of the wall as a one-dimensional problem (see Figure 5.9). There is a uniform heat source inside the wall generating $Q_g = 400 \, \text{W/m}^3$. We will model the temperature variation in the thickness direction using four one-dimensional finite elements.

The boundary conditions are: temperature at the leftmost node is prescribed, i.e., $T_1 = 200°C$, and the heat flux at the rightmost node vanishes, i.e., $Q_5 = 0$.

Since the elements are identical, the same element matrix equation can be used for all elements. The element matrix equations can be written as

Element 1: $100 \begin{bmatrix} 1 & -1 \\ -1 & 1 \end{bmatrix} \left\{ \begin{matrix} T_1 \\ T_2 \end{matrix} \right\} = \left\{ \begin{matrix} 50 \\ 50 \end{matrix} \right\} + \left\{ \begin{matrix} q_1^{(1)} \\ q_2^{(1)} \end{matrix} \right\}$

Element 2: $100 \begin{bmatrix} 1 & -1 \\ -1 & 1 \end{bmatrix} \left\{ \begin{matrix} T_2 \\ T_3 \end{matrix} \right\} = \left\{ \begin{matrix} 50 \\ 50 \end{matrix} \right\} + \left\{ \begin{matrix} q_2^{(2)} \\ q_3^{(2)} \end{matrix} \right\}$

Element 3: $100 \begin{bmatrix} 1 & -1 \\ -1 & 1 \end{bmatrix} \left\{ \begin{matrix} T_3 \\ T_4 \end{matrix} \right\} = \left\{ \begin{matrix} 50 \\ 50 \end{matrix} \right\} + \left\{ \begin{matrix} q_3^{(3)} \\ q_4^{(3)} \end{matrix} \right\}$

Element 4: $100 \begin{bmatrix} 1 & -1 \\ -1 & 1 \end{bmatrix} \left\{ \begin{matrix} T_4 \\ T_5 \end{matrix} \right\} = \left\{ \begin{matrix} 50 \\ 50 \end{matrix} \right\} + \left\{ \begin{matrix} q_4^{(4)} \\ q_5^{(4)} \end{matrix} \right\}$

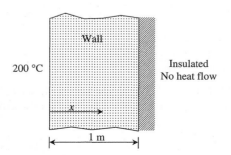

Figure 5.8 Heat transfer problem for insulated wall

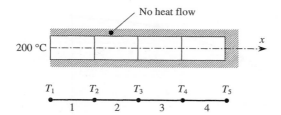

Figure 5.9 Finite element approximation of the wall

Note that the first term on the RHS is the thermal load from uniformly distributed heat source, which can be obtained using Eq. (5.28). Now, the conservation of energy at each node can be applied using Eq. (5.12). After replacing all heat flows by the element conduction equation, we obtain the assembled matrix equations as

$$
\begin{Bmatrix} Q_1 \\ Q_2 \\ Q_3 \\ Q_4 \\ Q_5 \end{Bmatrix} = \begin{Bmatrix} q_1^{(1)} \\ q_2^{(1)} + q_2^{(2)} \\ q_3^{(2)} + q_3^{(3)} \\ q_4^{(3)} + q_4^{(4)} \\ q_5^{(4)} \end{Bmatrix} = 100 \begin{bmatrix} 1 & -1 & 0 & 0 & 0 \\ -1 & 2 & -1 & 0 & 0 \\ 0 & -1 & 2 & -1 & 0 \\ 0 & 0 & -1 & 2 & -1 \\ 0 & 0 & 0 & -1 & 1 \end{bmatrix} \begin{Bmatrix} T_1 \\ T_2 \\ T_3 \\ T_4 \\ T_5 \end{Bmatrix} - \begin{Bmatrix} 50 \\ 100 \\ 100 \\ 100 \\ 50 \end{Bmatrix}
$$

Two boundary conditions are given. The temperature at Node 1 is fixed to 200°C, while heat flux at Node 5 is zero. Since the temperature is given at Node 1, we do not know how much heat needs to flow in order to maintain the constant temperature; i.e., Q_1 is unknown. There is no heat input for other nodes except for the contribution from the heat source; i.e., $Q_2 = Q_3 = Q_4 = Q_5 = 0$. Thus, after identifying known and unknown terms in the above equations, we obtain

$$
100 \begin{bmatrix} 1 & -1 & 0 & 0 & 0 \\ -1 & 2 & -1 & 0 & 0 \\ 0 & -1 & 2 & -1 & 0 \\ 0 & 0 & -1 & 2 & -1 \\ 0 & 0 & 0 & -1 & 1 \end{bmatrix} \begin{Bmatrix} 200 \\ T_2 \\ T_3 \\ T_4 \\ T_5 \end{Bmatrix} = \begin{Bmatrix} Q_1 + 50 \\ 100 \\ 100 \\ 100 \\ 50 \end{Bmatrix}
$$

Note that each row has only one unknown, either heat input or temperature. Since Q_1 is unknown, we want to remove the first row so that we only have temperature unknowns. We have

$$
100 \begin{bmatrix} -1 & 2 & -1 & 0 & 0 \\ 0 & -1 & 2 & -1 & 0 \\ 0 & 0 & -1 & 2 & -1 \\ 0 & 0 & 0 & -1 & 1 \end{bmatrix} \begin{Bmatrix} 200 \\ T_2 \\ T_3 \\ T_4 \\ T_5 \end{Bmatrix} = \begin{Bmatrix} 100 \\ 100 \\ 100 \\ 50 \end{Bmatrix}
$$

We cannot perform the "striking-the-column" procedure as we did in structural problems, because the prescribed temperature is not equal to zero. Observing that the first column will be multiplied by the prescribed temperature, we can move this column to the RHS to obtain

$$
100 \begin{bmatrix} 2 & -1 & 0 & 0 \\ -1 & 2 & -1 & 0 \\ 0 & -1 & 2 & -1 \\ 0 & 0 & -1 & 1 \end{bmatrix} \begin{Bmatrix} T_2 \\ T_3 \\ T_4 \\ T_5 \end{Bmatrix} = \begin{Bmatrix} 20100 \\ 100 \\ 100 \\ 50 \end{Bmatrix}
$$

Note the square matrix is not singular and we can solve for nodal temperatures:

$$ T_1 = 200°C, \quad T_2 = 203.5°C, \quad T_3 = 206°C, \quad T_4 = 207.5°C, \quad T_5 = 208°C $$

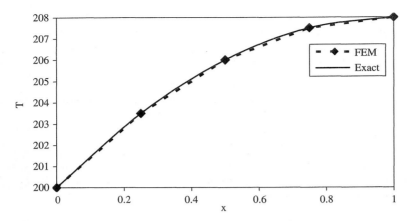

Figure 5.10 Temperature distribution along the wall thickness

The unknown heat flow Q_1 at Node 1 can be calculated from the first equation of the global conduction equations.

$$Q_1 + 50 = 100T_1 - 100T_2 = -350$$
$$\Rightarrow Q_1 = -400\,\text{W}$$

Thus, to maintain a constant temperature at the inner wall surface, 400 W of heat should be removed from the inside wall. Note this is also equal to the total hear generated.

Figure 5.10 plots the temperature variation in the wall. The exact solution is $T(x) = -8x^2 + 16x + 200$. Note that the finite element solution matches the exact solution at each node. But, due to linear interpolation the temperature within an element deviates from the exact one.

EXAMPLE 5.4   *Thermal Protection System of a Space Vehicle*

A thermal protection system for a space vehicle consists of a 0.1-m-thick outer ceramic foam and 0.1-m-thick inner metal foam. The thermal conductivities of the ceramic and metal foams, respectively, are 0.05 and 0.025 W/m/°K. A fluid is passed through the metal foam, which absorbs heat at the rate of 1,000 W/m^3. The metal foam is attached to the vehicle structure and is insulated. The outside temperature of the ceramic foam is estimated to be 1,000 °K. Use one-dimensional finite elements to determine the temperature distribution in the foams. What is the temperature of the vehicle structure? Calculate the heat entering the system due to aerodynamic heating.

SOLUTION   From the assumption that the temperature only varies in the thickness direction, it is possible to model a unit slice ($A = 1\,\text{m}^2$) of the wall as a one-dimensional problem. Since there is no distributed

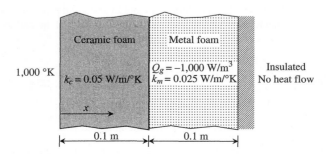

Figure 5.11 Heat transfer of a thermal protection system for a space vehicle

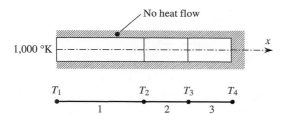

Figure 5.12 Finite element model of the thermal protection system

heat source in the ceramic foam, one finite element is enough. Due to heat absorption by fluid in the metal foam, more than one element is required for the metal foam. We will model the ceramic foam using one element, while two elements will be used for the metal foam as shown in Figure 5.12.

The element conduction matrix equations can be written as

Element 1:
$$\begin{bmatrix} 0.5 & -0.5 \\ -0.5 & 0.5 \end{bmatrix} \begin{Bmatrix} T_1 \\ T_2 \end{Bmatrix} = \begin{Bmatrix} q_1^{(1)} \\ q_2^{(1)} \end{Bmatrix}$$

Element 2:
$$\begin{bmatrix} 0.5 & -0.5 \\ -0.5 & 0.5 \end{bmatrix} \begin{Bmatrix} T_2 \\ T_3 \end{Bmatrix} = \begin{Bmatrix} -25 \\ -25 \end{Bmatrix} + \begin{Bmatrix} q_2^{(2)} \\ q_3^{(2)} \end{Bmatrix}$$

Element 3:
$$\begin{bmatrix} 0.5 & -0.5 \\ -0.5 & 0.5 \end{bmatrix} \begin{Bmatrix} T_3 \\ T_4 \end{Bmatrix} = \begin{Bmatrix} -25 \\ -25 \end{Bmatrix} + \begin{Bmatrix} q_3^{(3)} \\ q_4^{(3)} \end{Bmatrix}$$

Note that for Elements 2 and 3, the first term on the RHS is the contribution from uniformly distributed heat source, which can be obtained using Eq. (5.28). For assembly, the conservation of energy is applied at each node using Eq. (5.12). After replacing all heat flows by the element conduction equations, we can obtain the assembled matrix equations as

$$\begin{Bmatrix} Q_1 \\ Q_2 \\ Q_3 \\ Q_4 \end{Bmatrix} = \begin{Bmatrix} q_1^{(1)} \\ q_2^{(1)} + q_2^{(2)} \\ q_3^{(2)} + q_3^{(3)} \\ q_4^{(3)} \end{Bmatrix} = \begin{bmatrix} 0.5 & -0.5 & 0 & 0 \\ -0.5 & 1.0 & -0.5 & 0 \\ 0 & -0.5 & 1.0 & -0.5 \\ 0 & 0 & -0.5 & 0.5 \end{bmatrix} \begin{Bmatrix} T_1 \\ T_2 \\ T_3 \\ T_4 \end{Bmatrix} + \begin{Bmatrix} 0 \\ 25 \\ 50 \\ 25 \end{Bmatrix}$$

Two boundary conditions are given. The temperature at Node 1 is 1,000°K, while heat flux at Node 4 is zero. Since the temperature is given at Node 1, the heat flow is an unknown. There is no heat input for other nodes except for the contribution from the heat source. Thus, after identifying known and unknown terms from the above equation, we obtain

$$\begin{bmatrix} 0.5 & -0.5 & 0 & 0 \\ -0.5 & 1.0 & -0.5 & 0 \\ 0 & -0.5 & 1.0 & -0.5 \\ 0 & 0 & -0.5 & 0.5 \end{bmatrix} \begin{Bmatrix} 1000 \\ T_2 \\ T_3 \\ T_4 \end{Bmatrix} = \begin{Bmatrix} 0 \\ -25 \\ -50 \\ -25 \end{Bmatrix} + \begin{Bmatrix} Q_1 \\ 0 \\ 0 \\ 0 \end{Bmatrix}$$

Note that each row has only one unknown, either heat input or temperature. Since Q_1 is unknown, we want to remove the first row so that we only have temperature unknowns. In Chapter 2, this was called the "striking-the-row" procedure. We have

$$\begin{bmatrix} -0.5 & 1.0 & -0.5 & 0 \\ 0 & -0.5 & 1.0 & -0.5 \\ 0 & 0 & -0.5 & 0.5 \end{bmatrix} \begin{Bmatrix} 1000 \\ T_2 \\ T_3 \\ T_4 \end{Bmatrix} = \begin{Bmatrix} -25 \\ -50 \\ -25 \end{Bmatrix}$$

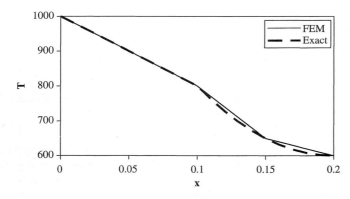

Figure 5.13 Temperature distribution in the thermal protection system

Observing that the first column will be multiplied by the prescribed temperature, we can move this column, after it is multiplied by the known temperature, to the RHS to yield

$$\begin{bmatrix} 1.0 & -0.5 & 0 \\ -0.5 & 1.0 & -0.5 \\ 0 & -0.5 & 0.5 \end{bmatrix} \begin{Bmatrix} T_2 \\ T_3 \\ T_4 \end{Bmatrix} = \begin{Bmatrix} 475 \\ -50 \\ -25 \end{Bmatrix}$$

Note that the matrix is not singular, and we can solve for nodal temperatures:

$$T_1 = 1,000°\text{K}, \ T_2 = 800°\text{K}, \ T_3 = 650°\text{K}, \ T_4 = 600°\text{K}$$

The unknown heat flow Q_1 at Node 1 can be calculated from the first equation of the global conduction equations.

$$Q_1 = 0.5T_1 - 0.5T_2 = 100\,\text{W}$$

Q_1 is the heat entering the system due to aerodynamic heating.

The exact solution can be obtained by integrating the governing differential equation in the two separate regions:

$$T(x) = \begin{cases} -2000x + 1000 & 0 \le x \le 0.1 \\ 20000x^2 - 8000x + 1400 & 0.1 \le x \le 0.2 \end{cases}$$

Figure 5.13 shows the temperature distributions in the ceramic and metal foams obtained from FEA and analytical method. The results from FEA are accurate in the ceramic foam. However, the FEA results for the metal foam are only approximate. The actual temperature variation is quadratic in x, whereas in FEA the temperature is piecewise linear.

5.5   CONVECTION BOUNDARY CONDITIONS

When a structure is surrounded by a fluid that has a different temperature from the structure, heat flow between the structure and fluid occurs and is called *convection*. Convection presents a special type of boundary conditions called the mixed boundary condition. The closest analogy to this in structural mechanics problems is that of a beam resting on an elastic spring (See Problem 4 in Section 4.8). The convection boundary condition contains an unknown temperature. For example, convective heat flow can be written as

$$q_h = hS(T^\infty - T) \tag{5.29}$$

where h is the convection coefficient, S is the area of the wet surface, and T^∞ is the surrounding fluid temperature, which is assumed to be constant. Note that the amount of heat flow is not prescribed. Rather, it is a function of surface temperature, which is unknown.

We will consider two different types of convection heat transfer. In the first case, the convection occurs at the end faces of the one-dimensional body. For example, the inner surface of the heat chamber in Example 5.3 can be exposed to a hot fluid, rather than prescribing a known temperature. The same applies to the outer surface of the ceramic foam in the thermal protection system in Example 5.4. In this case, the wet area S is the same as the cross-sectional area A. In the second case, the convection heat transfer occurs throughout the length of the one-dimensional body. For example, an electric wire immersed in water can transfer heat to the surrounding water. This type of convection is not a boundary condition in a strict sense. Rather it is treated as a distributed heat flow. Thus, it should be included in the governing differential equation as in Eq. (5.6). In this case, the wet area S is the perimeter times the length.

5.5.1   Convection on the boundary

For illustration purpose, consider two one-dimensional heat transfer elements, as shown in Figure 5.14. For simplicity, let us assume that both elements are identical in terms of geometry and material properties. Both ends of the system are under convective boundary conditions. An insulating wall can be an example for this type of problem.

Since both elements are identical, the element conduction equations can be written as

Element 1:

$$\frac{kA}{L}\begin{bmatrix} 1 & -1 \\ -1 & 1 \end{bmatrix}\begin{Bmatrix} T_1 \\ T_2 \end{Bmatrix} = \begin{Bmatrix} q_1^{(1)} \\ q_2^{(1)} \end{Bmatrix}$$

Element 2:

$$\frac{kA}{L}\begin{bmatrix} 1 & -1 \\ -1 & 1 \end{bmatrix}\begin{Bmatrix} T_2 \\ T_3 \end{Bmatrix} = \begin{Bmatrix} q_2^{(2)} \\ q_3^{(2)} \end{Bmatrix}$$

The balance of heat flow in Eq. (5.12) can be applied to all nodes in conjunction with convective heat flow in Eq. (5.29) as

Node 1:

$$q_1^{(1)} = h_1 A(T_1^\infty - T_1)$$

Node 2:

$$q_2^{(1)} + q_2^{(2)} = 0$$

Node 3:

$$q_3^{(2)} = h_3 A(T_3^\infty - T_3)$$

At Nodes 1 and 3, the amount of heat entering the element is equal to the heat transferred by convection, while at Node 2 the sum of heat entering Elements 1 and 2 is equal to zero.

T_1^∞ <.......> T_1 T_2 T_3 <.......> T_3^∞ **Figure 5.14** Finite element approximation of
h_1 1 2 h_3 the furnace wall

After substituting the element conduction equation in the balance of heat flow equations, we can assemble the global heat conduction equation as

$$\frac{kA}{L}\begin{bmatrix} 1 & -1 & 0 \\ -1 & 2 & -1 \\ 0 & -1 & 1 \end{bmatrix}\begin{Bmatrix} T_1 \\ T_2 \\ T_3 \end{Bmatrix} = \begin{Bmatrix} h_1 A(T_1^\infty - T_1) \\ 0 \\ h_3 A(T_3^\infty - T_3) \end{Bmatrix}$$

The above equation cannot be solved as it is because the RHS includes unknown nodal temperatures, T_1 and T_3. This is due to the convection boundary conditions at Nodes 1 and 3. To solve the above equations, we move the terms containing unknown nodal temperatures to the LHS. Then we obtain the following global equations:

$$\begin{bmatrix} \frac{kA}{L}+h_1 A & -\frac{kA}{L} & 0 \\ -\frac{kA}{L} & \frac{2kA}{L} & -\frac{kA}{L} \\ 0 & -\frac{kA}{L} & \frac{kA}{L}+h_3 A \end{bmatrix}\begin{Bmatrix} T_1 \\ T_2 \\ T_3 \end{Bmatrix} = \begin{Bmatrix} h_1 A T_1^\infty \\ 0 \\ h_3 A T_3^\infty \end{Bmatrix} \qquad (5.30)$$

The square matrix in the LHS is non-singular due to the addition of terms to the diagonal, which makes the matrix positive definite.

EXAMPLE 5.5 *Convection in the Furnace Wall*

A furnace wall, as shown in Figure 5.15, consists of two layers, firebrick and insulating brick. The thermal conductivities are $k_1 = 1.2$ W/m/°C for the firebrick and $k_2 = 0.2$ W/m/°C for the insulating brick. The temperature inside the furnace is $T_f = 1,500$°C and the convection coefficient at the inner surface is $h_i = 12$ W/m²/°C. The ambient temperature is $T_a = 20$°C and the convection coefficient at the outer surface is $h_o = 2.0$ W/m²/°C. The thermal resistance of the interface between firebrick and insulating brick can be neglected. Using two finite elements (one for each brick), determine the rate of heat loss through the outer wall, the temperature T_i at the inner surface, and the temperature T_o at the outer surface.

SOLUTION Assuming that the furnace is tall enough, the heat flux along the vertical direction can be ignored, and we can assume that the heat flow is one-dimensional in the thickness direction. It is then possible to model a unit slice ($A = 1$ m²) of the wall as a one-dimensional problem. We will model the temperature change in the thickness direction using two one-dimensional finite elements (see Figure 5.16).

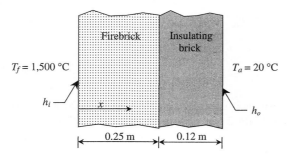

Figure 5.15 Heat transfer problem of a furnace wall

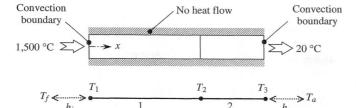

Convection boundary

No heat flow

Convection boundary

1,500 °C

x

20 °C

T_f $\quad$ h_i $\quad$ T_1 $\quad$ 1 $\quad$ T_2 $\quad$ 2 $\quad$ T_3 $\quad$ h_o $\quad$ T_a

Figure 5.16 Finite element approximation of the furnace wall

There are two convection boundaries: one at the inside of the wall and the other at the outer wall. Due to high temperature inside, heat will be transferred from inside to the wall and from wall to the outside. Using Eq. (5.30), we can construct the global matrix equations as

$$
\begin{bmatrix}
16.8 & -4.8 & 0 \\
-4.8 & 6.47 & -1.67 \\
0 & -1.67 & 3.67
\end{bmatrix}
\begin{Bmatrix}
T_1 \\
T_2 \\
T_3
\end{Bmatrix}
=
\begin{Bmatrix}
18,000 \\
0 \\
40
\end{Bmatrix}
$$

Note that the lengths and conductivities of the two elements are different. The matrix in the above equation is positive definite, and a unique solution (nodal temperatures) can be obtained as

$$\{\mathbf{T}\}^T = \{\, 1{,}411 \quad 1{,}190 \quad 552 \,\}°C$$

The surface temperature of the inner wall is $1{,}411°C$, and that of the outer wall is $552°C$. The rate of heat loss through the outer wall can be obtained from the convection boundary condition:

$$q_3^{(2)} = h_0(T_a - T_3) = -1054 \, \text{W/m}^2$$

The negative sign means the heat leaves the system.

The exact solution of the problem can be obtained by integrating the governing differential equation and applying the convection boundary conditions:

$$
T(x) = \begin{cases}
-883x + 1411 & 0 \le x \le 0.25 \\
-5300x + 2513 & 0.25 \le x \le 0.37
\end{cases}
$$

The finite element analysis results are identical to the exact solution.

5.5.2 Convection along the length of a rod

When a long rod is submerged into a fluid, convection occurs across the entire surface. This is different from the previous convection boundary condition in which convection occurs only at the end faces. As shown in Figure 5.17, the heat flow in convection is proportional to the perimeter of the cross-section multiplied by the length of the rod. In such a case, it is convenient to think of the convection heat flow as a distributed thermal load.

The governing differential equation is [see Eq. (5.6)]

$$\frac{d}{dx}\left(kA\frac{dT}{dx}\right) + AQ_g + hP(T^\infty - T) = 0, \quad 0 \le x \le L \tag{5.31}$$

where P is the perimeter of the cross-section, i.e., $P = 2(b + h)$. The approximate temperature in Eq. (5.14) is substituted in the above governing equation to obtain

$$\frac{d}{dx}\left(kA\frac{d\tilde{T}}{dx}\right) + AQ_g + hP(T^\infty - \tilde{T}) = R(x) \tag{5.32}$$

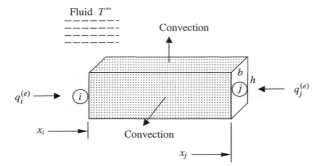

Figure 5.17 Heat conduction and convection in a long rod

where $R(x)$ is the error or the residual, as the approximation $\tilde{T}(x)$ will not satisfy the governing equation exactly. The above error is minimized by multiplying with by the weight function, $N_i(x)$, and integrating over the element length as follows:

$$\int_{x_i}^{x_j} \left(\frac{d}{dx}\left(kA\frac{d\tilde{T}}{dx}\right) + AQ_g + hP(T^\infty - \tilde{T}) \right) N_i(x)dx = 0 \tag{5.33}$$

We use integration by parts to obtain

$$kA\frac{d\tilde{T}}{dx}N_i(x)\bigg|_{x_i}^{x_j} - \int_{x_i}^{x_j}kA\frac{d\tilde{T}}{dx}\frac{dN_i}{dx}dx - \int_{x_i}^{x_j}hP\tilde{T}N_idx$$

$$= -\int_{x_i}^{x_j}AQ_gN_i(x)dx - \int_{x_j}hPT^\infty N_idx \tag{5.34}$$

Substituting for $\tilde{T}$ from Eq. (5.14) in the integrals on the LHS of Eq. (5.34) and rearranging the terms, we obtain

$$\int_{x_i}^{x_j}kA\left(T_i\frac{dN_i}{dx} + T_j\frac{dN_j}{dx}\right)\frac{dN_i}{dx}dx + \int_{x_i}^{x_j}hP(T_iN_i + T_jN_j)N_i\,dx$$

$$= \int_{x_i}^{x_j}(AQ_g + hPT^\infty)N_i\,dx - q(x_j)N_i(x_j) + q(x_i)N_i(x_i) \tag{5.35}$$

One may note that we have substituted the Fourier's Law $q = -kAdT/dx$ in the boundary terms in the above equation. Substituting for N_i and N_j and performing the integration, the above equation takes the form

$$\frac{kA}{L^{(e)}}(T_i - T_j) + hpL^{(e)}\left(\frac{T_i}{3} + \frac{T_j}{6}\right) = Q_i^{(e)} + q_i^{(e)} \tag{5.36}$$

where the thermal load at Node i due to the heat generation term Q_g and the convection term is given by

$$Q_i^{(e)} = \int_{x_i}^{x_j}(AQ_g + hPT^\infty)N_i(x)dx \tag{5.37}$$

and $Q_i^{(e)}$ is the heat flow across the cross-section at Node i into Element e. We can repeat the averaging procedure with the weight function N_j to obtain

$$\frac{kA}{L^{(e)}}(T_j - T_i) + hpL^{(e)}\left(\frac{T_i}{6} + \frac{T_j}{3}\right) = Q_j^{(e)} + q_j^{(e)} \qquad (5.38)$$

where

$$Q_j^{(e)} = \int_{x_i}^{x_j}(AQ_g + hPT^\infty)N_j(x)dx \qquad (5.39)$$

and $Q_j^{(e)}$ is the heat flow across the cross-section at Node j. Equations (5.36) and (5.38) can be combined as

$$\left[\frac{kA}{L^{(e)}}\begin{bmatrix} 1 & -1 \\ -1 & 1 \end{bmatrix} + \frac{hPL}{6}\begin{bmatrix} 2 & 1 \\ 1 & 2 \end{bmatrix}\right]\left\{\begin{matrix} T_i \\ T_j \end{matrix}\right\} = \left\{\begin{matrix} Q_i^{(e)} + q_i^{(e)} \\ Q_j^{(e)} + q_j^{(e)} \end{matrix}\right\}$$

or

$$\left[\left[\mathbf{k}_T^{(e)}\right] + \left[\mathbf{k}_h^{(e)}\right]\right]\{\mathbf{T}\} = \{\mathbf{Q}^{(e)}\} + \{\mathbf{q}^{(e)}\} \qquad (5.40)$$

Note that the first matrix on the LHS is the conductance matrix in Eq. (5.27), and the equivalent conductance matrix due to convective heat transfer across the periphery is given by

$$\left[\mathbf{k}_h^{(e)}\right] = \frac{hPL}{6}\begin{bmatrix} 2 & 1 \\ 1 & 2 \end{bmatrix} \qquad (5.41)$$

Using the balance of heat flow in Eq. (5.12), the above equations can be assembled to obtain the global matrix equations as

$$\boxed{\left[\left[\mathbf{K}_T\right] + \left[\mathbf{K}_h\right]\right]\{\mathbf{T}\} = \{\mathbf{Q}\}} \qquad (5.42)$$

Procedures for applying boundary conditions are identical to the previous section.

The thermal load vector on the RHS of Eq. (5.40) includes the contribution from the heat source and the convection term. When there is a uniformly distributed heat source Q_g, the thermal load in Eqs. (5.37) and (5.39) can be integrated to obtain

$$\left\{\mathbf{Q}^{(e)}\right\} = \left\{\begin{matrix} Q_i \\ Q_j \end{matrix}\right\} = \frac{AQ_gL^{(e)} + hPL^{(e)}T^\infty}{2}\left\{\begin{matrix} 1 \\ 1 \end{matrix}\right\} \qquad (5.43)$$

Note that the total amount of the thermal load is equally divided between the two nodes.

EXAMPLE 5.6 *Heat Flow in a Cooling Fin*

Determine the steady-state temperature distribution in a thin rectangular fin shown in Figure 5.18. The fin is 120 mm long-160 mm wide- and 1.25 mm thick. The inside wall is at a temperature of 330°C. Two side surfaces are insulated, while the top and bottom surfaces as well as the end surface are exposed to the air. The ambient air temperature is 30°C. Assume $k = 0.2$ W/mm/°C and $h = 2 \times 10^{-4}$ W/mm²/°C.

SOLUTION The problem can be treated as a one-dimensional problem because the thickness of the fin is small so that the temperature variation in the thickness direction can be neglected. A

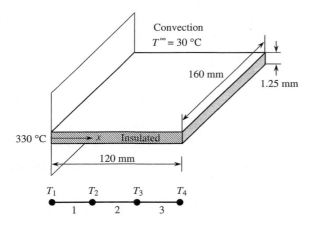

Figure 5.18 Heat flow through a thin fin and finite element model

one-dimensional finite element model using three linear elements can be used to calculate the temperature distribution. The temperature at the node 1 is prescribed, while the heat flux at node 4 is controlled by the convection boundary condition. In addition, there is convection heat transfer through the perimeter of the fin.

The conductance matrix and thermal load vector for all elements are identical and can be calculated from Eq. (5.40) as

$$[\mathbf{k}_T^{(e)}] + [\mathbf{k}_h^{(e)}] = \frac{0.2 \times 200}{40} \begin{bmatrix} 1 & -1 \\ -1 & 1 \end{bmatrix} + \frac{2 \times 10^{-4} \times 320 \times 40}{6} \begin{bmatrix} 2 & 1 \\ 1 & 2 \end{bmatrix}$$

and

$$\{\mathbf{Q}^{(e)}\} = \frac{2 \times 10^{-4} \times 320 \times 40 \times 30}{2} \begin{Bmatrix} 1 \\ 1 \end{Bmatrix}$$

Thus, the element conduction equation becomes

Element 1:
$$\begin{bmatrix} 1.8533 & -0.5733 \\ -0.5733 & 1.8533 \end{bmatrix} \begin{Bmatrix} T_1 \\ T_2 \end{Bmatrix} = \begin{Bmatrix} 38.4 \\ 38.4 \end{Bmatrix} + \begin{Bmatrix} q_1^{(1)} \\ q_2^{(1)} \end{Bmatrix}$$

Element 2:
$$\begin{bmatrix} 1.8533 & -0.5733 \\ -0.5733 & 1.8533 \end{bmatrix} \begin{Bmatrix} T_2 \\ T_3 \end{Bmatrix} = \begin{Bmatrix} 38.4 \\ 38.4 \end{Bmatrix} + \begin{Bmatrix} q_2^{(2)} \\ q_3^{(2)} \end{Bmatrix}$$

Element 3:
$$\begin{bmatrix} 1.8533 & -0.5733 \\ -0.5733 & 1.8533 \end{bmatrix} \begin{Bmatrix} T_3 \\ T_4 \end{Bmatrix} = \begin{Bmatrix} 38.4 \\ 38.4 \end{Bmatrix} + \begin{Bmatrix} q_3^{(3)} \\ q_4^{(3)} \end{Bmatrix}$$

The balance of heat flow in Eq. (5.12) can be applied for all nodes in conjunction with the convection heat flow in Eq. (5.29) as

Node 1:
$$q_1^{(1)} = Q_1$$

Node 2:
$$q_2^{(1)} + q_2^{(2)} = 0$$

Node 3:
$$q_3^{(2)} + q_3^{(2)} = 0$$

Node 4:
$$q_4^{(3)} = hA(T^\infty - T_4)$$

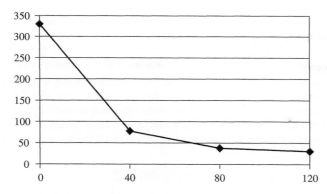

Figure 5.19 Temperature distribution in a thin fin

After substituting the element conduction equations, we obtain the global matrix equation as

$$
\begin{bmatrix}
1.853 & -.573 & 0 & 0 \\
-.573 & 3.706 & -.573 & 0 \\
0 & -.573 & 3.706 & -.573 \\
0 & 0 & -.573 & 1.853
\end{bmatrix}
\begin{Bmatrix}
T_1 \\ T_2 \\ T_3 \\ T_4
\end{Bmatrix}
=
\begin{Bmatrix}
38.4 + Q_1 \\
76.8 \\
76.8 \\
38.4 + hA(T^\infty - T_4)
\end{Bmatrix}.
$$

Note that the term containing the unknown temperature T_4 in the last equation needs to be moved to the LHS. After that, the following matrix equations is obtained:

$$
\begin{bmatrix}
1.853 & -.573 & 0 & 0 \\
-.573 & 3.706 & -.573 & 0 \\
0 & -.573 & 3.706 & -.573 \\
0 & 0 & -.573 & 1.893
\end{bmatrix}
\begin{Bmatrix}
330 \\ T_2 \\ T_3 \\ T_4
\end{Bmatrix}
=
\begin{Bmatrix}
38.4 + Q_1 \\
76.8 \\
76.8 \\
39.6
\end{Bmatrix}
$$

In the above equation, the prescribed temperature $T_1 = 330°C$ is substituted. Now eliminating the first row and moving the first column, after multiplying with T_1, to the RHS we obtain

$$
\begin{bmatrix}
3.706 & -.573 & 0 \\
-.573 & 3.706 & -.573 \\
0 & -.573 & 1.893
\end{bmatrix}
\begin{Bmatrix}
T_2 \\ T_3 \\ T_4
\end{Bmatrix}
=
\begin{Bmatrix}
265.89 \\
76.8 \\
39.6
\end{Bmatrix}
$$

The solution of this matrix equation yields the three nodal temperatures. By combining prescribed temperature, the entire nodal temperatures of the fin are determined as

$$
T_1 = 330°C, \; T_2 = 77.57°C, \; T_3 = 37.72°C, \; T_4 = 32.34°C
$$

Since the temperature interpolation within the element is linear, the temperature distribution of the fin can be plotted as shown in Figure 5.19. Due to convection across the lateral surfaces of the fin, the temperature drops dramatically along the x-axis.

5.6   EXERCISE

1. Consider heat conduction in a uniaxial rod surrounded by a fluid. The left end of the rod is at T_0. The free stream temperature is T^∞. There is convective heat transfer across the surface of the rod as well over the right end. The governing equation and boundary conditions are as follows:

$$
\frac{d}{dx}\left(kA\frac{dT}{dx}\right) + hP(T_\infty - T) = 0, \qquad 0 \leq x \leq L
$$

$$
\begin{cases}
T(0) = T_0, & x = 0 \\
k\dfrac{dT}{dx} = h(T_\infty - T), & x = L
\end{cases}
$$

where k is the thermal conductivity, h is the convection coefficient, and A and P are the area and circumference of the rod cross-section, respectively. Numerical values (SI units): $L = 0.3$, $k = 180$, $h = 12$, $T_0 = 700$, $T^\infty = 400$, $A = 10^{-4}$, $P = 4 \times 10^{-2}$. Use three finite elements to solve the problem. Use elements of lengths 0.05, 0.1, and 0.15, respectively.

2. Consider a heat conduction problem described in the figure. Inside the bar, heat is generated from a uniform heat source $Q_g = 10 \, \text{W/m}^3$, and the thermal conductivity of the material is $k = 0.1 \, \text{W/m/}^\circ\text{C}$. The cross-sectional area $A = 1 \, \text{m}^2$. When the temperatures at both ends are fixed to 0°C, calculate the temperature distribution using: (a) two equal-length elements and (b) three equal-length elements. Plot the temperature distribution along the bar and compare with the exact solution.

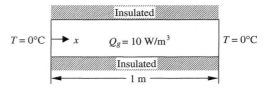

3. Repeat Problem 2 with $Q_g = 20x$.

4. Determine the temperature distribution (nodal temperatures) of the bar shown in the figure using two equal-length finite elements with cross-sectional area of $1 \, \text{m}^2$. The thermal conductivity is $10 \, \text{W/m}^\circ\text{C}$. The left side is maintained at 300°C. The right side is subjected to heat loss by convection with $h = 1 \, \text{W/m}^2{}^\circ\text{C}$ and $T_f = 30^\circ\text{C}$. All other sides are insulated.

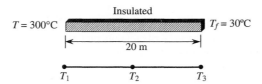

5. The nodal temperatures of heat conduction problem are given in the one-dimensional figure. Calculate the temperature at $x = 0.2$ using: (a) two two-node elements and (b) one three-node element.

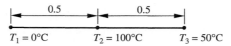

6. In order to solve one-dimensional steady-state heat transfer problem, one element with three-nodes is used. The shape functions and the conductivity matrix before applying boundary conditions are given.

$$\begin{cases} N_1(x) = 1 - 3x + 2x^2 \\ N_2(x) = 4x - 4x^2 \\ N_3(x) = -x + 2x^2 \end{cases} , \quad [K_T] = \begin{bmatrix} 1 & -2 & 1 \\ -2 & 4 & -2 \\ 1 & -2 & 2 \end{bmatrix}$$

(a) When the temperature at Node 1 is equal to 40°C and the heat flux of 80 W is provided at Node 3, calculate the temperature at $x = 1/4 \, \text{m}$.

(b) When the temperature at node 1 is equal to 40°C and the convection boundary condition is applied at node 3 with $h = 4 \, \text{W/m}^2/^\circ\text{C}$, $T_\infty = 100^\circ\text{C}$, calculate the temperature at $x = 1/4 \, \text{m}$.

(c) Instead of the previous boundary conditions, heat fluxes at Nodes 1 and 3 are given as Q_1 and Q_3, respectively. Can this problem be solved for the nodal temperatures? Explain your answer.

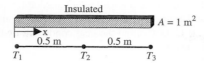

Insulated

$A = 1\,\text{m}^2$

x

0.5 m   0.5 m

T_1 T_2 T_3

7. The one-dimensional heat chamber in the figure is modeled using one element with three nodes. There is a uniform heat source inside the wall generating $Q = 300\,\text{W/m}^3$. The thermal conductivity of the wall is $k = 3\,\text{W/m°C}$. Assume that $A = 1\,\text{m}^2$ and $l = 1\,\text{m}$. The left end has a fixed temperature of $T = 20°C$, while the right end has zero heat flux.

   (a) Calculate the shape function $[\mathbf{N}] = [N_1, N_2, N_3]$ as a function of x.

   (b) Solve for nodal temperature $\{\mathbf{T}\} = \{\, T_1, T_2, T_3 \,\}^T$ using boundary conditions. Plot the temperature distribution in x-T graph.

   (c) Calculate the heat fluxes at $x = 0$ and $x = 1$ when the solution is $\{\mathbf{T}\} = \{\, 20,\; 58.25,\; 71 \,\}$.

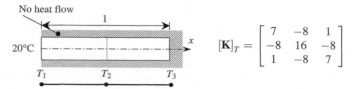

No heat flow

1

20°C

T_1 T_2 T_3

$$[\mathbf{K}]_T = \begin{bmatrix} 7 & -8 & 1 \\ -8 & 16 & -8 \\ 1 & -8 & 7 \end{bmatrix}$$

8. Consider heat conduction in a uniaxial rod surrounded by a fluid. The right end of the rod is attached to a wall and is at temperature T_R. One half of the bar is insulated as indicated. The free stream temperature is T_f. There is convective heat transfer across the un-insulated surface of the rod as well as over the left end face. Use two equal-length elements to determine the temperature distribution in the rod. Use the following numerical values: $L = 0.2\,\text{m}$, $k = 200\,\text{W/m/°C}$, $h = 36\,\text{W/m}^2\text{/°C}$, $T_R = 600\,°C$, $T_f = 300°C$, $A = 10^{-4}\,\text{m}^2$, and $P = 0.04\,\text{m}$.

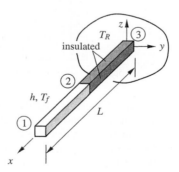

T_R ③

insulated

②

h, T_f

①

L

x

z

y

9. A well-mixed fluid is heated by a long iron plate of conductivity $k = 12\,\text{W/m/°C}$ and thickness $t = 0.12\,\text{m}$. Heat is generated uniformly in the plate at the rate $Q_g = 5,000\,\text{W/m}^3$. If the surface convection coefficient $h = 7\,\text{W/m}^2\text{/°C}$ and fluid temperature is $T_f = 45°C$, determine the temperature at the center of the plate T_c and the heat flow rate to the fluid q using three one-dimensional elements.

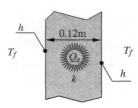

h

0.12m

T_f Q_g T_f

k h

10. A cooling fin of square cross-sectional area $A = 0.25 \times 0.25\,\text{m}^2$, length $L = 2\,\text{m}$, and conductivity $k = 10\,\text{W/m/°C}$ extends from a wall maintained at temperature $T_w = 100°\text{C}$. The surface convection coefficient between the fin and the surrounding air is $h = 0.5\,\text{W/m}^2\text{/°C}$, and the air temperature is $T_a = 20°\text{C}$. Determine the heat conducted by the fin and the temperature of the tip using five one-dimensional finite elements.

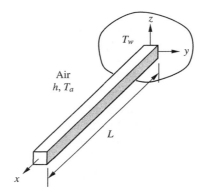

11. Find the heat transfer per unit area through the composite wall in the figure. Assume one-dimensional heat flow and there is no heat flow between B and C. The thermal conductivities are $k_A = 0.04\,\text{W/m/°C}$, $k_B = 0.1\,\text{W/m/°C}$ $k_C = 0.03\,\text{W/m/°C}$, and $k_D = 0.06\,\text{W/m/°C}$.

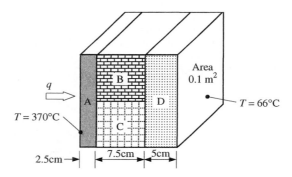

12. Consider a wall built up of concrete and thermal insulation. The outdoor temperature is $T_o = -17°\text{C}$, and the temperature inside is $T_i = 20°\text{C}$. The wall is subdivided into three elements. The thermal conductivity for concrete is $k_c = 1.7\,\text{W/m/°C}$, and that of the insulator is $k_i = 0.04\,\text{W/m/°C}$. Convection heat transfer is occurring at both surfaces with convection coefficients of $h_0 = 14\,\text{W/m}^2\text{/°C}$ and $h_i = 5.5\,\text{W/m}^2\text{/°C}$ Calculate the temperature distribution of the wall. Also, calculate the amount of heat flow through the outdoor surface.

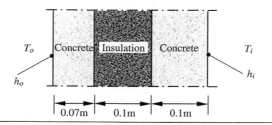

Chapter 6

Finite Elements for Plane Solids

6.1 INTRODUCTION

All real-life structures are three-dimensional. It is engineers who make the approximation as a one-dimensional (e.g., beam) or a two-dimensional (e.g., plate or plane solid) structure. In Chapter 1 we explained in detail the conditions under which such approximation could be made. When the stresses on a plane normal to one of the axes are approximately zero, then we say that the solid is in the state of plane stress. Similarly, when the corresponding strains are zero, the solid is in the state of plane strain. A two-dimensional solid is also called a plane solid. Some examples of plane solids are (1) a thin plate subjected to in-plane forces and (2) a very thick solid with constant cross-section in the thickness direction. In this chapter, we will discuss when an engineering problem can be assumed to be two-dimensional and how to solve such a problem using two-dimensional finite elements. We will introduce three different types of plane elements. Every element has its own characteristics. In order to use the finite element appropriately, thorough understandings of capabilities and limitations of each element are required.

In general, two-dimensional elasticity problems can be expressed by a system of coupled second-order partial differential equations. Based on the constraints imposed in the thickness direction, a two-dimensional problem can be considered as either a plane stress or plane strain problem. Although the two problems are different, the equations developed for plane stress problems can be used for plane strain case by modifying the elastic constants.

In the plane solid problem, the main variables are the displacements in the coordinate directions. After solving for the displacements, stresses and strains can be calculated from the derivatives of displacements. The displacements are calculated using the fact that the structure is in equilibrium when the total potential energy has its minimum value. This will yield a matrix equation similar to the beam problem in Chapter 4.

6.2 TYPES OF TWO-DIMENSIONAL PROBLEMS

6.2.1 Governing Differential Equations

In two-dimensional problems, the stresses and strains are independent of the coordinate in the thickness direction (usually the z-axis). By setting all the derivatives with respect to z-coordinate in Eq. (1.68) to zero, we obtain the governing differential equations for plane problems as

$$
\begin{cases}
\dfrac{\partial \sigma_{xx}}{\partial x} + \dfrac{\partial \tau_{xy}}{\partial y} + b_x = 0 \\[2ex]
\dfrac{\partial \tau_{xy}}{\partial x} + \dfrac{\partial \sigma_{yy}}{\partial y} + b_y = 0
\end{cases}
\tag{6.1}
$$

where b_x and b_y are the body forces per unit volume.

Let u and v be the displacement in the x- and y-directions, respectively. From Eqs. (1.39) through (1.45), the strain components in a plane solid are defined as

$$\varepsilon_{xx} = \frac{\partial u}{\partial x}, \quad \varepsilon_{yy} = \frac{\partial v}{\partial y}, \quad \gamma_{xy} = \left(\frac{\partial u}{\partial y} + \frac{\partial v}{\partial x}\right) \tag{6.2}$$

In addition, the stresses and strains are related by the following constitutive relation:

$$\begin{Bmatrix} \sigma_{xx} \\ \sigma_{yy} \\ \tau_{xy} \end{Bmatrix} = \begin{bmatrix} C_{11} & C_{12} & C_{13} \\ C_{21} & C_{22} & C_{23} \\ C_{31} & C_{32} & C_{33} \end{bmatrix} \begin{Bmatrix} \varepsilon_{xx} \\ \varepsilon_{yy} \\ \gamma_{xy} \end{Bmatrix} \Leftrightarrow \{\sigma\} = [C]\{\varepsilon\} \tag{6.3}$$

Substituting the stress-strain relations in Eq. (6.3) and strain-displacement relations in Eq. (6.2) in the equilibrium equations in Eq. (6.1), we obtain a pair of second-order partial differential equations in two variables, $u(x,y)$ and $v(x,y)$. The explicit form of the equations is available in textbooks on elasticity, e.g., Timoshenko and Goodier.[1]

The differential equation must be accompanied by boundary conditions. Two types of boundary conditions can be defined. The first one is the boundary in which the values of displacements are prescribed (*essential boundary condition*). The other is the boundary in which the tractions are prescribed (*natural boundary condition*). The boundary conditions can be formally stated as

$$\begin{aligned} \mathbf{u} &= \mathbf{g}, & \text{on } S_g \\ \boldsymbol{\sigma}\mathbf{n} &= \mathbf{T}, & \text{on } S_T \end{aligned} \tag{6.4}$$

where S_g and S_T, respectively, are the boundaries where the displacement and traction boundary conditions are prescribed. The objective is to determine the displacement field $u(x,y)$ and $v(x,y)$ that satisfies the differential equation (6.1) and the boundary conditions in Eq. (6.4). Now we will discuss the stress-strain relations in Eq. (6.3) for the two different plane problems.

6.2.2 Plane Stress Problems

Plane stress conditions exist when the thickness dimension (usually the z-direction) is much smaller than the length and width dimensions of a solid. Since stress at the two surfaces normal to the z-axis are zero, it is assumed that stresses in the normal direction are zero throughout the body; i.e., $\sigma_{zz} = \tau_{xz} = \tau_{yz} = 0$. In such a case, the structure can be modeled in two dimensions. An example of the plane stress problem is a thin plate or disk with applied in-plane forces (see Figure 6.1). If an out-of-plane force (e.g., transverse pressure in the z-direction) is applied, then the problem can be assumed plane stress only when the applied pressure load is much smaller than the in-plane stresses such as σ_{xx}.

- Non-zero stress components: σ_{xx}, σ_{yy}, τ_{xy}.
- Non-zero strain components: ε_{xx}, ε_{yy}, γ_{xy}, ε_{zz}.

For linear isotropic materials, the stress-strain relation can be written as (see Section 1.3)

$$\begin{Bmatrix} \sigma_{xx} \\ \sigma_{yy} \\ \tau_{xy} \end{Bmatrix} = \frac{E}{1 - v^2} \begin{bmatrix} 1 & v & 0 \\ v & 1 & 0 \\ 0 & 0 & \dfrac{1 - v}{2} \end{bmatrix} \begin{Bmatrix} \varepsilon_{xx} \\ \varepsilon_{yy} \\ \gamma_{xy} \end{Bmatrix} \tag{6.5}$$

$$\Leftrightarrow \{\sigma\} = [C_\sigma]\{\varepsilon\}$$

---

[1] S.P. Timoshenko and J.N. Goodier, Theory of Elasticity, McGraw-Hill, NY, 1984.

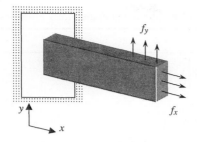

Figure 6.1 Thin plate with in-plane applied forces

where $[\mathbf{C}_\sigma]$ is the stress-strain matrix or elasticity matrix for the plane stress problem. It should be noted that the normal strain ε_{zz} in the thickness direction is not zero; it can be calculated from the following relation:

$$\varepsilon_{zz} = -\frac{v}{E}\left(\sigma_{xx} + \sigma_{yy}\right) \tag{6.6}$$

6.2.3 Plane Strain Problems

A state of plane strain will exist in a solid when the thickness dimension is much larger than other two dimensions. When the deformation in the thickness direction is constrained, the solid is assumed to be in a state of plane strain even if the thickness dimension is small. A proper assumption is that that strains with z subscript are zero i.e., $\varepsilon_{zz} = \gamma_{xz} = \gamma_{yz} = 0$. In such a case, it is sufficient to model a slice of the solid with unit thickness. Some examples of plane strain problems are the retaining wall of a dam and long cylinder such as a gun barrel (see Figure 6.2).

– Non-zero stress components: σ_{xx}, σ_{yy}, τ_{xy}, σ_{zz}.

– Non-zero strain components: ε_{xx}, ε_{yy}, γ_{xy}.

For linear isotropic materials, the stress-strain relations under plane strain conditions can be written as (see Section 1.3)

$$\begin{Bmatrix} \sigma_{xx} \\ \sigma_{yy} \\ \tau_{xy} \end{Bmatrix} = \frac{E}{(1+v)(1-2v)}\begin{bmatrix} 1-v & v & 0 \\ v & 1-v & 0 \\ 0 & 0 & \frac{1}{2}-v \end{bmatrix}\begin{Bmatrix} \varepsilon_{xx} \\ \varepsilon_{yy} \\ \gamma_{xy} \end{Bmatrix} \tag{6.7}$$

$$\Leftrightarrow \{\sigma\} = [\mathbf{C}_\varepsilon]\{\varepsilon\}$$

where $[\mathbf{C}_\varepsilon]$ is the stress-strain matrix for plane strain problem. It should be noted that the transverse stress σ_{zz} is not equal to zero in plane strain case, and it can be calculated from the following relation:

$$\sigma_{zz} = \frac{Ev}{(1+v)(1-2v)}\left(\varepsilon_{xx} + \varepsilon_{yy}\right) \tag{6.8}$$

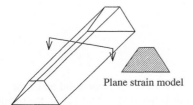

Plane strain model

Figure 6.2 Dam structure with plane strain assumption

Table 6.1   Material Property Conversion Between Plane Strain and Plane Stress Problems

From → To	E	ν
Plane strain → Plane stress	$E\left[1 - \left(\dfrac{\nu}{1+\nu}\right)^2\right]$	$\dfrac{\nu}{1+\nu}$
Plane stress → Plane strain	$\dfrac{E}{1 - \left(\dfrac{\nu}{1-\nu}\right)^2}$	$\dfrac{\nu}{1-\nu}$

6.2.4   Equivalence between Plane Stress and Plane Strain Problems

Although plane stress and plane strain problems are different by definition, they are quite similar from the computational viewpoint. Thus, it is possible to use the plane strain formulation and solve the plane stress problem. In such case, two material properties, E and ν, need to be modified. Similarly, it is also possible to convert the plane stress formulation into the plane strain formulation. Table 6.1 summarizes the conversion relations.

6.3   PRINCIPLE OF MINIMUM POTENTIAL ENERGY

Similar to the beam-bending problem in Chapter 4, the principle of minimum potential energy can be used to derive the finite element equations for the two-dimensional plane solid problems.

6.3.1   Strain Energy in a Plane Solid

Consider a plane elastic solid as illustrated in Figure 6.3. The *strain energy* is a form of energy that is stored in the solid due to the elastic deformation. Formally, it can be defined as

$$
\begin{aligned}
U &= \frac{1}{2} \iiint\limits_{\text{volume}} \{\varepsilon\}^T \{\sigma\} dV \\
&= \frac{h}{2} \iint\limits_{\text{area}} \{\varepsilon\}^T \{\sigma\} dA \\
&= \frac{h}{2} \iint\limits_{\text{area}} \{\varepsilon\}^T [\mathbf{C}] \{\varepsilon\} dA
\end{aligned}
\tag{6.9}
$$

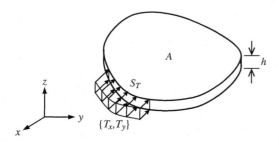

Figure 6.3 A plane solid under the distributed load $\{T_x, T_y\}$ on the traction boundary S_T

where h is the thickness of the plane solid ($h = 1$ for plane strain) and $[\mathbf{C}] = [\mathbf{C}_\sigma]$ for plane stress and $[\mathbf{C}] = [\mathbf{C}_\varepsilon]$ for plane strain. Since stress and strain are constant throughout the thickness, the volume integral is changed to area integral by multiplying by the thickness in the second line in Eq. (6.9). The linear elastic relation in Eq. (6.3) has been used in the last line of Eq. (6.9).

6.3.2   Potential Energy of Applied Loads

When a force acting on a body moves through a small distance, it loses its potential to do additional work, and hence its potential energy is given by the negative of product of the force and corresponding displacement. For example, when concentrated forces are applied to the solid, the potential energy becomes

$$V = -\sum_{i=1}^{ND} F_i q_i \tag{6.10}$$

where F_i is the i-th force, q_i is the displacement in the direction of the force, and ND is the total number of concentrated forces acting on the body. The negative sign indicates that the potential energy decreases as the force has expended some energy performing the work given by the product of force and corresponding displacement.

When distributed forces, such as a pressure load, act on the edge of a body (see Figure 6.3), the summation sign in the above expression is replaced by integration over the edge of the body as shown below:

$$
\begin{aligned}
V &= -h \int_{S_T} (T_x u + T_y v) \, dS \\
&= -h \int_{S_T} [u \quad v] \begin{Bmatrix} T_x \\ T_y \end{Bmatrix} dS \\
&\equiv -h \int_{S_T} \{\mathbf{u}\}^T \{\mathbf{T}\} dS
\end{aligned}
\tag{6.11}
$$

where T_x and T_y are the components of applied surface forces in the x- and y-direction, respectively.

If body forces (forces distributed over the volume) are present, work done by these forces can be computed in a similar manner. The gravitational force is an example of body force. In this case, integration should be performed over the volume. We will discuss this further, when we derive the finite element equations.

6.3.3   Total Potential Energy

As with the beam problem, the *potential energy* is defined as the sum of the strain energy and the potential energy of applied loads:

$$\Pi = U + V \tag{6.12}$$

where U is the strain energy and V is the potential energy of applied loads. The principle of minimum total potential energy states that of all possible displacement configurations of a solid/structure, the equilibrium configuration corresponds to the minimum total potential energy. That is, at equilibrium, we have

$$\frac{\partial \Pi}{\partial \{\mathbf{u}\}} = 0 \Rightarrow \frac{\partial \Pi}{\partial u_1} = 0, \; \frac{\partial \Pi}{\partial u_2} = 0 \cdots \frac{\partial \Pi}{\partial u_N} = 0 \tag{6.13}$$

where u_1, u_2, . . . , u_N are the displacements that define the deformed configuration of the body. In finite element analysis, the deformation of the body is defined in terms of the displacements of the nodes. In the following sections, we will use the principal of minimum total potential energy to derive finite element equations for different types of elements.

6.4 CONSTANT STRAIN TRIANGULAR (CST) ELEMENT

In finite element analysis, a plane solid can be divided into a number of contiguous elements. A simplest way of dividing a plane solid is to use triangular elements. Figure 6.4 shows a plane solid that is divided by triangular elements. Each element shares its edge and two corner nodes with an adjacent element, except for those on the boundary. The three vertices of a triangle are the nodes of that element as shown in Figure 6.4. The first node of an element can arbitrarily be chosen. However, the sequence of the nodes 1, 2, and 3 should be in the counter-clockwise direction. Each node has two displacements, u and v, respectively, in the x- and y-directions.

The displacements within the element are interpolated in terms of the nodal displacements using shape functions. In the polynomial approximation, the displacement has to be a linear function in x and y because displacement information is available only at three points (nodes), and a linear polynomial has three unknown coefficients. Since displacement is a linear function, strain and stress are constant within an element, and that is why a triangular element is called a *constant strain triangle element*.

6.4.1 Displacement and Strain Interpolation

The first step in deriving the finite element matrix equation is to interpolate the displacement function in terms of the nodal displacements. Let the x- and y-directional displacements be $u(x,y)$ and $v(x,y)$, respectively. Since the two-coordinates are perpendicular to each other (orthogonal), the $u(x,y)$ and $v(x,y)$ are independent of each other. Hence, $u(x,y)$ needs to be interpolated in terms of u_1, u_2, and u_3, and $v(x,y)$ in terms of v_1, v_2, and v_3. It is obvious that the interpolation function must be a three-term polynomial in x and y. Since we must have rigid body displacements (constant displacements) and constant strain terms in the interpolation function, the displacement interpolation must be of the following form:

$$\begin{cases} u(x, y) = \alpha_1 + \alpha_2 x + \alpha_3 y \\ v(x, y) = \beta_1 + \beta_2 x + \beta_3 y \end{cases} \tag{6.14}$$

where α's and β's are constants. In finite element analysis, we would like to replace the constants by the nodal displacements. Let us consider x-directional displacements, which

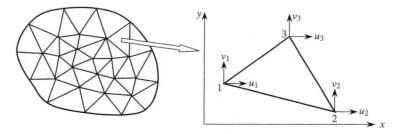

Figure 6.4 Constant strain triangular (CST) element

are u_1, u_2, and u_3. At Node 1, for example, x and y take the values of x_1 and y_1, respectively, and the nodal displacement is u_1. If we repeat this for the other two nodes, we obtain the following three simultaneous equations:

$$\begin{cases} u(x_1, y_1) \equiv u_1 = \alpha_1 + \alpha_2 x_1 + \alpha_3 y_1 \\ u(x_2, y_2) \equiv u_2 = \alpha_1 + \alpha_2 x_2 + \alpha_3 y_2 \\ u(x_3, y_3) \equiv u_3 = \alpha_1 + \alpha_2 x_3 + \alpha_3 y_3 \end{cases} \tag{6.15}$$

In matrix notation, the above equations can be written as

$$\begin{Bmatrix} u_1 \\ u_2 \\ u_3 \end{Bmatrix} = \begin{bmatrix} 1 & x_1 & y_1 \\ 1 & x_2 & y_2 \\ 1 & x_3 & y_3 \end{bmatrix} \begin{Bmatrix} \alpha_1 \\ \alpha_2 \\ \alpha_3 \end{Bmatrix} \tag{6.16}$$

If the three points (x_1, y_1), (x_2, y_2), and (x_3, y_3) are not on a straight line, then the inverse of the above coefficient matrix exists. Thus, we can calculate the unknown coefficients as

$$\begin{Bmatrix} \alpha_1 \\ \alpha_2 \\ \alpha_3 \end{Bmatrix} = \begin{bmatrix} 1 & x_1 & y_1 \\ 1 & x_2 & y_2 \\ 1 & x_3 & y_3 \end{bmatrix}^{-1} \begin{Bmatrix} u_1 \\ u_2 \\ u_3 \end{Bmatrix} = \frac{1}{2A} \begin{bmatrix} f_1 & f_2 & f_3 \\ b_1 & b_2 & b_3 \\ c_1 & c_2 & c_3 \end{bmatrix} \begin{Bmatrix} u_1 \\ u_2 \\ u_3 \end{Bmatrix} \tag{6.17}$$

where A is the area of the triangle and

$$\begin{cases} f_1 = x_2 y_3 - x_3 y_2, & b_1 = y_2 - y_3, & c_1 = x_3 - x_2 \\ f_2 = x_3 y_1 - x_1 y_3, & b_2 = y_3 - y_1, & c_2 = x_1 - x_3 \\ f_3 = x_1 y_2 - x_2 y_1, & b_3 = y_1 - y_2, & c_3 = x_2 - x_1 \end{cases} \tag{6.18}$$

The area A of the triangle can be calculated from

$$A = \frac{1}{2} \det \begin{vmatrix} 1 & x_1 & y_1 \\ 1 & x_2 & y_2 \\ 1 & x_3 & y_3 \end{vmatrix} \tag{6.19}$$

Note that the determinant in Eq. (6.19) is zero when three nodes are collinear. In such a case, the area of the triangular element is zero and we cannot uniquely determine the three coefficients.

A similar procedure can be applied for y-directional displacement $v(x, y)$, and the unknown coefficients β_i, $(i = 1, 2, 3)$ are determined using the following equation:

$$\begin{Bmatrix} \beta_1 \\ \beta_2 \\ \beta_3 \end{Bmatrix} = \frac{1}{2A} \begin{bmatrix} f_1 & f_2 & f_3 \\ b_1 & b_2 & b_3 \\ c_1 & c_2 & c_3 \end{bmatrix} \begin{Bmatrix} v_1 \\ v_2 \\ v_3 \end{Bmatrix} \tag{6.20}$$

After calculating α_i and β_i, the displacement interpolation can be written as

$$u(x, y) = \begin{bmatrix} N_1 & N_2 & N_3 \end{bmatrix} \begin{Bmatrix} u_1 \\ u_2 \\ u_3 \end{Bmatrix} \text{ and } v(x, y) = \begin{bmatrix} N_1 & N_2 & N_3 \end{bmatrix} \begin{Bmatrix} v_1 \\ v_2 \\ v_3 \end{Bmatrix} \tag{6.21}$$

where the shape functions are defined by

$$\begin{cases} N_1(x, y) = \dfrac{1}{2A}(f_1 + b_1 x + c_1 y) \\[2mm] N_2(x, y) = \dfrac{1}{2A}(f_2 + b_2 x + c_2 y) \\[2mm] N_3(x, y) = \dfrac{1}{2A}(f_3 + b_3 x + c_3 y) \end{cases} \tag{6.22}$$

Note that N_1, N_2, and N_3 are linear functions of x- and y-coordinates. Thus, interpolated displacement varies linearly in each coordinate direction.

To make the derivations simple, we would rewrite the interpolation relation in Eq. (6.21) in matrix form. Let $\{\mathbf{u}\} = \{u, v\}^T$ be the displacement vector at any point (x, y). The interpolation can be written in the matrix notation by

$$\{\mathbf{u}\} \equiv \begin{Bmatrix} u \\ v \end{Bmatrix} = \begin{bmatrix} N_1 & 0 & N_2 & 0 & N_3 & 0 \\ 0 & N_1 & 0 & N_2 & 0 & N_3 \end{bmatrix} \begin{Bmatrix} u_1 \\ v_1 \\ u_2 \\ v_2 \\ u_3 \\ v_3 \end{Bmatrix}$$

or

$$\boxed{\{\mathbf{u}(x, y)\} = [\mathbf{N}(x, y)]\{\mathbf{q}\}} \tag{6.23}$$

Equation (6.23) is the critical relation in finite element approximation. When a point (x, y) within a triangular element is given, the shape function $[\mathbf{N}]$ is calculated at this point. Then the displacement at this point can be calculated by multiplying this shape function matrix with the nodal displacement vector $\{\mathbf{q}\}$. Thus, if we solve for the nodal displacements, we can calculate the displacement everywhere in the element. Note that the nodal displacements will be evaluated using the principle of minimum total potential energy in the following section.

After calculating displacement within an element, it is possible to calculate the strain by differentiating the displacement with respect to x and y. From the expression in Eq. (6.23), it can be noted that the nodal displacements are constant, but the shape functions are functions of x and y. Thus, the strain can be calculated by differentiating the shape function with respect to the coordinates. For example, ε_{xx} can be written as

$$\varepsilon_{xx} \equiv \frac{\partial u}{\partial x} = \frac{\partial}{\partial x}\left(\sum_{i=1}^{3} N_i(x, y)u_i \right) = \sum_{i=1}^{3} \frac{\partial N_i}{\partial x}u_i = \sum_{i=1}^{3} \frac{b_i}{2A}u_i \tag{6.24}$$

Note that u_1, u_2, and u_3 are nodal displacements and they are independent of coordinate x. Thus, only the shape function is differentiated with respect to x. Similar calculation can be carried out for ε_{yy} and γ_{xy}. Using the matrix notation, we have

$$\{\varepsilon\} = \begin{Bmatrix} \dfrac{\partial u}{\partial x} \\[2mm] \dfrac{\partial v}{\partial y} \\[2mm] \dfrac{\partial u}{\partial y} + \dfrac{\partial v}{\partial x} \end{Bmatrix} = \frac{1}{2A} \begin{bmatrix} b_1 & 0 & b_2 & 0 & b_3 & 0 \\ 0 & c_1 & 0 & c_2 & 0 & c_3 \\ c_1 & b_1 & c_2 & b_2 & c_3 & b_3 \end{bmatrix} \begin{Bmatrix} u_1 \\ v_1 \\ u_2 \\ v_2 \\ u_3 \\ v_3 \end{Bmatrix} \equiv [\mathbf{B}]\{\mathbf{q}\} \tag{6.25}$$

It may be noted that the $[\mathbf{B}]$ matrix is constant and depends only on the coordinates of the three nodes of the triangular element. Thus, one can anticipate that if this element is used, then the strains will be constant over a given element and will depend only on nodal displacements. Hence, this element is called the **Constant Strain Triangular Element** or **CST** element.

The interpolation of displacement in Eq. (6.23) and the interpolation of strain in Eq. (6.25) are used for approximating the strain energy and potential energy of applied loads.

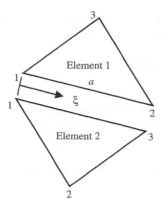

Figure 6.5 Two adjacent constant strain triangular (CST) elements

6.4.2 Properties of the CST Element

Before we derive the strain energy, it may be useful to study some interesting aspects of the CST element. Since the displacement field is assumed to be a linear function in x and y, one can show that the triangular element deforms into another triangle when forces are applied. Furthermore, an imaginary straight line drawn within an element before deformation becomes another straight line after deformation.

Let us consider the displacements of points along one of the edges of the triangle. Consider the points along the edge 1-2 of Element 1 in Figure 6.5. These points can be conveniently represented by a coordinate ξ. The coordinate $\xi = 0$ at Node 1 and $\xi = a$ at Node 2. Along this edge, x and y are related to ξ. By substituting this relation in the displacement functions, one can express the displacements of points on the edge 1-2 as a function of ξ. It can be easily shown that the displacement functions, for both u and v, must be linear in ξ, i.e.,

$$\begin{cases} u(\xi) = \gamma_1 + \gamma_2 \xi \\ v(\xi) = \gamma_3 + \gamma_4 \xi \end{cases}$$

where γ's are constants to be determined. Since the variation of displacement is linear, it might be argued that the displacements should depend only on u_1 and u_2, and not on u_3. Then, the displacement field along the edge 1-2 takes the following form:

$$\begin{cases} u(\xi) = \left(1 - \dfrac{\xi}{a}\right)u_1 + \dfrac{\xi}{a}u_2 = H_1(\xi)u_1 + H_2(\xi)u_2 \\[2ex] v(\xi) = \left(1 - \dfrac{\xi}{a}\right)v_1 + \dfrac{\xi}{a}v_2 = H_1(\xi)v_1 + H_2(\xi)v_2 \end{cases} \tag{6.26}$$

where H_1 and H_2 are shape functions defined along the edge 1-2 and a is the length of edge 1-2. One can also note that a condition called *inter-element displacement compatibility* is satisfied by triangular elements. This condition can be described as follows. After the loads are applied and the solid is deformed, the displacements at any point in an element can be computed from the nodal displacements of that particular element and the interpolation functions in Eq. (6.23). Consider a point on a common edge of two adjacent elements. This point can be considered as belonging to either of the elements. Then the nodes of either triangle can be used in interpolating the displacements of this point. However, one must obtain a unique set of displacements independent of the choice of the element. This can be true only if the displacements of the points depend only on the nodes common to both elements. In fact, this will be satisfied because of Eq. (6.26). Thus, the CST element satisfies the inter-element displacement compatibility.

EXAMPLE 6.1   *Interpolation in a Triangular Element*

Consider two triangular elements shown in Figure 6.6. The nodal displacements are given as $\{u_1, v_1, u_2, v_2, u_3, v_3, u_4, v_4\} = \{-0.1, 0, 0.1, 0, -0.1, 0, 0.1, 0\}$. Calculate displacements and strains in both elements.

SOLUTION   Element 1 has nodes 1-2-4. Then, using the nodal coordinates, we can derive the shape functions as shown below:

$$
\begin{array}{ccc}
x_1 = 0 & x_2 = 1 & x_3 = 0 \\
y_1 = 0 & y_2 = 0 & y_3 = 1 \\
f_1 = 1 & f_2 = 0 & f_3 = 0 \\
b_1 = -1 & b_2 = 1 & b_3 = 0 \\
c_1 = -1 & c_2 = 0 & c_3 = 1
\end{array}
$$

In addition, the area of the element is 0.5. Thus, from Eq. (6.22) the shape functions can be derived as

$$
\begin{aligned}
N_1(x, y) &= 1 - x - y \\
N_2(x, y) &= x \\
N_3(x, y) &= y
\end{aligned}
$$

Then, the displacements in Element 1 can be interpolated as

$$
u^{(1)}(x, y) = \sum_{I=1}^{3} N_I(x, y) u_I = 0.1(2x + 2y - 1)
$$

$$
v^{(1)}(x, y) = \sum_{I=1}^{3} N_I(x, y) v_I = 0.0
$$

Strains can be calculated from Eq. (6.25), or directly differentiating the above expressions for displacements, as

$$
\varepsilon_{xx}^{(1)} = \frac{\partial u^{(1)}}{\partial x} = 0.2
$$

$$
\varepsilon_{yy}^{(1)} = \frac{\partial v^{(1)}}{\partial y} = 0.0
$$

$$
\gamma_{xy}^{(1)} = \frac{\partial u^{(1)}}{\partial y} + \frac{\partial v^{(1)}}{\partial x} = 0.2
$$

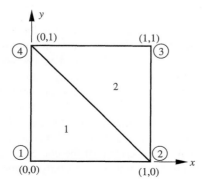

Figure 6.6 Interpolation of displacements in triangular elements

Element 2 connects Nodes 2-3-4. Thus, using the nodal coordinates, we can build the shape functions, as

$$
\begin{array}{lll}
x_1 = 1 & x_2 = 1 & x_3 = 0 \\
y_1 = 0 & y_2 = 1 & y_3 = 1 \\
f_1 = 1 & f_2 = -1 & f_3 = 1 \\
b_1 = 0 & b_2 = 1 & b_3 = -1 \\
c_1 = -1 & c_2 = 1 & c_3 = 0
\end{array}
$$

In addition, the area of the element is 0.5. Thus, from Eq. (6.22), the shape functions can be obtained as

$$
N_1(x, y) = 1 - y
$$
$$
N_2(x, y) = x + y - 1
$$
$$
N_3(x, y) = 1 - x
$$

Then, the displacements of Element 2 can be interpolated as

$$
u^{(2)}(x, y) = \sum_{i=1}^{3} N_i(x, y) u_i = 0.1(3 - 2x - 2y)
$$
$$
v^{(2)}(x, y) = \sum_{i=1}^{3} N_i(x, y) v_i = 0.0
$$

Strains can be calculated from Eq. (6.25), or directly differentiating the above expressions for displacements, as

$$
\varepsilon_{xx}^{(2)} = \frac{\partial u^{(2)}}{\partial x} = -0.2
$$
$$
\varepsilon_{yy}^{(2)} = \frac{\partial v^{(2)}}{\partial y} = 0.0
$$
$$
\gamma_{xy}^{(2)} = \frac{\partial u^{(2)}}{\partial y} + \frac{\partial v^{(2)}}{\partial x} = -0.2
$$

Note that the displacements are linear and the strains are constant in each element. From the given nodal displacements, it is clear that the top edge has strain $\varepsilon_{xx} = -0.2$, while the bottom edge has $\varepsilon_{xx} = 0.2$. The strain varies linearly along the y-coordinate. However, the triangular element cannot represent this change and provides constant values of $\varepsilon_{xx} = 0.2$ for Element 1 and $\varepsilon_{xx} = -0.2$ for Element 2. In general, if a plane solid is under the constant strain states, the CST element will provide accurate solutions. However, if the strain varies in the solid, then the CST element cannot represent it accurately. In such a case, many elements should be used to approximate it as a series of step functions. Note that the strains along the interface between two elements are discontinuous.

6.4.3   Strain Energy

Let us calculate the strain energy in a typical triangular element, say Element e. In Eq. (6.9), the strain energy of the plane solid was derived in terms of strains and the elasticity matrix [**C**]. Substituting for strains from Eq. (6.25), we obtain

$$
\begin{aligned}
U^{(e)} &= \frac{h}{2} \iint_A \{\varepsilon\}^T [\mathbf{C}] \{\varepsilon\} \, dA^{(e)} \\
&= \frac{h}{2} \{\mathbf{q}^{(e)}\}^T \iint_A \underset{6\times3}{[\mathbf{B}]^T} \underset{3\times3}{[\mathbf{C}]} \underset{3\times6}{[\mathbf{B}]} \, dA \{\mathbf{q}^{(e)}\} \\
&\equiv \frac{1}{2} \{\mathbf{q}^{(e)}\}^T \underset{6\times6}{[\mathbf{k}]^{(e)}} \{\mathbf{q}^{(e)}\}
\end{aligned}
\qquad (6.27)
$$

where $[\mathbf{k}^{(e)}]$ is the *element stiffness matrix* of the triangular element. The column vector $\mathbf{q}^{(e)}$ is the displacement of the three nodes that belong to the element. The dimension of $[\mathbf{k}^{(e)}]$ is 6×6. In the case of the triangular element, all entries in matrices $[\mathbf{B}]$ and $[\mathbf{C}]$ are constant and can be integrated easily. After integration, the element stiffness matrix takes the form

$$\boxed{[\mathbf{k}^{(e)}] = hA[\mathbf{B}]^T[\mathbf{C}][\mathbf{B}]} \qquad (6.28)$$

where A is the area of the plane element. Using the expression for $[\mathbf{B}]$ in Eq. (6.25) and stress-strain relation in Eq. (6.5), the element stiffness matrix can be calculated.

One may note that in the case of truss and frame elements, we used a transformation matrix $[\mathbf{T}]$ in deriving the element stiffness matrix. However, in the present case, we have used the global coordinate system in the derivation of $[\mathbf{k}^{(e)}]$, and that is the reason for not using a transformation matrix. In some cases, however, it is required to define the element in a local coordinate system. For example, if the material is not isotropic, it will have a specific directional property. In such a case, $[\mathbf{k}^{(e)}]$ is first derived in a local coordinate system and transformed to the global coordinates by multiplying by appropriate transformation matrices.

The strain energy of the entire solid is simply the sum of the element strain energies. That is,

$$U = \sum_{e=1}^{NE} U^{(e)} = \frac{1}{2}\sum_{e=1}^{NE}\{\mathbf{q}^{(e)}\}^T[\mathbf{k}^{(e)}]\{\mathbf{q}^{(e)}\} \qquad (6.29)$$

where NE is the number of elements in the model. The superscript (e) in $\mathbf{q}^{(e)}$ implies that it is the vector of displacements or degrees-of-freedom (DOFs) of Element e. The summation in the above equation leads to the assembling of the element stiffness matrices into the structural stiffness matrix.

$$U = \frac{1}{2}\{\mathbf{Q}_s\}^T[\mathbf{K}_s]\{\mathbf{Q}_s\} \qquad (6.30)$$

where $\{\mathbf{Q}_s\}$ is the column vector of all displacements in the model and $[\mathbf{K}_s]$ is the structural stiffness matrix obtained by assembling the element stiffness matrices.

6.4.4 Potential Energy of Concentrated Forces at Nodes

The next step is to calculate the potential energy of external forces. We will consider three different types of applied forces. The first type is concentrated forces at nodes. It may be noted that the element expects two forces, one in the x-direction and the other in the y-direction, at each node. In general, the potential energy of concentrated nodal forces can be written as

$$V = -\sum_{i=1}^{ND}(F_{ix}u_i + F_{iy}v_i) \equiv -\{\mathbf{Q}_s\}^T\{\mathbf{F}_N\} \qquad (6.31)$$

where $\{\mathbf{F}_N\} = \begin{bmatrix} F_{1x} & F_{1y} & \cdots & F_{NDx} & F_{NDy} \end{bmatrix}^T$ is the vector of applied nodal forces and ND is the number of nodes in the solid. The contribution of a particular node to the potential energy will be zero if no force is applied at the node, or the displacement of the node becomes zero. The above potential energy of concentrated forces does not include supporting reaction because the displacement at those nodes will be zero.

6.4.5 Potential Energy of Distributed Forces along Element Edges

The second type of applied force is the distributed force (traction) on the side surface of the plane solid. In the plane solid, the traction is assumed to be a constant through the thickness. Let the surface traction force $\{\mathbf{T}\} = \{T_x,\, T_y\}^T$ is applied on the element edge 1-2 as shown in Figure 6.7. The unit of the surface traction is Pa (N/m^2) or psi. Since the force is distributed along the edge, the potential energy of the surface traction force must defined in the form of integral as

$$V^{(e)} = -h \int_{S_T} \{\mathbf{u}(s)\}^T \{\mathbf{T}(s)\} ds = -\{\mathbf{d}\}^T h \int_{S_T} [\mathbf{H}(s)]^T \{\mathbf{T}(s)\} ds \tag{6.32}$$

where $\{u\}^T = [\, u(s) \quad v(s)\,]$ is the vector of displacements along the Edge 1-2, $\{T\}^T = [\, T_x(s) \quad T_y(s)\,]$ is the vector of applied tractions along the Edge 1-2, $\{d\}^T = [\, u_1 \quad v_1 \quad u_2 \quad v_2\,]$ is the vector of displacements of Nodes 1 and 2, and $[H] = \begin{bmatrix} H_1 & 0 & H_2 & 0 \\ 0 & H_1 & 0 & H_2 \end{bmatrix}$ is the matrix of shape functions defined in Eq. (6.26). The integration can be performed in a closed form if the specified surface tractions (T_x and T_y) are a simple function of s. We will modify Eq. (6.32) to include all the six DOFs of the element and rewrite as

$$V^{(e)} = -\{\mathbf{q}^{(e)}\}^T h \int_{S_T} [\mathbf{N}(s)]^T \{\mathbf{T}(s)\} ds = -\{\mathbf{q}^{(e)}\}^T \{\mathbf{f}_T^{(e)}\} \tag{6.33}$$

We have used the complete shape function matrix in Eq. (6.33):

$$[\mathbf{N}] = \begin{bmatrix} \dfrac{l-s}{l} & 0 & \dfrac{s}{l} & 0 & 0 & 0 \\[2mm] 0 & \dfrac{l-s}{l} & 0 & \dfrac{s}{l} & 0 & 0 \end{bmatrix} \tag{6.34}$$

If the last expression in Eq. (6.33) is examined carefully, it is possible to note that the force vector $\{\mathbf{f}_T\}$ is a nodal force vector that is equivalent to the distributed force applied on the edge of the element. This is also called the work-equivalent nodal force vector. For a constant surface traction T_x and T_y, we can calculate the equivalent nodal force, as

$$\{\mathbf{f}_T^{(e)}\} = h \int_0^l [\mathbf{N}]^T \{\mathbf{T}\} ds = h \int_0^l \begin{bmatrix} (l-s)/l & 0 \\ 0 & (l-s)/l \\ s/l & 0 \\ 0 & s/l \\ 0 & 0 \\ 0 & 0 \end{bmatrix} \begin{Bmatrix} T_x \\ T_y \end{Bmatrix} ds = \frac{hl}{2} \begin{Bmatrix} T_x \\ T_y \\ T_x \\ T_y \\ 0 \\ 0 \end{Bmatrix} \tag{6.35}$$

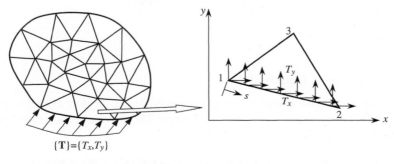

$\{\mathbf{T}\} = \{T_x, T_y\}$

Figure 6.7 Applied surface traction along edge 1-2

For the uniform surface traction force, the equivalent nodal forces are obtained by simply dividing the total force equally between the two nodes on the edge.

The potential energy of distributed forces of all elements whose edge belongs to the traction boundary S_T must be assembled to build the global force vector of distributed forces:

$$V = -\sum_{e=1}^{NS}\{\mathbf{q}^{(e)}\}^T\{\mathbf{f}_T^{(e)}\} = -\{\mathbf{Q}_s\}^T\{\mathbf{F}_T\} \tag{6.36}$$

where NS is the number of elements whose edge belongs to S_T.

6.4.6 Potential Energy of Body Forces

The body forces are distributed over the entire element (e.g., centrifugal forces, gravitational forces, inertia forces, magnetic forces). For simplification of the derivation, let us assume that a constant body force $\mathbf{b} = \{b_x, b_y\}^T$ is applied to the whole element. The potential energy of body force becomes

$$V^{(e)} = -h\iint_A [u \quad v]\begin{Bmatrix} b_x \\ b_y \end{Bmatrix} dA = -\{\mathbf{q}^{(e)}\}^T h\iint_A [\mathbf{N}]^T dA \begin{Bmatrix} b_x \\ b_y \end{Bmatrix}$$
$$\equiv -\{\mathbf{q}^{(e)}\}^T\{\mathbf{f}_b^{(e)}\} \tag{6.37}$$

where

$$\{\mathbf{f}_b^{(e)}\} = \frac{hA}{3}\begin{bmatrix} 1 & 0 \\ 0 & 1 \\ 1 & 0 \\ 0 & 1 \\ 1 & 0 \\ 0 & 1 \end{bmatrix}\begin{Bmatrix} b_x \\ b_y \end{Bmatrix} = \frac{hA}{3}\begin{Bmatrix} b_x \\ b_y \\ b_x \\ b_y \\ b_x \\ b_y \end{Bmatrix} \tag{6.38}$$

The resultant of body forces is hAb_x in the x-direction and hAb_y in the y-direction. Equation (6.38) equally distributes these forces to the three nodes. Similar to the distributed force, $\{\mathbf{f}_B\}$ is the equivalent nodal force that corresponds to the constant body force.

The potential energy of body forces of all elements must be assembled to build the global force vector of body forces:

$$V = -\sum_{e=1}^{NE}\{\mathbf{q}^{(e)}\}^T\{\mathbf{f}_b^{(e)}\} = -\{\mathbf{Q}_s\}^T\{\mathbf{F}_B\} \tag{6.39}$$

where NE is the number of elements.

6.4.7 Global Finite Element Equations

Since the strain energy and potential energy of applied forces are now available, let us go back to the potential energy of the triangular element. The discrete version of the potential energy becomes

$$\Pi = U + V = \frac{1}{2}\{\mathbf{Q}_s\}^T[\mathbf{K}_s]\{\mathbf{Q}_s\} - \{\mathbf{Q}_s\}^T\{\mathbf{F}_N + \mathbf{F}_T + \mathbf{F}_B\} \tag{6.40}$$

The principle of minimum potential energy in Chapter 3 states that the structure is in equilibrium when the potential energy is minimum. Since the potential energy in Eq. (6.40) is a quadratic form the displacement vector $\{\mathbf{Q}_s\}$, we can differentiate Π to obtain

$$\frac{\partial \Pi}{\partial \{\mathbf{Q}_s\}} = 0 \Rightarrow [\mathbf{K}_s]\{\mathbf{Q}_s\} = \{\mathbf{F}_N + \mathbf{F}_T + \mathbf{F}_B\} \tag{6.41}$$

The stationary condition of the potential energy yields the global finite element matrix equations.

The assembled structural stiffness matrix $[\mathbf{K}_s]$ is singular due to the presence of rigid body motion. After constructing the global matrix equation, the boundary conditions are applied by removing those DOFs that are fixed or prescribed. After imposing the boundary condition, the global stiffness matrix becomes non-singular and can be inverted to solve for the nodal displacements.

6.4.8   Calculation of Strains and Stresses

Once the nodal displacements are calculated, strains and stresses in individual elements can be calculated. First, the nodal displacement vector $\{\mathbf{q}^{(e)}\}$ for the element of interest needs to be extracted from the global displacement vector. Then, the strains and stresses in the element can be obtained from

$$\{\varepsilon\} = [\mathbf{B}]\{\mathbf{q}^{(e)}\} \tag{6.42}$$

and

$$\{\sigma\} = [\mathbf{C}]\{\varepsilon\} = [\mathbf{C}][\mathbf{B}]\{\mathbf{q}^{(e)}\} \tag{6.43}$$

where $[\mathbf{C}] = [\mathbf{C}_\sigma]$ for the plane stress problems and $[\mathbf{C}] = [\mathbf{C}_\varepsilon]$ for the plane strain problems.

As discussed before, stress and strain are constant within an element because the matrixes $[\mathbf{B}]$ and $[\mathbf{C}]$ are constant. This property can cause difficulties in interpreting the results of the finite element analysis. When two adjacent elements have different stress values, it is difficult to determine the stress value at the interface. Such discontinuity is not caused by the physics of the problem, but by the inability of the triangular element in describing the continuous change of stresses across element boundary. In fact, most finite elements cannot maintain continuity of stresses across the element boundary. Most programs average the stress at the element boundaries in order to make the stress look continuous. However, as we refine the model using smaller size elements, this discontinuity can be reduced. The following example illustrates discontinuity of stress and strain between two adjacent CST elements.

EXAMPLE 6.2   *Cantilevered Plate*

Consider a cantilevered plate as shown in Figure 6.8. The plate has the following properties: $h = 0.1$ in, $E = 30 \times 10^6$ psi, and $v = 0.3$. Model the plate using two CST elements to determine the displacements and stresses.

SOLUTION   This problem can be modeled as plane stress because the thickness of the plate is small compared to the other dimensions.

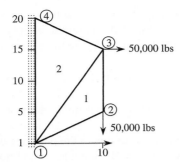

Figure 6.8 Cantilevered Plate

(1) Element 1: Nodes 1-2-3

Using nodal coordinates, we can calculate the constants defined in Eq. (6.18) as

$$
\begin{array}{lll}
x_1 = 0,\ y_1 = 0 & x_2 = 10,\ y_2 = 5 & x_3 = 10,\ y_3 = 15 \\
b_1 = y_2 - y_3 = -10 & b_2 = y_3 - y_1 = 15 & b_3 = y_1 - y_2 = -5 \\
c_1 = x_3 - x_2 = 0 & c_2 = x_1 - x_3 = -10 & c_3 = x_2 - x_1 = 10
\end{array}
$$

In addition, from the geometry of the element, the area of the triangle $A_1 = 0.5 \times 10 \times 10 = 50$. The matrix $[\mathbf{B}]$ in Eq. (6.25) and $[\mathbf{C}_\sigma]$ can be written as

$$
[\mathbf{B}] = \frac{1}{2A}
\begin{bmatrix}
b_1 & 0 & b_2 & 0 & b_3 & 0 \\
0 & c_1 & 0 & c_2 & 0 & c_3 \\
c_1 & b_1 & c_2 & b_2 & c_3 & b_3
\end{bmatrix}
$$

$$
= \frac{1}{100}
\begin{bmatrix}
-10 & 0 & 15 & 0 & -5 & 0 \\
0 & 0 & 0 & -10 & 0 & 10 \\
0 & -10 & -10 & 15 & 10 & -5
\end{bmatrix}
$$

and

$$
[\mathbf{C}_\sigma] = \frac{E}{1 - \nu^2}
\begin{bmatrix}
1 & \nu & 0 \\
\nu & 1 & 0 \\
0 & 0 & \tfrac{1}{2}(1 - \nu)
\end{bmatrix}
= 3.297 \times 10^7
\begin{bmatrix}
1 & .3 & 0 \\
.3 & 1 & 0 \\
0 & 0 & .35
\end{bmatrix}
$$

Using the above two matrices, the element stiffness matrix can be obtained as

$$
[\mathbf{k}^{(1)}] = hA[\mathbf{B}]^T[\mathbf{C}_\sigma][\mathbf{B}]
$$

$$
= 3.297 \times 10^6
\begin{bmatrix}
.5 & 0. & -.75 & .15 & .25 & -.15 \\
& .175 & .175 & -.263 & -.175 & .088 \\
& & 1.3 & -.488 & -.55 & .313 \\
& & & .894 & .338 & -.631 \\
& \text{Symmetric} & & & .3 & -.163 \\
& & & & & .544
\end{bmatrix}
$$

(2) Element 2: Nodes 1-3-4

By following similar procedures as for Element 1, the constants in Eq. (6.18) for Element 2 can be written as

$$
\begin{array}{lll}
x_1 = 0,\ y_1 = 0 & x_2 = 10,\ y_2 = 15 & x_3 = 0,\ y_3 = 20 \\
b_1 = y_2 - y_3 = -5 & b_2 = y_3 - y_1 = 20 & b_3 = y_1 - y_2 = -15 \\
c_1 = x_3 - x_2 = -10 & c_2 = x_1 - x_3 = 0 & c_3 = x_2 - x_1 = 10
\end{array}
$$

The area of Element 2 is twice that of Element 1: $A_2 = 0.5 \times 20 \times 10 = 100$. The strain-displacement matrix $[\mathbf{B}]$ can be obtained as

$$[\mathbf{B}] = \frac{1}{200} \begin{bmatrix} -5 & 0 & 20 & 0 & -15 & 0 \\ 0 & -10 & 0 & 0 & 0 & 10 \\ -10 & -5 & 0 & 20 & 10 & -15 \end{bmatrix}$$

By following the same procedure, the stiffness matrix for Element 2 can be computed as

$$[\mathbf{k}^{(2)}] = 3.297 \times 10^6 \begin{bmatrix} .15 & .081 & -.25 & -.175 & .1 & .094 \\ & .272 & -.15 & -.088 & .069 & -.184 \\ & & 1. & 0. & -.75 & .15 \\ & & & .35 & .175 & -.263 \\ & & & & .65 & -.244 \\ & \text{Symmetric} & & & & .447 \end{bmatrix}$$

(3) Global finite element matrix equations

The two element stiffness matrices are assembled to form the global stiffness matrix. Since there are four nodes, the model has eight DOFs; each node has two DOFs. Thus, the global matrix has a dimension of 8×8. After assembly, the global matrix equation can be written as

$$3.297 \times 10^6 \begin{vmatrix} .65 & .081 & -.75 & .15 & .0 & -.325 & .1 & .094 \\ & .447 & .175 & -.263 & -.325 & .0 & .069 & -.184 \\ & & 1.3 & -.488 & -.55 & .313 & .0 & .0 \\ & & & .894 & .338 & -.631 & .0 & .0 \\ & & & & 1.3 & -.163 & -.75 & .15 \\ & & & & & .894 & .175 & -.263 \\ & & & & & & .65 & -.244 \\ & \text{Symmetric} & & & & & & .447 \end{vmatrix} \begin{Bmatrix} u_1 \\ v_1 \\ u_2 \\ v_2 \\ u_3 \\ v_3 \\ u_4 \\ v_4 \end{Bmatrix} = \begin{Bmatrix} R_{x1} \\ R_{y1} \\ 0 \\ -50,000 \\ 50,000 \\ 0 \\ R_{x4} \\ R_{y4} \end{Bmatrix}$$

where R_{x1}, R_{y1}, R_{x4}, and R_{y4} are unknown reaction forces at nodes 1 and 4.

(4) Applying boundary conditions

The displacement boundary conditions are: $u_1 = v_1 = u_4 = v_4 = 0$. Thus, we remove first, second, seventh, and eighth, rows and columns. After removing those rows and columns we obtain the following reduced matrix equation:

$$3.297 \times 10^6 \begin{bmatrix} 1.3 & -.488 & -.55 & .313 \\ & .894 & .338 & -.631 \\ \text{Symmetric} & & 1.3 & -.163 \\ & & & .894 \end{bmatrix} \begin{Bmatrix} u_2 \\ v_2 \\ u_3 \\ v_3 \end{Bmatrix} = \begin{Bmatrix} 0 \\ -50,000 \\ 50,000 \\ 0 \end{Bmatrix}$$

Note that the global stiffness matrix of the above equation is non-singular and, therefore, the unique solution can be obtained.

(5) Solution

The above matrix equation can be solved for unknown nodal displacements:

$$u_2 = -2.147 \times 10^{-3}$$
$$v_2 = -4.455 \times 10^{-2}$$
$$u_3 = 1.891 \times 10^{-2}$$
$$v_3 = -2.727 \times 10^{-2}$$

(6) Strain and stress in Element 1

After calculating the nodal displacements, strain and stress can be calculated at the element level. First, the displacements for those nodes that belong to the element need to be extracted from the global nodal displacement vector. Since nodes 1, 2, and 3 belong to Element 1, the nodal displacements will be $\{\mathbf{q}\} = \{u_1,\, v_1,\, u_2,\, v_2,\, u_3,\, v_3\}^T = \{0,\, 0,\, -2.147 \times 10^{-3},\, -4.455 \times 10^{-2},$ $1.891 \times 10^{-2},\, -2.727 \times 10^{-2}\}^T$. Then, strain in Eq. (6.25) can be calculated using $\{\varepsilon\} = [\mathbf{B}]\{\mathbf{q}\}$

$$
\begin{Bmatrix} \varepsilon_{xx} \\ \varepsilon_{yy} \\ \gamma_{xy} \end{Bmatrix} = \frac{1}{100}
\begin{bmatrix}
-10 & 0 & 15 & 0 & -5 & 0 \\
0 & 0 & 0 & -10 & 0 & 10 \\
0 & -10 & -10 & 15 & 10 & -5
\end{bmatrix}
\begin{Bmatrix}
0 \\ 0 \\ -2.147 \times 10^{-3} \\ -4.455 \times 10^{-2} \\ 1.891 \times 10^{-2} \\ -2.727 \times 10^{-2}
\end{Bmatrix}
$$

$$
= \begin{Bmatrix} -1.268 \times 10^{-3} \\ 1.727 \times 10^{-3} \\ -3.212 \times 10^{-3} \end{Bmatrix}
$$

The stresses in the element are obtained from Eq. (6.5)

$$
\begin{Bmatrix} \sigma_{xx} \\ \sigma_{yy} \\ \tau_{xy} \end{Bmatrix} = 3.297 \times 10^7
\begin{bmatrix}
1 & .3 & 0 \\
.3 & 1 & 0 \\
0 & 0 & .35
\end{bmatrix}
\begin{Bmatrix} -1.268 \times 10^{-3} \\ 1.727 \times 10^{-3} \\ -3.212 \times 10^{-3} \end{Bmatrix}
= \begin{Bmatrix} -24,709 \\ 44,406 \\ -37,063 \end{Bmatrix} \text{psi}
$$

(7) Strains and stresses in Element 2

Element 2 has Nodes 1, 3, and 4. Thus, the nodal displacements will be $\{\mathbf{q}\} = \{u_1,\, v_1,\, u_3,\, v_3,\, u_4,\, v_4\}^T = \{0,\, 0,\, 1.891 \times 10^{-2},\, -2.727 \times 10^{-2},\, 0,\, 0\}^T$. Using the element displacements, the strains and stresses in the element can be obtained as

$$
\begin{Bmatrix} \varepsilon_{xx} \\ \varepsilon_{yy} \\ \gamma_{xy} \end{Bmatrix} = \frac{1}{200}
\begin{bmatrix}
-5 & 0 & 20 & 0 & -15 & 0 \\
0 & -10 & 0 & 0 & 0 & 10 \\
-10 & -5 & 0 & 20 & 10 & -15
\end{bmatrix}
\begin{Bmatrix}
0 \\ 0 \\ 1.891 \times 10^{-2} \\ -2.727 \times 10^{-2} \\ 0 \\ 0
\end{Bmatrix}
$$

$$
= \begin{Bmatrix} 1.891 \times 10^{-3} \\ 0 \\ -2.727 \times 10^{-3} \end{Bmatrix}
$$

and

$$
\begin{Bmatrix} \sigma_{xx} \\ \sigma_{yy} \\ \tau_{xy} \end{Bmatrix} = 3.297 \times 10^7
\begin{bmatrix}
1 & .3 & 0 \\
.3 & 1 & 0 \\
0 & 0 & .35
\end{bmatrix}
\begin{Bmatrix} 1.891 \times 10^{-3} \\ 0 \\ -2.727 \times 10^{-3} \end{Bmatrix}
= \begin{Bmatrix} 62,354 \\ 18,706 \\ -31,469 \end{Bmatrix} \text{psi}
$$

If the stresses in the two elements are examined, one can note that the stress value changes suddenly across the element boundary. For example, σ_{xx} in Element 1 is $-24,709$ psi, whereas in Element 2 it is 62,354 psi. Such a drastic change in stresses is an indicator that the finite element analysis results from the current model are not accurate and more elements are required.

From Example 6.2, we can conclude the following:

- Stresses are constant over the individual element.
- The solution is not accurate because there are large discontinuities in stresses across element boundaries.
- With only two elements, the mesh is very coarse and we obviously cannot expect very good results.

6.5   FOUR-NODE RECTANGULAR ELEMENT

6.5.1   Lagrange Interpolation for Rectangular Element

A rectangular element is composed of four nodes and eight DOFs (see Figure 6.9). It is a part of a plane solid that is composed of many rectangular elements. Each element shares its edge and two corner nodes with an adjacent element, except for those on the boundary. The four vertices of a rectangle are the nodes of that element, as shown in Figure 6.9. The first node of an element can arbitrarily be chosen. However, the sequence of Nodes 1, 2, 3, and 4 should be in the counter-clockwise direction. Each node has two displacements, u and v, respectively, in the x- and y-directions.

Since all edges are parallel to the coordinate directions, this element is not practical but useful as it is the basis for the quadrilateral element discussed in the following section. In addition, the behavior of the rectangular element is similar to that of the quadrilateral element. Shape functions can be calculated using procedures similar to that of CST element, but it is more instructive to use the Lagrange interpolation functions in x- and y-directions.

Consider the rectangular element in Figure 6.9. From the geometry, it is clear that $x_3 = x_2$, $y_4 = y_3$, $x_4 = x_1$, and $y_2 = y_1$. We will use a polynomial in x and y as the interpolation function. Since there are four nodes, we can apply four conditions and hence the polynomial should have four terms, as follows:

$$\begin{aligned} u &= \alpha_1 + \alpha_2 x + \alpha_3 y + \alpha_4 xy \\ v &= \beta_1 + \beta_2 x + \beta_3 y + \beta_4 xy \end{aligned} \tag{6.44}$$

Let us calculate unknown coefficients α_i using the x-directional displacement u:

$$\begin{cases} u_1 = \alpha_1 + \alpha_2 x_1 + \alpha_3 y_1 + \alpha_4 x_1 y_1 \\ u_2 = \alpha_1 + \alpha_2 x_2 + \alpha_3 y_2 + \alpha_4 x_2 y_2 \\ u_3 = \alpha_1 + \alpha_2 x_3 + \alpha_3 y_3 + \alpha_4 x_3 y_3 \\ u_4 = \alpha_1 + \alpha_2 x_4 + \alpha_3 y_4 + \alpha_4 x_4 y_4 \end{cases}$$

It is obvious that we need to invert the 4×4 matrix in order to calculate the interpolation coefficients.

Instead of matrix inversion method, we use the *Lagrange interpolation* method to interpolate u and v. The goal is to obtain the following expression:

$$u(x, y) = [N_1 \quad N_2 \quad N_3 \quad N_4] \begin{Bmatrix} u_1 \\ u_2 \\ u_3 \\ u_4 \end{Bmatrix} \tag{6.45}$$

where $N_1, \ldots, N_4$ are the interpolation functions. To do that, let us first consider displacement along edge 1-2 in Figure 6.9. Along edge 1-2, $y = y_1$ (constant); therefore shape functions must be functions of x only as shown below:

$$u_I(x, y_1) = [n_1(x) \quad n_2(x)] \begin{Bmatrix} u_1 \\ u_2 \end{Bmatrix} \tag{6.46}$$

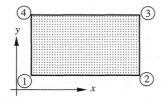

Figure 6.9 Four-node rectangular element

Using one-dimensional Lagrange interpolation formula the shape functions can be obtained as

$$n_1(x) = \frac{x - x_2}{x_1 - x_2}, \quad n_2(x) = \frac{x - x_1}{x_2 - x_1} \tag{6.47}$$

This is the same procedure that was used in Chapter 3. Next, since $y = y_3 = y_4$ along edge 4-3 in Figure 6.9, the displacement can be interpolated as

$$u_{II}(x, y_3) = [n_4(x) \quad n_3(x)] \begin{Bmatrix} u_4 \\ u_3 \end{Bmatrix} \tag{6.48}$$

Again from one-dimensional Lagrange interpolation formula, we have

$$n_4(x) = \frac{x - x_3}{x_4 - x_3}, \quad n_3(x) = \frac{x - x_4}{x_3 - x_4} \tag{6.49}$$

Equations (6.46) and (6.48) represent interpolation of displacements at the top and bottom of the element, respectively. So far, we have interpolated displacements in the x-direction only. Now, we can extend the interpolation in the y-direction between $u_I(x, y_1)$ and $u_{II}(x, y_3)$ using the same Lagrange interpolation method. By considering $u_I(x, y_1)$ and $u_{II}(x, y_3)$ as nodal displacements, we have the following interpolation formula:

$$u(x, y) = [n_1(y) \quad n_4(y)] \begin{Bmatrix} u_I(x, y_1) \\ u_{II}(x, y_3) \end{Bmatrix} \tag{6.50}$$

where

$$n_1(y) = \frac{y - y_4}{y_1 - y_4}, \quad n_4(y) = \frac{y - y_1}{y_4 - y_1} \tag{6.51}$$

are Lagrange interpolation in the y-direction. By substituting Eqs. (6.46) and (6.48) into Eq. (6.50), we have the following formula:

$$u(x, y) = [n_1(y) \quad n_4(y)] \begin{Bmatrix} [n_1(x) \quad n_2(x)] \begin{Bmatrix} u_1 \\ u_2 \end{Bmatrix} \\ [n_4(x) \quad n_3(x)] \begin{Bmatrix} u_4 \\ u_3 \end{Bmatrix} \end{Bmatrix} \tag{6.52}$$

Thus,

$$u(x, y) = [n_1(x)n_1(y) \quad n_2(x)n_1(y) \quad n_3(x)n_4(y) \quad n_4(x)n_4(y)] \begin{Bmatrix} u_1 \\ u_2 \\ u_3 \\ u_4 \end{Bmatrix} \tag{6.53}$$

Comparing the above expression with Eq. (6.45), we can define shape functions. $N_1, \ldots, N_4$. In the rectangular element, it is enough to use the coordinates of two nodes, because $x_1 = x_4$, $y_1 = y_2$, etc. We will use the coordinates of Nodes 1 and 3. Using the property that the area of the element is $A = (x_3 - x_1)(y_3 - y_1)$, we obtain

$$\begin{cases} N_1 \equiv n_1(x)n_1(y) = \dfrac{1}{A}(x_3 - x)(y_3 - y) \\[2mm] N_2 \equiv n_2(x)n_1(y) = -\dfrac{1}{A}(x_1 - x)(y_3 - y) \\[2mm] N_3 \equiv n_3(x)n_4(y) = \dfrac{1}{A}(x_1 - x)(y_1 - y) \\[2mm] N_4 \equiv n_4(x)n_4(y) = -\dfrac{1}{A}(x_3 - x)(y_1 - y) \end{cases} \tag{6.54}$$

Note that the *shape functions* for rectangular elements are the product of Lagrange interpolations in the two coordinate directions. Let us discuss the properties of the shape functions. It can be easily verified that $N_1(x, y)$ is:

- 1 at Node 1 and 0 at other nodes
- linear function of x along edge 1-2 and linear function of y along edge 1-4 (bilinear interpolation)
- zero along edges 2-3 and 3-4

Other shape functions have similar behavior. Because of these characteristics, the i-th shape function is considered associated with Node i of the element.

To make the derivations simple, we rewrite the interpolation relation in Eq. (6.45) in matrix form. Let $\{\mathbf{u}\} = \{u, v\}^T$ be the displacement vector at any point (x, y). The interpolation can be written using the matrix notation by

$$\{\mathbf{u}\} \equiv \begin{Bmatrix} u \\ v \end{Bmatrix} = \begin{bmatrix} N_1 & 0 & N_2 & 0 & N_3 & 0 & N_4 & 0 \\ 0 & N_1 & 0 & N_2 & 0 & N_3 & 0 & N_4 \end{bmatrix} \begin{Bmatrix} u_1 \\ v_1 \\ u_2 \\ v_2 \\ u_3 \\ v_3 \\ u_4 \\ v_4 \end{Bmatrix}$$

or,

$$\{\mathbf{u}\} = [\mathbf{N}]_{2\times8}\{\mathbf{q}\}_{8\times1} \tag{6.55}$$

Note that the dimension of the shape function matrix is 2×8.

EXAMPLE 6.3   *Shape Functions of a Rectangular Element*

A rectangular element is shown in Figure 6.10. By substituting the numerical values of nodal coordinates into the above shape function formulas, the expressions for shape functions for this rectangular element can be obtained as

$$N_1 = \frac{(3-x)(2-y)}{6} \qquad N_2 = \frac{x(2-y)}{6}$$

$$N_3 = \frac{xy}{6} \qquad N_4 = \frac{y(3-x)}{6}$$

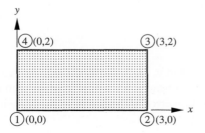

Figure 6.10 Four-node rectangular element

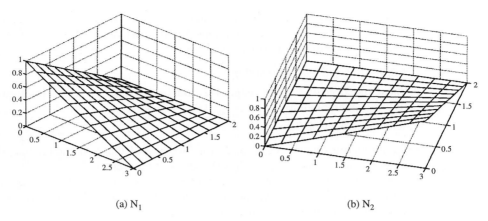

(a) N_1 (b) N_2

Figure 6.11 Three-dimensional surface plots of shape functions for a rectangular element

Three-dimensional plots of N_1 and N_2 are shown in Figure 6.11(a) and Figure 6.11(b).

Since the shape functions are given as function of x- and y-coordinates, we can use an approach similar to that of CST element to obtain the strain-displacement relations. Thus, the strain can be calculated by differentiating the shape function with respect to the coordinates. For example, ε_{xx} can be written as

$$\varepsilon_{xx} \equiv \frac{\partial u}{\partial x} = \frac{\partial}{\partial x}\left(\sum_{i=1}^{4} N_i(x,\,y)u_i\right) = \sum_{i=1}^{4} \frac{\partial N_i}{\partial x}u_i \qquad (6.56)$$

Note that u_1, u_2, u_3, and u_4 are nodal displacements and are independent of coordinate x. Thus, only the shape function is differentiated with respect to x. Similar calculation can be carried out for ε_{yy} and γ_{xy}. Then, we have

$$\{\varepsilon\} = \frac{1}{A}\begin{bmatrix} y-y_3 & 0 & y_3-y & 0 & y-y_1 & 0 & y_1-y & 0 \\ 0 & x-x_3 & 0 & x_1-x & 0 & x-x_1 & 0 & x_3-x \\ x-x_3 & y-y_3 & x_1-x & y_3-y & x-x_1 & y-y_1 & x_3-x & y_1-y \end{bmatrix}\begin{Bmatrix} u_1 \\ v_1 \\ u_2 \\ v_2 \\ u_3 \\ v_3 \\ u_4 \\ v_4 \end{Bmatrix}$$

$$\equiv [\mathbf{B}]\{\mathbf{q}\}$$

$$(6.57)$$

Note that the matrix $[\mathbf{B}]$ is a linear function of x and y. Thus, the strain will change linearly within the element. For example, ε_{xx} will vary linearly in y while constant with respect to x. Thus, the element will have an approximation error, if the actual strains vary in the x-direction.

6.5.2   Element Stiffness Matrix

The element stiffness matrix can be calculated from the strain energy of the element. By substituting for strains from Eq. (6.57) into the expression for strain energy in Eq. (6.9), we have

$$U^{(e)} = \frac{h}{2} \iint\limits_A \{\varepsilon\}^T [\mathbf{C}]\{\varepsilon\} dA^{(e)}$$

$$= \frac{h}{2} \{\mathbf{q}^{(e)}\}^T \iint\limits_A [\mathbf{B}]^T_{8\times3} [\mathbf{C}]_{3\times3} [\mathbf{B}]_{3\times8} \ dA \{\mathbf{q}^{(e)}\} \tag{6.58}$$

$$\equiv \frac{1}{2} \{\mathbf{q}^{(e)}\}^T [\mathbf{k}^{(e)}]_{8\times8} \{\mathbf{q}^{(e)}\}$$

where $[\mathbf{k}^{(e)}]$ is the element stiffness matrix. Calculation of the element stiffness matrix requires two-dimensional integration. We will discuss numerical integration in the next section. When the element is square and the problem is plane stress, analytical integration of the strain energy yields the following form of element stiffness matrix:

$$[\mathbf{k}^{(e)}] = \frac{Eh}{1-\nu^2} \times
\begin{vmatrix}
\frac{3-\nu}{6} & \frac{1+\nu}{8} & -\frac{3+\nu}{12} & \frac{-1+3\nu}{8} & \frac{-3+\nu}{12} & -\frac{1+\nu}{8} & \frac{\nu}{6} & \frac{1-3\nu}{8} \\
\frac{1+\nu}{8} & \frac{3-\nu}{6} & \frac{1-3\nu}{8} & \frac{\nu}{6} & \frac{1+\nu}{8} & \frac{-3+\nu}{12} & \frac{-1+3\nu}{8} & -\frac{3+\nu}{12} \\
-\frac{3+\nu}{12} & \frac{1-3\nu}{8} & \frac{3-\nu}{6} & -\frac{1+\nu}{8} & \frac{\nu}{6} & \frac{-1+3\nu}{8} & \frac{-3+\nu}{12} & \frac{1+\nu}{8} \\
\frac{-1+3\nu}{8} & \frac{\nu}{6} & -\frac{1+\nu}{8} & \frac{3-\nu}{6} & \frac{1-3\nu}{8} & -\frac{3+\nu}{12} & \frac{1+\nu}{8} & \frac{-3+\nu}{12} \\
\frac{-3+\nu}{12} & -\frac{1+\nu}{8} & \frac{\nu}{6} & \frac{1-3\nu}{8} & \frac{3-\nu}{6} & \frac{1+\nu}{8} & \frac{3+\nu}{12} & \frac{-1+3\nu}{8} \\
-\frac{1+\nu}{8} & \frac{-3+\nu}{12} & \frac{-1+3\nu}{8} & \frac{3+\nu}{12} & \frac{1+\nu}{8} & \frac{3-\nu}{6} & \frac{1-3\nu}{8} & \frac{\nu}{6} \\
\frac{\nu}{6} & \frac{-1+3\nu}{8} & \frac{-3+\nu}{12} & \frac{1+\nu}{8} & \frac{3+\nu}{12} & \frac{1-3\nu}{8} & \frac{3-\nu}{6} & -\frac{1+\nu}{8} \\
\frac{1-3\nu}{8} & -\frac{3+\nu}{12} & \frac{1+\nu}{8} & \frac{-3+\nu}{12} & \frac{-1+3\nu}{8} & \frac{\nu}{6} & \frac{1+\nu}{8} & \frac{3-\nu}{6}
\end{vmatrix}$$

$$\tag{6.59}$$

It is interesting to note that the element stiffness matrix does not depend on the actual element dimensions but it is a function of material properties (E and ν) and thickness h for a square element.

The strain energy of the entire solid can be obtained using Eq. (6.29), which involves the assembly process.

6.5.3   Potential Energy of Applied Loads

In the CST element, we discussed three different types of applied loads: concentrated forces at nodes, distributed forces along element edges, and body force. The first two types are independent of element used. Thus, the same forms in Eqs. (6.31) and (6.36) can be used for the potential energies of concentrated force and distributed force, respectively.

In the case of the body force, the element shape functions are used to calculate equivalent nodal forces. When a constant body force $\mathbf{b} = \{b_x, b_y\}^T$ acts on a rectangular element, the potential energy of body force becomes

$$V^{(e)} = -h \iint_A [u \quad v] \begin{Bmatrix} b_x \\ b_y \end{Bmatrix} dA = -\{\mathbf{q}^{(e)}\}^T h \iint_A [\mathbf{N}]^T dA \begin{Bmatrix} b_x \\ b_y \end{Bmatrix}$$

$$\equiv \{\mathbf{q}^{(e)}\}^T \{\mathbf{f}_b^{(e)}\} \tag{6.60}$$

where

$$\{\mathbf{f}_b^{(e)}\} = \frac{hA}{4} \begin{bmatrix} 1 & 0 \\ 0 & 1 \\ 1 & 0 \\ 0 & 1 \\ 1 & 0 \\ 0 & 1 \\ 1 & 0 \\ 0 & 1 \end{bmatrix} \begin{Bmatrix} b_x \\ b_y \end{Bmatrix} = \frac{hA}{4} \begin{Bmatrix} b_x \\ b_y \\ b_x \\ b_y \\ b_x \\ b_y \\ b_x \\ b_y \end{Bmatrix} \tag{6.61}$$

Equation (6.61) equally divides the total magnitude of the body force to the four nodes. $\{\mathbf{f}_b\}$ is the equivalent nodal force that corresponds to the constant body force. The potential energy of body forces of all elements must be assembled to build the global force vector of body forces as in Eq. (6.39).

Using the principle of minimum total potential energy in Eq. (6.41), a similar global matrix equation for the rectangular elements can be obtained. Applying boundary conditions and solving the matrix equations are identical to the CST element. After solving for nodal displacements, strains and stresses in each element can be calculated using Eqs. (6.42) and (6.43), respectively.

EXAMPLE 6.4   *Simple Shear Deformation of a Square Element*

A square element shown in Figure 6.12 is under a simple shear deformation. Material properties are given as $E = 10\,\text{GPa}$, $\nu = 0.25$, and thickness is $h = 0.1\,\text{m}$. When distributed force $f = 100\,\text{kN/m}^2$ is applied horizontally at the top edge, calculate stress and strain components. Compare the results with the exact solution.

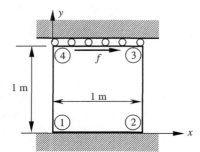

Figure 6.12 A square element under a simple shear condition

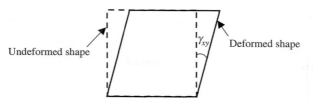

Undeformed shape γ_{xy} Deformed shape

Figure 6.13 Simple shear
deformation of a square element

SOLUTION   Since the problem consists of one element, we do not need assembly process. The element
has eight DOFs: $\{Q_s\} = \{u_1, v_1, u_2, v_2, u_3, v_3, u_4, v_4\}^T$. From the boundary condition given in
Figure 6.12, only two DOFs are non-zero: u_3 and u_4. Thus, from the element stiffness matrix given
in Eq. (6.59), all fixed DOFs are deleted to obtain

$$[\mathbf{K}] = \frac{Eh}{1-v^2}\begin{bmatrix} \dfrac{3-v}{6} & -\dfrac{3+v}{12} \\[2mm] -\dfrac{3+v}{12} & \dfrac{3-v}{6} \end{bmatrix}\begin{matrix} u_3 \\ u_4 \end{matrix} = 10^8\begin{bmatrix} 4.88 & -2.88 \\ -2.88 & 4.88 \end{bmatrix}\begin{matrix} u_3 \\ u_4 \end{matrix}$$

The total distributed load of 10,000 N at the top edge will be equally divided into two nodes: 4 and
3. Thus, the global matrix equation becomes

$$10^8\begin{bmatrix} 4.88 & -2.88 \\ -2.88 & 4.88 \end{bmatrix}\begin{Bmatrix} u_3 \\ u_4 \end{Bmatrix} = \begin{Bmatrix} 5,000 \\ 5,000 \end{Bmatrix}$$

The above equation can be solved for unknown nodal displacements, as $u_3 = u_4 = 0.025\,\text{mm}$. Then,
from (6.57), the strain components can be obtained, as

$$\{\varepsilon\} = \begin{bmatrix} y-1 & 0 & 1-y & 0 & y & 0 & -y & 0 \\ 0 & x-1 & 0 & -x & 0 & x & 0 & 1-x \\ x-1 & y-1 & -x & 1-y & x & y & 1-x & -y \end{bmatrix}\begin{Bmatrix} 0 \\ 0 \\ 0 \\ 0 \\ 2.5\times 10^{-5} \\ 0 \\ 2.5\times 10^{-5} \\ 0 \end{Bmatrix}$$

$$= \begin{Bmatrix} 0 \\ 0 \\ 2.5\times 10^{-5} \end{Bmatrix}$$

Note that the shear strain is the only non-zero strain. Thus, the rectangular element can accurately
represent the simple shear condition. Figure 6.13 shows the deformed shape of the solid. Note that
the deformation is magnified for the illustration purpose.

Using the stress-strain relation in Eq. (6.5) for plane stress, the stress components can be
obtained, as

$$\begin{Bmatrix} \sigma_{xx} \\ \sigma_{yy} \\ \tau_{xy} \end{Bmatrix} = \frac{10^{10}}{1-0.25^2}\begin{bmatrix} 1 & 0.25 & 0 \\ 0.25 & 1 & 0 \\ 0 & 0 & 0.375 \end{bmatrix}\begin{Bmatrix} 0 \\ 0 \\ 2.5\times 10^{-5} \end{Bmatrix} = \begin{Bmatrix} 0 \\ 0 \\ 10^5 \end{Bmatrix}\text{Pa}$$

Since distributed force $f = 10\,\text{kN/m}^2$ is applied at the top edge, the above shear stress is exact.

EXAMPLE 6.5   ***Pure Bending Deformation of a Square Element***

A pure bending condition can be achieved by applying a couples in the case of a beam (see Chapter
4). For the plane solid, the effect of a couple can be achieved by applying equal forces in opposite

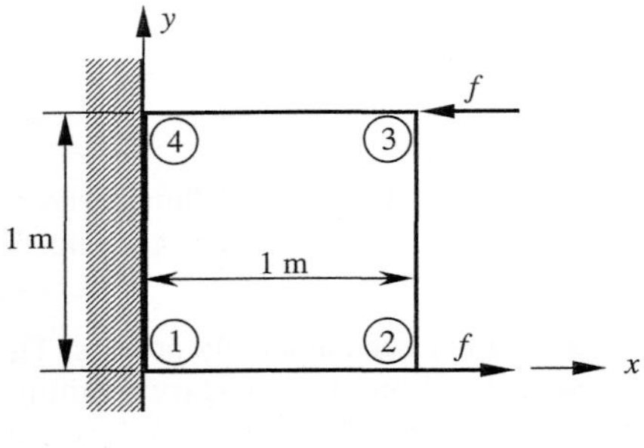

Figure 6.14 A square element under pure bending condition

directions. A square element shown in Figure 6.14 is under a pure bending condition. Material properties are given as $E = 10\,\text{GPa}$, $\nu = 0.25$, and thickness is $h = 0.1\,\text{m}$. When an equal and opposite force $f = 100\,\text{kN}$ is applied at Nodes 2 and 3, calculate stress and strain components. Compare the results with the exact solutions from the beam theory.

SOLUTION

(a) Analytical solution: If we consider the above plane solid as a cantilevered beam, the moment of inertia $I = 8.333 \times 10^{-3}\,\text{m}^4$ and the applied couple $M = 100\,\text{kN} \cdot \text{m}$. Thus, the maximum stress will occur at the bottom edge with the magnitude of

$$(\sigma_{xx})_{\max} = -\frac{M\left(-\dfrac{L}{2}\right)}{I} = 6.0\,\text{MPa}$$

where L is the length of the element. The minimum stress will occur at the top edge with the same magnitude, but in compression. Since the stress varies linearly along the y-coordinate, we have

$$\sigma_{xx} = 6.0(1 - 2y)\,\text{MPa}$$

All other stress components are zero.

(b) Finite element solution: Since we use only one element, we do not need assembly process. The element has eight DOFs: $\{\mathbf{Q}_s\} = \{u_1, v_1, u_2, v_2, u_3, v_3, u_4, v_4\}^T$. From the boundary condition given in Figure 6.14, only four DOFs are non-zero: u_2, v_2, u_3, and v_3. Thus, from the stiffness matrix of a square element given in Eq. (6.59), all fixed DOFs are deleted to obtain

$$[\mathbf{K}] = \frac{Eh}{1-\nu^2}
\begin{bmatrix}
\dfrac{3-\nu}{6} & -\dfrac{1+\nu}{8} & \dfrac{\nu}{6} & \dfrac{-1+3\nu}{8} \\[2mm]
-\dfrac{1+\nu}{8} & \dfrac{3-\nu}{6} & \dfrac{1-3\nu}{8} & \dfrac{3+\nu}{12} \\[2mm]
\dfrac{\nu}{6} & \dfrac{1-3\nu}{8} & \dfrac{3-\nu}{6} & \dfrac{1+\nu}{8} \\[2mm]
\dfrac{-1+3\nu}{8} & -\dfrac{3+\nu}{12} & \dfrac{1+\nu}{8} & \dfrac{3-\nu}{6}
\end{bmatrix}$$

$$= 10^8
\begin{bmatrix}
4.89 & -1.67 & 0.44 & -0.33 \\
-1.67 & 4.89 & 0.33 & -2.89 \\
0.44 & 0.33 & 4.89 & 1.67 \\
-0.33 & -2.89 & 1.67 & 4.89
\end{bmatrix}
\begin{matrix} u_2 \\ v_2 \\ u_3 \\ v_3 \end{matrix}$$

Using the applied nodal forces, the global matrix equation becomes

$$10^8
\begin{bmatrix}
4.89 & -1.67 & 0.44 & -0.33 \\
-1.67 & 4.89 & 0.33 & -2.89 \\
0.44 & 0.33 & 4.89 & 1.67 \\
-0.33 & -2.89 & 1.67 & 4.89
\end{bmatrix}
\begin{Bmatrix} u_2 \\ v_2 \\ u_3 \\ v_3 \end{Bmatrix}
=
\begin{Bmatrix} 100,000 \\ 0 \\ -100,000 \\ 0 \end{Bmatrix}$$

The above equation can be solved for unknown nodal displacements as

$$u_2 = 0.4091 \text{ mm}, \quad v_2 = 0.4091 \text{ mm}$$
$$u_3 = -0.4091 \text{ mm}, \quad v_3 = 0.4091 \text{ mm}$$

Then, from Eq. (6.57), the strain components can be obtained as

$$\{\varepsilon\} = \begin{bmatrix} y-1 & 0 & 1-y & 0 & y & 0 & -y & 0 \\ 0 & x-1 & 0 & -x & 0 & x & 0 & 1-x \\ x-1 & y-1 & -x & 1-y & x & y & 1-x & -y \end{bmatrix} \begin{Bmatrix} 0 \\ 0 \\ 0.4091 \\ 0.4091 \\ -0.4091 \\ 0.4091 \\ 0 \\ 0 \end{Bmatrix} \times 10^{-3}$$

$$= \begin{Bmatrix} 0.4091 \times 10^{-3}(1-2y) \\ 0 \\ 0.4091 \times 10^{-3}(1-2x) \end{Bmatrix}$$

Using the stress-strain relation in Eq. (6.5) for plane stress, the stress components can be obtained as

$$\begin{Bmatrix} \sigma_{xx} \\ \sigma_{yy} \\ \tau_{xy} \end{Bmatrix} = \frac{10^{10}}{1-0.25^2} \begin{bmatrix} 1 & 0.25 & 0 \\ 0.25 & 1 & 0 \\ 0 & 0 & 0.375 \end{bmatrix} \begin{Bmatrix} 0.4091 \times 10^{-3}(1-2y) \\ 0 \\ 0.4091 \times 10^{-3}(1-2x) \end{Bmatrix}$$

$$= \begin{Bmatrix} 4.364(1-2y) \\ 1.091(1-2y) \\ 1.636(1-2x) \end{Bmatrix} \text{MPa}$$

The deformed shape of the element is shown in the figure below.

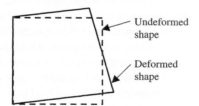

Undeformed shape

Deformed shape

Figure 6.15 Pure bending deformation of a square element

In a plane solid, the applied couple produces a curvature, but the rectangular element is unable to produce deformation corresponding to the curvature because the displacement can only change linearly within the element. The rectangular shape deforms to the trapezoidal shape, and as a result, non-zero shear stress is produced. Note that the maximum stress $(\sigma_{xx})_{\max}$ is only 73% (4.364/6.0) of the exact solution. In addition, σ_{yy} and τ_{xy} have non-zero values. The applied couple is supported by other stress components, σ_{yy} and τ_{xy}, and as a result the element shows smaller $(\sigma_{xx})_{\max}$. In a sense, the element shows a *stiff* behavior.

6.6 FOUR-NODE ISO-PARAMETRIC QUADRILATERAL ELEMENT

As discussed in Section 6.5, the rectangular element is limited in practical applications due to its inability to represent irregular geometries. The four-node quadrilateral element, shown in Figure 6.16, can overcome such limitations. The four-node iso-parametric finite element is one of the most commonly used elements in engineering applications.

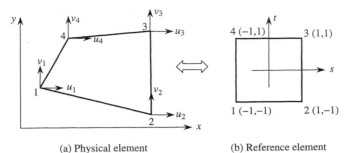

Figure 6.16 Four-node quadrilateral element for plane solids

(a) Physical element   (b) Reference element

The element consists of four nodes and two DOFs at each node. Since the geometry of the element is irregular, it is convenient to introduce a *reference element* and use a mapping relation between the physical element and the reference element. The term *iso-parametric* comes from the fact that the same interpolation scheme is used for interpolating both displacement and geometry.

6.6.1 Iso-parametric Mapping

The physical element in Figure 6.16 is a general quadrilateral shape. However, all interior angles should be less than 180 degrees. The order of node numbers is the same as that of the rectangular element: starting from one corner and moving in the counter-clockwise direction. Each node has two DOFs: u and v. Thus, the element has a total of eight DOFs.

Since different elements have different shapes, it would not be a trivial task if the interpolation functions need to be developed for individual elements. The interpolation functions must satisfy the inter-element displacement compatibility condition discussed earlier in the context of triangular elements. Instead, the concept of mapping to the reference element will be used. The physical element in Figure 6.16(a) will be mapped into the reference element shown in Figure 6.16(b). The physical element is defined in x-y coordinates, while the reference element is defined in s-t coordinates. The reference element is a square element and has the center at the origin. Although the physical element can have the first node at any corner, the reference element always has the first node at the lower-left corner $(-1, -1)$.

The interpolation functions are defined in the reference element so that different elements have the same interpolation function. The only difference is the mapping relation between the two elements. Since the reference element is of square shape, the Lagrange interpolation for rectangular elements can be used. Using (6.54), the interpolation or shape functions can be written in s-t coordinates as

$$\begin{cases} N_1(s, t) = \dfrac{1}{4}(1 - s)(1 - t) \\ N_2(s, t) = \dfrac{1}{4}(1 + s)(1 - t) \\ N_3(s, t) = \dfrac{1}{4}(1 + s)(1 + t) \\ N_4(s, t) = \dfrac{1}{4}(1 - s)(1 + t) \end{cases} \tag{6.62}$$

Since the above shape functions are Lagrange interpolation functions, they satisfy the same properties as those of the rectangular element. Thus, N_1 is equal to unity at Node 1 and zero at other nodes, etc.

Note that in the CST and rectangular elements, the shape functions are used to inter-polate displacements within the element. In the quadrilateral element, the shape functions are also used for mapping between the physical element and the reference element. The quadrilateral element is defined by the coordinates of four corner nodes. These four corner nodes are mapped into the four corner nodes of the reference element. In addition, every point in the physical element is also mapped into a point in the reference element. The mapping relation is one-to-one such that every point in the reference element also has a mapped point in the physical element. Thus a physical point (x, y) is a function of reference point (s, t). A relation between (x, y) and (s, t) can be derived using the same shape functions as

$$x(s, t) = [N_1(s, t) \quad N_2(s, t) \quad N_3(s, t) \quad N_4(s, t)] \begin{Bmatrix} x_1 \\ x_2 \\ x_3 \\ x_4 \end{Bmatrix}$$

$$y(s, t) = [N_1(s, t) \quad N_2(s, t) \quad N_3(s, t) \quad N_4(s, t)] \begin{Bmatrix} y_1 \\ y_2 \\ y_3 \\ y_4 \end{Bmatrix}$$

(6.63)

It can be easily checked that at Node 1, for example, $(s, t) = (-1, -1)$ and $N_1 = 1$, $N_2 = N_3 = N_4 = 0$. Thus, we have $x(-1, -1) = x_1$ and $y(-1, -1) = y_1$, i.e., Node 1 in the physical element is mapped into Node 1 in the reference element. The above mapping relation is called *iso-parametric mapping* because the same shape functions are used for interpolating geometry as well as displacements.

The above mapping relation is explicit in terms of x and y, which means that when s and t are given, x and y can be calculated explicitly. The reverse relation is not straightforward. However, the following example explains how s and t can be calculated for a given x and y.

EXAMPLE 6.6   *Iso-Parametric Mapping*

Consider a quadrilateral element of the trapezoidal shape shown in Figure 6.17. Using the iso-parametric mapping method calculate: (a) the physical coordinates of point A (0.5, 0.5), and (b) the reference coordinate of point B (1, 2).

SOLUTION

(a)   At point A, $(s, t) = (0.5, 0.5)$. The values of the shape functions at A are

$$N_1\left(\tfrac{1}{2}, \tfrac{1}{2}\right) = \frac{1}{16}, \quad N_2\left(\tfrac{1}{2}, \tfrac{1}{2}\right) = \frac{3}{16}, \quad N_3\left(\tfrac{1}{2}, \tfrac{1}{2}\right) = \frac{9}{16}, \quad N_4\left(\tfrac{1}{2}, \tfrac{1}{2}\right) = \frac{3}{16}$$

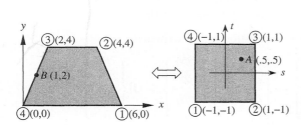

Figure 6.17 Mapping of a quadrilateral element

Thus, the physical coordinate becomes

$$x\left(\tfrac{1}{2}, \tfrac{1}{2}\right) = \sum_{I=1}^{4} N_I\left(\tfrac{1}{2}, \tfrac{1}{2}\right) x_I = \frac{1}{16} \cdot 6 + \frac{3}{16} \cdot 4 + \frac{9}{16} \cdot 2 + \frac{3}{16} \cdot 0 = 2.25$$

$$y\left(\tfrac{1}{2}, \tfrac{1}{2}\right) = \sum_{I=1}^{4} N_I\left(\tfrac{1}{2}, \tfrac{1}{2}\right) y_I = \frac{1}{16} \cdot 0 + \frac{3}{16} \cdot 4 + \frac{9}{16} \cdot 4 + \frac{3}{16} \cdot 0 = 3$$

Thus, the reference point $(s, t) = (0.5, 0.5)$ is mapped into the physical point $(x, y) = (2.25, 3.0)$.

(b) At point B, $(x, y) = (1, 2)$. From the iso-parametric mapping relation, we have

$$x = 1 = \sum_{I=1}^{4} N_I(s, t) x_I = \tfrac{1}{4}(1 - s)(1 - t) \cdot 6 + \tfrac{1}{4}(1 + s)(1 - t) \cdot 4$$
$$+ \tfrac{1}{4}(1 + s)(1 + t) \cdot 2 + \tfrac{1}{4}(1 - s)(1 + t) \cdot 0$$
$$= st - 2t + 3$$

$$y = 2 = \sum_{I=1}^{4} N_I(s, t) y_I = \tfrac{1}{4}(1 - s)(1 - t) \cdot 0 + \tfrac{1}{4}(1 + s)(1 - t) \cdot 4$$
$$+ \tfrac{1}{4}(1 + s)(1 + t) \cdot 4 + \tfrac{1}{4}(1 - s)(1 + t) \cdot 0$$
$$= 2 + 2s$$

From the above two relations, we obtain $(s, t) = (0, 1)$. Note that the above results will not be the same, if the sequence of node numbers in the physical element is changed.

6.6.2 Jacobian of Mapping

The idea of using the reference element is convenient because it is unnecessary to build different shape functions for different elements. The same shape functions can be used for all elements. However, it has its own drawbacks. The strain energy in the plane solid element requires the derivative of displacement, i.e., strains. As can be seen in Eq. (6.25), the strains are defined as derivatives of displacements. In the case of CST and rectangular elements, the shape functions could be differentiated directly because the nodal displacements are explicit functions of x and y. For those elements, the derivatives of the shape functions can be easily obtained because they are defined as a function of physical coordinates (x, y). However, in the case of the quadrilateral element, the shape functions are defined in the reference coordinates. Thus, differentiation with respect to the physical coordinates is not straightforward. In this case, we use a Jacobian relation and the chain rule of differentiation. From the fact that $s = s(x, y)$ and $t = t(x, y)$, we can write the derivatives of N_I as follows:

$$\frac{\partial N_I}{\partial s} = \frac{\partial N_I}{\partial x} \frac{\partial x}{\partial s} + \frac{\partial N_I}{\partial y} \frac{\partial y}{\partial s}$$

$$\frac{\partial N_I}{\partial t} = \frac{\partial N_I}{\partial x} \frac{\partial x}{\partial t} + \frac{\partial N_I}{\partial y} \frac{\partial y}{\partial t}$$

Using the matrix form, the above equation can be written as

$$\left\{ \begin{array}{c} \dfrac{\partial N_I}{\partial s} \\[2ex] \dfrac{\partial N_I}{\partial t} \end{array} \right\} = \begin{bmatrix} \dfrac{\partial x}{\partial s} & \dfrac{\partial y}{\partial s} \\[2ex] \dfrac{\partial x}{\partial t} & \dfrac{\partial y}{\partial t} \end{bmatrix} \left\{ \begin{array}{c} \dfrac{\partial N_I}{\partial x} \\[2ex] \dfrac{\partial N_I}{\partial y} \end{array} \right\} = [\mathbf{J}] \left\{ \begin{array}{c} \dfrac{\partial N_I}{\partial x} \\[2ex] \dfrac{\partial N_I}{\partial y} \end{array} \right\} \tag{6.64}$$

where [**J**] is the *Jacobian matrix* and its determinant is called the Jacobian. By inverting the Jacobian matrix, the desired derivatives with respect to x and y can be obtained:

$$
\left\{\begin{array}{c} \dfrac{\partial N_I}{\partial x} \\[2mm] \dfrac{\partial N_I}{\partial y} \end{array}\right\} = [\mathbf{J}]^{-1} \left\{\begin{array}{c} \dfrac{\partial N_I}{\partial s} \\[2mm] \dfrac{\partial N_I}{\partial t} \end{array}\right\} = \frac{1}{|\mathbf{J}|} \left[\begin{array}{cc} \dfrac{\partial y}{\partial t} & -\dfrac{\partial y}{\partial s} \\[2mm] -\dfrac{\partial x}{\partial t} & \dfrac{\partial x}{\partial s} \end{array}\right] \left\{\begin{array}{c} \dfrac{\partial N_I}{\partial s} \\[2mm] \dfrac{\partial N_I}{\partial t} \end{array}\right\}
\tag{6.65}
$$

where $|\mathbf{J}|$ is the *Jacobian* and is defined by

$$
|\mathbf{J}| = \frac{\partial x}{\partial s}\frac{\partial y}{\partial t} - \frac{\partial x}{\partial t}\frac{\partial y}{\partial s}
\tag{6.66}
$$

Since iso-parametric mapping is used, the above Jacobian can be obtained by differentiating the relation in Eq. (6.63) with respect to s and t. For example,

$$
\frac{\partial x}{\partial s} = \sum_{I=1}^{4} \frac{\partial N_I}{\partial s} x_I = \frac{1}{4}(-x_1 + x_2 + x_3 - x_4) + \frac{t}{4}(x_1 - x_2 + x_3 - x_4)
$$

$$
\frac{\partial x}{\partial t} = \sum_{I=1}^{4} \frac{\partial N_I}{\partial t} x_I = \frac{1}{4}(-x_1 - x_2 + x_3 + x_4) + \frac{s}{4}(x_1 - x_2 + x_3 - x_4)
$$

A similar expression can be obtained for $\partial y/\partial s$ and $\partial y/\partial t$ by replacing x_i with y_i. Note that $\partial x/\partial s$ is the function of t only, while $\partial x/\partial t$ is the function of s only.

As seen from Eq. (6.65), the derivative of the shape function cannot be obtained if the Jacobian is zero anywhere in the element. In fact, the mapping relation between (x, y) and (s, t) is not valid if the Jacobian is zero or negative anywhere in the element $(-1 \le s, t \le 1)$.

The Jacobian plays an important role in evaluating the validity of mapping as well as the quality of the quadrilateral element. The fundamental requirement is that every point in the reference element should be mapped into the interior of the physical element, and vice versa. When an interior point in (s, t) coordinates is mapped into an exterior point in the (x, y) coordinates, the Jacobian becomes negative. If multiple points in (s, t) coordinates are mapped into a single point in (x, y) coordinates, the Jacobian becomes zero at that point. Thus, it is important to maintain the element shape so that the Jacobian is positive everywhere in the element.

EXAMPLE 6.7   *Jacobian of Mapping*

Check the validity of iso-parametric mapping for the two elements shown in Figure 6.18.

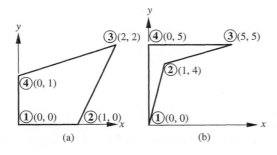

Figure 6.18 Four-node quadrilateral elements

SOLUTION

(a) Nodal coordinates:

$$x_1 = 0, x_2 = 1, x_3 = 2, x_4 = 0$$
$$y_1 = 0, y_2 = 0, y_3 = 2, y_4 = 1$$

– Iso-parametric mapping:

$$x = \sum_{I=1}^{4} N_I x_I = N_2 + 2N_3 = \frac{1}{4}(3 + 3s + t + st)$$

$$y = \sum_{I=1}^{4} N_I y_I = 2N_3 + N_4 = \frac{1}{4}(3 + s + 3t + st)$$

– Jacobian:

$$[\mathbf{J}] = \begin{bmatrix} \dfrac{\partial x}{\partial s} & \dfrac{\partial y}{\partial s} \\ \dfrac{\partial x}{\partial t} & \dfrac{\partial y}{\partial t} \end{bmatrix} = \frac{1}{4}\begin{bmatrix} 3+t & 1+t \\ 1+s & 3+s \end{bmatrix}$$

$$|\mathbf{J}| = \frac{1}{4}[(3+t)(3+s) - (1+t)(1+s)] = \frac{1}{2} + \frac{1}{8}s + \frac{1}{8}t$$

Thus, it is clear that $|\mathbf{J}| > 0$ for $-1 \le s \le 1$ and $-1 \le t \le 1$. Figure 6.19 shows constant s and t lines. Since all lines are within the element boundary, the mapping is valid.

(b) Nodal coordinates:

$$x_1 = 0, x_2 = 1, x_3 = 5, x_4 = 0$$
$$y_1 = 0, y_2 = 4, y_3 = 5, y_4 = 5$$

– Iso-parametric mapping:

$$x = \sum_{I=1}^{4} N_I x_I = \frac{1}{2}(1+s)(3+2t)$$

$$y = \sum_{I=1}^{4} N_I y_I = \frac{1}{2}(7 + 2s + 3t - 2st)$$

– Jacobian:

$$|\mathbf{J}| = \frac{1}{4}(5 - 10s + 10t)$$

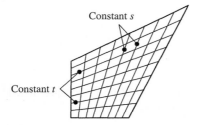

Constant s

Constant t

Figure 6.19 Iso-parametric lines of a quadrilateral element

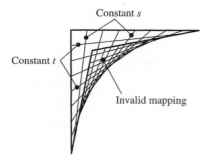

Constant s

Constant t

Invalid mapping

Figure 6.20 An example of invalid mapping

Note that $|\mathbf{J}| = 0$ at $5 - 10s + 10t = 0$; i.e., $s - t = 1/2$. The mapping illustrated in Figure 6.20 clearly shows that the mapping is invalid. Some points in the reference element are mapped into the outside of the physical element.

In the practical sense, maintaining positive Jacobian is not enough because of the numerical nature. For example, when the Jacobian is small, i.e., $|\mathbf{J}| \ll 1$, calculation of stress and strain is not accurate and the integration of the strain energy will losen its accuracy. The small value of the Jacobian occurs when the element shape is far from the rectangular one. To avoid problems due to badly shaped elements, it is recommended that the inside angles in quadrilateral elements be $> 15°$ and $< 165°$, as illustrated in Figure 6.21.

6.6.3   Interpolation of Displacements and Strains

As we explained in iso-parametric mapping above, the same shape functions are used for interpolating displacements. Similar to the rectangular element the quadrilateral element also has eight DOFs. Then the displacements within the element can be interpolated as

$$
\begin{Bmatrix} u \\ v \end{Bmatrix} = \begin{bmatrix} N_1 & 0 & N_2 & 0 & N_3 & 0 & N_4 & 0 \\ 0 & N_1 & 0 & N_2 & 0 & N_3 & 0 & N_4 \end{bmatrix} \begin{Bmatrix} u_1 \\ v_1 \\ u_2 \\ v_2 \\ u_3 \\ v_3 \\ u_4 \\ v_4 \end{Bmatrix} = [\mathbf{N}]\{\mathbf{q}\} \qquad (6.67)
$$

where the same shape functions in Eq. (6.62) are used for interpolation. The difference between the previous two elements (CST and rectangular elements) and the quadrilateral element is that the interpolation is done in the reference coordinates (s, t). However, the behavior of the element is similar to that of the rectangular element because both of them are based on the bilinear Lagrange interpolation

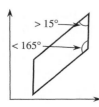

$> 15°$

$< 165°$

Figure 6.21 Recommended ranges of internal angles in a quadrilateral element

Now we derive the strain-displacement relationship for the quadrilateral element. To make the following matrix operation convenient, we first reorder the strain components into the derivatives of displacements, as

$$\{\varepsilon\} = \begin{Bmatrix} \varepsilon_{xx} \\ \varepsilon_{yy} \\ \gamma_{xy} \end{Bmatrix} = \begin{Bmatrix} \partial u/\partial x \\ \partial v/\partial y \\ \partial u/\partial y + \partial v/\partial x \end{Bmatrix} = \begin{bmatrix} 1 & 0 & 0 & 0 \\ 0 & 0 & 0 & 1 \\ 0 & 1 & 1 & 0 \end{bmatrix} \begin{Bmatrix} \partial u/\partial x \\ \partial u/\partial y \\ \partial v/\partial x \\ \partial v/\partial y \end{Bmatrix}$$

As we discussed above, the derivatives of displacements cannot be obtained directly. Instead, we use the inverse Jacobian relation so that the derivatives of displacements are written in terms of the reference coordinates. Thus, we have

$$\begin{Bmatrix} \dfrac{\partial u}{\partial x} \\[2mm] \dfrac{\partial u}{\partial y} \end{Bmatrix} = \frac{1}{|\mathbf{J}|} \begin{bmatrix} \dfrac{\partial y}{\partial t} & -\dfrac{\partial y}{\partial s} \\[2mm] -\dfrac{\partial x}{\partial t} & \dfrac{\partial x}{\partial s} \end{bmatrix} \begin{Bmatrix} \dfrac{\partial u}{\partial s} \\[2mm] \dfrac{\partial u}{\partial t} \end{Bmatrix}$$

$$\begin{Bmatrix} \dfrac{\partial v}{\partial x} \\[2mm] \dfrac{\partial v}{\partial y} \end{Bmatrix} = \frac{1}{|\mathbf{J}|} \begin{bmatrix} \dfrac{\partial y}{\partial t} & -\dfrac{\partial y}{\partial s} \\[2mm] -\dfrac{\partial x}{\partial t} & \dfrac{\partial x}{\partial s} \end{bmatrix} \begin{Bmatrix} \dfrac{\partial v}{\partial s} \\[2mm] \dfrac{\partial v}{\partial t} \end{Bmatrix}$$

Writing the two equations together, we have

$$\begin{Bmatrix} \partial u/\partial x \\ \partial u/\partial y \\ \partial v/\partial x \\ \partial v/\partial y \end{Bmatrix} = \frac{1}{|\mathbf{J}|} \begin{bmatrix} \partial y/\partial t & -\partial y/\partial s & 0 & 0 \\ -\partial x/\partial t & \partial x/\partial s & 0 & 0 \\ 0 & 0 & \partial y/\partial t & -\partial y/\partial s \\ 0 & 0 & -\partial x/\partial t & \partial x/\partial s \end{bmatrix} \begin{Bmatrix} \partial u/\partial s \\ \partial u/\partial t \\ \partial v/\partial s \\ \partial v/\partial t \end{Bmatrix}$$

The strains can now be expressed as

$$\begin{Bmatrix} \varepsilon_{xx} \\ \varepsilon_{yy} \\ \gamma_{xy} \end{Bmatrix} = \frac{1}{|\mathbf{J}|} \begin{bmatrix} 1 & 0 & 0 & 0 \\ 0 & 0 & 0 & 1 \\ 0 & 1 & 1 & 0 \end{bmatrix} \begin{bmatrix} \partial y/\partial t & -\partial y/\partial s & 0 & 0 \\ -\partial x/\partial t & \partial x/\partial s & 0 & 0 \\ 0 & 0 & \partial y/\partial t & -\partial y/\partial s \\ 0 & 0 & -\partial x/\partial t & \partial x/\partial s \end{bmatrix} \begin{Bmatrix} \partial u/\partial s \\ \partial u/\partial t \\ \partial v/\partial s \\ \partial v/\partial t \end{Bmatrix}$$

$$\equiv [\mathbf{A}] \begin{Bmatrix} \partial u/\partial s \\ \partial u/\partial t \\ \partial v/\partial s \\ \partial v/\partial t \end{Bmatrix}$$

where $[\mathbf{A}]$ is a 3×4 matrix. The derivatives of the displacements with respect to s and t can be obtained by differentiating $u(s,t)$ and $v(s,t)$ in Eq. (6.67), which involves the derivatives of the shape functions:

$$\begin{Bmatrix} \partial u/\partial s \\ \partial u/\partial t \\ \partial v/\partial s \\ \partial v/\partial t \end{Bmatrix} = \frac{1}{4} \begin{bmatrix} -1+t & 0 & 1-t & 0 & 1+t & 0 & -1-t & 0 \\ -1+s & 0 & -1-s & 0 & 1+s & 0 & 1-s & 0 \\ 0 & -1+t & 0 & 1-t & 0 & 1+t & 0 & -1-t \\ 0 & -1+s & 0 & -1-s & 0 & 1+s & 0 & 1-s \end{bmatrix} \begin{Bmatrix} u_1 \\ v_1 \\ u_2 \\ v_2 \\ u_3 \\ v_3 \\ u_4 \\ v_4 \end{Bmatrix}$$

$$\equiv [\mathbf{G}]\{\mathbf{q}\}$$

where the dimension of matrix $[\mathbf{G}]$ is 4×8. The strain-displacement matrix $[\mathbf{B}]$ can now be written as follows:

$$\begin{Bmatrix} \varepsilon_{xx} \\ \varepsilon_{yy} \\ \gamma_{xy} \end{Bmatrix} = [\mathbf{A}] \begin{Bmatrix} \partial u/\partial s \\ \partial u/\partial t \\ \partial v/\partial s \\ \partial v/\partial t \end{Bmatrix} = [\mathbf{A}][\mathbf{G}]\{\mathbf{q}\} \equiv [\mathbf{B}]\{\mathbf{q}\} \tag{6.68}$$

where $[\mathbf{B}]$ is a 3×8 matrix. The explicit expression of $[\mathbf{B}]$ is not readily available because the matrix $[\mathbf{A}]$ involves inverse of Jacobian matrix. However, for given reference coordinate (s, t), it can be calculated using Eq. (6.68). Note that the strain-displacement matrix $[\mathbf{B}]$ is not constant as in CST elements. Thus, the strains and stresses within an element vary as a function of s and t coordinates.

EXAMPLE 6.8 *Interpolation using Quadrilateral Element*

For a rectangular element shown in Figure 6.22, displacements at four nodes are given by $\{u_1, v_1, u_2, v_2, u_3, v_3, u_4, v_4\} = \{0.0, 0.0, 1.0, 0.0, 2.0, 1.0, 0.0, 2.0\}$. Calculate displacement and strain at point $(s, t) = (1/3, 0)$.

SOLUTION When the reference coordinate $(s, t) = (1/3, 0)$, the shape functions become

$$N_1 = \frac{1}{6}, \qquad N_2 = \frac{1}{3}, \qquad N_3 = \frac{1}{3}, \qquad N_4 = \frac{1}{6}$$

Using Eq. (6.67), the displacements can be interpolated, as

$$\begin{cases} u = \sum_{I=1}^{4} N_I u_I = \frac{1}{6} \cdot 0 + \frac{1}{3} \cdot 1 + \frac{1}{3} \cdot 2 + \frac{1}{6} \cdot 0 = 1 \\ v = \sum_{I=1}^{4} N_I v_I = \frac{1}{6} \cdot 0 + \frac{1}{3} \cdot 0 + \frac{1}{3} \cdot 1 + \frac{1}{6} \cdot 2 = \frac{2}{3} \end{cases}$$

In order to calculate strains, we need the derivatives of the shape functions. First, we calculate the derivatives with respect to the reference coordinates, as

$$\begin{cases} \dfrac{\partial N_1}{\partial s} = -\dfrac{1}{4}(1-t) = -\dfrac{1}{4} \\[2mm] \dfrac{\partial N_2}{\partial s} = \dfrac{1}{4}(1-t) = \dfrac{1}{4} \\[2mm] \dfrac{\partial N_3}{\partial s} = \dfrac{1}{4}(1+t) = \dfrac{1}{4} \\[2mm] \dfrac{\partial N_4}{\partial s} = -\dfrac{1}{4}(1+t) = -\dfrac{1}{4} \end{cases} \qquad \begin{cases} \dfrac{\partial N_1}{\partial t} = -\dfrac{1}{4}(1-s) = -\dfrac{1}{6} \\[2mm] \dfrac{\partial N_2}{\partial t} = -\dfrac{1}{4}(1+s) = -\dfrac{1}{3} \\[2mm] \dfrac{\partial N_3}{\partial t} = \dfrac{1}{4}(1+s) = \dfrac{1}{3} \\[2mm] \dfrac{\partial N_4}{\partial t} = \dfrac{1}{4}(1-s) = \dfrac{1}{6} \end{cases}$$

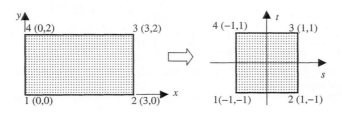

Figure 6.22 Mapping of a rectangular element

In addition, the Jacobian matrix can be calculated using Eq. (6.65), as

$$
\begin{cases}
\dfrac{\partial x}{\partial s} = -\dfrac{1}{4}\cdot 0 + \dfrac{1}{4}\cdot 3 + \dfrac{1}{4}\cdot 3 - \dfrac{1}{4}\cdot 0 = \dfrac{3}{2} \\[2mm]
\dfrac{\partial y}{\partial s} = -\dfrac{1}{4}\cdot 0 + \dfrac{1}{4}\cdot 0 + \dfrac{1}{4}\cdot 2 - \dfrac{1}{4}\cdot 2 = 0 \\[2mm]
\dfrac{\partial x}{\partial t} = -\dfrac{1}{6}\cdot 0 - \dfrac{1}{3}\cdot 3 + \dfrac{1}{3}\cdot 3 + \dfrac{1}{6}\cdot 0 = 0 \\[2mm]
\dfrac{\partial y}{\partial t} = -\dfrac{1}{6}\cdot 0 - \dfrac{1}{3}\cdot 0 + \dfrac{1}{3}\cdot 2 + \dfrac{1}{6}\cdot 2 = 1
\end{cases}
$$

$$
[\mathbf{J}] = \begin{bmatrix} \dfrac{\partial x}{\partial s} & \dfrac{\partial y}{\partial s} \\[2mm] \dfrac{\partial x}{\partial t} & \dfrac{\partial y}{\partial t} \end{bmatrix} = \begin{bmatrix} \dfrac{3}{2} & 0 \\[1mm] 0 & 1 \end{bmatrix}, \quad [\mathbf{J}]^{-1} = \begin{bmatrix} \dfrac{2}{3} & 0 \\[1mm] 0 & 1 \end{bmatrix}
$$

The Jacobian is positive, and the mapping is valid at this point. In fact, the Jacobian matrix is constant throughout the element. Note that the Jacobian matrix only has diagonal components, which means that the physical element is a rectangle. The horizontal dimension of the physical element is 1.5 times that of the reference element, and the vertical dimension is the same.

Using the inverse Jacobian matrix and the derivatives of the shape functions, we can calculate the following:

$$
\begin{cases}
\dfrac{\partial N_I}{\partial s} = \dfrac{\partial N_I}{\partial x}\dfrac{\partial x}{\partial s} + \dfrac{\partial N_I}{\partial y}\dfrac{\partial y}{\partial s} \\[2mm]
\dfrac{\partial N_I}{\partial t} = \dfrac{\partial N_I}{\partial x}\dfrac{\partial x}{\partial t} + \dfrac{\partial N_I}{\partial y}\dfrac{\partial y}{\partial t}
\end{cases}
\Rightarrow
\begin{Bmatrix} \dfrac{\partial N_I}{\partial s} \\[2mm] \dfrac{\partial N_I}{\partial t} \end{Bmatrix} = [\mathbf{J}] \begin{Bmatrix} \dfrac{\partial N_I}{\partial x} \\[2mm] \dfrac{\partial N_I}{\partial y} \end{Bmatrix}
$$

$$
\begin{Bmatrix} \dfrac{\partial N_I}{\partial x} \\[2mm] \dfrac{\partial N_I}{\partial y} \end{Bmatrix} = [\mathbf{J}]^{-1} \begin{Bmatrix} \dfrac{\partial N_I}{\partial s} \\[2mm] \dfrac{\partial N_I}{\partial t} \end{Bmatrix} = \begin{bmatrix} \dfrac{2}{3} & 0 \\[1mm] 0 & 1 \end{bmatrix} \begin{Bmatrix} \dfrac{\partial N_I}{\partial s} \\[2mm] \dfrac{\partial N_I}{\partial t} \end{Bmatrix} = \begin{Bmatrix} \dfrac{2}{3}\dfrac{\partial N_I}{\partial s} \\[2mm] \dfrac{\partial N_I}{\partial t} \end{Bmatrix}
$$

Using the derivatives of the shape functions, the strains can be calculated using Eq. (6.68), as

$$
\varepsilon_{xx} = \dfrac{\partial u}{\partial x} = \sum_{I=1}^{4} \dfrac{\partial N_I}{\partial x} u_I = \sum_{I=1}^{4} \dfrac{2}{3}\dfrac{\partial N_I}{\partial s} u_I
$$

$$
= \dfrac{2}{3}\left(-\dfrac{1}{4}\cdot 0 + \dfrac{1}{4}\cdot 1 + \dfrac{1}{4}\cdot 2 - \dfrac{1}{4}\cdot 0 \right) = \dfrac{1}{2}
$$

$$
\varepsilon_{yy} = \dfrac{\partial v}{\partial y} = \sum_{I=1}^{4} \dfrac{\partial N_I}{\partial y} v_I = \sum_{I=1}^{4} \dfrac{\partial N_I}{\partial t} v_I
$$

$$
= -\dfrac{1}{6}\cdot 0 - \dfrac{1}{3}\cdot 0 + \dfrac{1}{3}\cdot 1 + \dfrac{1}{6}\cdot 2 = \dfrac{2}{3}
$$

$$
\gamma_{xy} = \dfrac{\partial u}{\partial y} + \dfrac{\partial v}{\partial x} = \sum_{I=1}^{4}\left(\dfrac{\partial N_I}{\partial y} u_I + \dfrac{\partial N_I}{\partial x} v_I \right) = \sum_{I=1}^{4}\left(\dfrac{\partial N_I}{\partial t} u_I + \dfrac{2}{3}\dfrac{\partial N_I}{\partial s} v_I \right)
$$

$$
= -\dfrac{1}{6}\cdot 0 - \dfrac{1}{3}\cdot 1 + \dfrac{1}{3}\cdot 2 + \dfrac{1}{6}\cdot 0 - \dfrac{1}{4}\cdot 0 + \dfrac{1}{4}\cdot 0 + \dfrac{1}{4}\cdot 1 - \dfrac{1}{4}\cdot 2 = \dfrac{1}{12}
$$

The reader can verify that the same results could have been obtained using the formulas in Eq. (6.57) derived for rectangular elements.

6.6.4   Finite Element Matrix Equation

As in the case of the CST element, the element stiffness matrix can be calculated from the strain energy of the element. By substituting for strains from Eq. (6.68) into the strain energy in Eq. (6.9) we have

$$U^{(e)} = \frac{h}{2} \iint\limits_{A} \{\varepsilon\}^{T} [\mathbf{C}] \{\varepsilon\} \, dA^{(e)}$$

$$= \frac{h}{2} \{\mathbf{q}^{(e)}\}^{T} \iint\limits_{A} \underset{8\times3}{[\mathbf{B}]}^{T} \underset{3\times3}{[\mathbf{C}]} \underset{3\times8}{[\mathbf{B}]} \, dA \{\mathbf{q}^{(e)}\} \qquad (6.69)$$

$$\equiv \frac{1}{2} \{\mathbf{q}^{(e)}\}^{T} \underset{8\times8}{[\mathbf{k}^{(e)}]} \{\mathbf{q}^{(e)}\}$$

where $[\mathbf{k}^{(e)}]$ is the element stiffness matrix. Calculation of the element stiffness matrix has two challenges. First, the integration domain is a general quadrilateral shape, and, second, the displacement-strain matrix $[\mathbf{B}]$ is written in (s, t) coordinates. Thus, the integration in Eq. (6.69) is not trivial. Using the idea of mapping the physical element into the reference element, we can perform the integration in Eq. (6.69) in the reference element. Since the reference element is a square and is defined in (s, t) coordinates, the above two challenges can be resolved simultaneously. Again, the Jacobian plays an important role in transforming the integral to the reference element. Let us consider an infinitesimal area dA of the physical element is mapped into an infinitesimal rectangle $ds \cdot dt$ in the reference element. Then, the relation between the two areas becomes

$$dA = |\mathbf{J}| ds dt \qquad (6.70)$$

Thus, the element stiffness matrix in the reference element can be written as

$$[\mathbf{k}^{(e)}] = h \iint\limits_{A} [\mathbf{B}]^{T} [\mathbf{C}] [\mathbf{B}] \, dA \equiv h \int\limits_{-1}^{1} \int\limits_{-1}^{1} [\mathbf{B}]^{T} [\mathbf{C}] [\mathbf{B}] |\mathbf{J}| \, ds dt \qquad (6.71)$$

Although the integration has been transformed to the reference element, still the integration in Eq. (6.71) is not trivial because the integrand cannot be written down as an explicit function of s and t. Note that the matrix $[\mathbf{B}]$ includes the inverse of the Jacobian matrix. Thus, it is going to be extremely difficult, if not impossible, to integrate Eq. (6.71) analytically. However, since the integral domain is a square, numerical integration can be used to calculate the element stiffness matrix. Numerical integration methods using Gauss quadrature, which is the most popular method, will be discussed in the following section. Similar to the other elements, the strain energy of entire solid can be obtained using Eq. (6.29), which involves the assembly process.

The potentials of applied loads can be obtained by following a similar procedure as the CST and rectangular elements. The potential energy of concentrated forces and distributed forces will be the same as that of the CST element. The potential energy of the body force can be calculated using Eq. (6.60), except that the transformation in Eq. (6.70) should be used so that the integration be performed in the reference element. For rectangular element, the uniform body force yields the equally divided nodal forces. In the case of the quadrilateral element, however, the work-equivalent nodal forces will not divide the body force equally because the Jacobian is not constant within the element. The numerical integration can be used for integrating Eq. (6.60).

Using the principle of minimum total potential energy in Eq. (6.41), a similar global matrix equation for the quadrilateral elements can be obtained. Applying boundary conditions and solving the matrix equations are identical to the CST element. After solving for

Table 6.2 Gauss Quadrature Points and Weights

NG	Integration points (s_i)	Weights (w_i)	Exact polynomial degree
1	0.0	2.0	1
2	±0.5773502692	1.0	3
3	±0.7745966692	0.5555555556	5
	0.0	0.8888888889	
4	±0.8611363116	0.3478546451	7
	±0.3399810436	0.6521451549	
5	±0.9061798459	0.2369268851	9
	±0.5384693101	0.4786286705	
	0.0	0.5688888889	

nodal displacements, strain and stress in an element can be calculated using Eqs. (6.68) and (6.43), respectively.

6.7 NUMERICAL INTEGRATION

As discussed before, it is not trivial to analytically integrate the element stiffness matrix and body force for the quadrilateral element. Although there are many numerical integration methods available, Gauss quadrature is the preferred method in the finite element analysis because it requires fewer function evaluations compared to other methods. We will explain the one-dimensional Gauss quadrature first.

In the Gauss quadrature, the integrand is evaluated at predefined points (called Gauss points). The sum of these integrand values, multiplied by integration weights (called Gauss weight), provides an approximation to the integral:

$$I = \int_{-1}^{1} f(s)\, ds \approx \sum_{i=1}^{n} w_i f(s_i) \tag{6.72}$$

where n is the number of Gauss points, s_i the Gauss points, w_i the Gauss weights, and $f(s_i)$ the function value at the Gauss point s_i. The locations of Gauss points and weights are derived in such a way that with n points, a polynomial of degree $2n-1$ can be integrated exactly. Note that the integral domain is normalized, i.e., $[-1, 1]$. The Gauss quadrature performs well when the integrand is a smooth function. Table 6.2 shows the locations of the Gauss points and corresponding weights.

EXAMPLE 6.9 *Numerical Integration*

Evaluate the following integral using Gauss quadrature with 1~4 integration points. Compare the numerical integration results with the analytical integration.

$$I = \int_{-1}^{1} (8x^7 + 7x^6)\, dx$$

SOLUTION It can be easily verified that the exact integral will yield $I = 2$. Now we assume that the exact integral is unknown and calculate its approximate value using Gauss quadrature.

(a) 1-point integral:

$$s_1 = 0, \quad f(s_1) = 0, \quad w_1 = 2$$
$$I = w_1 f(s_1) = 2 \times 0 = 0$$

Obviously, the one-point integral is not accurate.

(b) 2-point integral:

$$s_1 = -.577, \quad f(s_1) = 8(-.577)^7 + 7(-.577)^6 = .0882 \quad w_1 = 1$$
$$s_2 = .577, \quad f(s_2) = 8(.577)^7 + 7(.577)^6 = .4303, \qquad w_2 = 1$$
$$I = w_1 f(s_1) + w_2 f(s_2) = .0882 + .4303 = .5185$$

2-point integral still has a large error because it is accurate only up to the third-order polynomial.

(c) 3-point integral:

$$s_1 = -.7746, \quad f(s_1) = .17350, \quad w_1 = .5556$$
$$s_2 = 0.0, \qquad f(s_2) = 0.0 \qquad w_2 = .8889$$
$$s_3 = .7746, \qquad f(s_3) = 2.8505, \quad w_3 = .5556$$
$$I = w_1 f(s_1) + w_2 f(s_2) + w_3 f(s_3) = .5556(.17350 + 2.8505) = 1.6800$$

(d) 4-point integral:

$$s_1 = -.8611, \quad f(s_1) = .0452, \quad w_1 = .3479$$
$$s_2 = -.3400, \quad f(s_2) = .0066, \quad w_2 = .6521$$
$$s_3 = .3400, \qquad f(s_3) = .0150, \quad w_3 = .6521$$
$$s_3 = .8611, \qquad f(s_3) = 5.6638, \quad w_3 = .3479$$
$$I = w_1 f(s_1) + w_2 f(s_2) + w_3 f(s_3) + w_4 f(s_4) = 2.0$$

Note that the 4-point integral is exact up to seventh-order polynomials. Since the given problem is seventh-order polynomial, the numerical integration is exact.

Two-dimensional Gauss integration formulas can be obtained by combining two one-dimensional Gauss quadrature formulas as shown below:

$$I = \int_{-1}^{1}\int_{-1}^{1} f(s,t)\, ds\, dt$$

$$\approx \int_{-1}^{1} \sum_{i=1}^{m} w_i f(s_i, t)\, dt \qquad (6.73)$$

$$= \sum_{j=1}^{n}\sum_{i=1}^{m} w_i w_j f(s_i, t_j)$$

where s_i and t_j are Gauss points, m is the number of Gauss points in s direction, n is the number of Gauss points in t direction, and w_i and w_j are Gauss weights. The total number of Gauss points becomes $m \times n$. Figure 6.23 shows few commonly used integration formulas.

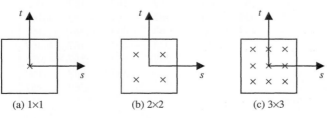

(a) 1×1   (b) 2×2   (c) 3×3

Figure 6.23 Gauss integration points in two-dimensional parent elements

The element stiffness matrix in Eq. (6.71) can be evaluated using 2×2 Gauss integration formulas:

$$[\mathbf{k}^{(e)}] = h \int_{-1}^{1} \int_{-1}^{1} [\mathbf{B}]^T [\mathbf{C}][\mathbf{B}]|\mathbf{J}| \, ds dt$$

$$\approx h \sum_{i=1}^{2} \sum_{j=1}^{2} w_i w_j [\mathbf{B}(s_i, t_j)]^T [\mathbf{C}][\mathbf{B}(s_i, t_j)]|\mathbf{J}(s_i, t_j)| \tag{6.74}$$

EXAMPLE 6.10   *Numerical Integration of Element Stiffness Matrix*

Calculate the element stiffness matrix of the square element shown in Figure 6.24 using (a) 1×1 Gauss quadrature and (b) 2×2 Gauss quadrature. Compare the numerically integrated element stiffness matrix with the exact one calculated using Eq. (6.59). Assume plane stress with thickness $h = 0.1$ m, Young's modulus $E = 10$ GPa, and Poisson's ratio $\nu = 0.25$.

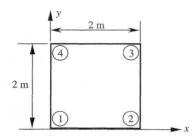

Figure 6.24 Numerical integration of a square element

SOLUTION   Since the element size is the same as that of the reference element, the Jacobian matrix becomes the identity matrix. Thus, from Eq. (6.68), the displacement-strain matrix [**B**] becomes

$$[\mathbf{B}] = \frac{1}{4} \begin{bmatrix} -1+t & 0 & 1-t & 0 & 1+t & 0 & -1-t & 0 \\ 0 & -1+s & 0 & -1-s & 0 & 1+s & 0 & 1-s \\ -1+s & -1+t & -1-s & 1-t & 1+s & 1+t & 1-s & -1-t \end{bmatrix}$$

(a) 1×1 Gauss quadrature uses one-point integration at $(s, t) = (0, 0)$ with weight equal to 4. The [**B**] matrix at this point becomes

$$[\mathbf{B}] = \frac{1}{4} \begin{bmatrix} -1 & 0 & 1 & 0 & 1 & 0 & -1 & 0 \\ 0 & -1 & 0 & -1 & 0 & 1 & 0 & 1 \\ -1 & -1 & -1 & 1 & 1 & 1 & 1 & -1 \end{bmatrix}$$

Then, the numerical integration of the element stiffness matrix becomes

$$[\mathbf{k}_1] \approx h w_1 w_1 [\mathbf{B}(0,0)]^T [\mathbf{C}][\mathbf{B}(0,0)]$$

$$= 10^9 \begin{bmatrix} .367 & .167 & -.167 & -.033 & -.367 & -.167 & .167 & .033 \\ & .367 & .033 & .167 & -.167 & -.367 & -.033 & -.167 \\ & & .367 & -.167 & .167 & -.033 & -.367 & .167 \\ & & & .367 & .033 & -.167 & .167 & -.367 \\ & & & & .367 & .167 & -.167 & -.033 \\ & & & & & .367 & .033 & .167 \\ & & & & & & .367 & -.167 \\ & & \text{Symmetric} & & & & & .367 \end{bmatrix}$$

(b) For 2×2 Gauss quadrature, we need four integration points and weights are unit.

Integration point	s	t
1	$-.5773502692$	$-.5773502692$
2	$+.5773502692$	$-.5773502692$
3	$+.5773502692$	$+.5773502692$
4	$-.5773502692$	$+.5773502692$

Then, the numerical integration of the element stiffness matrix in Eq. (6.74) becomes

$$[\mathbf{k}_2] \approx h \sum_{i=1}^{2} \sum_{j=1}^{2} w_i w_j [\mathbf{B}(s_i, t_j)]^T [\mathbf{C}] [\mathbf{B}(s_i, t_j)] |\mathbf{J}(s_i, t_j)|$$

$$= 10^9 \begin{bmatrix} .489 & .167 & -.289 & -.033 & -.244 & -.167 & .044 & .033 \\ & .489 & .033 & .044 & -.167 & -.244 & -.033 & -.289 \\ & & .489 & -.167 & .044 & -.033 & -.244 & .167 \\ & & & .489 & .033 & -.289 & .167 & -.244 \\ & & & & .489 & .167 & -.289 & -.033 \\ & & & & & .489 & .033 & .044 \\ & & \text{Symmetric} & & & & .489 & -.167 \\ & & & & & & & .489 \end{bmatrix}$$

Using the exact stiffness in Eq. (6.59), we can find that the element stiffness matrix obtained from 2×2 Gauss quadrature is exact.

In general, the 2×2 Gauss quadrature is not exact for quadrilateral elements. The exact results in the above example occur because the element shape is a square.

6.7.1   Lower-Order Integration and Extra Zero-Energy Modes

It is important that the proper order of Gauss quadrature should be used. Otherwise, the element may show undesirable behavior. One of the well-known phenomena of lower-order integration is *extra zero energy modes*. The zero-energy mode is the deformation of an element without changing its strain energy. In plane solids, there are three types of deformations (more precisely, motions) that do not change the strain energy: x-translation, y-translation, and z-rotation. Figure 6.25 illustrates these modes. Since the relative locations of nodes do not change, the stress and strain of the elements are zero, and the strain energy remains constant. In finite element analysis, these modes should be fixed by applying displacement boundary conditions. Otherwise, the stiffness matrix will be singular and there will be no unique solution.

While the zero-energy modes in Figure 6.25 are proper modes, there are improper modes, called extra zero-energy modes, which often occurs when an element is underintegrated. For example, a square element is integrated using 1×1 Gauss quadrature

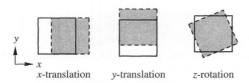

x-translation y-translation z-rotation

Figure 6.25 Three rigid-body modes of plane solids

y

x

Figure 6.26 Two extra zero-energy modes of plane solids

there will be two extra zero-energy modes in addition to the three rigid-body modes. Figure 6.26 illustrates the two extra zero-energy modes of plane solids. It is clear that the element is being deformed but the centroid (the quadrature point) of the element does not experience any deformation and hence the strain energy remains constant. In other words, the element will deform without having externally applied forces, which is a numerical artifact. Thus, the extra zero-energy modes must be removed in order to obtain meaningful deformation.

The most common way of checking the extra zero-energy modes is using eigen values of the stiffness matrix. For a plane solid, the number of zero eigen values must be equal to three. However, the element stiffness matrix with 1×1 integration will have five zero eigen values corresponding to five zero energy modes shown in Figure 6.25 and Figure 6.26. In the following example, we will show another method of checking the extra zero-energy modes.

EXAMPLE 6.11   *Extra Zero-Energy Modes*

Consider two stiffness matrices of the square element in Example 6.10: $[\mathbf{k}_1]$ for 1×1 integration and $[\mathbf{k}_2]$ for 2×2 integration. When nodal displacements are given as $\{\mathbf{q}\}^T = \{0.1, 0, -0.1, 0, 0.1, 0, -0.1, 0\}$, check the reaction forces and determine if the stiffness matrix has extra zero-energy mode.

SOLUTION

(a) For $[\mathbf{k}_1]$ (1×1 integration), the reaction force can be calculated by multiplying the stiffness matrix with the nodal displacements as

$$[\mathbf{k}_1]\{\mathbf{q}\} = 10^9 \begin{bmatrix} .367 & .167 & -.167 & -.033 & -.367 & -.167 & .167 & .033 \\ & .367 & .033 & .167 & -.167 & -.367 & -.033 & -.167 \\ & & .367 & -.167 & .167 & -.033 & -.367 & .167 \\ & & & .367 & .033 & -.167 & .167 & -.367 \\ & & & & .367 & .167 & -.167 & -.033 \\ & & & & & .367 & .033 & .167 \\ & \text{Symmetric} & & & & & .367 & -.167 \\ & & & & & & & .367 \end{bmatrix} \begin{Bmatrix} 0.1 \\ 0.0 \\ -0.1 \\ 0.0 \\ 0.1 \\ 0.0 \\ -0.1 \\ 0.0 \end{Bmatrix} = \begin{Bmatrix} 0 \\ 0 \\ 0 \\ 0 \\ 0 \\ 0 \\ 0 \\ 0 \end{Bmatrix}$$

No force is required to deform the element. Thus, the $[\mathbf{k}_1]$ matrix has extra zero-energy mode.

(b) For $[\mathbf{k}_2]$ (2×2 integration), the reaction force can be calculated by multiplying the stiffness matrix with the nodal displacements as

$$[\mathbf{k}_2]\{\mathbf{q}\} = 10^9 \begin{bmatrix} .489 & .167 & -.289 & -.033 & -.244 & -.167 & .044 & .033 \\ & .489 & .033 & .044 & -.167 & -.244 & -.033 & -.289 \\ & & .489 & -.167 & .044 & -.033 & -.244 & .167 \\ & & & .489 & .033 & -.289 & .167 & -.244 \\ & & & & .489 & .167 & -.289 & -.033 \\ & & & & & .489 & .033 & .044 \\ & \text{Symmetric} & & & & & .489 & -.167 \\ & & & & & & & .489 \end{bmatrix} \begin{Bmatrix} 0.1 \\ 0 \\ -0.1 \\ 0 \\ 0.1 \\ 0 \\ -0.1 \\ 0 \end{Bmatrix} = 10^7 \begin{Bmatrix} 4.89 \\ 0 \\ -4.89 \\ 0 \\ 4.89 \\ 0 \\ -4.89 \\ 0 \end{Bmatrix}$$

Non-zero nodal forces are required to deform the element. Thus, the $[\mathbf{k}_2]$ matrix does not have extra zero-energy mode corresponding to the given deformation.

6.8 PROJECT

Project 6.1 – Accuracy and Convergence Analysis of a Cantilever Beam

In this project, we want to compare the finite element results of plane solid elements with those of uniaxial bar and beam elements. Consider a cantilever beam shown in Figure 6.27 under horizontal and transverse forces at the tip. The beam has a square cross-section of $0.1\,\text{m} \times 0.1\,\text{m}$, length of $L = 1\,\text{m}$, Young's modulus $E = 207\,\text{GPa}$, and Poisson's ratio $\nu = 0.3$.

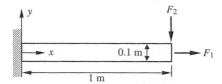

Figure 6.27 Cantilever beam model

Part I

(a) Consider the case of $F_1 = 100\,\text{N}$ and $F_2 = 0$. Solve the problem using a uniaxial bar element to find the elongation $u(x)$. Calculate ε_{xx} and σ_{xx}. Assume that $\sigma_{yy} = \sigma_{zz} = \tau_{xy} = \tau_{yz} = \tau_{xz} = 0$. Compare the results with analytical solution.

(b) Consider the case of $F_1 = 0$ and $F_2 = 500\,\text{N}$. Solve the problem using a beam element to find the deflection $w(x)$. Calculate ε_{xx} and σ_{xx}. Assume that $\sigma_{yy} = \sigma_{zz} = \tau_{xy} = \tau_{yz} = \tau_{xz} = 0$. Compare the results with analytical solution. Plot σ_{xx} as a function of y at $x = L/2$.

Part II

(a) Consider the case of $F_1 = 100\,\text{N}$ and $F_2 = 0$. Solve the problem using: (i) 20 CST elements and (ii) 10 rectangular elements to find the elongation $u(x)$. Calculate ε_{xx} and σ_{xx}. Compare the results with those from Part I. Explain the results using interpolation scheme.

(b) Consider the case of $F_1 = 0$ and $F_2 = 500\,\text{N}$. Solve the problem using: (i) 20 CST elements and (ii) 10 rectangular elements to find the deflection $w(x)$. Calculate ε_{xx} and σ_{xx}. Compare the results with those of Part I. Explain the results using interpolation scheme.

(c) Consider the case of $F_1 = 0$ and $F_2 = 500\,\text{N}$. Perform convergence study by gradually decreasing element size and show the deflection and stress converge to the exact solution.

Project 6.2 – Design of a Torque-arm

A torque arm shown in Figure 6.28 is under horizontal and vertical loads transmitted from a shaft at the right hole, while the left hole is fixed. Assume: Young's modulus $= 206.8\,\text{GPa}$, Poisson's ratio $= 0.29$, and thickness $= 1.0\,\text{cm}$.

1. Provide a preliminary analysis result that can estimate the maximum von Mises stress.

2. Using plane stress elements, carry out finite element analysis for the given loads. Clearly state all assumptions and simplifications that you adopted in modeling.

All dimensions in cm

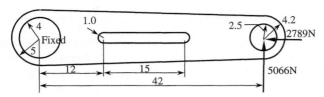

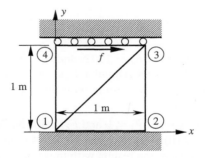

Figure 6.28 Dimensions of torque arm model

Carry out convergence study and determine the size of elements for a reasonably accurate solution.

6.9   EXERCISE

1. Repeat Example 6.2 with the following element connectivity:

   Element 1: 1–2–4
   Element 2: 2–3–4

   Does the different element connectivity change the results?

2. Solve Example 6.2 using one of the finite element programs given in the Appendix.

3. Using two CST elements, solve the simple shear problem depicted in the figure and determine whether the CST elements can represent the simple shear condition accurately. Material properties are given as $E = 10\,\text{GPa}$, $v = 0.25$, and thickness is $h = 0.1\,\text{m}$. The distributed force $f = 100\,\text{kN/m}^2$ is applied at the top edge.

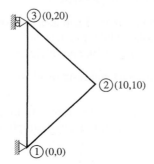

4. Solve Example 6.4 using one of finite element programs in the Appendix.

5. A structure shown in the figure is modeled using one triangular element. Plane strain assumption is used.

   (a) Calculate the strain-displacement matrix [**B**].

   (b) When nodal displacements are given by $\{u_1, v_1, u_2, v_2, u_3, v_3\} = \{0, 0, 2, 0, 0, 1\}$, calculate element strains.

6. Calculate the shape function matrix [**N**] and strain-displacement matrix [**B**] of the triangular element shown in the figure.

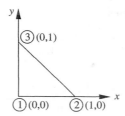

7. The coordinate of the nodes and corresponding displacements in a triangular element are given in the table. Calculate the displacement u and v and strains ε_{xx}, ε_{yy}, and γ_{xy} at the centroid of the element given by the coordinates (1/3, 1/3)

Node	x (m)	y (m)	u (m)	v (m)
1	0	0	0	0
2	1	0	0.1	0.2
3	0	1	0	0.1

8. A 2 m × 2 m × 1 mm square plate with $E = 70$ GPa and $\nu = 0.3$ is subjected to a uniformly distributed load as shown in Figure (a). Due to symmetry, it is sufficient to model one quarter of the plate with artificial boundary conditions, as shown in Figure (b). Use two triangular elements to find the displacements, strains, and stresses in the plate. Check the answers using simple calculations from mechanics of materials.

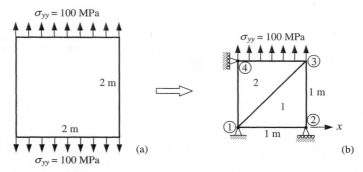

9. A beam problem under the pure bending moment is solved using CST finite elements, as shown in the figure. Assume $E = 200$ GPa and $\nu = 0.3$. The thickness of the beam is 0.01 m. To simulate the pure bending moment, two opposite forces $F = \pm 100,000$ N are applied at the end of the beam. Using any available finite element program, calculate the stresses in the beam along the neutral axis and top and bottom surfaces. Compare the numerical results with the elementary theory of beam. Provide an element stress contour plot for σ_{xx}.

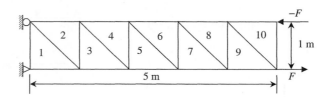

10. For a rectangular element shown in the figure, displacements at four nodes are given by $\{u_1, v_1, u_2, v_2, u_3, v_3, u_4, v_4\} = \{0.0, 0.0, 1.0, 0.0, 2.0, 1.0, 0.0, 2.0\}$. Calculate displacement (u, v) and strain ε_{xx} at point $(x, y) = (2, 1)$.

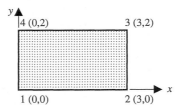

11. Six rectangular elements are used to model the cantilevered beam shown in the figure. Sketch the graph of σ_{xx} along the top surface that a finite element analysis would yield. There is no need to actually solve the problem, but use your knowledge of shape functions for rectangular elements.

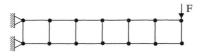

12. A rectangular element as shown in the figure is used to represent a pure bending problem. Due to the bending moment M, the element is deformed as shown in the figure with displacement $\{\mathbf{q}\} = \{u_1, v_1, u_2, v_2, u_3, v_3, u_4, v_4\}^T = \{-1, 0, 1, 0, -1, 0, 1, 0\}^T$.

(a) Write the mathematical expressions of strain component $\varepsilon_{xx}, \varepsilon_{yy}$, and γ_{xy}, as functions of x and y.

(b) Does the element satisfy pure bending condition? Explain your answer.

(c) If two CST elements are used by connecting nodes 1–2–4 and 4–2–3, what will be ε_{xx} along line A-B?

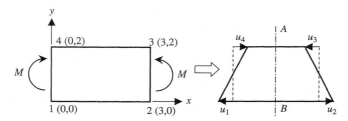

13. Five rectangular elements are used to model a plane beam under pure bending. The element in the middle has nodal displacements, as shown in the figure. Using the bilinear interpolation scheme, calculate the shear strain along the edge AB and compare it with the exact shear strain.

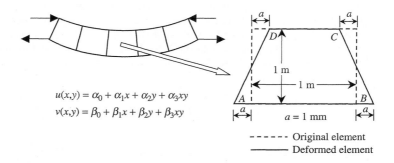

$$u(x,y) = \alpha_0 + \alpha_1 x + \alpha_2 y + \alpha_3 xy$$
$$v(x,y) = \beta_0 + \beta_1 x + \beta_2 y + \beta_3 xy$$

- - - - - - Original element
———— Deformed element

14. A uniform beam is modeled by two rectangular elements with thickness b. Qualitatively, and without performing calculations, plot σ_{xx} and τ_{xy} along the top edge from A to C, as predicted by FEA. Also, plot the exact stresses according to beam theory.

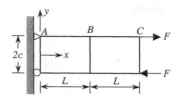

15. A beam problem under the pure bending moment is solved using five rectangular finite elements, as shown in the figure. Assume $E = 200\,\text{GPa}$ and $v = 0.3$ are used. The thickness of the beam is 0.01 m. To simulate the pure bending moment, two opposite forces $F = \pm 100,000\,\text{N}$ are applied at the end of the beam. Using a commercial FE program, calculate strains in the beam along the bottom surface. Draw graphs of ε_{xx} and γ_{xy}, with x-axis being the beam length. Compare the numerical results with the elementary theory of beam. Provide an explanation for the differences, if any. Is the rectangular element stiff or soft compared to the CST element?

Normally, a commercial finite element program provides stress and strain at the corners of the element by averaging with stresses at the adjacent elements. Thus, you may use nodal displacement data from FE code to calculate strains along the bottom surface of the element. Calculate the strains at about 10 points in each element for plotting purpose. Make sure that the commercial program uses the standard Lagrange shape function.

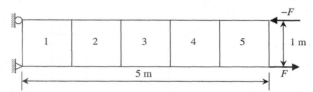

Repeat the above procedure when an upward vertical force of 200,000 N is applied at the tip of the beam. Use boundary conditions similar to the clamped boundary conditions of a cantilevered beam.

16. The quadrilateral element shown in the figure has the nodal displacement of $\{u_1, v_1, u_2, v_2, u_3, v_3, u_4, v_4\} = \{-1, 0, -1, 0, 0, 1, 0, 1\}$.

   (a) Find the (s, t) reference coordinates of point A (0.5, 0) using iso-parametric mapping method.

   (b) Calculate the displacement at point B whose reference coordinate is $(s, t) = (0, -0.5)$

   (c) Calculate the Jacobian matrix $[\mathbf{J}]$ at point B.

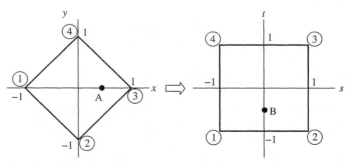

Physical Element   Reference Element

17. A four-node quadrilateral element is defined as shown in the figure.

   (a) Find the coordinates in the reference element corresponding to $(x, y) = (0, 0.5)$.

   (b) Calculate the Jacobian matrix as a function of s and t.

   (c) Is the mapping valid? Explain your answer.

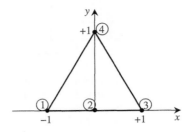

18. A quadrilateral element in the figure is mapped into the reference element.

   (a) A point P has a coordinate $(x, y) = (½, y)$ in the physical element and $(s, t) = (-½, t)$ in the parent element. Find the y and t coordinates of the point using iso-parametric mapping.

   (b) Calculate the Jacobian matrix at the center of the element.

   (c) Is the mapping valid? Explain your answer.

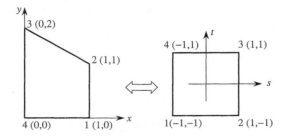

19. Consider the plane stress four-node element shown below. Its global node numbers are shown in the figure. The coordinates of the nodes in the global x-y coordinate systems is shown next to each node.

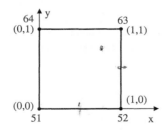

The element connectivity is as follows:

Element #	Local Node 1	Local Node 2	Local Node 3	Local Node 4
27	51	52	63	64

Nodal   displacement   vector $= \{\mathbf{q}\}^T = \{u_{51},\, v_{51},\, u_{52},\, v_{52},\, u_{63},\, v_{63},\, u_{64},\, v_{64}\} = \{0,\, 0,\, 0.1,\, 0,\, 0.1,\, 0.1,\, 0,\, 0\}$.

(a) Determine the displacement at the point $(x, y) = (0.75, 0.75)$ by interpolating the nodal displacements.

(b) Compute the Jacobian matrix at the point in (b).

(c) Compute strain ε_{yy} at the center of the element.

20. A linearly varying pressure p is applied along the edge of the four-node element shown in the figure. The finite element method converts the distributed force into an equivalent set of nodal forces $\{\mathbf{F}^e\}$ such that

$$\int_S \mathbf{u}^T \mathbf{T} dS = \{\mathbf{q}^{(e)}\}^T \{\mathbf{F}^{(e)}\}$$

where $\mathbf{T}$ is the applied traction (force per unit area) and $\mathbf{u}$ is the vector of displacements. Since the applied pressure is normal to the surface (in the x-direction), the traction can be expressed as $\mathbf{T} = \{p, 0\}^T$ where p can be expressed as $p = p_0(t + 1)/2$, $t = -1$ at Node 1 and $t = +1$ at Node 4. The length of the edge is L. Integrate the left-hand side of the above equation to compute the work-equivalent nodal forces $\{\mathbf{F}^{(e)}\}$ when $\{\mathbf{q}^{(e)}\}^T = \{u_1,\, v_1,\, u_2,\, v_2,\, u_3,\, v_3,\, u_4,\, v_4\}$.

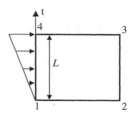

21. Determine the Jacobian matrix for the following isoparametric elements. If the temperature at the nodes of both elements are $\{T_1, T_2, T_3, T_4\} = \{100, 90, 80, 90\}$, compute the temperature at the midpoint of the element and at the midpoint of the edge between connecting Nodes 1 and 4.

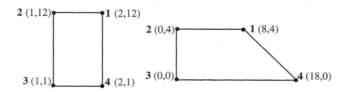

22. Integrate the following function using one-point and two-point numerical integration (Gauss quadrature). The exact integral is equal to 2. Compare the accuracy of the numerical integration with the exact solution.

$$I = \int_0^{\pi} \sin(x)\, dx$$

23. A six-node finite element as shown in the figure is used for approximating the beam problem.

(a) Write the expressions of displacements $u(x,y)$ and $v(x,y)$ in terms of polynomials with unknown coefficients. For example, $u(x, y) = a_0 + a_1 x + \cdots$.

(b) Can this element represent the pure bending problem accurately? Explain your answer. Bending moment M is applied at the edge 2-3.

(c) Can this element represent a uniformly distributed load problem accurately? The distributed load q is applied along the edge 4-6-3.

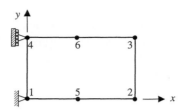

24. Consider a quadrilateral element shown in the figure below. The nodal temperatures of the element are given as $\{T_1, T_2, T_3, T_4\} = \{80, 40, 40, 80\}$.

(a) Compute the expression of the temperature T along the line ξ that connects Nodes 3 and 1. For example, $T = 3 + 5\xi + 3\xi^2 + \cdots$. You can assume that $\xi = 0$ at Node 3 and $\xi = 1$ at Node 1. Plot the graph of $T(\xi)$ with respect to ξ.

(b) Compute the temperature gradient $\partial T / \partial x$ at the center of the element.

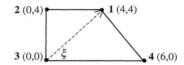

Chapter 7

Finite Element Procedures and Modeling

Finite element method (FEM) is one of the numerical methods of solving differential equations that describe many engineering problems. The FEM, originated in the area of structural mechanics, has been extended to other areas of solid mechanics and later to other fields such as heat transfer, fluid dynamics, and electro-magnetism. In fact, FEM has been recognized as a powerful tool for solving partial differential equations and integro-differential equations, and in the near future, it may become the numerical method of choice in many engineering and applied science areas. One of the reasons for its popularity is that the method results in computer programs versatile in nature that can solve many practical problems with the least amount of training. Obviously, there is a danger in using computer programs without properly understanding the assumptions and limitations of the method, which is the objective of this textbook.

In the previous chapters, we developed a variety of finite elements and studied their application in solving problems in solid and structural mechanics. Different elements were used for different types of problems. When a structural problem is given, it is important to understand the following steps: (1) creation of the FE model of the given problem; (2) applying the boundary conditions and the loads; (3) solution of the finite element matrix equations; and (4) interpretation and verification of the FE results. In this chapter, we will learn some of the formal procedures of solving structural problems using finite elements and various modeling techniques. We will limit our interest to calculation of deflections and stresses. However, these procedures can be extended to problems in other engineering disciplines such as heat transfer.

7.1 FINITE ELEMENT ANALYSIS PROCEDURES

Finite element analysis involves dividing the structure into a set of contiguous elements. This process is called discretization. Each element has a simple shape such as a line, a triangle, or a rectangle, and is connected to other elements by sharing ''nodes.'' The unknowns for each element are the displacements at the nodes. These are also called degrees of freedom. Displacement boundary conditions and applied loads are then specified. The element level matrix equations are assembled to form global level equations. The global matrix equations are solved for the unknown displacements, given the forces and boundary conditions. From the displacements at the nodes, strains and then stress in each element are calculated. However, in practice there are many difficulties in solving the real-life problems using finite elements. For example, when a physical problem is given, engineers should know how to model the problem using finite elements; what kind of

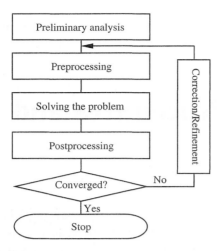

Figure 7.1 Finite element analysis procedures

elements and how many elements should be used; how the boundary conditions and loads should be specified; and how to interpret the results. Experience definitely plays an important role in some of the above steps. In this section, we will discuss some procedures that will be of some help to the analyst.

In general, the finite element analysis procedures can be divided into four stages: preliminary analysis, preprocessing, solution of equations, and postprocessing. In the following subsections, we will discuss these four stages of finite element procedures. In many cases, one may not be able to obtain a satisfactory solution from a single analysis. The model may have errors, or the accuracy of the solution may not be satisfactory. When the model has errors, it needs to be corrected and the procedures should be repeated. If the accuracy of the solution is not satisfactory, the model needs to be refined and the procedures are repeated until the solution converges. Figure 7.1 illustrates the sequence of finite element procedures.

7.1.1 Preliminary Analysis

Preliminary analysis is one of the most important parts of the finite element analysis, but it is often ignored by many engineers. Preliminary analysis will provide an insight into the problem at hand and predict proper behavior of the model. At this stage, the given problem is idealized and analytical methods are used to obtain an approximate solution. The analytical procedures include for example, drawing free-body diagrams of different components and analyzing force equilibrium and applying simplified, strain-displacement relations, and stress-strain relations. Obtaining analytical solutions of practical problems is often not feasible, but in many cases the problem can be simplified. For example three-dimensional or two-dimensional structures can be approximated as bars or beams. The goal is not solving the problem with precision, but predicting the level of displacements and stresses as well as the locations of their maximum values. Before performing any numerical analysis, the engineer should at least know the range of the solution and the expected location of critical points. If the finite element solutions are far away from the results of the preliminary analysis, there must be a good explanation of the discrepancy. In many cases, small mistakes such as use of inconsistent units, may cause large errors.

EXAMPLE 7.1 *Preliminary Analysis of a Plane Frame*

Consider a plane frame shown in Figure 7.2. Using the analytical method calculate the stress at point C and the maximum normal stress that the uniformly distributed load produces in the member.

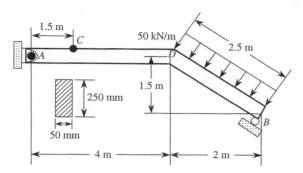

Figure 7.2 Frame structure under uniformly distributed load

SOLUTION The state of stress at point C can be determined by using the principle of superposition. The stress distribution due to each loading (axial, shear, and bending) is first determined, and then every contribution is superposed to obtain the final stress distribution. The principle of superposition can be used because of a linear relationship between the stress and loads.

Let us calculate the reaction forces of the member using the free-body diagram, as shown below.

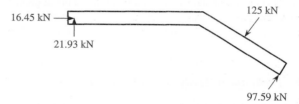

To find forces and moment at point C, let us examine the horizontal segment of the member. From the three equilibrium conditions we have

$$\sum F_X = 0 \quad \Rightarrow \quad 16.45 - N = 0, \quad N = 16.45\,\text{kN}$$

$$\sum F_Y = 0 \quad \Rightarrow \quad 21.93 - V = 0, \quad V = 21.93\,\text{kN}$$

$$\sum M = 0 \quad \Rightarrow \quad -21.93x + M = 0, \quad M = 21.93x\,\text{kN} \cdot \text{m}$$

Normal Force: The normal stress caused by N at C is a compressive uniform stress.

$$\sigma_{\text{normal\_force}} = -\frac{16.45\,\text{kN}}{0.05 \times 0.25\,\text{m}^2} = -1.32\,\text{MPa}$$

Shear Force: Since point C is located at the top of the beam, no shear stress exists at point C.
Bending Moment: Since point C is located 1.5 m from point A and 125 mm from the neutral axis, the normal stress cause by bending moment is compressive.

$$\sigma_{bending\_moment} = -\frac{32.89\ kN \cdot m \times 0.125\ m}{0.05 \times 0.25^3/12\ m^4} = -63.15\ MPa$$

Superposition:
Thus, the stress at point C is obtained using superposition as

$$\sigma_C = -1.32\ MPa - 63.15\ MPa = -64.5\ MPa$$

As the bending stress will increase along the horizontal segment of the member, the maximum normal stress for the horizontal segment will occur at point D, where $x = 4m$.

$$\sigma_{normal\_force} = -\frac{16.45\ kN}{0.05 \times 0.25\ m^2} = -1.32\ MPa$$

$$\sigma_{bending\_moment} = -\frac{87.72\ kN \cdot m \times 0.125\ m}{0.05 \times 0.25^3/12\ m^4} = -168.4\ MPa$$

$$\sigma_D = -1.32\ MPa - 168.4\ MPa = -169.7\ MPa$$

However, it is unclear whether the maximum stress occurs at point D. It needs to be verified using finite element analysis.

EXAMPLE 7.2 *Preliminary Analysis of a Plate with a Hole*

Figure 7.3 shows a plate with a hole under uniaxial tension load. Because of the hole, stress concentration occurs at the hole edge. Using stress concentration table and analytical method calculate the maximum stress.

SOLUTION The first step of analysis is to estimate the stress concentration factor using the analytical method. From the assumption that the stress is constant in the cross-section, the nominal stress can be calculated, as

$$\sigma_{nominal} = \frac{P}{A} = \frac{300}{(2 - 0.75) \times 0.25} = 960\ psi$$

To calculate the stress concentration factor at the hole, we need to calculate the geometric factor, which is the ratio between the diameter of the hole and the width of the plate. From Figure 7.3, the geometric factor becomes $d/D = 0.75/2 = 0.375$. The stress concentration factor corresponding to

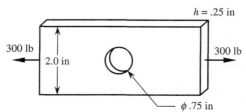

Figure 7.3 Plate with a hole

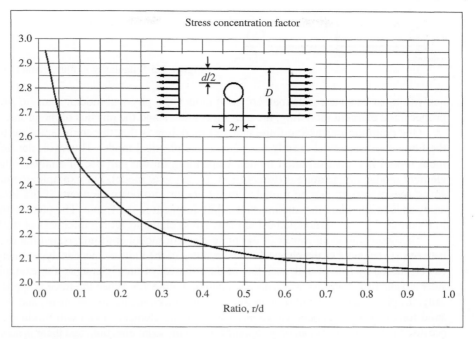

Figure 7.4 Stress concentration factor of plate with a hole

the geometric factor of 0.375 can be obtained from the graph in Fig. 7.4, which is $K = 2.17$. Then, the maximum stress occurs at the top part of the hole and its value becomes

$$\sigma_{\text{MAX}} = K\sigma_{\text{nominal}} = 2.17 \times 960 = 2083.2 \, \text{psi}$$

Preliminary analysis is important in that it helps to understand the physical problem better. Based on preliminary analysis results, engineers can plan a modeling strategy in preprocessing.

7.1.2   Preprocessing

Preprocessing is the stage of preparing a model for finite element analysis. It includes discretizing the structure into elements, as well as specifying displacement boundary conditions and applied loads. At this stage, the engineer considers the following modeling-related issues:

(a) Modeling a physical problem using finite elements

(b) The types and number of elements that should be used

(c) Applying displacement boundary conditions

(d) Applying external loads

We will discuss each issue in the following subsections.

A. Modeling a Physical Problem

It is important to understand the difference between the physical model and the finite element model. The finite element model is not a replication of the physical model but a mathematical representation of the physical model. The purpose of finite element analysis is to analyze mathematically the behavior of a physical model. In other words, the

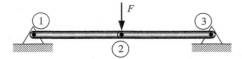

Figure 7.5 Singularity in finite element model

analysis must be an accurate mathematical model of a physical model. Thus, to perform a proper finite element modeling, the user must understand the physics of the problem.

One common mistake in finite element modeling is that the engineers want to make the finite element model exactly the same as the physical model. However, the finite element model is a goal-oriented model. Depending on the purpose of analysis, the finite element model can be quite different from the physical model. For example, the complex space rocket system can be modeled using one or two beam elements if the interest of analysis is to calculate the maximum bending moment in the frame. The plate with a hole in Example 7.2 can be modeled using plane stress elements with a thickness of 0.25 in.

The other important aspect of the finite element model is to understand the difference between the behavior of the physical model and that of the finite element analysis. Many errors can be caused by lack of understanding of the behavior of finite elements. For example, consider a plane truss with two elements, as shown in Figure 7.5. When the force is applied vertically, the physical model will support this force by deforming the two elements and producing axial forces in the members. However, a linear finite element model will produce an error because the two elements can only support force in the axial direction, and there is no stiffness in the vertical direction. Actually the global stiffness matrix will be singular for this case.

This aspect is clear from the following assembled matrix equation:

$$\frac{EA}{L}\begin{bmatrix} 1 & 0 & -1 & 0 & 0 & 0 \\ 0 & 0 & 0 & 0 & 0 & 0 \\ -1 & 0 & 2 & 0 & -1 & 0 \\ 0 & 0 & 0 & 0 & 0 & 0 \\ 0 & 0 & -1 & 0 & 1 & 0 \\ 0 & 0 & 0 & 0 & 0 & 0 \end{bmatrix}\begin{Bmatrix} u_1 \\ v_1 \\ u_2 \\ v_2 \\ u_3 \\ v_3 \end{Bmatrix} = \begin{Bmatrix} R_{1x} \\ R_{1y} \\ 0 \\ -F \\ R_{3x} \\ R_{3y} \end{Bmatrix} \tag{7.1}$$

After applying displacement boundary conditions (striking the first, second, fifth, and sixth rows and columns), the global matrix equation becomes

$$\frac{EA}{L}\begin{bmatrix} 2 & 0 \\ 0 & 0 \end{bmatrix}\begin{Bmatrix} u_2 \\ v_2 \end{Bmatrix} = \begin{Bmatrix} 0 \\ -F \end{Bmatrix} \tag{7.2}$$

It is clear that the matrix is singular, and we cannot solve for unknown nodal displacements. More precisely, we cannot solve the problem using linear finite element analysis in which the initial geometry is used to calculate the response of the structure. In the advanced finite element analysis,[1] the response of the structure can be calculated by following the deformation of the structure. Then, since the element stiffness is a function of deformation, the problem becomes nonlinear and can be solved iteratively.

Units: Use of proper units is fundamental in engineering analysis, but often neglected by many engineers. In general, there are no embedded units in the finite element model. It is the engineer's responsibility to use consistent units throughout the entire analysis procedure. Due to the numerical nature of finite element analysis, non-standard units are often employed. For example, the length unit ''meter'' in the SI system is too big for finite

---

[1] Bathe, 1996, Finite element procedures, Prentice-Hall, Upper Saddle River, 1996

element analysis if the magnitude of displacements is in the order of microns. In such cases it is common to use millimeters as the unit of length. On the other hand, the pressure unit Pascal in the SI system is too small for engineering applications. For example, the Young's modulus of steel is about 2×10^{11} Pascal. Mega-Pascal unit is often used for describing pressure. It is recommended that the user select the proper units for the various quantities before beginning the modeling.

Automatic mesh generation: In the second step, the physical model is approximated using the finite element model, which is composed of nodes and elements. For a simple model, it is relatively straightforward to create individual nodes by specifying its location and to define elements by connecting nodes. For a complex model, however, it would be laborious to define thousands of nodes and elements manually. Fortunately, many commercial preprocessing programs have mesh generation capability so that nodes and elements are automatically generated using graphical user interface. In such cases, the user first defines a solid model, which is similar to the physical model. Then, nodes and elements can be automatically created on the edges, surfaces, and volumes of the solid model. Bar and beam elements can be generated on edges, thin-shelled elements on surfaces, and solid elements on volumes. Sometimes, automatic mesh generation using solid model may not be enough to completely represent the physical model. In such a case, manual creation of nodes and elements are often combined with automatic mesh generation.

Using a solid model: The solid modeling technique starts from computer-aided design (CAD) programs, but now many commercial preprocessing programs have a similar capability. Many CAD programs also have a preprocessing capability. A solid model is a geometric representation of the physical model and can be used for various purposes. In general, a solid model consists of points, lines/curves, areas/surfaces, and volumes. Figure 7.6 shows the solid model of plate with a hole in Example 7.2. The preprocessing programs can either create the solid model or import it from CAD programs. Once the model is available, the user can create nodes and elements on the solid model.

Mesh control: To create nodes and elements automatically on the solid model, the user must provide mesh parameters that define the size and type of elements and other attributes. The user can also control the global element size, local element size, and curvature-based element size. These parameters are chosen based on the engineering knowledge. For example, the user may want to have a fine mesh in the vicinity of a hole where there is stress concentration. In that case the local element size is set smaller than the global element size. Then, the program will create small-sized elements near the hole and gradually increase to the global element size away from that point. The curvature-based element size control is good to represent curves. Figure 7.7 shows quadrilateral elements generated on the solid model in Figure 7.6 using the global element size of 0.1 and 0.2 in.

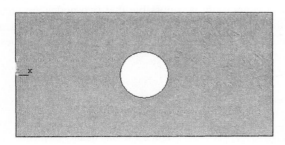

Figure 7.6 Solid model of plate with a hole

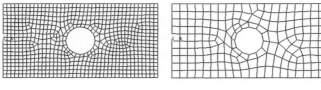

(a) Element size = 0.1   (b) Element size = 0.2

Figure 7.7 Automatically generated elements in a plate with hole

Note that the element size is a general guideline; not all elements have the same size. In addition, the circular hole is discretized using piecewise linear segments because the element edge is linear.

Mesh quality: Mesh quality is an important criterion in creating elements. A good quality mesh is a recipe for success in finite element analysis. As we learned in Chapter 6, the accuracy of analysis depends on the shape of elements.

(a) The first quality criterion is the shape of elements. For quadrilateral elements, the element perform best when the shape is a rectangle because the Jacobian matrix is diagonal and constant (see Section 6.6). When an element is distorted too much such as shown in Figure 7.8(a), numerical integration becomes inaccurate and the Jacobian is close to zero. Although we cannot make all elements rectangular, the angle between adjacent edges should be close to 90 degrees.

(b) The second quality criterion is aspect ratio. The square element has the aspect ratio of one, and, thus, it is perfect. Large-aspect ratio elements should be avoided. They perform worse when the shape is not rectangular as shown in Figure 7.8(b).

(c) The third criterion is related to the element size. In general, the element size is related the dimensions of the model. However, quick transition from small elements to large elements should be avoided as shown in Figure 7.9. This often happens when the ratio between the global and local element sizes are too large.

(d) One misconception in controlling element size is to use small elements at high stress region. However, elements perform equally well for representing low and high stresses. Rather, small-sized elements must be used in regions where the stress gradients are large.

Checking the mesh: After nodes and elements are created, it is important to check the created model for any possible errors or defects. Although there are many possible errors, we will discuss three common mesh errors. The first one is duplicated nodes. All interior edges and faces of elements must be shared with other elements. This can automatically be achieved when the nodes on the common edge/surface are shared with adjacent elements. However, during the mesh generation, it is possible that two nodes are created at the same location and each of them belongs to different elements. Due to graphical illusion, two elements seem to be connected, but during the assembly process, these two

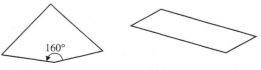

160°

(a) Distorted element   (b) Large aspect ratio   **Figure 7.8** Bad quality elements

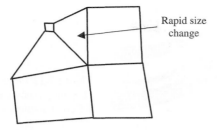

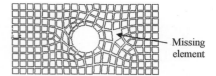

Figure 7.10 Shrink plot of elements to find missing elements

Figure 7.9 Quick transition of element size

elements are considered separate. This can cause an artificial crack in the model. Many preprocessing programs provide a capability to check "free surface," which plots the edges/surfaces that are not shared with adjacent elements. If the free surface occurs at an unintended location, the user needs to check duplicated nodes and merge them.

The second type of mesh error is missing elements. This error often occurs when elements are created manually. In the regular graphical display, it is not easy to find one or two missing elements. Many preprocessing programs can plot elements with shrunk size. As can be seen in Figure 7.10, the missing element can be easily identified in the shrink plot.

The third type of mesh error is mismatched boundary. Consider the three elements shown in Figure 7.11. Element E_1 is defined by connecting Nodes 1-2-3-4, E_2 by Nodes 4-5-7-6, and E_3 by Nodes 5-3-8-7. Graphically, these three elements look connected each other, but mathematically, there exists a crack between Nodes 3 and 4. To explain this, let us consider that all nodes have zero displacement except for Node 5, which has positive x-translation. Since the displacements of each element are interpolated by nodal displacements, E_1 has zero displacements. Thus, the edge along Nodes 3-4 has zero displacement, too. In E_2, however, since Node 5 has nonzero displacement, the edge along Nodes 4-5 has non-zero displacements. Thus, a gap develops between E_1 and E_2. The same is true for E_1 and E_3. To connect two elements, all nodes on the same edge must be shared.

Material properties: Although we only considered isotropic, linear elastic materials in this text, a variety of material models can be used in finite element analysis, including anisotropic materials, composite materials, nonlinear elastic materials, elasto-plastic materials, and visco-elastic materials. The interested users are referred to advanced textbooks for a detailed discussion of these materials. For isotropic, linear elastic materials, the following three material properties are often used: Young's modulus, shear modulus, and Poisson's ratio. However, it is normally required to provide two material constants, because the third can be calculated from other two.[2] Sometimes, it is necessary to provide

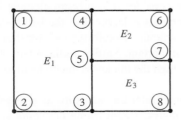

Figure 7.11 Error in element connection

---

[2] For an example, see Eq. (1.56) in Chapter 1.

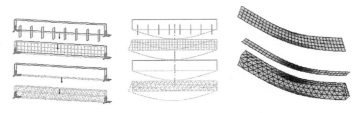

Figure 7.12 Finite element modeling using different element types

the yield strength or failure stress of the material in order to check the safety status of the finite element model under given loading conditions. It is important to note that the units of material constants should be consistent with the units used in the finite element model.

B. Choosing Element Types and Size

There is no unique way of modeling the given problem using finite elements. Different models and hence different element types can be used for solving the same problem. However, that does not mean that any element and model can be used. An important issue is that the user should understand the capability of the elements and models so that a proper element is used.

Selection of element type is one of the most important steps. The same part can be modeled using different types of elements. For example, the rectangular structure in Figure 7.12 can be modeled using beam, plane stress, shell, or solid elements. Although all different elements can be used to solve the same problem, each element has different characteristics.

Solid element: In general, the best way of modeling a structure is using solid elements, for this can represent structural details, such as sectional properties of components, filet and rounded corners, and detailed joint geometry. However, the number of elements required for modeling a structure using solid element can increase so quickly that it may not be computationally feasible. For example, let us consider that a sheet metal component is being modeled using solid elements. As we discussed in Chapter 6, several elements will be required in the thickness direction in order to capture local bending effects accurately, and the other dimensions of the elements would have to be kept small so that the aspect ratios of the elements are acceptable. Thus, it is not feasible to model many thin-wall structures with solid elements.

Shell/plate element: This element is not studied in this text but can be found in more advanced finite element textbooks. Shell/plate elements were originally developed to efficiently represent thin sheets or plates, both flat and curved surfaces. They model the structure as a two-dimensional plane with thickness of the structure. They include out-of-plane bending effects in their fundamental formulation, as well as transverse shear, tension, and compression in the plane. Conceptually, it can be considered as a beam element in two-coordinate directions. The required number of elements is larger than the beam elements but smaller than the solid elements. This element particularly performs well for thin-walled structures where bending and in-plane forces are important. Similar to the beam element, this element also cannot predict stress variation through the thickness due to local loading.

Beam/frame element: The beam/frame elements model the structure as a one-dimensional line with appropriate cross-sectional geometry. Beam/frame elements are even simpler and more efficient when structures employ beam-like details. As shown in Chapter 4, the beam element has translational and rotational DOFs at each node. The beam element is good for predicting the overall deflection and bending moments of a slender

Table 7.1   Different types of finite elements

Element	Name
○—○	1D linear element
△	2D triangular element
▭	2D rectangular element
◁	3D tetrahedral element
▱	3D hexahedral element

member. There are occasions when beams will be more fully represented as shells or solids in order to examine in detail how they are behaving or interacting with the structure where they are connected to other parts. Structural steel tubing and rolled sections can sometimes be simplified as beam elements. However, the beam element is limited in predicting stresses due to local stress concentration at the point of applied load or at a junction not modeled due to the simplification in shapes.

Table 7.1 shows different types of finite elements. Beam and bar elements belong to one-dimensional linear elements, whereas plane solid and shell/plate elements to two-dimensional triangular and rectangular elements. Tetrahedral and hexahedral elements can be used to model a three-dimensional solids.

Element order: The elements that are introduced in the previous chapters are categorized as linear elements. In general, higher-order elements are also available. Most-frequently used elements are listed as follows.

(a) Linear elements: two-node bar, three-node triangular, four-node quadrilateral, four-node tetrahedral, eight-node hexahedral elements

(b) Quadratic elements: three-node bar, six-node triangular, eight-node quadrilateral, 10-node tetrahedral, 20-node hexahedral elements

(c) Cubic elements: four-node bar, nine-node triangular, 12-node quadrilateral, 16-node tetrahedral, 32-node hexahedral elements

Linear elements have two nodes along each edge, parabolic elements have three, and cubic elements have four nodes along an edge. In general, a higher-order element is more accurate than a lower-order element because the former has more nodes than the latter.

Element size: Choosing a proper element size is extremely important in obtaining good results, and yet there is no systematic method available that will help in determining the proper element size *a priori*. If the mesh is too coarse, finite element analysis results can contain large errors. If the mesh is too fine, solving the problem becomes computationally expensive due to excessive number of DOFs. The proper mesh size depends on the problem at hand and the user's experience. The results from the preliminary analysis can be a guide to estimate the accuracy of the obtained finite element analysis results. However,

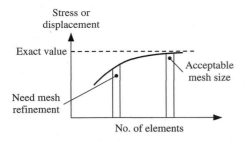

Figure 7.13 Convergence of finite element analysis results

this is only possible when the preliminary analysis results are accurate. In addition, the mesh does not have to be refined everywhere. Often critical regions require a fine mesh, and the other regions can have relatively coarse mesh.

Two common methods to determine if the current mesh is appropriate are (1) error analysis and (2) convergence study. The error analysis is based on error between original stress and averaged stress. If the error is larger than a threshold, it indicates that the current mesh is not acceptable and needs refinements. We will discuss error estimation in Section 7.2.

The convergence study is a powerful tool when there is no analytical solution available. Since we do not know the exact solution in most cases, it is impossible to know how accurate the current analysis result is even if we used thousands of elements or more. After a convergence study, we can have a confidence in the analysis results. The convergence study consists of two sets of meshes: one is the original mesh and the other is the mesh with twice as many elements in critical regions. If the two meshes yield nearly the same results, then the mesh is probably adequate. If the two meshes yield substantially different results, then further mesh refinement might be required. Figure 7.13 shows a typical output from the convergence study. The vertical axis represents the function of interest, such as displacement, stress, and temperature. The horizontal axis shows the number of elements used for each of the trial solutions. The number of DOFs can also be used. In general, the number of elements is doubled with successive trial solutions.

The convergence study in Figure 7.13 can go one step further to calculate the convergence rate. In this case, we have to calculate the function of interest at three different meshes. Let h_1, h_2, and h_3 be the sizes of elements; they are ordered such that $h_1 > h_2 > h_3$; i.e., h_3 represents the finest mesh. It is recommended that $h_1 = 2h_2 = 4h_3$. To calculate the convergence rate, the function of interest, u, is calculated at each of the mesh. Then the ratio in difference in the functions of interest is defined as

$$\frac{\|u_{h_3} - u_{h_2}\|}{\|u_{h_3} - u_{h_1}\|} \approx \left(\frac{h_2}{h_1}\right)^{\alpha} \tag{7.3}$$

where α is the *convergence rate*. It indicates how fast the solution will converge to the exact one.

One important question in the convergence study is the degree of accuracy the designer should seek. This is a very practical question and yet difficult to answer because in many cases we do not know the exact solution. In the above equations, the most accurate solution, u_{h_3}, allows us to estimate the error in the previous two less accurate solutions, u_{h_1} and u_{h_2}. Since u_{h_3} is close to the exact solution, the differences between it and the earlier solutions are really estimates of the errors in the earlier solutions. With those estimates, we can select the element size that gives us the least accurate solution that is still acceptable.

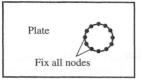

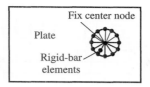

Figure 7.14 Applying displacement boundary conditions at a hole in a plate

C. Applying Displacement Boundary Conditions

Once the finite element model is built, displacement boundary conditions need to be implemented. It is important to note that the finite element model should be properly restrained from rigid body motions, both translation and rotation. Otherwise a unique solution is not possible. Special techniques are available to analyze a finite element model under rigid-body motion, but in general, all deformations are supposed to produce strains and thus stresses. Otherwise, the global stiffness matrix will be singular and will indicate error when solving the matrix equations. On the other hand, if the model is restrained at too many places, the analysis results will be different from the actual response of the physical model. For example, consider a plate with a circular hole that is connected to the other parts of the structure through a pin as shown in Figure 7.14. Depending on what actually happens, two different ways of modeling can be possible. If the pin is tight so that the plate is not allowed to rotate, all the nodes along the circumference of the hole can be fixed. The second scenario is as follows. The pin is not tight so that the plate can rotate, but it is not allowed to move in the x- and y-directions. In this case, we can create a node at the center of the hole and connect all the nodes in the circumference of the hole with the center node using rigid-bar elements. Then the x- and y-translations of the center node can be fixed.

In the previous section, we mentioned that the errors in finite element solutions will be reduced as the mesh of the model is refined. However, the errors in boundary condition will not reduce no matter how much one refines the model. Any unexplained high stress may be due to a wrong boundary condition. Thus, it is important to check if the displacement boundary conditions are properly implemented.

Errors in the boundary conditions are sometimes subtle; therefore, they are not easy to identify. For example, consider two plane trusses as shown in Figure 7.15. Since the problem is two-dimensional, overall three DOFs need to be fixed in order to eliminate all rigid-body motions. In Figure 7.15(a), Node 1 is fixed in both directions and Node 3 is fixed in the y-direction. Thus, all three DOFs are fixed. However, the model is not free from rigid-body motions because it is possible to move Node 3 in x-direction, although the movement is infinitesimal. Thus, in Case (a), the stiffness matrix will be singular and the matrix equation cannot be inverted to solve for displacements. On the other hand, the truss in Figure 7.15(b) is properly restrained and the global stiffness matrix will be positive definite.

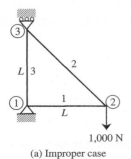

(a) Improper case

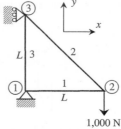

(b) Proper case

Figure 7.15 Applying displacement boundary conditions on truss

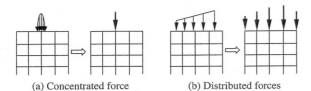

(a) Concentrated force (b) Distributed forces

Figure 7.16 Concentrated and distributed forces in a finite element model

When the elements are generated from the solid model, it is often possible to apply the boundary conditions directly to the solid model. This depends on preprocessing programs. In that case, the program automatically converts the boundary conditions into equivalent nodal forces and displacements. The advantage of applying boundary conditions in the solid model is when the problem is solved multiple times with different mesh densities. In that case, it will not be necessary to redefined boundary conditions for different meshes.

D. Applying External Forces

In practice, forces may be applied to a machine or a structure through complex mechanisms. Forces from one part are transferred to the other part through the contact pressure in the interface. However, if the region of interest is far from the location at which the force is applied, the complex mechanisms that transmit the force could be approximated using simpler ones. In most cases the mechanisms can be completely eliminated and the forces can be applied directly to the structure. In such cases it is important to understand that the results in the vicinity of the force will not be accurate. In general, three different types of forces are applied to the finite element model: (1) concentrated forces at nodes, (2) distributed forces on the surface or edge, and (3) body forces. The body forces and surface forces usually vary over the volume or area in which they are acting.

Applying a concentrated force: In theory, if a concentrated force is applied at a point, the stress at that point becomes infinitely large because the area is zero. In reality, there is no way to apply a force at a point. All forces are distributed in a region. The concentrated force in finite elements is an idealization of distributed forces in a small region. When the region is relatively small compared to the size of element, it can be idealized as a concentrated force at a point, as shown in Figure 7.16(a). On the other hand, when the region is larger than the size of element, it can be treated as distributed forces, as shown in Figure 7.16(b). Note that the distributed forces are converted to the work-equivalent nodal forces. In fact, all applied forces must be converted to the equivalent nodal forces because the right-hand side of finite element matrix equations is the vector of nodal forces.

The effect of this idealization is limited to the immediate vicinity of the points of application of the force. If the interest region is relatively far from the force location, the stress distribution may be assumed independent of the actual mode of application of the force (St. Vernant's principle). Figure 7.17 shows the distribution of stress due to a concentrated force at the top. In the section near the top, the stress is concentrated at the

$\sigma_{min} = 0.973\sigma_{ave}$
$\sigma_{max} = 1.027\sigma_{ave}$

$\sigma_{min} = 0.668\sigma_{ave}$
$\sigma_{max} = 1.387\sigma_{ave}$

$\sigma_{min} = 0.198\sigma_{ave}$
$\sigma_{max} = 2.575\sigma_{ave}$

Figure 7.17 Stress distribution due to concentrated force

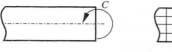
(a) Beam element

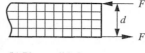
(b) Plane solid elements

Figure 7.18 Applying a couple to different element types

center. However, when the distance from the force application point is the same as the member cross-sectional dimension, the stress is almost uniformly distributed.

In the finite element model, the concentrated force can only be applied to a node. Thus, it is important to make sure that a node exists at the location of the force. If a concentrated force needs to be applied at the location where no node exists, it can be approximated by calculating equivalent nodal forces at the surrounding nodes using the shape functions (for example, see Problem 19 in Chapter 4). Even if the concentrated force is applied at a node, the stress at that point will not be singular but very large because the finite element distributes the effect of the force throughout the element.

Applying a couple to a plane solid: Sometimes, different methods should be used to apply the same type of force. For example, consider applying a couple to a structure (Figure 7.18). The structure can be modeled either using beam elements or plane solid elements. In the case of beam elements shown in Figure 7.18(a), it is straightforward to apply a couple because it is a force corresponding to the rotational DOF. At each node of a beam element, a transverse force and couple can be applied. However, in the case of plane solid elements, shown in Figure 7.18(b), we can only apply x- and y-directional forces. Since there is no rotational DOF, a couple cannot be applied directly as a nodal force. In this case, a pair of equal and opposite forces has to be applied and the forces will be separated by a distance d such that the equivalent couple $C = F \cdot d$. Of course, the effect will be different in the vicinity of the forces, but according to St. Vernant's principle, the local effect of force application method will disappear at a short distance.

Force transmitted by a shaft: In mechanical system, a force in a part is transferred to the other part through connections, such as shafts and joints. One can model both the part and shaft and apply contact condition between them. Then the applied force at the shaft is transferred to the part through the contact pressure at the interface. However, the problem becomes nonlinear and solving the problem will be very complicated. If the stress distribution of a part due to the transferred force is main interest, it is unnecessary to model the shaft. Rather, the force in the shaft can directly be applied to the surface of the hole. At this point, it is necessary to approximate the distribution of force to the surface. Figure 7.19 explains two commonly used approximations. Note that only half the surface of the shaft is in contact with the plate and the maximum contact pressure occurs at the center. The first method assumes that contact pressure distribution is elliptic, that is, given by the equation of an ellipse. The magnitude of the maximum distribution is calculated from the fact that the total force supported by the contact pressure is the same as the force transmitted by the shaft. The second method approximates the effect of shaft using bar

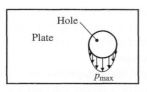

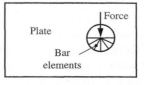

Figure 7.19 Modeling a shaft force using assumed pressure and bar elements

elements that connect the center of shaft to one side of the hole surface. Since the shaft is often more stiffer than the plate, a large value of the axial stiffness is used for the bar elements, or sometimes the bar elements are assumed to be rigid.

EXAMPLE 7.3 *Displacement Boundary Conditions for a Plate Model*

For the plate with a hole in Example 7.2, apply displacement boundary conditions and external load such that it is under uniaxial tension.

SOLUTION The displacement boundary conditions are given in Figure 7.20. All nodes on the left edge are fixed in x-direction. However, there exists a rigid-body motion in y-direction. Thus, the node on the center of the left edge is fixed in both x- and y-directions. Since the model is plane solid, the three rigid-body motions are fixed.

A uniform pressure is applied on the right edge. The magnitude of the pressure is 600 psi, which is equivalent to the 300 lb in Example 7.2. Figure 7.20 shows the equivalent nodal forces corresponding to the uniform pressure. Note that the nodal forces are equal except for the two nodes at the top and bottom. As discussed in Chapter 6, the uniform pressure force is equally divided into two end nodes in a rectangular element. Thus, after assembly, two end nodes have half of the force compared to nodes inside.

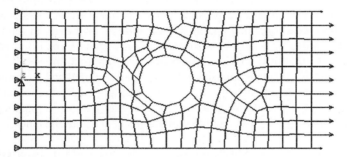

Figure 7.20 Displacement and forces of the plate model

7.1.3 Solution Techniques

In this stage, individual element stiffness matrices and the vector of nodal forces are assembled and solved for unknown DOFs. After solving the global matrix equation, two different types of solutions are produced:

(a) Nodal DOF solutions – these are primary unknowns.

(b) Derived solutions – such as stresses and strains in individual element.

The solutions will be used in the postprocessing stage to display and interpret the results. The primary solutions are available at each node, whereas the secondary or derived solutions are usually calculated at the integration points of individual elements.

Although we only discussed static analysis and steady-state heat conduction analysis, many different types of analyses are available, such as static, buckling or stability analysis, heat transfer, potential flow, dynamics, and nonlinear analysis. The engineer should understand the purpose of these analyses and expected outputs.

In most finite element analysis programs, the solution stage is transparent to the user, such that constructing element stiffness matrices, assembling the global stiffness matrix, constructing vector of applied forces, deleting rows and columns, and solving the matrix equations are performed automatically using the information that the user provided in the preprocessing stage. Due to this fact, most users do not pay attention to this step. However, most failures in finite element analysis procedures occur in this step. Although the failure is usually caused by errors or wrong information from the preprocessing stage, the effects of these errors can only be detected while solving the matrix equation. Thus, it is important to understand what types of errors or wrong information in the preprocessing stage can cause the problems in the solution stage.

Singularities: One of the most common problems in the solution stage is singularity in the global stiffness matrix. As we learned in the previous chapters, the element stiffness matrices and the structural stiffness matrix are singular. However, after applying displacement boundary conditions (striking rows and columns), the global stiffness matrix becomes positive definite and can be inverted. If singularity happens, the matrix solver cannot calculate nodal DOFs and the solution stage stops with an error message. A singularity may also indicate existence of an indeterminate or non-unique solution. Mathematically, as we learned in Chapter 0, when the coefficient matrix is positive definite, there always exists an inverse of the matrix and the equation has a unique solution. When the matrix is singular, the determinant of the matrix is zero, and the inverse of the matrix cannot be obtained. The following conditions may cause singularities in the global stiffness matrix:

(a) Insufficient/wrong displacement boundary conditions. These are the most common mistakes/errors that cause singularity in the matrix. The displacement boundary conditions should be such that all rigid body modes are eliminated.

(b) Negative values of material properties, such as Young's modulus or density. Zero value of the Poisson's ratio will not cause singularity.

(c) Unconstrained joints. The element usage may cause singularities. For example, most commercial programs do not have one-dimensional or two-dimensional truss. Three-dimensional truss element is supposed to cover these elements. However, when the three-dimensional truss element is used to solve for one-dimensional bar problems, the displacements in y- and z-axis for all nodes must be fixed even if no forces are applied in those directions. Some commercial programs automatically fix these DOFs.

(d) Coincident nodes causing cracks in the model. As explained in the preprocessing stage, a modeling error such as coincident nodes separates elements and can cause a rigid-body motion.

(e) Large difference in stiffness. When two different materials are used in the model, if the difference in stiffness is too large, the stiffness of the weaker material is considered as zero numerically. This happens because all data are stored with limited number of significant digits. When the ratio between Young's modulus is greater than 10^6, most finite element programs have singularity errors.

(f) Irregular node numbering could also lead to singular stiffness matrices. For example, when using triangular elements it is important to number the nodes in the counter clockwise direction. Otherwise, the area of the triangle will be negative and the stiffness matrix may not be positive definite.

Multiple load conditions: In this text, we only discussed the case where the RHS of the matrix equation is a single-column vector, and the solution is also a single-column vector. This corresponds to a situation in which the structure is under a single loading condition.

However, the engineer may want to evaluate the safety of the structure under different loading conditions. For example, bicycle design in Section 4.7 of Chapter 4 must satisfy two loading conditions, vertical bending and horizontal impact loads, but not simultaneously. The safety of a building structure involves the wind load and the seismic load. Of course, it is possible to solve the finite element equation multiple times, but it is possible to solve multiple loading conditions more efficiently. Let us consider the case that we want to calculate the responses of the system under N different load conditions. In this case, the global matrix equation can be written as

$$[\mathbf{K}][\tilde{\mathbf{Q}}] = [\tilde{\mathbf{F}}] \qquad (7.4)$$

where

$$[\tilde{\mathbf{Q}}] = [\mathbf{Q}_1 \, \mathbf{Q}_2 \, \cdots \, \mathbf{Q}_N]$$
$$[\tilde{\mathbf{F}}] = [\mathbf{F}_1 \, \mathbf{F}_2 \, \cdots \, \mathbf{F}_N]$$

Vector $\{\mathbf{Q}_1\}$ is the nodal responses corresponding to the nodal forces $\{\mathbf{F}_1\}$ and $\{\mathbf{Q}_2\}$ to $\{\mathbf{F}_2\}$, etc. This is different from solving the matrix equation N times. In fact, it is much more efficient than N finite element analyses. In general, solving a matrix equation can be decomposed into two stages. In the first stage, which is the most time-consuming part, the stiffness matrix is factorized using LU decomposition. The stiffness matrix is expressed as the product of lower and upper triangular matrices. However, factorizing the stiffness matrix is independent of the force vectors. Once the stiffness matrix is factorized, solving for $\{\mathbf{Q}_I\}$ for a given $\{\mathbf{F}_I\}$ is computationally inexpensive. Thus, it is very efficient to decompose the stiffness matrix once and solve for multiple solutions.

Restarting solution stage: Occasionally, the user may need to restart an analysis after the initial result has been obtained. For example, the user may want to add more load conditions to the analysis. In this case, restarting the solution can be an effective tool. Let us assume that the matrix equation with initial load condition has been solved. Then the LU-factorized stiffness matrix is stored in the disk space because this part is independent of the applied loads and is the most time-consuming part in the solution stage. When the user wants to restart the analysis with different load conditions, it is easier to construct the new vector of global forces and solve for unknown DOFs using the already factorized stiffness matrix. This procedure seems attractive, but it has its own drawbacks. It requires storing the factorized stiffness matrix in the disk space. Thus, the user trades off between saving computational time and disk space.

7.1.4   Post-processing

After building the model and obtaining the solution, the user may want to review the analysis results and evaluate the performance of the structure. Postprocessing provides tools to display and interpret the results. It is probably the most important step in the analysis, because the purpose of analysis is to understand the effect of applied loads to the structure and to evaluate the quality of the mesh. The most important task of postprocessing is to interpret results. This stage requires knowledge and experience in mechanics. Through postprocessing, the user can check any discrepancy between the preliminary analysis results and the finite element analysis results.

Deformed shape display: After solving the finite element matrix equation, the values of all nodal DOFs are available. For structural problems, the nodal DOFs are displacement or temperature. For a simple model, it is easy to interpret the deformation of the structure using nodal displacements. For a complicated model, however, it is not straightforward to

Figure 7.21 Deformed shape of the plate model

interpret the deformation using numerical data. Most postprocessing programs provide plotting capability of deformed geometry of the model using graphical user interface. The deformed shape plot is a strong tool in understanding the structural behavior. Many modeling errors can often be found from the deformed shape plot. Through the deformed shape plot, the user can verify whether the displacement boundary conditions and external forces are correctly applied.

Note that the magnitude of deformation for linear elastic material is usually in the range of microns or millimeters. Thus, it is difficult to recognize the deformation in a regular plot. For this reason, the deformation is often magnified such that it is visible. Many postprocessing programs automatically scale the deformation such that the maximum deformation is approximately 5% to 10% of the model size. Figure 7.21 shows magnified deformation of the plate model.

Contour display: The displacements are vectors and hence a deformed shape of the mesh is the best way to depict the deformations. On the other hand, contour plots are suitable for scalar quantities. Although stress at a point is a matrix quantity, a single component of stress, say σ_{xx}, can be considered as a scalar and the stress variation throughout the model can be displayed using a contour plot. Finite element analysis outputs include stress results for all elements. Using the contour plot of the stress, the user can understand the distribution of the stress in the structure and identify the most critical locations. In ductile materials yielding is governed by von Mises stress and hence it makes sense to have a contour plot of von Mises stress in order to understand the behavior of the structure.

The plate model in Figure 7.22 shows that the maximum stress occurs at the top and bottom of the hole, which was expected from the preliminary analysis. From the preliminary analysis in Example 7.2, the stress concentration at these locations was estimated to be 2,083 psi, but the finite element analysis yields 2,209 psi, which is approximately 6% higher than the preliminary analysis results. Note that the stress results in Figure 7.22 are obtained with the mesh of size 0.2 in.

There are several reasons for mismatch in stress between preliminary analysis and finite element analysis. First, the stress concentration factor in the preliminary analysis is

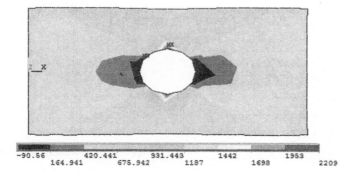

-90.56   420.441   931.443   1442   1953
      164.941   675.942   1187   1698   2209

Figure 7.22 Contour plot of σ_{xx} of the plate model (element size = 0.2 in)

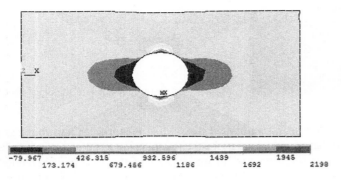

Figure 7.23 Contour plot of σ_{xx} in the refined model (element size = 0.1 in)

based on approximate reading of the stress concentration factor graph. In addition, the stress in Figure 7.22 may not be accurate because of extrapolation and coarse mesh size. In finite element analysis, the stress value is most accurate at the Gauss quadrature points. The stresses at the nodes are obtained by extrapolating stresses at the integration points and also by averaging the stresses from adjacent elements. To see the effect of mesh size, the initial mesh size of 0.2 in is refined to 0.1 in. Figure 7.23 shows stress contour plot with the refined mesh size. Note that the maximum stress is reduced to 2,198 psi. The maximum stress only changes by 0.5% from the coarse mesh stress. Thus, the stress results from finite element analysis can be considered to be reasonably accurate.

Stress averaging: One of the challenging tasks in postprocessing is to graphically display analysis results accurately. Since most contour–plotting algorithms are based on nodal values, it is necessary to calculate stress values at nodes. In the case of displacements, they are continuous between elements and hence accuracy is not an issue. In the case of stresses, however, they are continuous within the element but discontinuous across element boundaries. Thus, it is necessary to perform averaging on calculated values to produce a contour plot. In addition, the stress results are available at the Gauss quadrature points, and the values at nodes are obtained through extrapolation. Thus, extrapolation and averaging can affect the accuracy of stress results in the contour plot. For exact stress values, it is better to use the numerical values at the Gauss quadrature points. Figure 7.24 shows stress averaging for one-dimensional elements. Stress values are calculated at Gauss quadrature points (in this case, two per element). Then the stress within the element is calculated using the interpolation function. The stress values at the node that connects two elements are in general different. Then the averaged nodal stress is calculated at that node. If the difference in stress at the connecting node is small, the error due to averaging is small. The difference between actual and averaged stress values are often used as a criterion of accuracy in analysis results.

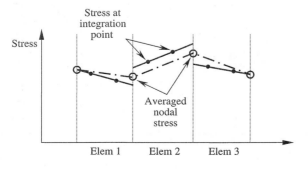

Figure 7.24 Averaging stresses at nodes

7.2   FINITE ELEMENT MODELING TECHNIQUES

There are many modeling techniques in finite element analysis. They depend on particular finite element programs. In the following subsections, we summarize a small list of modeling techniques.

7.2.1   Model Abstraction

One of common misunderstandings in finite element modeling is that the finite element model should be exactly the same as the physical model. Thus, many engineers waste their time by including more details during the preprocessing phase. However, small details that are unimportant to the analysis should not be included in the model, since they will only make the model more complicated than necessary. It would be better to gain insight from several simple models than to spend time making a single-detail model. Figure 7.25 shows a wheel cover model. If the purpose of analysis is to estimate bending/torsional stiffness, the model shown in the figure is too detailed.

For some structures, however, small features such as fillets or holes can be locations of maximum stress and might be quite important, depending on the purpose of analysis. In such a case, the structural details are important and should be included in the model. The user must have an adequate understanding of the structure's expected behavior in order to make proper decisions concerning how much detail should be included in the model. For example, if the purpose of analysis is to estimate the maximum deflection of a bridge structure, it might be good enough to model the bridge using beam elements. However, if the purpose is to predict the stress concentration at the joints or connections, then it is necessary to model holes and pins. It is important to perform a trade off between details of results and modeling and computational costs. For example, when the size of a hole is

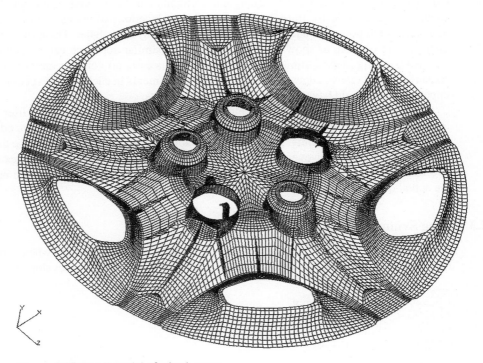

Figure 7.25 Detail model of wheel cover

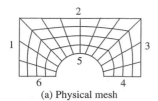

(a) Physical mesh

(b) Topological mesh

Figure 7.26 Mesh generation using mapping

less than 1% of the entire dimension of the model, it would be better to ignore it in modeling. Instead, nominal stress values in conjunction with stress concentration factors can be used to predict the effects of the hole on the overall behavior of the structure.

7.2.2 Free Meshing vs. Mapped Meshing

In general, there are two ways of creating a mesh on the solid model: free meshing and mapped meshing. In free meshing, the user provides a general guideline of meshing and the preprocessing programs will make the mesh according to it. Thus, there are not many things that the user needs to do. On the other hand, in mapped meshing, the user provides detailed instructions of how the mesh should be created. In two dimensions, all surfaces are divided into topologically four-sided quadrilaterals. The user then specifies how many elements will be generated in each side of the quadrilateral. Consider a surface shown in Figure 7.26. It consists of six lines/curves. To make a mapped mesh, these six curves are divided into four groups. For example, we can group curves 1, 2, and 3 into one so that the surface consists of three single curves (4, 5, and 6) and one composite curve (1-2-3). Now, let us consider that four elements are assigned to curves 4 and 6 and twelve elements to curve 5 and the composite curve. Note that the opposite side should have the same number of elements. Then, 4×12 rectangular elements can be created in the topological square in Figure 7.26(b). In the physical mesh, the locations of nodes can be found by assuming the topological square is morphed into the physical shape. Then, quadrilateral elements are created, as shown in Figure 7.26(a). A similar approach can be applied for mapped meshing in three-dimensional volume. In this case, all volumes are grouped into six surfaces and twelve edges. Then the number of elements in each edge is specified. The opposite sides should have the same number of elements.

There are advantages and disadvantages in both meshing methods. More user action is required in mapped mesh, whereas more complex computer algorithms need to be implemented in free meshing. In general, the mapped mesh looks better because the grid looks more regular. However, the quality of elements cannot be assured. For example, two meshes are compared in Figure 7.27. Even if the mapped mesh looks more regular, the actual quality should be measured from the level of distortion.

7.2.3 Using Symmetry

Exploiting symmetry in finite element analysis was considered important when computational capabilities used to be limited. As computer hardware has developed rapidly,

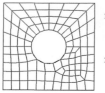

mapped mesh

free mesh

Figure 7.27 Mapped and free meshes

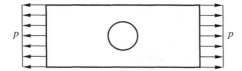

Figure 7.28 Plate with a uniform pressure

symmetry in reducing modeling effort is less used than before. Consider a plate with a hole, as shown in Figure 7.28. The plate is under uniform traction at the both ends. As is clear in the model, there is no displacement boundary condition. Either traction is applied on boundaries or they are free boundaries. If these boundary conditions are used as such the stiffness matrix will be singular because the rigid-body motion is not removed in the model. Although it is possible to fix arbitrary points, we can use symmetry to remove the rigid-body motion. Note that the geometry as well as the applied load is symmetric with respect to both the x- and y-axes. Thus, the results of the finite element analysis must be symmetric.

In general, symmetry of the problem implies that geometry, applied loadings, and boundary conditions are symmetric. If non-symmetric loads are applied on the symmetric geometry, the problem is not symmetric. A symmetry boundary condition means that out-of-plane translations and in-plane rotations are set to zero along the line of symmetry. For the structure shown in Figure 7.29, the geometry and applied load are symmetric with respect to the vertical plane indicated by dotted lines. In such a case, only a half of the structure needs to be modeled with appropriate symmetry boundary conditions on the plane of symmetry. If more than one symmetry plane exists, the model can further be reduced.

The symmetric model in Figure 7.30(a) removes the x-translational and z-rotational rigid-body motions, but still has y-translational rigid-body motion. (Remember that a plane solid has three rigid-body motions). Thus, the symmetric model in Figure 7.30(a) cannot be solved as it is. We can further apply additional symmetry with respect to the x-axis, as shown in Figure 7.30(b). Now all rigid-body motions are fixed and the $\frac{1}{4}$ finite element model can be used.

When a concentrated force is applied along the symmetric line, the force must be halved in the symmetric model.

In some cases, a structure could be considered symmetric, but for a few minor details that disrupt symmetry. In such cases those conditions that prevent symmetry could be removed as long as the user understands the effect of such deliberate changes. The user must weigh the gain in model simplification against the cost in reduced accuracy when deciding whether to deliberately ignore unsymmetrical features of an otherwise symmetric structure.

7.2.4 Connecting Beam with Plane Solids

For a complex system, it is often necessary to use different types of elements in order to model the physical problem. These different types of elements are connected to each other by sharing common nodes. For some types of elements, this connection is trivial, but for others, it is not. The issue is that not all elements have the same DOFs at the node. For example,

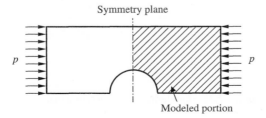

Figure 7.29 Uniform pressure

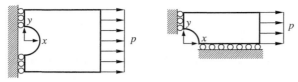

(a) One symmetric plane   (b) Two symmetric planes

Figure 7.30 Symmetric model of a plate with a hole

quadrilateral elements can be connected to triangular elements without any modification because both types have the same type of DOFs at the node. However, this is not the case when we want to connect plane solid elements with beam/frame elements. The plane solid element has x- and y-translation DOFs at the node, but the plane frame element has x- and y-translation as well as z-rotational DOFs at the node. When these two types of elements are connected, the rotational DOF of the frame is not constrained and, thus, the frame element will experience rigid-body motion; i.e., there is a singularity in the model (see Figure 7.31).

When a plane solid is connected to a frame as shown in Figure 7.31, it means these two are connected through welding. Thus, the rotation of the frame should be related to the rotation of the plane solid. However, the plane solid element does not have rotational DOF. In fact a special technique will be required to apply a couple to the plane solid elements, as shown in Figure 7.18. There are two possible modeling techniques to achieve this connection effect. In Figure 7.32(a), the frame element is extended to inside of the plane solid. Thus, the rotation of the beam can cause deformation of the solid. This technique is useful when the mesh in the solid is parallel to the axial direction of the frame, such as mapped mesh. The second method is illustrated in Figure 7.32(b), in which the following constraint equation is imposed between the nodes on the solid and the rotational DOF of the frame so that the rotation of the frame can cause deformation in the solid:

$$u_1 + h\theta_2 - u_3 = 0 \qquad (7.5)$$

The constraint is equivalent to the welding joint in the interface. This may not be a perfect fix because this approach will result in high local stresses. Similar approach could be used to connect a three-dimensional solid to shell elements.

7.2.5 Modeling Bolted Joints

Many mechanical parts are connected using bolted joints. A frequent question is how detailed a bolted joint should be modeled in finite element analysis. One of the most complex modeling schemes will be using all three-dimensional representations of bolts and parts and connecting them through contact constraints. However, this approach will not be practical unless the purpose of the analysis is to calculate the local stresses in the bolt. In addition, this approach will be very difficult to solve because: (1) the model size will be huge due to detailed representation of small features; (2) the problem becomes nonlinear due to the contact constraints; and (3) there will be rigid-body motion, if an initial gap exists between the bolt and parts. Thus, we are looking for practical and simpler modeling technique for bolted joints.

Plane solid
element

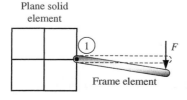

Frame element

Figure 7.31 Singularity in connecting a plane solid with a frame

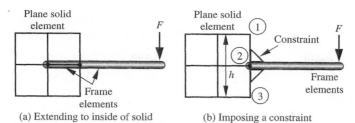

Figure 7.32 Connecting a plane solid with frame

Nodal coupling: The structural parts that are connected through the bolted joints are in general thin plates. In addition, the size of the bolt hole is much smaller that the dimension of the parts. Hence, the bolt hole in the part can be ignored in modeling. Thus, the first simplified approach is to represent the bolted (or riveted) connection of overlapping shell structures by locating a node on each surface at the location of the bolt. The two nodes have to be located at the same x-, y-, and z-locations in space. This is an approximation because physically it is impossible to locate two plates in the same location due to finite thickness of the plates. Then a constraint is imposed so that these two nodes are tied such that they have the same displacements. Most finite element programs support a nodal coupling capability. It will be desirable to tie two of the three rotations as well. The only rotation that is free is that around an axis perpendicular to the planes of elements (around the axis of the bolt).

When bolt holes are ignored in modeling, the user needs to do additional work to examine the stresses in the vicinity of the bolt holes, using the stress concentration factor to find the allowable net stress, bearing force, and total force in that zone.

Using rigid link: To remove the approximation of locating two nodes in the same location, the two plates are positioned properly in space and then a rigid element is used to link pairs of nodes. The rotational DOF around the axis of the bolt must be free at the ends of the rigid element. Instead of the rigid element, it is possible to use a very stiff bar element to connect the two points.

Bolt preload: When bolts are used to tighten two plates, there exists a preload in the bolts. The previous two methods will not be able to include bolt preload. When a preload needs to be modeled, a frame element with "initial strain" can be used. The intended preload must exist BEFORE the structure is loaded. In addition, the finite element model becomes much more complicated because the preload must be squeezing two surfaces together, which means surface contact elements must be in use between separated plate element surfaces. If the bolts are not overloaded when the structure is loaded, the bolt preload will be nearly unchanged when the structure is loaded. The use of friction coefficient at the contact surfaces needs to be considered

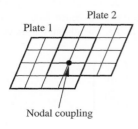

Nodal coupling

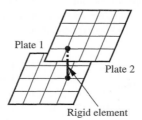

Rigid element

Figure 7.33 Modeling bolted joints

carefully, for it can reduce or prevent completely the shear loading on the bolts. It would be conservative to assume frictionless contact because the bolts will take all the loads.

EXAMPLE 7.4 *Bolted Joint*

Two identical beams of length 1 m are clamped at one end and connected by a bolt, as shown in Figure 7.34. An upward transverse force of 240 N is applied at the bolted joint. Assume $EI = 1000\,\text{Nm}^2$. Use two beam elements to determine the deflection at the bolted joint. Compare the deflection with that of Example 4.6.

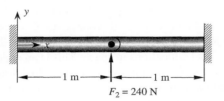

1 m   1 m

$F_2 = 240$ N

Figure 7.34 Finite element models of stepped cantilevered beam

SOLUTION   To make a bolted joint, two beam elements must be separated. Thus, the finite element model has four nodes, as shown in the figure below.

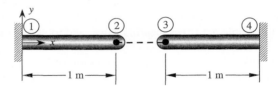

1 m   1 m

The element stiffness matrices of the two elements will be the same, although the row and column addresses will be different. Using the formula in Eq. (4.52), we can write the element stiffness matrices as

$$[\mathbf{k}^{(1)}] = 1000 \begin{array}{c} \\ \\ \\ \\ \end{array} \begin{bmatrix} \begin{array}{cccc} v_1 & \theta_1 & v_2 & \theta_2 \end{array} \\ \begin{array}{cccc} 12 & 6 & -12 & 6 \\ 6 & 4 & -6 & 2 \\ -12 & -6 & 12 & -6 \\ 6 & 2 & -6 & 4 \end{array} \end{bmatrix} \begin{array}{c} v_1 \\ \theta_1 \\ v_2 \\ \theta_2 \end{array}$$

$$[\mathbf{k}^{(2)}] = 1000 \begin{array}{c} \\ \\ \\ \\ \end{array} \begin{bmatrix} \begin{array}{cccc} v_3 & \theta_3 & v_4 & \theta_4 \end{array} \\ \begin{array}{cccc} 12 & 6 & -12 & 6 \\ 6 & 4 & -6 & 2 \\ -12 & -6 & 12 & -6 \\ 6 & 2 & -6 & 4 \end{array} \end{bmatrix} \begin{array}{c} v_3 \\ \theta_3 \\ v_4 \\ \theta_4 \end{array}$$

Assembling the element stiffness matrices, we can obtain the 8×8 structural stiffness matrix $[\mathbf{K}_s]$. However, it is possible to apply the boundary conditions in the element level so that zero deflections and rotations can be deleted from the above element stiffness matrices. Since Nodes 1 and 4 are fixed, we can only keep those DOFs corresponding to Nodes 2 and 3. The global matrix equation is then obtained as

$$1000 \begin{bmatrix} 12 & -6 & 0 & 0 \\ -6 & 4 & 0 & 0 \\ 0 & 0 & 12 & 6 \\ 0 & 0 & 6 & 4 \end{bmatrix} \begin{Bmatrix} v_2 \\ \theta_2 \\ v_3 \\ \theta_3 \end{Bmatrix} = \begin{Bmatrix} 120 \\ 0 \\ 120 \\ 0 \end{Bmatrix} \tag{7.6}$$

In Eq. (7.6), we applied 120 N of vertical loads at Nodes 2 and 3. Since we applied displacement boundary conditions, the above global matrix equation is positive definite and we can solve for unknown nodal DOFs. However, the solution will be for the case when Nodes 2 and 3 are not connected, and thus it does not represent the physical problem. Now, we want to apply the constraint that corresponds to the bolted joint. Due to the joint, the vertical deflection at Nodes 2 and 3 will be identical, but two rotations will be independent if we assume the joint is frictionless. Thus, the only constraint equation that we need is $v_2 = v_3$, which can be written in terms of unknown DOFs, as

$$[1 \quad 0 \quad -1 \quad 0]\begin{Bmatrix} v_2 \\ \theta_2 \\ v_3 \\ \theta_3 \end{Bmatrix} = \{0\} \tag{7.7}$$

We impose the above constraint by replacing the third equation in Eq. (7.6) with the above equation. Note that we have multiplied Eq. (7.7) by a factor 1000 throughout. Now, we have

$$1000\begin{bmatrix} 12 & 6 & 0 & 0 \\ 6 & 4 & 0 & 0 \\ 1 & 0 & -1 & 0 \\ 0 & 0 & 6 & 4 \end{bmatrix}\begin{Bmatrix} v_2 \\ \theta_2 \\ v_3 \\ \theta_3 \end{Bmatrix} = \begin{Bmatrix} 120 \\ 0 \\ 0 \\ 0 \end{Bmatrix} \tag{7.8}$$

The above matrix equation is solved for unknown nodal DOFs, to yield

$$v_2 = v_3 = 0.04, \ \theta_2 = 0.06, \ \theta_3 = -0.06$$

Due to the constraint, it is expected that the vertical deflections at both nodes are identical. Due to symmetry in geometry, the rotations are equal and opposite sign. Note that the vertical deflection at Node 2 is eight times larger than that of welded joint in Example 4.6.

7.2.6   Patch Test

As we have discussed in the previous chapters, finite element solutions are often an approximation of the exact one. We also discussed that a more accurate solution can be obtained by reducing element size. In general, we expect that finite element solutions converge to the exact solution as the mesh is refined. Although this is true for most cases, it does not always apply. Some specially designed elements do not converge to the exact solution even if the mesh is gradually refined. To systematically address the convergent behavior of elements, we will discuss the requirements and methods of testing it.

For one-dimensional bars and two-dimensional plane solids, the strain energy involves the integral of first-order derivatives, from which we can deduce the following two requirements:

1. **Compatibility:** Displacements must be continuous across element boundaries— no gaps in materials. The derivatives of displacements, strains, must be continuous within the element, but not necessarily across element boundaries.

2. **Completeness:** The element should be able to represent rigid-body motions and constant strain conditions. For example, when an element is under rigid-body translation, there should be no strains developed by that motion.

If an element satisfies the above two requirements, it is called a *conforming* or *compatible* element. Elements that violate these conditions are called non-conforming or incompatible elements. When an element is conforming, the solution converges monotonically as the mesh is refined.

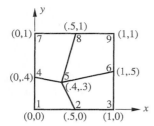

Figure 7.35 A patch of quadrilateral elements

A compatible element may become incompatible if a lower-order Gauss quadrature rule is used than is necessary for numerical integration of stiffness matrix. The extra zero-energy modes in Section 6.7 are examples of incompatibility. Thus, not only the element formulation, but also numerical integration can cause it.

To guarantee the convergence of the solution, the element must pass a test, called *patch test*, which is based on the requirements that the element should be able to represent rigid-body motions as well as constant strain conditions. The patch test is a simple test to determine if an element satisfies the basic convergence requirements. The patch test can find not only problems in the basic formulation but also problems in numerical integration or the computer implementation. The test procedure is as follows.

Consider a set of finite elements, called a patch, with at least one interior node, such as shown in Figure 7.35. It is important to make sure that the shape of the elements is not regular because some errors in the formulation may not appear in the elements with regular shape. Now we will discuss three different patch tests using these elements.

1. **Rigid-body motion test:** The first test checks if the elements can represent rigid-body motions. The displacements of all nodes on the boundary are prescribed corresponding to the rigid-body motion. With these prescribed displacements as boundary conditions, the displacements at the interior nodes are calculated. Since there are no applied loads, the calculated displacements at the interior nodes must be consistent with the rigid-body motion. If this in fact is the case, the element is said to pass the rigid-body patch test. For example, for the patch in Figure 7.35, consider x-translation of the boundary nodes i.e., $u_1 = u_2 = u_3 = u_4 = u_6 = u_7 = u_8 = u_9 = 1$, and the y-displacement of these nodes are all zero. Using these displacements, the finite element matrix equations are solved for unknown displacements at Node 5. To pass the rigid-body patch test, the computed horizontal displacement at Node 5 should be $u_5 = 1$ and $v_5 = 0$. Note that all three rigid-body motion cases (two translations and one rotation) must be verified before concluding the element passes the rigid-body patch test.

2. **Constant strain patch test:** The second test checks if the elements can represent constant strain conditions. Since strains are derivatives of displacements, constant strain conditions can be produced by linear displacements. First, the linearly varying displacements are applied to the nodes on the boundary and the displacements at the interior nodes are calculated. If the calculated displacements satisfy the constant strain conditions (i.e., linearly varying displacements), the element passes the constant strain patch test. For example, for the patch in Figure 7.35, consider the condition of $\varepsilon_{xx} = 1$. This strain condition can be achieved by prescribing displacements of nodes at the boundary given by $u(x) = x$ and $v(x) = 0$; i.e., nodal x-displacement is the same as its x-coordinate. In order to pass the test, the computed x-displacement at Node 5 must be equal to its x-coordinates; i.e., $u_5 = 0.4$ and $v_5 = 0.0$. Again, several different linear displacements must be tried before concluding that the element passes the patch test.

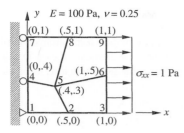

Figure 7.36 Generalized patch test for constant σ_{xx}

3. **Generalized patch test:** In the previous two patch tests, only displacements are specified along the boundary. Thus, this form of the test cannot verify errors associated with the applied loads at the boundary. A more general form of patch test is proposed to check the element formulation and implementation more thoroughly. In the generalized form, the patch of element is supported by minimum number of boundary conditions to prevent rigid body motion. For the remainder of the boundary nodes, a set of loads consistent with constant stress state in the element is applied. The computed stress state in the elements should obviously match with the assumed stress state. The procedure is illustrated for the patch in Figure 7.36. The objective is to test if the element can represent uniform stress of $\sigma_{xx} = 1$ Pa. For the minimum displacement boundary conditions, we apply $u_1 = v_1 = u_4 = u_7 = 0$. A distributed load is then applied on the right edge to produce a constant σ_{xx}. This distributed load must be converted to the equivalent nodal forces. The patch can now be analyzed using a finite element code. If the assumed stresses and displacements are recovered, the element passes the patch test. The generalized patch test requires a little more work but tests the implementation more thoroughly. Furthermore, the test does not need an interior node and therefore can be performed even on a single element.

Note that the patch test numerically tries to test whether linear and constant terms are present in the finite element approximation. If the shape functions are derived from an assumed polynomial and the analytical integration is performed, there is no need to go through the patch test. However, if shape functions are written by intuition, it may not be obvious if the necessary terms have been included. The numerical integration scheme and introduction of additional shape functions make the behavior of elements more complicated. The patch test has therefore become a standard tool in structural problems to test if the element has the necessary convergence properties.

Patch test for two-dimensional problems involves consideration of several cases. For example, Table 7.2 summarizes various cases of patch tests.

Table 7.2 Patch tests for plane solids ($E = 1$ GPa, $\nu = 0.3$)

Patch test type	u	v
x-translational rigid-body motion	1	0
y-translational rigid-body motion	0	1
xy-diagonal rigid-body motion	1	1
Constant $\sigma_{xx} = 200$ MPa	$0.2x$	$-0.06y$
Constant $\sigma_{yy} = 200$ MPa	$-0.06x$	$0.2y$
Constant $\tau_{xy} = 100$ MPa	$0.26y$	0

EXAMPLE 7.5   *Patch Test for Bar Elements*

In one-dimensional bar elements, there is only one rigid-body motion test (x-translation) and one constant strain test (ε_{xx} = constant). Consider a patch of two bar elements shown in Figure 7.37 with $E = 10\,\text{GPa}$, $A = 10^{-6}\,\text{m}^2$. Show that the element passes the two patch tests using the following shape functions for Element (i):

$$N_1(x) = \frac{x_{i+1} - x}{L^{(i)}}, \quad N_2(x) = \frac{x - x_i}{L^{(i)}}$$

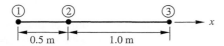

         Figure 7.37 Patch test for bar elements

SOLUTION   Element stiffness matrices

Element 1: $[\mathbf{k}^{(1)}] = 10^4 \begin{bmatrix} 2 & -2 \\ 2 & -2 \end{bmatrix}$

Element 2: $[\mathbf{k}^{(2)}] = 10^4 \begin{bmatrix} 1 & -1 \\ 1 & -1 \end{bmatrix}$

Using the standard assembly procedure, the equations for the patch can be written as follows.

$$10^4 \begin{bmatrix} 2 & -2 & 0 \\ -2 & 3 & -1 \\ 0 & -1 & 1 \end{bmatrix} \begin{Bmatrix} u_1 \\ u_2 \\ u_3 \end{Bmatrix} = \begin{Bmatrix} F_1 \\ F_2 \\ F_3 \end{Bmatrix}$$

(a) Rigid-body motion test: In the rigid-body motion test, the displacements at the boundary nodes are prescribed. In this case, we can specify $u_1 = u_3 = 1$ and need to check if $u_2 = 1$. By applying displacement boundary conditions, we have

$$10^4 \begin{bmatrix} 2 & -2 & 0 \\ -2 & 3 & -1 \\ 0 & -1 & 1 \end{bmatrix} \begin{Bmatrix} 1 \\ u_2 \\ 1 \end{Bmatrix} = \begin{Bmatrix} R_1 \\ 0 \\ R_3 \end{Bmatrix}$$

where R_1 and R_3 are unknown reaction force in order to impose the displacement boundary conditions. From the second equation, we have

$$10^4(-2 \cdot 1 + 3 \cdot u_2 - 1 \cdot 1) = 0 \quad \Rightarrow \quad u_2 = 1$$

Thus, the element passes the patch test for rigid-body motion.

(b) Constant strain test: In order to produce the constant strain, the linear displacements are assumed. We prescribe $u(x) = x$ for the nodes at the boundaries; i.e., $u_1 = 0$ and $u_3 = 1.5$. We expect $u_2 = 0.5$ in order to pass the patch test. Then, the matrix equation becomes

$$10^4 \begin{bmatrix} 2 & -2 & 0 \\ -2 & 3 & -1 \\ 0 & -1 & 1 \end{bmatrix} \begin{Bmatrix} 0 \\ u_2 \\ 1.5 \end{Bmatrix} = \begin{Bmatrix} R_1 \\ 0 \\ R_3 \end{Bmatrix}$$

From the second equation, we have

$$10^4(-2 \cdot 0 + 3 \cdot u_2 - 1 \cdot 1.5) = 0 \quad \Rightarrow \quad u_2 = 0.5$$

Thus, the element passes the patch test for constant strain.

7.2.7   Estimating Errors

Error estimation is important in order to check the accuracy of the current analysis results. It is also used as a criterion of mesh refinement. There are many error estimation methods, but we will discuss error estimation based on stress results. We will explain the error estimation using uniaxial stress case. However, the same idea can be generalized to three-dimensional stress state. Let σ be the stress at a node (usually extrapolated from the stress at the Gauss quadrature points) and σ^* be the continuous, averaged stress at node location. If analysis results are accurate, there should no significant difference between σ and σ^*. Thus, the error estimation is based on the difference between calculated stress and averaged stress.

$$\sigma_E = \sigma - \sigma^* \tag{7.9}$$

Since we are interested in the error in the entire model, we use the strain energy. The objective is to make the strain energy from σ_E small compared to the strain energy from the original stress. We define the two strain energies as

$$U = \sum_{e=1}^{NE} \int_{V^{(e)}} \frac{\sigma^2}{2E} dV \tag{7.10}$$

$$U_E = \sum_{e=1}^{NE} \int_{V^{(e)}} \frac{\sigma_E^2}{2E} dV \tag{7.11}$$

Note that the strain energy is written in terms of stress not strain. Equation (7.10) is regular strain energy of a structure under deformation, while Eq. (7.11) is the contribution from the error in stresses. Then, the accuracy is measured as a ratio between U_E and U. In particular, since strain energy contains square of stresses, we define

$$\eta = \sqrt{\frac{U_E}{U + U_E}} \tag{7.12}$$

It is suggested that the current mesh size is considered to be appropriate, if $\eta \approx 0.05$. The error estimate is often used for the adaptive mesh refinement. The mesh is gradually refined until η in Eq. (7.12) is less than 0.05.

7.3   PROJECT

Figure 7.38 shows a rough design of a bracket that carries a load at the circular hole whose diameter is 40 mm and is attached to the wall at the other end. Optimize its design to minimize weight subject to constraints so that it should not deform plastically. The final

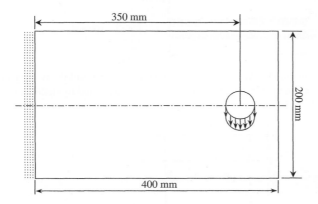

Figure 7.38 Design domain and boundary condition for the bracket

design must fit within the $400 \times 200\,\text{mm}^2$ box shown in Figure 7.38. The part is cast and then machined. Therefore, it is preferred that it is not less than 10 mm wide anywhere. The bracket is made of Aluminum 6061, which has Young's modulus $E = 69\,\text{GPa}$, Poisson's ratio $v = 0.3$, and yield strength $\sigma_Y = 378\,\text{MPa}$. Assume that the bracket has a uniform thickness of 10 mm. Use the safety factor of 2.0. The bracket has to support a resultant maximum load $F = 15{,}000\,\text{N}$. Write a project report describing the objective, preliminary analysis, finite element analysis results, and conclusion of your project. Carry out the analysis using both triangular and quadrilateral elements and turn in plots (mesh showing boundary conditions, deformed shape, stress plots) to justify the validity of your design.

7.4   EXERCISE

1. Solve the frame in Example 7.1 using one of the finite element programs given in the Appendix. Draw the axial force, bending moment, and shear force diagram. Compare the maximum stress value and its location from the preliminary analysis.

2. Consider a cantilevered beam shown in the figure. Solve the problem using: (a) five beam elements, (b) 4×20 plane stress solid elements, and (c) $4 \times 4 \times 20$ hexahedral elements. Compare the maximum stress and tip deflection. Assume $E = 72\,\text{GPa}$ and $v = 0.3$.

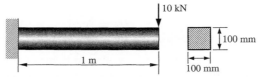

3. Consider a simply supported beam shown in the figure. Solve the problem using plane stress solids. Perform a convergence study for displacement at the center and maximum tensile stress. Increase the mesh size by a factor of two starting from an initial mesh size of 2×10. Compare the results with the exact solution. Assume $E = 72\,\text{GPa}$ and $v = 0.3$.

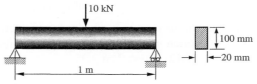

4. Consider a cantilevered beam shown in the figure. Solve the problem using plane stress solids. Perform a convergence study for displacement at the tip and maximum tensile stress. Increase the mesh size by a factor of two starting from an initial mesh size of 2×10. Compare the results with exact solution. Assume $E = 72\,\text{GPa}$ and $v = 0.3$.

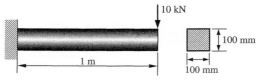

5. A cantilevered beam is modeled using: (a) two rectangular elements and (b) two quadrilateral elements. Compare the accuracy of analysis results with exact stress and tip deflection. Assume $E = 72\,\text{GPa}$ and $v = 0.3$.

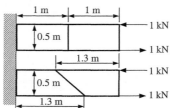

6. Using symmetric properties, draw an equivalent, simplified geometry of the structure shown in the figure with appropriate boundary conditions. Can the original geometry be solved using FEM? Can the simplified geometry be solved using FEM? Explain your answers.

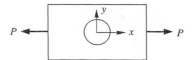

7. Using symmetric properties, draw an equivalent, simplified geometry of the beam shown in the figure with appropriate boundary conditions and applied loads.

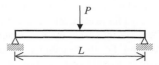

8. Using symmetry, find deflection, bending moment, and shear force in a continuous beam shown in the figure when $F = 8\,\text{kN}$ is applied at the center. Assume $E = 200\,\text{GPa}$ and $I = 10^5\,\text{mm}^4$.

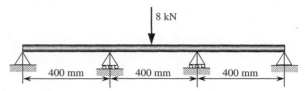

9. Using symmetry, find deflection, bending moment, and shear force in a continuous beam shown in the figure. Assume $E = 200\,\text{GPa}$ and $I = 10^5\,\text{mm}^4$.

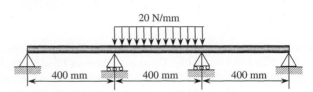

10. Consider a plane structure shown in the figure. Using symmetric modeling, draw the smallest geometry, and corresponding forces and boundary conditions that can provide the same analysis results as the original problem. Consider two cases:

(a) when P_x and P_y are different

(b) when $P_x = P_y = P$

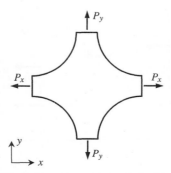

11. In one-dimensional bar elements, there is only one rigid-body motion test (x-translation) and one constant strain test ($\varepsilon_{xx} = $ constant). Consider a patch of two bar elements shown in the figure with $E = 10\,\text{GPa}$, $A = 10^{-6}\,\text{m}^2$. Check if the element passes the two patch tests using the following shape functions for Element (i):

$$N_1(x) = \frac{x_{i+1}^2 - x^2}{x_{i+1}^2 - x_i^2}, \quad N_2(x) = \frac{x^2 - x_i^2}{x_{i+1}^2 - x_i^2}$$

12. Patch test is often used to ensure that the solutions from the finite element method converge to the exact solution as the finite element mesh is refined. In the generalized form of the patch test, the patch of elements is supported by a minimum number of boundary conditions to prevent rigid body displacement. For the remainder of the boundary nodes, a set of load consistent with constant stress state in the element is applied. The computed stress state should obviously match with the assumed stress state. Let a structure be approximated by four plane stress finite elements with thickness $= 0.1$, as shown in the figure with $E = 1000$ and $\nu = 0.3$.

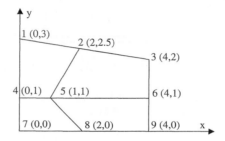

(a) For all boundary nodes (1, 2, 3, 4, 6, 7, 8, 9), apply displacement boundary conditions of $u = 0.2x$ and $v = -0.06y$. Using a commercial FE software, check the analysis results with the analytical solution of a constant stress field: $\sigma_{xx} = 200$, $\sigma_{yy} = 0$, and $\tau_{xy} = 0$. You need to make sure that incompatible shape functions are removed in the analysis. Provide a discussion of patch test results.

(b) Instead of applying displacement boundary condition, apply distributed traction forces that represent a constant stress field $\sigma_{xx} = 200$. Convert the distributed traction forces into work-equivalent nodal forces. In the figure shown below, draw your nodal force vectors showing their directions and magnitudes. Carry out FE analysis with these nodal forces and verify whether all elements have the constant stress field.

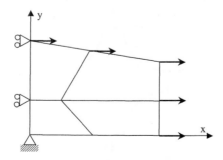

13. A rectangular solid under plane-stress conditions is modeled using four finite elements as shown in the figure. It has been proved that the finite elements pass the patch test. (a) When a

constant displacement $u = 0.01$ m is applied to nodes 3, 6, and 9, calculate the displacement $\{u, v\}$ at node 5, whose initial coordinates are $(0.4, 0.4)$. (b) Explain how passing the patch test is important for the element's performance.

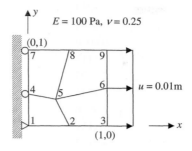

14. A cantilever beam can be analyzed using the three different finite element models shown in the figure.

    (a) Rank these three models based on which one you expect to give results closest to the analytic solution for the displacement at the tip of the beam (point of application of load). Give a brief explanation for the ranking you assign.

    (b) Plot the normal stress distribution you expect to compute at a section near the center of the beam for each model. Explain why you expect these plots.

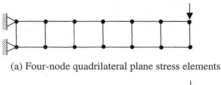

(a) Four-node quadrilateral plane stress elements

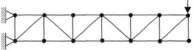

(b) Three-node triangular plane stress elements

(c) Single two-node hermite beam element

15. A thick cylinder is under the internal pressure $p = 1,000$ psi as illustrated in the figure. Young's modulus is 10^6 psi and Poisson's ratio is 0.2. Assuming plane strain conditions, calculate nodal displacements, element stress components, principal stresses, and von Mises stress. Solve 1/4 model using symmetry boundary conditions. Nodal coordinates and element connectivity are given in the following table.

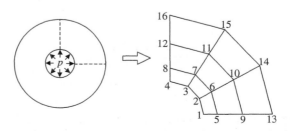

Node	x	y
1	2.0000	0.0
2	1.7052	1.0349
3	1.0349	1.7015
4	0.0000	2.0000
5	3.1111	0.0000
6	2.8105	1.7734
7	1.7734	2.8105
8	0.0000	3.1111
9	5.1111	0.0000
10	4.5120	2.8083
11	2.8083	4.5120
12	0.0000	5.1111
13	8.0000	0.0000
14	6.8061	4.1394
15	4.1394	6.8061
16	0.0000	8.0000

Element	Node 1	Node 2	Node 3	Node 4
1	1	5	6	2
2	2	6	7	3
3	3	7	8	4
4	5	9	10	6
5	6	10	11	7
6	7	11	12	8
7	9	13	14	10
8	10	14	15	11
9	11	15	16	12

Chapter 8

Structural Design Using Finite Elements

8.1 INTRODUCTION

Finite element analysis is concerned with determining the response (displacements and stresses) of a given structure for a given set of loads and boundary conditions. It is an analysis procedure in which the structure – its geometry, material properties, boundary conditions and loads – is well defined, and the goal is to determine its response. On the other hand, engineering design is a process of synthesis in which parts are put together to build a structure that will perform a given set of functions satisfactorily. Analysis is very systematic and can be taught easily. Design is an iterative process. Clearly, analysis is one of the several steps in the design process because we use analysis to evaluate the adequacy of the design. In this chapter, we will discuss the use of finite element analysis in the process of designing a structure.

There are two general approaches to design: creative design and adaptive design. The former is concerned with creating a new structure or machine that does not exist, whereas the latter is concerned with modifying an existing design. Although analysis techniques such as FEA play a crucial role, the designer's experience and creative ideas are important for the former. Adaptive design is an evolutionary process, and is encountered much more frequently in practice. For example, how many times does an automotive company design a new car from scratch? The majority of an engineer's work concentrates on improving the existing vehicle so that the new car will be more comfortable, more durable, safer and fuel-efficient. In this chapter, we will discuss the role finite element analysis in the adaptive design process.

Structural design is a procedure to improve or enhance the performance of a structure by changing its parameters. *Performances* can be quite general in engineering fields and can include the weight, stiffness or compliance; the fatigue life; noise and vibration levels; safety, etc. However, the performance does not include such aesthetic measure as attractiveness. The performance metrics are measurable quantities. We are especially interested in the performances that are obtained from finite element analysis results.

In structural design, two different types of performances are often considered. The first type is related to the criteria that the system must satisfy. As long as the performance satisfies the criteria, its level is not important. In engineering design, this type of performance is called a *constraint*. For example, the allowable stress is often used as a constraint in structural design so that the stresses are less than the allowable stress. The second type, called *goal*, is the performance that the engineer wants to improve as much as possible. The total weight of the structure or cost of manufacturing is an example of goal.

System parameters are variables that the engineer can change during the design process. For example, the thickness of a vehicle body panel can be changed to improve the stiffness of the vehicle. The cross-section of a beam can be changed in designing a bridge structure. System parameters that can be changed during the design process are called *design variables*. The design variables can include the plate thickness, cross-sectional dimensions, location and size of cutouts, and shape of the structure. In this chapter, we will learn how to change the design variables to improve the goal while satisfying system constraints.

A set of design variables that satisfies the constraints is called a *feasible design*, while a set that does not satisfy constraints is called an *infeasible design*. It is difficult to determine whether a current design is feasible unless the structural problem is analyzed. For complicated structural problems, it may not be easy to choose appropriate design constraints so that the feasible region is not empty.

Since this chapter is only an introduction to structural design, we will first present conventional design approaches using safety margin and intuitive design in the first two sections. Design parameterization in Section 8.4 deals with the definition of design variables. The topics of parametric study and sensitivity analysis in Section 8.5 investigate the effect of change in design variables on the performance. Section 8.6 introduces structural optimization, which is a mathematical tool to find the best design.

8.2   SAFETY MARGIN IN DESIGN

In Chapter 1, we defined the concept of factor of safety and its application in determining if a structure is safe or not. For a given loading condition, the structural analysis determines the stresses, and the ratio between the failure stress and the actual stress is defined as the factor of safety. In this section, the same concept is used for the purpose of design; i.e., we want to design a structure that has a given level of safety factor.

8.2.1   Factor of Safety

In engineering design, the basic strength-based approach is through a factor of safety. For example, a structure should satisfy the following constraint:

$$\sigma_i(\mathbf{x}) \leq \sigma_{\text{allowable}} \tag{8.1}$$

where $\sigma_i(\mathbf{x})$ is the applied stress calculated to act at a point $\mathbf{x}$ in the i-th structural component, and $\sigma_{\text{allowable}}$ is the *allowable stress* for the component. The allowable stress is usually defined in the structural design code. It is derived from material strength (ultimate strength or failure strength, etc) but reduced by a factor S_F:

$$\sigma_{\text{allowable}} = \sigma_F / S_F \tag{8.2}$$

where S_F is the *factor of safety*, which considers the effect of material variability, experimental errors, etc. The structure is considered failed when any part of the structure reaches the allowable stress. The actual failure, however, depends on how well $\sigma_i(\mathbf{x})$ represents the actual stress in the real structure and how well $\sigma_{\text{allowable}}$ represents the actual material failure. The stress calculated from finite element analysis may be differ from the actual stress due to various errors and also various assumptions made in the model, and the ultimate strength might be beyond linear region of the material behavior. From the viewpoint of safety, these errors are not important as long as Eq. (8.1) is a conservative safety measure. Conservativeness is the main difference between analysis and design. As long as the analysis result is conservative, the accuracy in design is not as important as in analysis.

Although we explained the factor of safety using the strength constraint, the above safety constraint can be applied to different types of performances. The above safety constraint can be restated that the response (R_i) of a member should be less than the capacity (C_i) of the member material. In the case of strength constraint, the applied stress is the response, while the ultimate strength is the capacity. Then, we can extend Eq. (8.1) to general performances as

$$R_i(\mathbf{x}) \le \frac{C_i}{S_F} \tag{8.3}$$

When multiple loads are applied simultaneously, the response R_i should be the combination of the effects of all loads. Typically,

$$R_i = \sum_{j=1}^{N_L} R_i(F_j) \tag{8.4}$$

where F_j is the jth load and N_L is total number of applied loads.

Note that the safety constraint in Eq. (8.3) is applied to a point $\mathbf{x}$ of ith member, and it must be satisfied everywhere. In practice, the location at which the response is critical is considered.

An alternative and useful measure of safety is the safety margin, which measures the excess capacity compared with the response; thus

$$Z_i = C_i - R_i \tag{8.5}$$

In the case of the strength constraint, the member can afford additional Z_i stress before it fails.

From a different viewpoint, we can define a sufficiency factor S_i as a ratio of the allowable capacity to the response. Then, the safety constraint can be restated, as

$$\frac{C_i}{S_F R_i} = S_i \ge 1 \tag{8.6}$$

For example, a sufficiency factor 0.8 means that R_i has to be multiplied by 0.8 or C_i is divided by 0.8 so that the sufficiency factor increases to one. In other words, this means that R_i has to be decreased by 20% $(1 - 0.8)$ or C_i has to be increased by 25% $(1/0.8 - 1)$ in order to achieve the required safety. The sufficient factor is automatically normalized in terms of the capacity. The sufficiency factor is useful in estimating the resources needed to achieve the required safety factor. For example, if the current design has the sufficiency factor of 0.8, then this indicates that maximum stresses must be lowered by 20% to meet the safety. This permits the engineers to readily estimate the weight required to reduce stresses to a given level.

8.2.2 Load Factor

Instead of reducing the capacity by the factor of safety, it is possible to increase the applied loads, which is basic idea of the load factor. The *load factor* λ is the factor by which a set of loads acting on the structure must be multiplied just enough to cause the structure to fail. Commonly, the loads are taken as those acting on the structure during service conditions. For ductile materials the strength of the structure is determined from the idealized elastic-perfectly plastic material model. For brittle materials the tensile strength is used as the allowable stress.

A set of working loads F_j are applied to the structure and the structure is safe. For a given collapse mode (i.e., for a given ultimate strength), the structure is considered to have "failed" or collapsed when the capacity C_i is related to the factored load λF_j by

$$R_i(\lambda \mathbf{F}) = \mathbf{C_i} \tag{8.7}$$

where $\mathbf{F}$ is the vector of all applied loads. The load factor is the scale of the applied loads such that the response becomes equal to the capacity. If proportional loading is assumed, R_i is a linear function and the load factor can be taken out of parentheses. Then, Eq. (8.7) can be written in the following form:

$$\frac{C_i}{R_i(\mathbf{F})} = \lambda \tag{8.8}$$

In the structural analysis viewpoint, the load factor is the ratio between capacity and response. From the design viewpoint, the structure should be designed to satisfy a given level of load factor.

Clearly, there is much similarity in formulation between the factor of safety and the load factor as measures of structural safety. The difference is the reference level at which the two measures operate: the first at the level of working loads level, the second at the level of collapse loads.

EXAMPLE 8.1   *Factor of Safety*

Consider a cantilevered beam shown in Figure 8.1 with $E = 2.9 \times 10^4$ ksi. The width w is taken as 2.25 inches, while the height h is considered a design variable. The design should satisfy the factor of safety 1.5. (a) When the allowable tip displacement is $D_{\text{allowable}} = 2.5$ in, determine the height of the beam. (b) When the failure stress of the material is 40 ksi, determine the height of the beam.

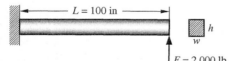

$F = 2,000$ lb   **Figure 8.1** Cantilevered beam design

SOLUTION

(a) We use one element to model the beam. From Chapter 4 the matrix equation can be written as

$$\frac{EI}{L^3}\begin{bmatrix} 12 & 6L & -12 & 6L \\ 6L & 4L^2 & -6L & 2L^2 \\ -12 & -6L & 12 & -6L \\ 6L & 2L^2 & -6L & 4L^2 \end{bmatrix} \begin{Bmatrix} v_1 = 0 \\ \theta_1 = 0 \\ v_2 \\ \theta_2 \end{Bmatrix} = \begin{Bmatrix} R_1 \\ C_1 \\ F \\ 0 \end{Bmatrix} \tag{8.9}$$

where R_1 and C_1 are the supporting force and couple at the wall. After deleting the first and second columns and rows, we obtain

$$\frac{EI}{L^3}\begin{bmatrix} 12 & -6L \\ -6L & 4L^2 \end{bmatrix}\begin{Bmatrix} v_2 \\ \theta_2 \end{Bmatrix} = \begin{Bmatrix} F \\ 0 \end{Bmatrix} \tag{8.10}$$

The solutions of the above equation becomes

$$v_2 = \frac{4FL^3}{Ewh^3}, \quad \theta_2 = \frac{6FL^2}{Ewh^3}$$

Since the allowable deflection is given, we do not need safety factor. Thus, we have

$$v_2 = \frac{4FL^3}{Ewh^3} = D_{\text{allowable}} \Rightarrow h = \sqrt[3]{\frac{4FL^3}{EwD_{\text{allowable}}}} = 3.66 \, \text{in}$$

(b) The bending moment can be calculated from the second row of Eq. (8.9). The supporting moment at the wall is

$$C_1 = \frac{EI}{L^3}[6Lv_1 + 4L^2\theta_1 - 6Lv_2 + 2L^2\theta_2] = -FL$$

Thus, the bending moment at the wall is $M = FL$, which is consistent with the result from elementary mechanics. Then, the maximum stress at the wall becomes

$$\sigma_{\max} = \frac{M\frac{h}{2}}{I} = \frac{6FL}{wh^2}$$

Since the safety factor is given as 1.5, the failure stress is divided by the safety factor and compared to the maximum stress to calculate height, as

$$\frac{6FL}{wh^2} = \frac{\sigma_F}{S_F} \Rightarrow h = \sqrt{\frac{6FLS_F}{w\sigma_F}} = 4.47 \, \text{in}$$

Thus, the strength requirement is severer than the displacement requirement. Note that in this case we can use load factor $\lambda = 1.5$ instead of the factor of safety because Part (b) has only one load case and one performance.

8.3   INTUITIVE DESIGN: FULLY-STRESSED DESIGN

In Example 8.1 we were able to determine the height of the beam that satisfies the stress or deflection constraint because the constraint was explicitly written in terms of the design variable. For more complex structures, it is impractical to have this explicit relationship. Instead, we can evaluate the values of performances at given design variables. In such a case, an iterative process can be used to find the best design. Starting from initial values of design variables, we can update the design variables according to the calculated performances. In the case of the beam design problem, for example, if the stress is too high, we increase the height gradually. We can repeat this process until the stress becomes just right. The abovementioned process is called *intuitive design*.

When structures are subject only to stress and minimum gage constraints, the *fully-stressed design* (FSD) can be very useful to find the best design. The basic concept is as follows:

> For the best design, each member of the structure that is not at its minimum gage is fully stressed under at least one of the design load conditions.

This basic concept implies that we should remove material from members that are not fully stressed unless prevented by minimum gage constraints. This appears reasonable, but it is based on an implicit assumption that the primary effect of adding or removing material from a structural member is to change the stresses in that member. If this assumption is not true, that is, if adding material to one part of the structure can have large effects on the stresses in other parts of the structure, we may want to have members that are not fully stressed because they help to relieve stresses in other members.

For *statically determinate* structures, the assumption that adding material to a member influences primarily the stresses in that member is correct. In fact, without inertia or thermal loads there is no effect at all on stresses in other members. Therefore, we can expect that the FSD criterion will hold at the minimum weight design for such structures. However, for statically indeterminate structures, the minimum weight design may not be fully stressed. In most cases of a structure made of a single material, there is a fully stressed design near the optimum design, and so the method has been extensively used for metal structures. The FSD method may not do as well when several materials are used.

The FSD technique is usually complemented by a resizing algorithm based on the assumption that the load distribution in the structure is independent of member sizes. That is, the stress in each member is calculated, and then the member is resized to bring the stresses to their allowable values, assuming that the loads carried by members remained constant (this is logical since the FSD criterion is based on a similar assumption). For example, for truss structures, where the design variables are often cross-sectional areas, the force in any member is $\sigma \cdot A$ where σ is the axial stress and A the cross-sectional area. Assuming that $\sigma \cdot A$ is constant leads to the following stress ratio resizing technique:

$$A_{\text{new}} = A_{\text{old}} \left| \frac{\sigma}{\sigma_{\text{allowable}}} \right| \tag{8.11}$$

which gives the resized area A_{new} in terms of the current area A_{old}, the current stress σ, and the allowable stress $\sigma_{\text{allowable}}$. For a statically determinate truss, the assumption that member forces are constant is exact, and Eq. (8.11) will bring the stress in each member to its allowable value. If the structure is not statically determinate, Eq. (8.11) has to be applied repeatedly until convergence to any desired tolerance is achieved. Also, if A_{new} obtained by Eq. (8.11) is smaller than the minimum gage, the minimum gage is selected rather than the value given by Eq. (8.11).

Equation (8.11) works well for truss structures and under the assumption that the member force does not change significantly as the cross-sections change. In the case of beam element, we can obtain a similar stress ratio resizing technique. In that case, we use the section modulus. The maximum stress in the beam occurs at the top or bottom of the cross-section. Thus,

$$\sigma = \frac{M \frac{h}{2}}{I} = \frac{M}{S} \tag{8.12}$$

where S is the section modulus. Assuming that the bending moment remains constant during the design process, we can obtain the following stress ratio resizing technique:

$$S_{\text{new}} = S_{\text{old}} \left| \frac{\sigma}{\sigma_{\text{allowable}}} \right| \tag{8.13}$$

Note that the difference compared to the truss structure is the use of section modulus rather than cross-sectional area. Again, if the bending moment remains constant (i.e., statically determinate), then Eq. (8.13) will yield the fully-stress design in one iteration.

EXAMPLE 8.2   *Fully-Stressed Design of a Cantilevered Beam*

Consider the cantilevered beam in Figure 8.1. Let the initial height of the cross-section be 3.5 in. Using fully-stressed design, calculate the new height so that the maximum stress is equal to $\sigma_{\text{allowable}}$.

SOLUTION   First, we calculate the section modulus and the maximum stress at the initial design:

$$S_{\text{old}} = \frac{2I}{h} = \frac{wh^2}{6} = \frac{2.25 \times 3.5^2}{6} = 4.594 \, \text{in}^3$$

$$\sigma_{\max} = \frac{M}{S_{\text{old}}} = 43.537 \, \text{ksi}$$

Thus, the new section modulus can be obtained using the stress ratio resizing technique, as

$$S_{\text{new}} = S_{\text{old}} \frac{\sigma_{\max}}{\sigma_{\text{allowable}}} = 4.594 \times \frac{43.537}{26.667} = 7.5 \, \text{in}^3$$

Thus, the new height can be obtained from the definition of the section modulus, as

$$S_{\text{new}} = \frac{wh^2}{6} \Rightarrow h = \sqrt{\frac{6S_{\text{new}}}{w}} = 4.47 \, \text{in}$$

Note that the solution is identical with that of Example 8.1. Thus, the fully-stressed design converges to the optimum design in one iteration for the statically determinate system.

EXAMPLE 8.3   *Fully-Stressed Design of Three-Bar Truss*

The goal of three-bar truss in Figure 8.2 is to find the cross-sectional areas so that the weight of the truss is minimal while the stresses in all members are less than the yield strength with the factor of safety 2.0. Perform one iteration using fully-stressed design. The current cross-sectional areas are $b_1 = b_2 = b_3 = 10 \, \text{mm}^2$. Use Length $L = 1 \, \text{m}$, Young's modulus 80 GPa, and yield stress $\sigma_Y = 250 \, \text{MPa}$.

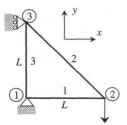

1,000 N   **Figure 8.2** Three-bar truss for fully-stressed design

SOLUTION   The element table is shown below.

Element	First Node i	Second Node j	AE/L	l	m
1	1	2	Eb_1	1	0
2	2	3	$Eb_2/\sqrt{2}$	$-1/\sqrt{2}$	$1/\sqrt{2}$
3	1	3	Eb_3	0	1

Using the above element table, the global element equations can be obtained after applying displacement boundary conditions:

$$10^5 \begin{bmatrix} 10.83 & -2.83 & 2.83 \\ -2.83 & 2.83 & -2.83 \\ 2.83 & -2.83 & 10.83 \end{bmatrix} \begin{Bmatrix} u_2 \\ v_2 \\ v_3 \end{Bmatrix} = \begin{Bmatrix} 0 \\ -1000 \\ 0 \end{Bmatrix} \qquad (8.14)$$

By solving the above equation we have the following unknown nodal degrees of freedom (DOFs):

$$u_2 = -1.25 \, \text{mm}, \quad v_2 = -6.04 \, \text{mm}, \quad v_3 = -1.25 \, \text{mm} \qquad (8.15)$$

The stress in each element can be found from

Element 1: $\qquad \sigma^{(1)} = \dfrac{E}{L^{(1)}} \left[l^{(1)}(u_2 - u_1) - m^{(1)}(v_2 - v_1) \right] = -100 \, \text{MPa}$

Element 2: $\qquad \sigma^{(2)} = \dfrac{E}{L^{(2)}} \left[l^{(2)}(u_3 - u_2) - m^{(2)}(v_3 - v_2) \right] = 141.4 \, \text{MPa}$

Element 3: $\qquad \sigma^{(3)} = \dfrac{E}{L^{(3)}} \left[l^{(3)}(u_3 - u_1) - m^{(3)}(v_3 - v_1) \right] = -100 \, \text{MPa}$

Now, using the stress ratio test, the new area of each element can be obtained, as

Element 1: $\qquad b_{\text{new}}^{(1)} = b_{\text{new}}^{(1)} \left| \dfrac{\sigma^{(1)}}{\sigma_{\text{allowable}}} \right| = 10 \left| \dfrac{-100}{250/2} \right| = 8 \, \text{mm}^2$

Element 2: $\qquad b_{\text{new}}^{(2)} = b_{\text{new}}^{(2)} \left| \dfrac{\sigma^{(2)}}{\sigma_{\text{allowable}}} \right| = 10 \left| \dfrac{141.4}{250/2} \right| = 11.31 \, \text{mm}^2$

Element 3: $\qquad b_{\text{new}}^{(3)} = b_{\text{new}}^{(3)} \left| \dfrac{\sigma^{(3)}}{\sigma_{\text{allowable}}} \right| = 10 \left| \dfrac{-100}{250/2} \right| = 8 \, \text{mm}^2$

Using the new element areas, the new global element matrix equation can be obtained after applying displacement boundary conditions, as

$$10^5 \begin{bmatrix} 9.6 & -3.2 & 3.2 \\ -3.2 & 3.2 & -3.2 \\ 3.2 & -3.2 & 9.6 \end{bmatrix} \begin{Bmatrix} u_2 \\ v_2 \\ v_3 \end{Bmatrix} = \begin{Bmatrix} 0 \\ -1000 \\ 0 \end{Bmatrix}$$

By solving the above equation, we have the following unknown nodal DOFs:

$$u_2 = -1.6 \, \text{mm}, \quad v_2 = -6.2 \, \text{mm}, \quad v_3 = -1.6 \, \text{mm}$$

The stress in each element can be found from

Element 1: $\qquad \sigma^{(1)} = \dfrac{E}{L^{(1)}} \left[l^{(1)}(u_2 - u_1) - m^{(1)}(v_2 - v_1) \right] = -125 \, \text{MPa}$

Element 2: $\qquad \sigma^{(2)} = \dfrac{E}{L^{(2)}} \left[l^{(2)}(u_3 - u_2) - m^{(2)}(v_3 - v_2) \right] = 125 \, \text{MPa}$

Element 3: $\qquad \sigma^{(3)} = \dfrac{E}{L^{(3)}} \left[l^{(3)}(u_3 - u_1) - m^{(3)}(v_3 - v_1) \right] = -125 \, \text{MPa}$

Now, all three members are in the allowable stress. Thus, the fully-stressed design converged to the optimum design in one iteration.

8.4 DESIGN PARAMETERIZATION

In the previous sections, we discussed simple design processes that can be applied to design variables that are related to cross-sectional geometry. However, when more complicated structures, such as plane solids or three-dimensional solids, are considered, it may not be trivial to define design variables. In this section, we discuss various types of design variables and procedures to define them.

Selecting design variables is called design parameterization. The design variables are assumed to vary during the design process. In some cases, it is relatively simple to choose them from analysis parameters and to vary their values. In other cases, however, it may not be easy because changing the design variables involves modifying finite element mesh. Based on their role in finite element analysis, we will discuss three different types of design variables.

Material property design variables: In structural analysis, material properties are used as a parameter. Young's modulus and Poisson's ratio, for example, are required in the linear elastic problem. If these material properties are subject to change, then they are called *material property design variables*. These kinds of design variables do not appear in regular design problems, since in most cases material properties are presumed to be constant. Analysis using constant material properties is called a deterministic approach. Another approach assumes that material properties are not constant but distributed within certain ranges. This is called probabilistic approach and is more practical, since tests with different material properties usually yield different results. In this case, material properties are no longer considered to be constant and can therefore be used as design variables.

Sizing design variables: Sizing design variables are related to geometric parameters of a structure, and they are often called parametric design variables. For example, most automotive and airplane parts are made from plate/shell components. It is natural that engineers want to vary the thicknesses (or gauge) of the plate/shell to reduce the weight of the vehicle. In that case, the plate thicknesses are sizing design variables. During structural analysis, the thicknesses are considered as a parameter. The sizing design variable is similar to the material property design variable in the sense that both variables change analysis parameters, not the geometry of the structure.

   Another important type of sizing design variable is the cross-sectional geometry of bars and beams. Figure 8.3 provides some examples of the shapes and parameters that define these cross-sections. In the structural analysis of bars, for example, the cross-sectional area is required in the axial rigidity. If a rectangular cross-section is used, then the area would be defined as $A = b \times h$. Thus, two parameters, b and h, can be considered design variables. Note that these variables contribute to the cross-sectional area for bars and the moment of inertia for beams.

Shape design variables: While material properties and sizing design variables are related to the parameters of structural analysis, shape design variables are related to the structure's geometry. The difficulty in this approach is that the shape of the structure does not explicitly appear as a parameter. Although the design variables in Figure 8.3 determine the cross-sectional shape, they are not shape design variables, since these cross-sectional shapes are considered parameters in structural analysis. However, the length of bars or beams should be treated as a shape design variable. Usually, the shape design variable defines the domain of integration in structural analysis. Thus, it is not convenient to extract shape design variables from a structural model and to use them as sizing design variables.

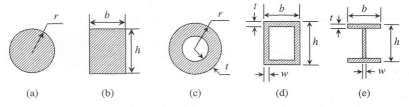

(a)   (b)   (c)   (d)   (e)

Figure 8.3 Sizing design variables for cross-sections of bars and beams; (a) Solid circular cross section; (b) Rectangular cross section; (c) Circular tube; (d) Rectangular tube; (e) I-section

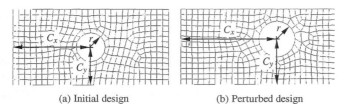

(a) Initial design   (b) Perturbed design

Figure 8.4 Shape design variables in plate with a hole

Consider a rectangular block with a hole, as shown in Figure 8.4. The location and size of the hole is determined by the geometric values of C_x, C_y, and r, which are shape design variables. Different values of shape design variables yield different structural shapes. However, these shape design variables do not explicitly appear in structural analysis. If finite element analysis is used to perform structural analysis, then the shape design variables change the mesh, as shown in Figure 8.4. Note that the mesh remains constant in the case of material properties and sizing design variables. Thus, the shape design problem is more difficult to solve than the sizing design problem.

It is important to note that inappropriate parameterization can lead to unacceptable shapes. This includes not only design variables, but also the range of designs. For example, the ranges in C_x, C_y, and r should be limited such that the hole remains inside of the rectangle in Figure 8.4.

Shape design parameterization describes the boundary shape of a structure as a function of design variables. There are many different methods, but we will only discuss two methods: iso-parametric mapping method and solid model-based parameterization. Both methods have advantages and disadvantages. The first method works well with the mapped mesh in which the topological mesh remains constant throughout the design process. Only the physical mesh changes according to shape designs. For example, let us consider changing the radius of the structure shown in Figure 8.5. The initial mesh is generated using the mapped mesh. Then when the radius is changed, the topology of the mesh remains unchanged, but due to change in physical dimension, the geometries of elements change accordingly. This type of parameterization is convenient because the number of elements and nodes will not change during the design process. For example, if the user specifies the performance as displacement at a node or stress at an element, it is easy to track the location and value of the performance. However, mesh distortion will be the bottleneck of this method when the changes in design variables are large. Initially well-shaped elements will eventually be distorted as design variables change from their initial values.

The second method parameterizes the shape designs on the solid model. This method assumes that the preprocessing program has solid modeling and automatic mesh generation capabilities. When a solid model is generated, the user provides dimensions to geometric features, such as fillets, holes, and cutouts. It is also assumed that the preprocessing has a capability of automatically updating the solid model when the values of dimensions are changed. Most CAD software has these capabilities. As shown in Figure 8.4, dimensions of solid models are usually selected as shape designs. It is unnecessary to select all dimensions. Only those dimensions that are supposed to change are considered design

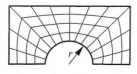

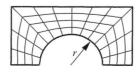

(a) Initial mesh   (b) Perturbed mesh

Figure 8.5 Design perturbation using iso-parametric mapping method

variables. For given values of design, the dimensions of the solid model are fixed, and finite elements are automatically generated on the solid model [refer to Figure 8.4(a)]. When design variables are changed, the solid model is updated to reflect the new designs and finite elements are regenerated on the new model [refer to Figure 8.4(b)]. This type of parameterization can reduce mesh distortion problems because the preprocessing program will generate free mesh for the given designs. However, small changes in design may end up generating a completely different mesh. Thus, it is difficult to track the performance value at a particular location. Particularly, numerical/discretization errors may dominate in finding the trend of performance.

8.5   PARAMETRIC STUDY – SENSITIVITY ANALYSIS

8.5.1   Parameter Study

Once the design variables are determined, structural analysis (i.e., finite element analysis) can be carried out to calculate performances. When the design variables are changed, we expect different values of the performances. Often we can estimate if a performance will improve or not based on our knowledge of mechanics. However, it would be nontrivial to estimate how much the performance will change. The parametric study investigates the effect of design variables on the performance. Usually, it changes one design variable at a time and plots the performance changes in a graph.

For example, the cantilevered beam in Example 8.1 has two design variables: w (width) and h (height) of the cross-section. Let each design variable (or parameter) have three levels. Then, Table 8.1 shows nine cases for the parametric study. Since there are only two parameters, it is possible to plot the results as a three-dimensional surface as shown in Figure 8.6. Note that the maximum stress decreases as both the width and height of the beam increase. In addition, the maximum stress is higher than the allowable stress when the height and width are small. Thus, a feasible design can be chosen from the acceptable region.

8.5.2   Sensitivity Analysis

The parametric study is a useful tool to provide quantitative behavior of the performance according to design changes. However, when the number of design variables becomes large, the parameter study can be expensive. In the case of two design variables with three levels, nine analyses were required. Sometimes, multiple parameters are varied simultaneously in order to reduce the number of analyses, but still this can be expensive. In addition, when the performance changes rapidly, the parametric study cannot capture the local change unless the interval is small enough.

Table 8.1   Parametric study of a cantilevered beam

w (in)	h (in)	σ_{max} (ksi)
2.0	4.0	37.5
2.0	4.5	29.6
2.0	5.0	24.0
2.5	4.0	30.0
2.5	4.5	23.7
2.5	5.0	19.2
3.0	4.0	25.0
3.0	4.5	19.8
3.0	5.0	16.0

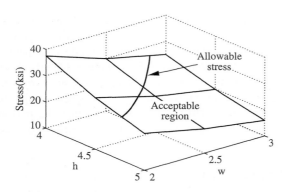

Figure 8.6 Parametric study plot for the cantilevered beam

In many cases, we cannot afford to perform the parametric study with all variables. In addition, we do not need to find the performance changes throughout the entire range. It is often enough to find the effect of design variables in the vicinity of the current design point. In such a case, design sensitivity analysis can be used to effectively find it. *Design sensitivity analysis computes the rate of performance change with respect to design variables.* In conjunction with structural analysis, design sensitivity analysis generates a critical information, gradient, for design optimization. Obviously, the performance is presumed to be a continuous function of the design, at least in the neighborhood of the current design point.

Explicit dependence on design: In general, a structural performance measure depends on the design. For example, a change in the cross-sectional area of a beam would affect the structural weight. This type of dependence is simple, if the expression of weight in terms of the design variables is known. For example, the weight of a beam with a circular cross-section can be expressed as

$$W(r) = \pi r^2 l \qquad (8.16)$$

where r is the radius and l is the length of the beam. If the radius is a design variable, then the design sensitivity of W with respect to r would be

$$\frac{dW}{dr} = 2\pi r l \qquad (8.17)$$

This type of function is *explicitly dependent* on the design, since the function can be explicitly written in terms of that design. Consequently, only algebraic manipulation is involved and no finite element analysis is required to obtain the design sensitivity of an explicitly dependent performance.

Implicit dependence on design: However, in most cases, a structural performance does not explicitly depend on the design. For example, when the stress in a beam is considered as a performance, there is no simple way to express the design sensitivity of stress explicitly in terms of the design variable r. In the linear elastic problem, the stress of the structure is determined from the displacement, which is a solution to the finite element analysis. Thus, the sensitivity of stress $\sigma(\mathbf{q})$ can be written as

$$\frac{d\sigma}{dr} = \frac{d\sigma}{d\mathbf{q}} \cdot \frac{d\mathbf{q}}{dr} \qquad (8.18)$$

where $\mathbf{q}$ is the matrix of nodal DOFs of the beam element. Since the expression of stress as a function of displacement is known, $d\sigma/d\mathbf{q}$ can easily be obtained. The only difficulty is the computation of dq/dr, which is the displacement sensitivity with respect to the design variable r.

When a design engineer wants to compute the design sensitivity of performance such as stress $\sigma(\mathbf{q})$ in Eq. (8.18), structural analysis (finite element analysis, for example) has presumably already been carried out. We will use symbol b for generic design variable, and the nodal DOFs $\mathbf{q}$ is a part of the global DOF vector $\{\mathbf{Q}\}$. Assume that the structural problem is governed by the following linear algebraic equation

$$[\mathbf{K}(b)]\{\mathbf{Q}\} = \{\mathbf{F}(b)\} \tag{8.19}$$

Equation (8.19) is a matrix equation of finite elements. Suppose the explicit expressions of $[\mathbf{K}(b)]$ and $\{\mathbf{F}(b)\}$ are known and differentiable with respect to design variable b. Since the stiffness matrix $[\mathbf{K}(b)]$ and load vector $\{\mathbf{F}(b)\}$ depend on the design b, solution $\{\mathbf{Q}\}$ also depends on the design b. However, it is important to note that this dependency is implicit, which is why we need to develop a design sensitivity analysis methodology. As shown in Eq. (8.18), dq/db must be computed using the governing equation of Eq. (8.19). This can be achieved by differentiating Eq. (8.19) with respect to b as

$$[\mathbf{K}]\left\{\frac{d\mathbf{Q}}{db}\right\} = \left\{\frac{d\mathbf{F}}{db}\right\} - \left[\frac{d\mathbf{K}}{db}\right]\{\mathbf{Q}\} \tag{8.20}$$

Assuming that the explicit expressions of $[\mathbf{K}(b)]$ and $\{\mathbf{F}(b)\}$ are known, $[d\mathbf{K}/db]$ and $\{d\mathbf{F}/db\}$ can be evaluated. Thus, if solution $\{\mathbf{Q}\}$ in Eq. (8.19) is known, then $\{d\mathbf{Q}/db\}$ can be computed from Eq. (8.20), which can then be substituted into Eq. (8.18) to compute $d\sigma/db$. Note that the stress is *implicitly dependent* on the design through nodal DOFs $\mathbf{q}$.

When more than one design variable is defined, the above sensitivity equation must be solved for each design variable. Thus, the sensitivity analysis can be expensive when the problem has a large number of design variables. However, the sensitivity equation (8.20) uses the same stiffness matrix with the original finite element analysis. The difference is on the RHS. As we discussed in Chapter 7, Eq. (8.20) is similar to the finite element analysis with multiple load cases. The RHS of Eq. (8.20) can be considered a pseudo-force vector. The best way of solving Eq. (8.20) might be to construct the RHS of Eq. (8.20) for different design variables and to use the restart procedure. Then, the sensitivity equation can be solved with the stiffness matrix that is already factorized during the finite element analysis. Thus, the computational cost in solving Eq. (8.20) is usually less than 5% of that of finite element analysis.

In general, it can be assumed that the general performance H depends on the design explicitly and implicitly. That is, the performance measure H is presumed to be a function of design b, and nodal DOFs $\mathbf{q}(b)$ as

$$H = H(\mathbf{q}(b), b) \tag{8.21}$$

The sensitivity of H can thus be expressed as

$$\frac{dH(\mathbf{q}(b), b)}{db} = \left.\frac{\partial H}{\partial b}\right|_{\mathbf{q}=\text{const}} + \left.\frac{\partial H}{\partial \mathbf{q}}\right|_{b=\text{const}} \cdot \frac{d\mathbf{q}}{db} \tag{8.22}$$

The only unknown term in Eq. (8.22) is $d\mathbf{q}/db$, which can be obtained from Eq. (8.20). When $[d\mathbf{K}/db]$ and $\{d\mathbf{F}/db\}$ are not available, we can calculate the sensitivity using a finite difference method as follows.

Finite difference method: The easiest way to compute sensitivity information of the performance is by using the finite difference method. Different designs yield different analysis results and, thus, different performance values. The finite difference method actually computes design sensitivity of performance by evaluating performance at different stages in the design process. If b is the current design, then the analysis results provide the value of performance measure $H(b)$. In addition, if the design is perturbed to $b + \Delta b$, where Δb represents a small change in the design, then the sensitivity of $H(b)$ can be approximated as

$$\frac{dH}{db} \approx \frac{H(b + \Delta b) - H(b)}{\Delta b} \tag{8.23}$$

Equation (8.23) is called the *forward difference method* since the design is perturbed in the direction of $+\Delta b$. If $-\Delta b$ is substituted in Eq. (8.23) for Δb, then the equation is defined as the *backward difference method*. Additionally, if the design is perturbed in both directions, such that the design sensitivity is approximated by

$$\frac{dH}{db} \approx \frac{H(b + \Delta b) - H(b - \Delta b)}{2\Delta b} \tag{8.24}$$

then the equation is defined as the *central difference method*.

The advantage of the finite difference method is obvious. If structural analysis can be performed and the performance can be obtained as a result of structural analysis, then the expressions in Eqs. (8.23) and (8.24) are virtually independent of the problem types considered. Consequently, this method is still popular in engineering design.

However, sensitivity computation costs become the dominant concern in the design process. If n represents the number of designs, then $n + 1$ analyses have to be carried out for the forward and backward difference methods, and $2n + 1$ analyses are required for the central difference method. Unlike the sensitivity analysis described in Eq. (8.20), we cannot use the restart procedure in the finite difference method because the stiffness matrix at the perturbed design is different from that of the original design. For modern practical engineering applications the cost of structural analysis is rather expensive. Hence, this method is not feasible for large-scale problems containing many design variables.

Another major disadvantage of the finite difference method is the accuracy of its sensitivity results. In Eq. (8.23), accurate results can be expected when Δb approaches zero. Figure 8.7 shows some sensitivity results using the finite difference method. The tangential slope of the curve at b_0 is the exact sensitivity value. Depending on perturbation size, we can see that sensitivity results are quite different. For a mildly nonlinear performance measure, relatively large perturbation provides a reasonable estimation of sensitivity results. However, for highly nonlinear performances, a large perturbation yields completely inaccurate results. Thus, the determination of perturbation size greatly affects the sensitivity result. And even though it may be necessary to choose a very small perturbation, numerical noise becomes dominant for a too-small perturbation size. That is, with a too-small perturbation, no reliable difference can be found in the analysis

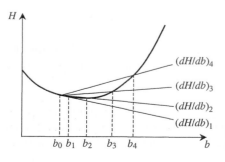

Figure 8.7 Influence of step size in forward finite difference method

results. For example, if up to five digits of significant numbers are valid in a structural analysis, then any design perturbation in the finite difference that is smaller than the first five significant digits cannot provide meaningful results. As a result, it is very difficult to determine design perturbation sizes that work for all problems.

EXAMPLE 8.4   *Sensitivity Analysis of a Cantilevered Beam*

Consider the cantilevered beam in Example 8.1. At the optimum design with strength constraint, we have $w = 2.25$ in and $h = 4.47$ in. Calculate the sensitivity of the tip displacement with respect to the height of the beam. Compare the sensitivity results with the exact sensitivity.

$$\left.\frac{dv_2}{dh}\right|_{\text{exact}} = -\frac{12FL^3}{Ewh^4} = -\frac{12 \times 2,000 \times 100^3}{2.9 \times 10^7 \times 2.25 \times 4.47^4} = -0.921 \tag{8.25}$$

SOLUTION   The design sensitivity of nodal DOFs can be obtained by differentiating the finite element matrix equation in Eq. (8.9) with respect to the design. Consider height $b = h$ of the cross-sectional dimension as a design variable. The design sensitivity equation is given in Eq. (8.20). To solve the design sensitivity equation, we need to calculate the RHS of Eq. (8.20). Since the applied load $\{\mathbf{F}\}$ is independent of the design, the first term $\{d\mathbf{F}/db\} = \{0\}$. The stiffness matrix in Eq. (8.9) depends on design through the moment of inertia $I = wh^3/12$. Thus, it can be differentiated with respect to design. After multiplying with $\{\mathbf{Q}\}$, we have

$$\left[\frac{d\mathbf{K}}{db}\right]\{\mathbf{Q}\} = \frac{F}{4Lh}\begin{bmatrix} 12 & 6L & -12 & 6L \\ 6L & 4L^2 & -6L & 2L^2 \\ -12 & -6L & 12 & -6L \\ 6L & 2L^2 & -6L & 4L^2 \end{bmatrix}\begin{Bmatrix} 0 \\ 0 \\ 4L \\ 6 \end{Bmatrix} = \frac{F}{4h}\begin{Bmatrix} -12 \\ -12L \\ 12 \\ 0 \end{Bmatrix} \tag{8.26}$$

Thus, the RHS of Eq. (8.20) can be computed as

$$\left\{\frac{d\mathbf{F}}{db}\right\} - \left[\frac{d\mathbf{K}}{db}\right]\{\mathbf{Q}\} = \frac{F}{4h}\begin{Bmatrix} 12 \\ 12L \\ -12 \\ 0 \end{Bmatrix} \tag{8.27}$$

Then, the design sensitivity equation can be obtained as

$$\frac{EI}{L^3}\begin{bmatrix} 12 & 6L & -12 & 6L \\ 6L & 4L^2 & -6L & 2L^2 \\ -12 & -6L & 12 & -6L \\ 6L & 2L^2 & -6L & 4L^2 \end{bmatrix}\begin{Bmatrix} dv_1/db = 0 \\ d\theta_1/db = 0 \\ dv_2/db \\ d\theta_2/db \end{Bmatrix} = \frac{F}{4h}\begin{Bmatrix} 12 \\ 12L \\ -12 \\ 0 \end{Bmatrix} \tag{8.28}$$

The first two rows and columns are deleted due to zero displacement boundary conditions. Note that this part is the same as the original finite element analysis. When a displacement is fixed, the sensitivity is also zero. After applying displacement boundary conditions, we have

$$\frac{EI}{L^3}\begin{bmatrix} 12 & -6L \\ -6L & 4L^2 \end{bmatrix}\begin{Bmatrix} dv_2/db \\ d\theta_2/db \end{Bmatrix} = \frac{F}{4h}\begin{Bmatrix} -12 \\ 0 \end{Bmatrix} \tag{8.29}$$

The above equation can be solved for the unknown nodal DOFs. Now we have

$$\frac{dv_2}{db} = -\frac{12FL^3}{Ewh^4}, \quad \frac{d\theta_2}{db} = -\frac{18FL^2}{Ewh^3} \tag{8.30}$$

Note that dv_2/db is the same with Eq. (8.25). Thus, the sensitivity we calculated is exact. Note that in differentiating the stiffness matrix in Eq. (8.26), only the moment of inertia I was differentiated. The basic form of the matrix remains unchanged. This type of design variable is called a sizing design variable.

EXAMPLE 8.5 *Finite Difference Sensitivity of a Three-Bar Truss*

Calculate the sensitivity of the vertical displacement (v_2) at Node 2 in the three-bar truss problem in Example 8.3 with respect to b_2. Compare the accuracy of calculated sensitivity with the finite difference sensitivity. Use 1.0% perturbation size for the forward finite difference.

SOLUTION Consider the three-bar truss example shown in Figure 8.2. The finite element matrix equation, after applying displacement boundary conditions, is given in Eq. (8.14), along with the nodal solutions in Eq. (8.15). To build the sensitivity equation, we need to calculate the RHS of Eq. (8.20). Since the applied load $\{F\}$ is independent of the design, the first term $\{dF/db\} = \{0\}$. Out of three element stiffness matrices, only $[\mathbf{k}^{(2)}]$ depends on design b_2. Thus,

$$\left[\frac{d\mathbf{k}^{(1)}}{db_2}\right] = \left[\frac{d\mathbf{k}^{(3)}}{db_2}\right] = [\mathbf{0}], \quad \left[\frac{d\mathbf{k}^{(2)}}{db_2}\right] = \frac{E}{2L^{(2)}}\begin{bmatrix} 1 & -1 & -1 & 1 \\ -1 & 1 & 1 & -1 \\ -1 & 1 & 1 & -1 \\ 1 & -1 & -1 & 1 \end{bmatrix}\begin{array}{c} u_2 \\ v_2 \\ u_3 \\ v_3 \end{array}$$

These three matrices are assembled in the same way with the stiffness matrix and then multiplied by the vector of nodal displacements to obtain

$$\left\{\frac{dF}{db_2}\right\} - \left[\frac{d\mathbf{K}}{db_2}\right]\{\mathbf{Q}\} = -2.828 \times 10^{10}\begin{bmatrix} 0 & 0 & 0 & 0 & 0 & 0 \\ 0 & 0 & 0 & 0 & 0 & 0 \\ 0 & 0 & 1 & -1 & -1 & 1 \\ 0 & 0 & -1 & 1 & 1 & -1 \\ 0 & 0 & -1 & 1 & 1 & -1 \\ 0 & 0 & 1 & -1 & -1 & 1 \end{bmatrix}\begin{Bmatrix} 0 \\ 0 \\ -.0013 \\ -.0060 \\ 0 \\ -.0013 \end{Bmatrix} = \begin{Bmatrix} 0 \\ 0 \\ -10^8 \\ 10^8 \\ 10^8 \\ -10^8 \end{Bmatrix}$$

Then, the design sensitivity equation, after applying displacement boundary conditions, becomes

$$10^5\begin{bmatrix} 10.83 & -2.83 & 2.83 \\ -2.83 & 2.83 & -2.83 \\ 2.83 & -2.83 & 10.83 \end{bmatrix}\begin{Bmatrix} du_2/db_2 \\ dv_2/db_2 \\ dv_3/db_2 \end{Bmatrix} = \begin{Bmatrix} -10^8 \\ 10^8 \\ -10^8 \end{Bmatrix} \tag{8.31}$$

The solution of the above sensitivity equation yields

$$\frac{du_2}{db_2} = 0, \frac{dv_2}{db_2} = 353.55, \frac{dv_3}{db_2} = 0 \tag{8.32}$$

Thus, the change in the cross-sectional area of member 2 will only change the vertical displacement of Node 2.

Let us compute the design sensitivity of v_2 by using the finite difference method. The original displacements in Eq. (8.15) are saved as $H(b_2) = -6.036$ mm. Then design b_2 is perturbed by 1.0%; i.e., $b_2 = 10.1$ mm. A new global matrix equation is produced with new design, as

$$10^5\begin{bmatrix} 10.86 & -2.86 & 2.86 \\ -2.86 & 2.86 & -2.86 \\ 2.86 & -2.86 & 10.86 \end{bmatrix}\begin{Bmatrix} u_2 \\ v_2 \\ v_3 \end{Bmatrix} = \begin{Bmatrix} 0 \\ -1000 \\ 0 \end{Bmatrix} \tag{8.33}$$

Note that the matrix is slightly different from the one in Eq. (8.14). By solving the above equation we have the following unknown nodal DOFs:

$$u_2 = -1.25 \text{ mm}, v_2 = -6.00 \text{ mm}, v_3 = -1.25 \text{ mm} \tag{8.34}$$

Note that u_2 and v_3 did not change, which is consistent from the zero sensitivity in Eq. (8.32). With the vertical displacement at Node 2, we have the performance at the perturbed design, $H(b_2 + \Delta b_2) = -6.001$ mm. From the finite difference sensitivity formula in Eq. (8.23), we have

$$\frac{dH}{db_2} \approx \frac{H(b_2 + \Delta b_2) - H(b_2)}{\Delta b_2} = \frac{-6.001 \times 10^{-3} + 6.306 \times 10^{-3}}{0.1 \times 10^{-5}} = 350.05 \tag{8.35}$$

Note that the finite difference sensitivity in Eq. (8.35) is slightly different from the one in Eq. (8.32). This is because the influence of finite perturbation size. When 0.1% perturbation is used, the finite difference sensitivity becomes 353.2, which is much closer to the one in Eq. (8.32). However, as we can see in Eqs. (8.15) and (8.33), the difference in stiffness matrix is small. Thus, it is required to maintain high accuracy in matrix solution in order to have a small perturbation size.

8.6   STRUCTURAL OPTIMIZATION

The purpose of many structural design problems is to find the best design among many possible candidates. As discussed in Section 8.1, at least one possible candidate should exist within a feasible design region that satisfies problem constraints. Every design in the feasible region is an acceptable design, even if it is not the best one. The best design is usually the one that minimizes (or maximizes) the objective function (goal) of the design problem. Thus, the goal of the design optimization problem is to find the design that minimizes the objective function among all feasible designs. Unfortunately, no mathematical theory exists that can find the global optimum design for general nonlinear functions. In this section, simple optimization methods are briefly introduced. However, this brief discussion is by no means the complete treatment of optimization methods. For a more detailed treatment, refer to Haftka and Gurdal[1] or Arora.[2]

Most gradient-based optimization algorithms are based on mathematical programming methods, which require performance values and sensitivity information at given design variables. For a given design variable that defines the structural model, structural analysis provides the values of the objective and constraint functions to the algorithm. Gradients or design sensitivities of the objective and constraint functions must also be supplied to the optimization algorithm. Then, the optimization algorithm calculates the best possible design of the problem. In this section, we will introduce how the optimization problem can be formulated and how it can be solved using graphical and mathematical programming methods.

8.6.1   Optimization Problem Formulation

An important step in optimization is to transcribe the verbal statement of optimization problem into a well-defined mathematical statement. In the verbal statement, the goal of the optimization problem is to find the best design that minimizes (or maximizes) the objective function (or cost function) under the given constraints by changing design parameters. Examples of objective function in structural analysis are the weight, stiffness or compliance; the fatigue life; the noise level; amplitude and frequency of vibration; safety, etc. Constraints are similar to objective functions, but they are given as limits so that a design must satisfy them in order to be a candidate. Design variables are parameters or geometry of the structure that can be varied during design. Figure 8.8 shows the flow chart of the design optimization process.

Three-step problem formulation: To formulate the optimization problem, it is important to properly define three components – design variables, objective, and constraint functions.

1. Identification and precise definition of design variables. As we discussed in Section 8.4, different types of design variables are available. Some designs only

[1] Haftka, R. T. and Z. Gurdal, *Elements of Structural Optimization*, Kluwer Academic Publishers, Dordrecht, 1992

[2] Arora, J. S., *Introduction to Optimum Design*, Elsevier Academic Press, New York, 2004

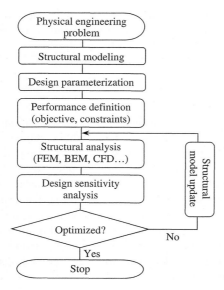

Figure 8.8 Structural design optimization procedure

change some parameters in the finite element analysis, but some designs change the finite element model itself. It is important to make sure that the finite element model, including properties and parameters, must completely be defined with a given set of design variables. For example, the cross-section in Figure 8.9(a) requires four parameters to completely define it: w, h, t_1, and t_2. Of course, it is unnecessary to choose all four as design variables. We can fix t_1 and t_2 and define w and h as design variables. The important thing is that once the values of design variables are given, the model should be unique. Also, it is important to provide the lower and upper bounds of design variables and to make sure that the finite element analysis can be carried out successfully over the entire range of design variables. When finite element analysis cannot be carried out for a given design, the objective function or constraint is often assigned with very large values so that the optimum design occurs somewhere else.

Another important aspect of selecting design variables is 'independence'. All design variables must be independent of each other. If design variables are not independent, they may not be able to generate the model, and the optimization algorithm may fail. Figure 8.9(b) shows an example of improperly defined design variables. If w, h, b, t_1, and t_2 are defined as design variables, then the following relation exists: $h = b + 2t_1$. It is possible to define six design variables and provide this relation as a constraint, but it will make the optimization problem unnecessarily complicated. In addition, selection of design variables is not unique.

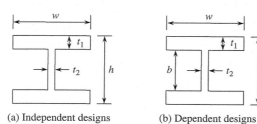

(a) Independent designs   (b) Dependent designs

Figure 8.9 Design parameters for beam cross-section

In Figure 8.9(b), for example, we can choose w, h, t_1, and t_2 as independent design variables, or w, b, t_1, and t_2.

2. Defining objective function (cost function): Once design variables are defined, the next step is to define the objective function of the problem. The objective function defines the ranking of different designs. The goal of optimization is to find the design that has the best ranking in terms of the objective function. It is obvious that the objective function must depend on design variables. The absolute magnitude of objective function is not important. A constant can be added or multiplied to the objective function without changing the optimization result.

When more than one objective function is involved, it is possible to combined them using weights as

$$F(\mathbf{b}) = \sum_{I=1}^{N_{OBJ}} w_I f_I(\mathbf{b})$$ (8.36)

where $f_I(\mathbf{b})$ is an individual objective function and w_I is the corresponding weight. The weights must be chosen based on importance of each objective function. Different optimum designs will be expected for different sets of weights. When the weights are unclear, the multi-objective optimization problem can be solved, which will provide all possible sets of optimum designs.[3]

3. Identification and definition of constraints. Unlike the objective functions, constraints do not rank the designs, but they validate whether the design is feasible or not. Once the design is determined to be feasible, it is not important how much safety margin the design has. In general, there are two types of constraints: equality and inequality. Equality constraints provide relations between design variables or impose conditions that a usable design must satisfy. When the relation is linear, it is possible to remove one design variable. Because of that, they are considered strong constraints. In general, however, it is not easy to remove a design variable when the equality constraint is nonlinear. Inequality constraints are more popular in structural problems and provide limits of performances. For example, the maximum stress of usable design should be less than the allowable stress of the material. As long as the maximum stress is lower than the allowable stress, it is not important how low the stress is.

Some constraints limit the lower or upper bounds of design variables. These constraints are called *side constraints*. Since it is relatively easy to impose the lower or upper bounds of design variables, these constraints are often treated separately.

Usually, an optimization problem has one objective function with many constraints. In general, there is no limit in the number of inequality constraints. For example, one can choose stresses of all elements and make them less than the allowable stress. In such a case, many elements will have stress much less than the allowable stress, and those elements will be ignored during optimization until their values become close to the allowable stress. However, the number of equality constraints should be less than that of design variables. This is obvious from the fact that we can theoretically remove one design variable from each equality constraint.

---

[3] Arora, J. S., *Introduction to Optimum Design*, Elsevier Academic Press, New York, 2004

Standard form: The above description of three components in optimization needs to be written in mathematical form so that there will be no ambiguity in the problem definition. The standard form of design optimization problem can be written as

$$\text{minimize} \quad f(\mathbf{b})$$

$$\text{subject } to \quad g_i(\mathbf{b}) \leq 0, \quad i = 1, \cdots, N$$

$$h_j(\mathbf{b}) = 0, \quad j = 1, \cdots, M \tag{8.37}$$

$$b_l^L \leq b_l \leq b_l^U, \quad l = 1, \cdots, K$$

where $\mathbf{b} = \{b_1, b_2, \ldots, b_K\}^T$ is the vector of design variables, $f(\mathbf{b})$ is the objective function, $g_i(\mathbf{b})(i = 1, \ldots, N)$ is inequality constraints, $h_j(\mathbf{b})(j = 1, \ldots, M)$ is equality constraints, and b_l^L and b_l^U are, respectively, the lower and upper bounds of design variables. Note that the objective function is minimized, and the inequality constraints are written in "less than or equal to" form.

The standard form in Eq. (8.37) is to find the minimum value of the objective function within the region that satisfies constraints. Thus, it is convenient to define the following feasible set S:

$$S = \{\mathbf{b}|g_i(\mathbf{b}) \leq 0, i = 1, \cdots N, \quad h_j(\mathbf{b}) = 0, j = 1, \cdots M\} \tag{8.38}$$

Then the optimization problem is to find the minimum $f(\mathbf{b})$ in the feasible set S. Once an optimization problem is written in the standard form, solving the optimization problem is independent of applications. Whether it is a structural or financial problem, the same optimization technique can be used to solve it.

EXAMPLE 8.6   *Beer Can Design Formulation*[4]

Formulate a standard optimization problem for a beer can design problem. We want to design the beer can that can hold at least a specific amount of beer and meet other design requirements. The goal is to minimize the manufacturing cost, as they are produced in billions. Since the cost can be related directly to the surface area of the sheet metal used, it is reasonable to minimize the sheet metal required to fabricate the can. Fabrication, handling, aesthetic, and shipping considerations impose the following restrictions on the size of the can:

1. The can is required to hold at least 400 ml of fluid.
2. The diameter of the can should be no more than 8 cm. In addition, it should not be less than 3.5 cm.
3. The height of the can should be no more than 18 cm and no less than 8 cm.

Following the three-step procedure, write a standard form of the optimization problem.

SOLUTION   By following three-step procedure, we first choose design variables. To design the volume and surface area of the can, it is necessary to choose the height of the can as well as either the radius or diameter of the can. Thus, we choose $\mathbf{b} = \{D, H\}^T$ from Figure 8.10 as design variables.

The objective function will be the surface area of the can, which is the amount of sheet metal used. Since the surface area is composed of top, bottom, and side, we have

$$f(D, H) = \pi DH + \frac{\pi}{2}D^2$$

---

[4] Arora, J. S., *Introduction to Optimum Design*, Elsevier Academic Press, New York, 2004

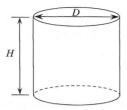

Figure 8.10 Design of a beer can

The three constraints also can be written in terms of design variables. Thus, the standard form of optimization problem becomes

$$\text{minimize} \quad f(\mathbf{b}) = \pi D H + \frac{\pi}{2} D^2 \quad \text{cm}^2$$

$$\text{subject to} \quad 400 - \frac{\pi}{4} D^2 H \leq 0 \tag{8.39}$$

$$3.5 \leq D \leq 8 \qquad \text{cm}$$

$$8.0 \leq H \leq 18 \qquad \text{cm}$$

It is important to distinguish two different types of minima in order to understand the nature of optimum design that we can obtain using practical methods.

Global minimum: In optimization problems, the best design is the one that yields the smallest value of objective function within the feasible set, which is called a global minimum. Formally, a point $\mathbf{b}^*$ is called a global minimum for $f(\mathbf{b})$ if

$$f(\mathbf{b}^*) \leq f(\mathbf{b}) \tag{8.40}$$

for all $\mathbf{b} \in S$. The global minimum is the eventual goal of optimization. It is easy to define the global minimum, but unfortunately, it is not trivial to find one. If $f(\mathbf{b})$ is continuous and the set S is closed and bounded, then there is a global minimum. However, there is no mathematical method to find it. In addition, it is possible that the problem may have multiple global minima. Figure 8.11 shows an objective function f as a function of design b. The point b_3 corresponds to the global minimum because $f(b_3)$ is smallest within the feasible set. In this case, a unique global minimum exists.

Local minimum: Instead of the global minimum, a local minimum is the one that we can afford to find. There are many mathematical theories to find it. Formally, a point $\mathbf{b}^*$ is called a local minimum for $f(\mathbf{b})$ if

$$f(\mathbf{b}^*) \leq f(\mathbf{b}) \tag{8.41}$$

for all $\mathbf{b} \in S$ in a small neighborhood of $\mathbf{b}^*$. The small neighborhood means for an arbitrary small Δ, $b^* - \Delta \leq b \leq b^* + \Delta$ in the case of single design variable. If we can find

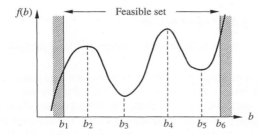

Figure 8.11 Local and global minima of a function

any Δ so that f satisfies Eq. (8.41), then it is a local minimum. In Figure 8.11, b_1, b_3, and b_5 are local minima. Note that b_3 is a local minimum as well as a global minimum.

8.6.2   Graphical Optimization

After the optimization problem is written in the standard form, it can be solved using numerical methods. Before we discuss the various numerical methods to solve the optimization problem, let us consider simple cases in which we can solve the optimization problem graphically. When the optimization problem has only one or two design variables, graphical methods can be used to solve the problem. Graphical methods are often more expensive than other numerical methods because they require a large number of evaluations of the objective functions and constraints. However, they help engineers to visualize the design space and to understand the nature of the design problem.

The first step in graphical optimization is to set up the graphical domain using the side constraints. The range of design variables defines all possible combination of the two design variables. Since any design out of the side constraints is not feasible, it is unnecessary to plot functions in infeasible region. The feasible region is typically referred to as the *design space*. The next step is to plot the objective function and constraints on the graph. In the case of constraints, it is possible to draw a boundary curve for each of the constraints. An inequality constraint will define the feasible region as one side (i.e., constraint is negative) of the curve, while an equality constraint will define the feasible region along the curve of the constraint. The difference between the equality and inequality constraints is in the interpretation of the design space.

Once feasible set is clearly defined using constraints, the next step is to plot the objective function. Unlike the constraints, the objective function does not have a fixed value. One way of representing the objective function is to plot contour lines. They are similar to equal-temperature or equal-pressure curves in weather maps. They can be created by setting the objective function being equal to a constant. This will provide an equation for one of the curves. Along this curve, the objective function has the same value. By gradually increasing or decreasing the constant, we can plot the contour curves. Sometimes, different colors are used for different constant values so that the values of the objective function can be easily found.

The final step of graphical optimization is to inspect the optimal solution. The optimal solution is the point in the feasible set that has the smallest value of the objective function. The optimal design may locate on the constraint curves or inside the feasible set. The latter is rarely encountered in structural optimization because most problems have important constraints that limit the objective function. If the lowest contour line intersects the boundary of one or more inequality constraints, then some or all of these constraints dictate the location of the optimal solution and are called active constraints.

EXAMPLE 8.7   *Graphical Optimization of Beer can Problem*

Find the optimum design of the beer can problem in Example 8.6 using a graphical method.

SOLUTION   Figure 8.12 depicts the optimization problem. Since two constraints are side constraints, we plot the graph between lower and upper bounds of design variables. Then the only constraint that we need to take care of is the volume constraint, which is plotted as a solid curve. Then, we can identify the feasible set as shown in Figure 8.12. Now we gradually reduce the objective function and plot them in dashed curves. When the value of objective function is less than 300, there is no

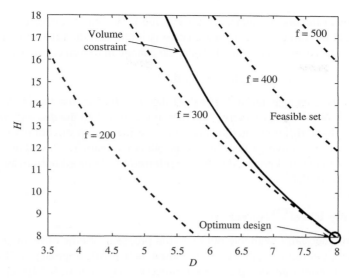

Figure 8.12 Graphical optimization of the beer can problem

feasible design. In fact, the optimum design occurs when $H = 8$ cm and the volume constraint is just satisfied, which means the inequality becomes equality. Thus, we can calculate the diameter at the optimum design, as

$$400 - \frac{\pi}{4} D^2 H = 0$$

From the above relation, we can solve for the diameter; $D = 7.98$ cm. Thus, we have

$$\mathbf{b}^* = \{D^*, H^*\}^T = \{7.98, 8\}^T$$

At the optimum design, the objective function becomes $f(\mathbf{b}^*) = 300.53$ cm^2.

8.6.3   Numerical Methods

As we discussed before, there is no mathematical method to find a global minimum, except for very simple cases. Almost all numerical methods provide a local minimum. They are basically starting from an initial design and finding a new design that reduces the objective function while satisfying all constraints. The method repeats until there is no design in the vicinity that can reduce the objective function further.

Basic algorithm: Most of gradient-based optimization algorithms take the following steps:

1. Start with initial design $\mathbf{b}^{(0)}$ and iteration counter $K = 0$.
2. Evaluate function values and their gradients.
3. Using information from Step 2, determine $\Delta \mathbf{b}^{(K)}$.
4. Check for termination.
5. Update design.

$$\mathbf{b}^{(K+1)} = \mathbf{b}^{(K)} + \Delta \mathbf{b}^{(K)} \tag{8.42}$$

6. Increase $K = K + 1$ and go to Step 2.

Change in design: In the above algorithm, the change in design is further decomposed into two steps. First, the direction of design change is found and then the amount of design change in that direction is determined. Thus, the change in design can be written as

$$\Delta \mathbf{b}^{(K)} = \alpha_K \, \mathbf{d}^{(K)} \tag{8.43}$$

where a_K is called the *step size* and $\mathbf{d}^{(K)}$ is called the *search direction* vector. Since the search direction also reduces the objective function, it is also called the *descent direction*. Various algorithms have different methods to calculate the search direction so that the optimization problem can converge fast. In this introductory chapter, we will not present different algorithms in detail. Interested users are referred to advanced optimization textbooks by Haftka and Gurdal.[5]

8.6.4 Optimization Using Excel™ Solver

There are many commercially available optimization programs. Some of them are based on mathematical programming, while others are based on heuristic approaches. However, in this chapter, we demonstrate the optimization process using Microsoft® Excel spreadsheet. We explain the process using a simple structural example.

Problem definition: Consider the minimum weight design of the four bar truss shown in Figure 8.13. For the sake of simplicity, we assume that members 1 through 3 have the same area A_1 and member 4 has an area A_2. The constraints are limits on the stresses in the members and on the vertical displacement at the right end of the truss. Under the specified loading, the member forces and the vertical displacement δ at the end are found to be

$$P^{(1)} = 5F, \ P^{(2)} = -F, \ P^{(3)} = 4F, \ P^{(4)} = -2\sqrt{3}F$$

$$\delta = v_4 = \frac{6FL}{E}\left(\frac{3}{A_1} + \frac{\sqrt{3}}{A_2}\right)$$

We assume the allowable stresses in tension and compression to be $8.74 \times 10^{-4}E$ and $4.83 \times 10^{-4}E$, respectively, and limit the displacement to be no greater than $3 \times 10^{-3}L$. First, convert the design variables into non-dimensional ones as

$$b_1 = 10^{-3}\frac{A_1 E}{F}, \quad b_2 = 10^{-3}\frac{A_2 E}{F}$$

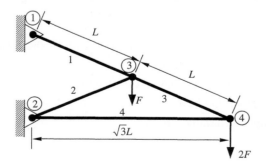

Figure 8.13 Minimum weight design of four-bar truss

---

[5] Haftka, R. T. and Z. Gurdal, Elements of structural optimization, Kluwer Academic Publishers, New York, 1992

Then, the minimum weight design subject to stress and displacement constraints can be formulated as

$$\text{Minimize} \quad f(\mathbf{b}) = 3b_1 + \sqrt{3}b_2$$

$$\text{subject to} \quad g_1 = \frac{18}{b_1} + \frac{6\sqrt{3}}{b_2} - 3 \leq 0$$

$$g_2 = 5.73 - b_1 \leq 0$$

$$g_3 = 7.17 - b_2 \leq 0$$

Loading the Solver Add-in: Before using *Solver*, the user must first load the *Solver* add-in into the memory. When installing Excel, the user was given the option of installing the add-ins that ship with Excel. If the add-ins were installed, *Solver* can be loaded into the memory using *Add-Ins* submenu in *Tools*. If Add-Ins is not installed, it must be installed first in order to use *Solver*. To load *Solver* into the memory, follow these steps:

1. Select *Tools* on the main menu. Figure 8.14 shows submenus of *Tools*.

2. If *Solver* submenu is shown on submenus of *Tools*, then *Solver* is already loaded into the memory. Skip the next steps.

3. If *Solver* submenu is not shown, click *Add-Ins* on the submenu. Figure 8.15 shows the currently installed *Add-ins*.

4. From the list of installed add-ins, check *Solver Add-In*.

5. Choose *OK* or press Enter. Now, *Solver* submenu should appear in *Tools* as shown in Figure 8.14.

Setting up the example problem: To use *Solver* in Excel worksheets, the optimization problem must be defined first. The user must specify the cells corresponding to design variables, objective function, and constraints. With *Solver*, the objective and constraint

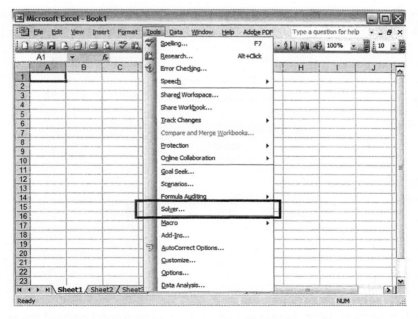

Figure 8.14 List of submenu in Tools menu (Solver appears in Tools menu)

Figure 8.15 Add-in dialog box with installed *Solver* Add-In

cells should be based on formulas in terms of design variable cells. Thus, if the values of design variables cells are changed, the objective and constraints are changed accordingly. Therefore, to set up the problem, determine which of the cells will be used as the objective and constraints and make sure that they contain formulas. The worksheet shown in Figure 8.16 illustrates the example problem that the Solver will use.

In Figure 8.16, cells C4 and C5 are selected for design variables b_1 and b_2, respectively (*Changing Cell* in *Solver Parameters* dialog box). The objective function is formulated in cell C7 (*Target Cell* in *Solver Parameters* dialog box). The constraint equations g_1, g_2, and g_3 are formulated in cells C10, C11, and C12, respectively.

1. Type text portions in cells A1, A3, A7, A9, B4, B5, B10, B11, and B12, as shown in Figure 8.16. These are unnecessary for *Solver* but can help the user understand the meaning of each cell.

2. Select cell C4 and type 10. This is an arbitrary initial value of design variable b_1.

3. Select cell C5 and type 10. This is an arbitrary initial value of design variable b_2.

4. Click C7 and key in objective function formulas as follows: `=3*C4+sqrt(3)*C5`

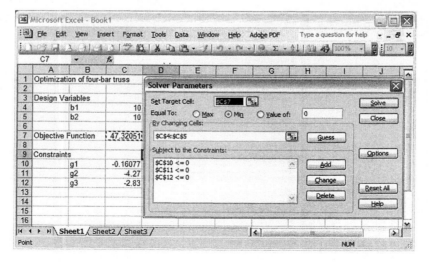

Figure 8.16 Excel worksheet for minimum weight design of the four-bar truss and Solver Parameters dialog box

Figure 8.17 Add Constraint dialog box

5. Click C10 and key in constraint equation formulas g_1 as follows:
   =18/C4+6*sqrt(3)/C5—3

6. Click C11 and key in constraint equation formulas g_2 as follows: =5.73-C4

7. Click C12 and type constraint equation formulas g_3 as follows: =7.17-C5

Running Solver: After the locations and formulas of design variables, objective, and constraints are set up, follow these steps to run *Solver*:

1. Select *Solver* in *Tools* to start the solve add-in. The *Solver Parameter* dialog box will be displayed as shown in Figure 8.16.

2. Indicate the cell that contains the objective function formula in the *Set Target Cell* text box. The objective cell can be directly type is as C7 or can be selected using the cell selection method in Excel.

3. In the *Equal To* section of the dialog box, select *Min* button as the objective function will be minimized.

4. In the *By Changing Cells* text box, indicate the cell or range of cells that will be used as design variables. In this case, choose C4 and C5.

5. To specify constraints, select the *Add* button to add each constraint to the problem. Figure 8.17 shows the *Add Constraint* dialog box.

6. To create a constraint, specify the cell containing the formula on which the constraint is based in the *Cell Reference* text box (for example, C10 for the first constraint). Click the drop-down arrow to display the list of constraint operators, and select the appropriate operator (choose $\Leftarrow$ symbol). In the final text box, enter the value the constraint must meet (type in 0). Choose the *Add* button to add the current constraint to the problem and create another, or choose *OK* to add the constraint and return to the *Solver Parameters* dialog box. The three constraints should appear in the *Subject to the Constraints* list box in the *Solver Parameters* dialog box (See Figure 8.16).

7. Choose *Options* button in the *Solver Parameters* dialog box to control maximum computing time, iteration, convergence, etc. The *Solver Options* dialog box will be displayed as shown in Figure 8.18. Choose appropriate settings in the *Solver*

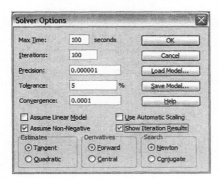

Figure 8.18 Solver Options dialog box

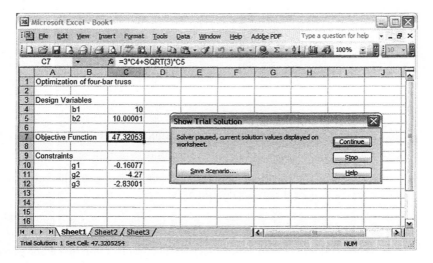

Figure 8.19 Show Trial Solution dialog box

Options dialog box. Check *Assume Non-Negative* and *Show Iteration Results* buttons and click *OK* button. The *Solver Parameters* dialog box in Figure 8.16 will be redisplayed.

8. Click *Solve* button to start the Solver. The *Solver* begins calculating the optimal solutions. Intermediate solution values are displayed on the worksheet, and *Show Trial Solution* dialog box, as shown in Figure 8.19 will appear.

9. The intermediate solution can be stored by selecting *Save Scenario* button in *Show Trial Solution* dialog box. In order to continue the optimization, click the *Continue* button. When *Solver* finds a solution, the *Solver Results* dialog box appears, as shown in Figure 8.20. Note that the values of objective function, deign variables, and constraints are changed. Select *Keep Solver Solution* to use the

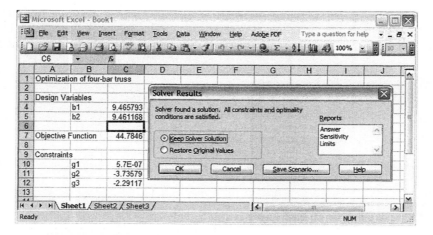

Figure 8.20 Solver Results dialog box

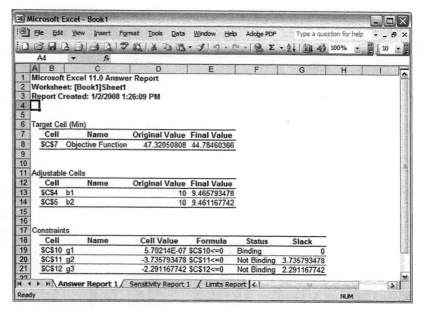

Figure 8.21 Answer Report worksheet

offered Solutions. If *Restore Original Values* button is selected, then the worksheet will return to the original values. Figure 8.20 also shows the worksheet after the *Solver* has found the solutions for the problem.

Creating Solver Reports: Solver can generate reports summarizing the results of its solutions. There are three types of reports: *Answer*, *Sensitivity*, and *Limit* reports. The *Answer* report shows the original and final values for the target cell (objective function) and the adjustable cells (design variables), as well as the status of each constraint. The *Sensitivity* report shows the sensitivity of each element of the solution to changes in input cells or constraints. The *Limits* report shows the upper and lower values of the design variables within the specified constraints. To create a report, select the reports from the list that appears in the *Solver Results* dialog box, and choose *OK*. Excel creates the reports in a separate sheet. Figure 8.21 shows *Answer Report* of the four-bar truss example.

8.7   PROJECT: DESIGN OPTIMIZATION OF A BRACKET

A bracket shown in Figure 8.22 has the following properties: Young's modulus $E = 2.068 \times 10^{11}\,\text{N/m}^2$, Poisson's ratio $\nu = 0.29$, density $\rho = 7.82 \times 10^3\,\text{kg/m}^3$, thickness = 3 mm. A horizontal force $F_x = 15,000\,\text{N}$ is applied at the center of the upper hole, and the two bottom holes are fixed to the ground. The design goal is to minimize the mass of the bracket, while the maximum stress is less than 800 MPa.[6]

---

[6] Bennett and Botkin, *AIAA Journal*, Vol. 23, 1985, pp. 458–464

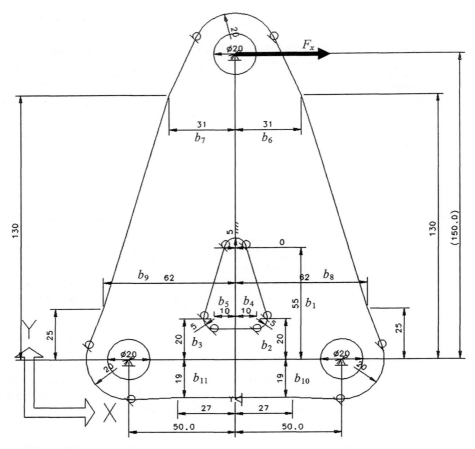

Figure 8.22 Geometry of a bracket (unit mm)

(a) In the report, clearly state the units for mass, length, force, stress.

(b) Using a CAD tool, create a solid mode with appropriate dimensions as shown in Figure 8.22. It is important to define dimensions and relations so that the solid mode can update properly when the values in dimensions are changed.

Table 8.2   Lower and upper bounds of design parameters (unit mm)

Design	Name	Lower bound	Initial value	Upper bound
b_1	Slot height	54	55	120
b_2	Slot Vr 1	9	20	21
b_3	Slot Vr 2	9	20	21
b_4	Slot Hr 1	9	10	30
b_5	Slot Hr 2	9	10	30
b_6	Out Hr 1	17	31	32
b_7	Out Hr 2	17	31	32
b_8	Out Hr 3	43	62	63
b_9	Out Hr 4	43	62	63
b_{10}	Bottom 1	1	19	21
b_{11}	Bottom 2	1	19	21

(c) Provide a plot of the finite element model that includes boundary conditions and applied force. Use proper modeling techniques in Chapter 7 to approximate the load application method and the displacement boundary conditions.

(d) Carry out parametric study by changing design b_1 (55, 60, 65, 70, 75). Provide mass and maximum stress plots as functions of design b_1.

(e) Carry out design optimization with design boundaries given in Table 8.2 to minimize the mass, while the maximum stress is less than 800 MPa. Provide an optimum geometry plot and optimum stress plot. Provide history of design goal, stress constraints, and design parameters.

8.8 EXERCISE

1. Determine the height of the beam in Example 8.1 when the load factor of $\lambda = 2.0$ is used with the failure stress of 40 ksi.

2. Determine the height of the beam in Example 8.1 so that the safety margin is 10 ksi with the failure stress of 40 ksi.

3. A two-dimensional truss shown in the figure is made of aluminum with Young's modulus $E = 80$ GPa and failure stress $\sigma_Y = 150$ MPa. Determine the minimum cross-sectional area of each member so that the truss is safe with safety factor 1.5.

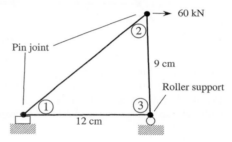

4. Consider a stepped beam modeled using two beam elements. The cross-sections are circular. Use Young's modulus $E = 80$ GPa, yield stress $\sigma_Y = 250$ MPa, and $L = 1$ m. When $F_2 = F_3 = 1,000$ N, calculate the minimum diameters of two sections so that the beam does not fail with safety margin of 100 MPa.

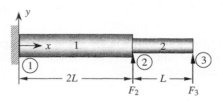

5. A cantilever beam of length 1 m is subjected to a uniformly distributed load $p(x) = p_0 = 12,000$ N-m and a clockwise couple 5,000 N-m at the tip. The load factors for the distributed load and couple are, respectively, 1.5 and 2.0. When the cross-section is circular, calculate the minimum diameter. Use Young's modulus 80 GPa and yield strength 250 MPa.

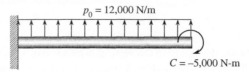

6. The frame shown in the figure is clamped at the left end and supported on a hinged roller at the right end. An axial force P and a couple C act at the right end. The load factor for the axial force is 1.5 and that of the couple is 2.0. Determine the radius of the circular cross-section. Assume the following numerical values: $L = 1\,\text{m}$, $E = 80\,\text{GPa}$, $P = 15,000\,\text{N}$, $C = 1,000\,\text{Nm}$, $\sigma_Y = 250\,\text{MPa}$.

7. All members of the truss shown in the figure initially have a circular cross-section with diameter of 2 in. Using one of the finite element analysis programs in Appendix, calculate the minimum diameter of each member using fully stressed design. Assume that Young's modulus $E = 10^4\,\text{psi}$, yield stress $\sigma_Y = 500\,\text{psi}$, and safety factor $S_F = 1.5$. For the zero-force member, use the smallest diameter of $d = 0.1\,\text{in}$.

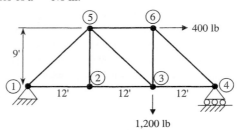

8. Consider a two-bar structure in the figure with Young's modulus $E = 100\,\text{GPa}$, yield stress $\sigma_Y = 250\,\text{MPa}$, and $F = 10,000\,\text{N}$. Design variables are $b_1 = $ area of section AB and $b_2 = $ area of section BC. Starting from the initial design of $b_1 = 1 \times 10^{-4}\,\text{m}^2$ and $b_2 = 2 \times 10^{-4}\,\text{m}^2$, perform the fully stressed design to obtain minimum cross-sectional areas.

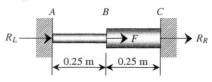

9. Repeat Problem 8 with the initial design of $b_1 = 2 \times 10^{-4}\,\text{m}^2$ and $b_2 = 1 \times 10^{-4}\,\text{m}^2$. Compare the results with those of Problem 8.

10. For the clamped beam shown in the figure, two design variables are defined, as $I_1 = b_1$ and $I_2 = b_2$. Using the finite element method and sensitivity analysis, calculate the sensitivity of the vertical displacement v_2 with respect to b_1 and b_2. Use the following values for the current design: $E = 30 \times 10^6\,\text{psi}$, $b_1 = 0.1\,\text{in}^4$, and $b_2 = 0.05\,\text{in}^4$. Compare the results with the exact sensitivity.

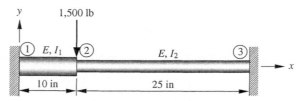

11. Calculate the sensitivity of the vertical displacement v_2 in Problem 10 using forward finite difference method with perturbation size 1%. Compare the results with the exact sensitivity.

12. Consider a simply supported beam of length $L = 1$ m subjected to a uniformly distributed transverse load $p_0 = 100$ N-m. The cross-section is rectangular with width $w = 0.01$ m and height $h = 0.02$ m. Calculate the sensitivity of the vertical displacement at the center with respect to h. Use one finite element with Young's modulus $= 80$ GPa. Compare the sensitivity with finite difference method with perturbation size 1%.

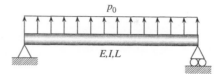

p_0

E,I,L

13. Repeat Problem 12 to calculate the sensitivity of maximum tensile stress with respect to h.

14. A cantilevered beam shown in the figure is under a couple of 500 lb · in at the end. The optimization problem is to find a design that minimizes the cross-sectional area, while the maximum stress is less than 2,000 psi. The thicknesses of the flange and the web of the cross-section are fixed with $t = 0.1$ in. The design variables are the width w and the height h of the cross-section. Determine graphically the optimal design. The width and the height are constrained to remain in the range $0.1 \leq w \leq 10$ in and $0.2 \leq h \leq 10$ in.

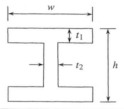

w

t_1

t_2

h

APPENDIX

Appendix A

Finite Element Analysis Using Pro/Engineer

A.1 INTRODUCTION

Pro/ENGINEER is a three-dimensional product design tool that promotes practices in design while ensuring compliance with industry and company standards. Integrated, parametric, three-dimensional CAD/CAM/CAE solutions allow engineers to design products fast, while improving innovation and quality to ultimately create exceptional products. Although the main function of Pro/ENGINEER is a solid modeler, it also provides various engineering analyses tools such as kinematic analysis, structural and thermal analysis. In this chapter, we will introduce structural finite element modeling and analysis using Pro/ENGINEER Mechanica, which is a CAE (Computer Aided Engineering) tool that enables engineers to simulate the physical behavior of a model and to understand and improve the mechanical performance of the design. It can directly calculate stresses, deflections, frequencies, heat transfer paths, and other factors, showing how the model will behave in a test laboratory or in the real world.

The Pro/ENGINEER Mechanica product line features two modules—Structure and Thermal—each of which solves for a different family of mechanical behaviors. Structure focuses on the structural integrity of the model, while Thermal evaluates heat-transfer characteristics.

Mechanica is available in two basic modes—integrated mode and independent mode. In integrated mode, engineers perform all Mechanica functions within Pro/ENGINEER environment. This version of the product offers the convenience and power of Pro/ENGINEER's parametric feature-creation technology coupled with the full range of Mechanica's solution software. However, some specific capabilities of Mechanica cannot be used in the integrated mode. In independent mode, engineers work in a separate user interface, developing the model from imported geometry or geometry that is created using Mechanica's geometry-creation facilities. In this Chapter we will use the integrated mode in performing finite element analysis.

A.2 GETTING START

To access Pro/Engineer program on Windows systems, choose **Start > All Programs > PTC > Pro Engineer > Pro Engineer**. This command will open Pro/Engineer window as shown in Figure A.1. The Pro/Engineer window is composed of different areas. The drawing area shows part geometry and finite elements as well as results of simulation. Once a part is created, the Navigation window will show part tree. Under the main menu, various toolbars are shown, which collect frequently used commands. Since no part is created, most commands are inactive. Usable commands will be activated depending on the current status of the model. In addition, locations of toolbars can be moved.

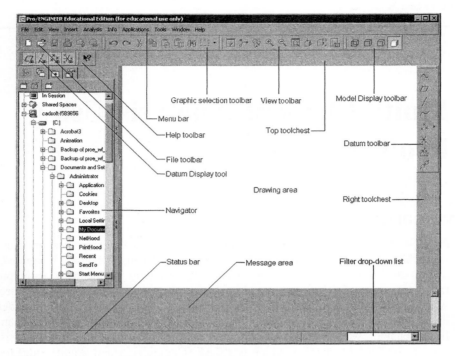

Figure A.1 Layout of Pro/Engineer main window

In order to finish Pro/Engineer, choose **File** > **Exit** command. Then, **Confirmation** window will show up. Choose **Yes** to finish the program.

Pro/Engineer has its full documents in help menu. In the main menu, choose **Help** > **Help Center**. Then Pro/Engineer Help windows will show up. In this documentation, the user can find all manuals and tutorials available in Pro/Engineer.

A.3   PLATE WITH A HOLE ANALYSIS

A.   Introduction

This example problem illustrates the use of Pro/Engineer for a simple static analysis. You will learn how to build a solid model of a plate with a hole, to create finite elements, to perform the analysis, and to examine the results suing Pro/Engineer.

Figure A.2 shows a plate with a hole. The plate is fixed on the left edge and a uniform pressure is applied on the right edge. The magnitude of pressure is chosen such that the total applied force is 300 lb. The purpose of finite element analysis is to estimate the stress concentration along the hole. Even if the model is three-dimensional solid, we will use plane stress elements because stress along the thickness direction is negligible.

B.   Creating the Part

In this section, a simple plate with a hole is constructed using Pro/Engineer solid model. This model will be used for finite element analysis.

1. Start Pro/Engineer and change working folder
   File > **Set Working Directory**
      Select or create a directory in which a part will be created. Note Pro/Engineer use the term ''directory'' which is identical to ''folder'' in windows system.

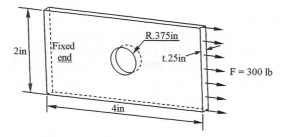

Figure A.2 Plate with a hole model

2. Create a new part
   File > New > Part > part name "PlaneStress"

3. Set up a unit to inch (pound f)
   Edit > Set up > Unit > Inch Pound Second (IPS) > Set > OK (Changing model unit) > Close (Unit Manager) > Done (Menu manager)

4. Specify part creation method
   Insert > Extrude > Placement (in the message area) > Define
   We will sketch the section of the part and extrude it to make a solid. This is the most common way of creating a part. Sketch window will appear

5. Select Sketch Plane
   In the drawing area, three datum planes and a coordinate system are shown. Select the **FRONT** plane and click **Sketch** button in the Sketch window.
   The viewpoint is changed such that the **FRONT** plane is the sketch plane. The sketch menu appears on the right side of the main window.

6. Create 4″ × 2″ rectangular section
   Select **Sketch > Rectangle** and pick the origin of the coordinate system and pick another point in the drawing area to create a rectangle. The size of the rectangle is not important yet. Click the **middle mouse button** finish creating the rectangle. The same action can be done by selecting rectangle symbol in the toolbar on the right side of the drawing area. Now the size of the rectangle will be shown as dimensions.
   Double click each dimension text and change their values so that the size of rectangle be 4″ × 2″ as shown in the figure below. You can move the dimension texts by dragging them in the drawing area.

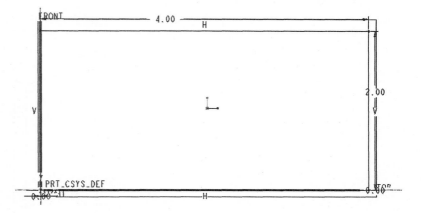

7. Create ∅ 0.75 circle at the center of the rectangle.

    Select **circle icon** on the right side toolbar. Select the center location close to the center of the rectangle and click another point to define the circle. Click the **middle mouse button** to finish creating the circle. Change dimensions in the circle so that the location is in the center of the rectangle and its diameter is 0.75 in. Now the section should look like the figure below.

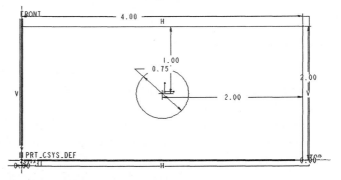

8. Finish creating the section

    Click the **checkmark** (continue with the current section) on the right toolbar to finish creating the section. The drawing toolbars on the right will disappear.

9. Extrude the section to create a solid

    In the message window (below the **placement** button), type in 0.25 for the depth value; click extrude direction so that the part is extruded in backward; and click **checkmark** to extrude the section by 0.25 in.

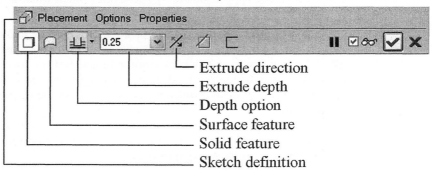

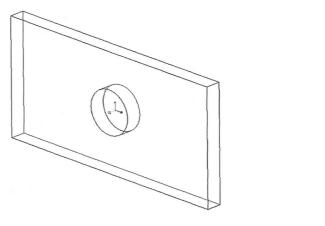

10. Save the solid model and exit
File > Save > file name "PLANESTRESS.PRT"
File > Exit > Yes
   Check whether file "planestress.prt.1" in your working folder exists.

C.   Finite Element Analysis (Integrated Mode)

In the integrated mode, the finite element analysis is performed within Pro/Engineer environment. In the integrate mode, the Mechanica is considered as a sub-menu system of Pro/Engineer.

1. Start Pro/Engineer, change the working directory, and open file "planestress.prt"
File > Set Working Directory
File > Open > planestress.prt

2. Start Mechanica Application
Application > Mechanica > Continue > Structure > OK

3. Idealization
Insert > Midsurface > New > Constant > Select top surface > Hold control-key and select bottom surface > OK > Done/Return
   Finite element model does not have to be exactly the same as the solid model, as long as its mathematical behavior is same. Idealization simplifies the solid model to a plate because displacement and stress are constant through the thickness. In order to see the created mid-surface,
Insert > Midsurface > Compress > Shell & Solid > Done > Done/Return > Done/Return
   We are creating plane stress element with the thickness the same as the solid

4. Displacement boundary condition
Insert > Displacement constraints
   In the Constraint window, type in **displBC** for Name, choose **Edge(s)/Curve(s)** for Reference and select **Edge (1)** in the figure below. Select **OK** to close the Constraint window.
   ConstraintSet1 is the name of default set that can include multiple displacement BCs.

5. Applied load
Insert > Force/Moment Load
   In the Force/Moment Load window, type in **Traction** for Name, choose **Edge(s)/Curve(s)** for Reference and select **Edge (2)** in the figure below. Type in **X = 300** in force column. Select **OK** to close the Force/Moment Load window.

LoadSet1 is the name of default set that can include multiple loads.

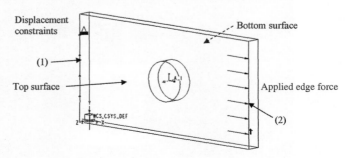

6. Material Data
   Properties > Materials
   In Material window, choose **STEEL** and click **triple arrow** to add the material to the model. Click **OK** to close the Material window.
   Assign the material to the part by clicking **Properties > Material Assignment > OK**. Since PlaneStress is only part and STEEL is only material, we accept defaults.

7. Create Analyses
   Analyses > Mechanica Analyses/Studies...
   In **Analyses and Design Studies** window, select File > **New Static**

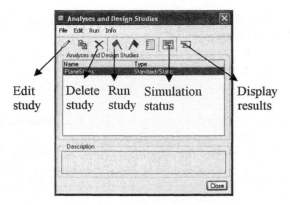

Edit study   Delete study   Run study   Simulation status   Display results

   In **Static Analysis Definition** window, type in ''**PlaneStress**'' for Name, highlight ''**ConstraintSet1**'' and ''**LoadSet1**''. Select Multi-pass adaptive with Polynomial Order, Minimum = 1, Maximum = 9, and Percent convergence = 5% → OK → Close

In pro/Mechanica, the order of interpolation is increased progressively until the solution converges.

8. Run the simulation
   In Analyses and Design Studies window, select ''PlaneStress'' in the box and click **Run study** icon. The finite element analysis will start in background. The status of analysis can be found by clicking **Simulation status** icon.

9. Post-processing (Stress contour plot)
   Once the analysis is finished, the analysis results can be plotted by clicking **Display results** icon. **Result Window Definition** windows will appear. The default values in the window plot the von Mises stress contour. Click **OK and Show** to plot with default values.

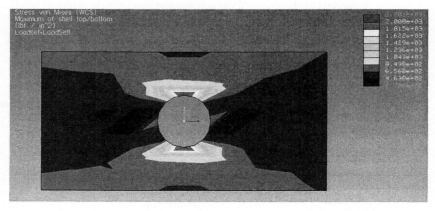

10. Post-processing (Convergence graph)

In addition to the stress contour plot, it is possible to plot a convergence graph of strain energy. Before creating the graph, close the stress contour window by choosing **File > Exit Results > No**

In the **Analysis and Design Studies** window, choose **Display results** icon again. **Result Window Definition** window will appear. Select **Graph** for Display type and **Measure** for Quantity. Click the **Measure** icon below Measure, and select **Strain energy** when the Measure window appears. Select **OK** to close the Measure window and **OK and Show** to display the graph.

The graph shows that the strain energy of the model converges to 0.0148 as higher-order polynomials are used for elements.

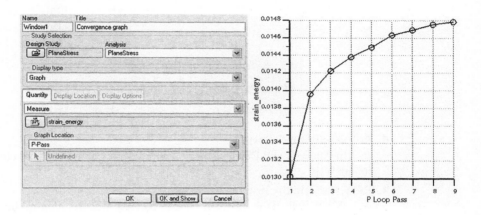

Close the convergence graph by selecting **File > Exit Results > No**

A.4   DESIGN SENSITIVITY ANALYSIS/ PARAMETER STUDY

One of the powerful capabilities in finite element analysis is to easily change the design parameters and to perform what-if study without building the actual part. The purpose of this section is to identify how the performance of the model changes according to the design parameters.

1. Design Sensitivity Study Definition

Select **Analysis > Mechanica Analysis/Studies** from the main menu. In the Analysis and Design Studies window, select **File > New Sensitivity Design Study**. Sensitivity Study Definition windows will appear. Type in Sensitivity Analysis for Name and select PlaneStress in Analysis section.

2. Design Variables

In the same window, click **Select dimension from model** icon and select the part. Then, dimensions of the part will appear. Select height dimension of the plate. Change the name of the variable to "Height". Change the start and end value of the height dimension to 1.85 and 3.0, respectively.

Select the diameter dimension using the same way and provide 0.5 for start and 0.9 for end. Change the name of the variable to "Diameter". Click **OK** to close Sensitivity Study Definition window.

Name
SensitivityAnalysis
Description

Type
Global Sensitivity

Analyses
PlaneStress (Static)
regenerate (Model Regeneration Only)

4

Variables

le	Current	Start	End	Units
	2	1.85	3	in
	0.75	0.5	0.9	in

Steps 10

Options...

OK   Cancel

3. Run sensitivity analysis

In Analysis and Design Studies window, click **Start Run** icon to run sensitivity analysis.

4. Review sensitivity results

When analysis is finished, choose **Review Results** icon in the Analysis and Design Study window.

In the Result Window Definition window, type in ''Sensitivity w.r.t. Height'' in Title. Note that SensitivityAnalysis and PlaneStress are chosen by default in Design Study section. Click Measure icon below Measure section and choose max_stress_vm in the Measure window. Click **OK** to close it. Select Height design variable (or d1) in the Graph Location section and click **OK and Show** button to plot the parameter study results.

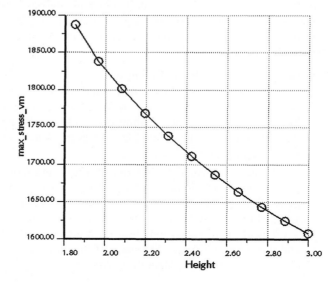

Note that the von Mises stress decreases as the height of the plate increases. A similar graph can be plotted for the diameter design variable.

A.5   DESIGN OPTIMIZATION

Design optimization is automatic design process to find what would be the best design given limitations and conditions. First, we have to define what the goal (measure) is and what the constraints (measure) are. Also, the design parameters need to have the lower and upper bounds that the engineer wants to change. All designs that are covered by the lower and upper bounds are called the design domain. Then, the optimization program sequentially changes the design parameters and finds the design that can minimize (or maximize) the goal while satisfying all constraints within the design domain.

1. Optimization Study Definition

    Select **Analysis** > **Mechanica Analysis/Studies** from the main menu. In the Analysis and Design Studies window, select **File** > **New Optimization Design Study**. Optimization Study Definition windows will appear. Type in ''Optimization'' for Name.

2. Define optimization problem

    Note that minimize total_mass is chosen for the goal by default.

    We want to limit the maximum von Mises stress less than 2500 lbf/in^2. In order to do that, click **Measure** icon on the right side of Design Limits section, and choose max_stress_vm in the Measure window. Type in ''2500'' in Value column so that the design limit looks like max_stress_vm < 2500.

    By following the same procedure in the sensitivity study, define two design variables (height and diameter). For diameter, use minimum (0.5), initial (0.75), and maximum (0.9). For height, use minimum (1.85), initial (2), and maximum (3).

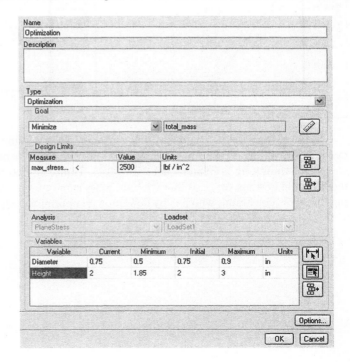

3. Launch the optimization program

In Analysis and Design Studies window, click **Start Run** icon to run sensitivity analysis.

4. Plot the design history

When analysis is finished, choose **Review Results** icon in the Analysis and Design Study window.

In Result Window Definition window, type in ''Optimization'' for Name. Note that Optimization and PlaneStress are already chosen for Design Study and Analysis, respectively. Select **Graph** in the Display type section and **Measure** in the Quantity section. Using the Measure icon below Measure, choose total_mass as the quantity of plot. Click **OK and Show** to plot the history of the total mass.

In order to insert maximum stress history, choose **Insert > Result window** from the main menu. Result Window Definition window will appear. Type in ''History2'' for Name and choose Optimization in Design Study section. Note that PlaneStress will be automatically added in Analysis because it is the only analysis in Optimization. Choose **Graph** in the Display type section and **Measure** in the Quantity section. Using the Measure icon below Measure, choose max_stress_vm as the quantity of plot. Click **OK and Show** to plot the history of the total mass.

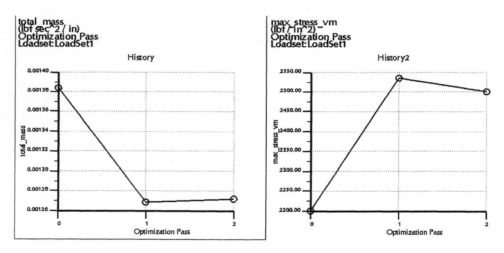

In the history figure, the goal (total mass) is reduced significantly at the first design cycle. However, as can be found in the stress history, the new design slightly violates the stress constraints (it is greater than 2500 psi!). Thus, the optimization program slightly modifies the designs twice so that the total mass cannot be reduce further without violating the stress constraint, which is by definition an optimum design.

Appendix B

Finite Element Analysis Using NEi Nastran

B.1 INTRODUCTION

NEi Nastran is engineering analysis and simulation software developed by Noran Engineering, Inc. NEi Nastran is a general purpose finite element analysis tool with an integrated graphical user interface and model Editor which is used to analyze linear and nonlinear stress, dynamics, and heat transfer characteristics of structures and mechanical components. A complimentary copy of NEi Nastran student edition is included with this book, which is composed of two modules: NEi Fusion and NEi Nastran. NEi Fusion combines a finite element modeler, pre- and post-processing capabilities, and Nastran solvers to create an analysis package for small and medium size problems. It is good for engineers who need simulation for product development, virtual testing, design validation, and quality assurance. Parts and assemblies can be analyzed for a spectrum of structural and thermal loadings. In this chapter, we will present two tutorials using integrated NEi Fusion/NEi Nastran software and several example problems in the text using NEi Nastran alone.

B.2 GETTING START

To access NEi Fusion program on Windows systems, choose **Start** > **All Programs** > **NEi Fusion** > **NEi Fusion v1.2** (Note that the version number v1.2 may be different for other versions). The startup screen will only show **New** and **Open** icons in the standard toolbar. In order to see main windows, choose **File** > **New** in the main menu. Choose **Part** and click **OK** in the New Document window. This command will open NEi Fusion window as shown in Figure B.1. The NEi Fusion window is composed of different areas. The graphic area shows geometries and finite elements as well as results of simulation. The main menu contains primary NEi Fusion functions. The standard toolbars collects commands that are frequently used. The Tab Menu on the left of the screen provides model related functions. In order to finish NEi Fusion, choose **File** > **Exit** command. NEi Fusion has its full documents as well as tutorials in help menu.

B.3 PLATE WITH A HOLE ANALYSIS

A. Introduction

This example problem illustrates the use of NEi Fusion/NEi Nastran for a simple static analysis. You will learn how to build a solid model of a plate with a hole using NEi Fusion

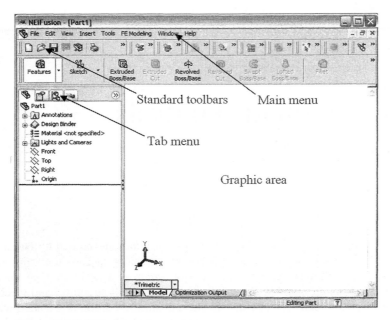

Figure B.1 NEi Fusion main windows

pre- and post-processor, perform the analysis with NEi Nastran, and examine the results with NEi Fusion.

Figure B.2 shows a plate with a hole. The plate is fixed on the left edge and a uniform pressure is applied on the right edge. The magnitude of pressure is chosen such that the total applied force is 300 lb. The purpose of finite element analysis is to estimate the stress concentration along the hole. Even if the model is three-dimensional solid, we will use plane stress elements because stress along the thickness direction is negligible.

B. Creating the Part

1. Create a new part by selecting **File> New** in the main menu. Choose **Part** and click **OK** in the New Document window.

2. Click **Tools> Options...** to open System Options window. Click **Document Properties** tab and select **Units** on the left-hand side tree. Change the units to **IPS (inch, pound, second)** in Unit system. Click **OK** to close Document Properties window.

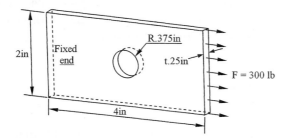

Figure B.2 Plate with a hole model

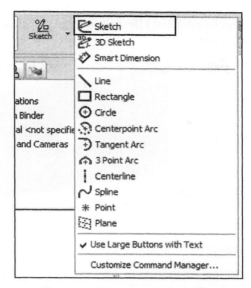

3. In the standard toolbar, select pull-down button of **Sketch** and choose Sketch in the pull-down menu. Select **Front** plane in the graphic window as a sketch plane. Note that the Front plane is x-y plane in the global coordinate system.

4. Draw a rectangle using rectangle icon in the toolbar. It is unnecessary to be a precise dimension yet. Draw a circle inside the rectangle using circle icon in the toolbar.

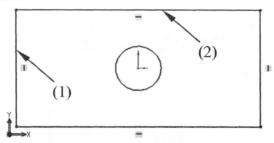

5. The vertical and horizontal line symbols on the rectangle represent that these lines are either vertical or horizontal. Choose the left vertical line (1) of the rectangle. On the Tab menu area, Line Properties window will appear. Change the starting coordinates to x = −2 and y = 1 and length = 2. Choose the top horizontal line (2) of the rectangle. Change the starting coordinates to x = −2 and y = 1 and length = 4. Now the rectangle is properly located.

6. Choose the circle and make its center locations (0, 0) and radius 0.375.

7. Click **Exit Sketch** icon in the standard toolbar to leave the sketch mode.

8. In the standard toolbar, select pull-down button of **Feature** and choose **Extrude Boss/Base** command.

9. In the Extrude window, change thickness = 0.25 in and click **OK** button to generate a solid.

10. Save the model by selecting **File** > **Save** menu. Provide ''PlateHole.SLDPRT'' as the name of the part.

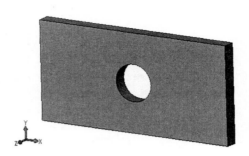

C.   Pre-Processing the Model

Using the solid model that we created in the previous section, we will apply (1) material properties, (2) physical properties, (3) boundary conditions, (4) external forces, and (5) mesh the geometry. These steps are called pre-processing and will be done with the NEi - Fusion program.

Switching to NEi Fusion

1. The NEi Fusion Modeler tab view looks as shown below. Choose the NEi Fusion tab at the top of the tree view to switch to the NEi Fusion environment.

NEi Fusion tab

2. The NEi Fusion tree view looks as shown in the figure below.

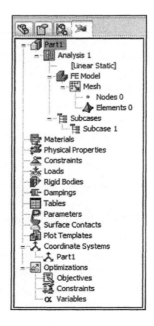

Select Materials

1. Right click on **Materials** in the entity list, and select **New** to create a new material.
2. In the Material window, type in **Material1** in Name section.
3. Change the Young's modulus **E** = 2.8E7 and Poisson's ratio $\nu = 0.3$. Leave shear modulus **G** empty as it will be calculated using previous two constants.
4. Click **OK** (checkmark icon) to leave material section.

Apply the Constraints

1. We will create the mesh on the back face of the plate (This is the plane whose z-coordinates are zero). Change the view to **Hidden Line** using the icon in the standard toolbar.

2. Right-click on **Constraints** in the Entity List and select **New**.
3. In Constraints window, type in **Constraint1** in Name section.
4. Select the left edge of the back face (1) as shown in the figure below. Click the **checkmark** icon to add the constraint to the finite element model.

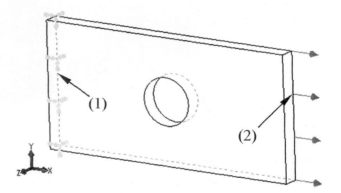

Apply the Load

1. The load is a 300 lbf force applied at the right edge of the plate in x-direction. Right-click on **Loads** in the Entity List and select **New**.
2. In Load window, type in **Load1** in Name section.
3. Select the right edge of the back face (2) as shown in the figure. Type in 300 in **Fx** section. Click the **checkmark** icon to add the load to the finite element model.

Note that the constraint and load are defined before generating finite elements. The nodal constraints and forces will be calculated when the model is meshed using finite elements. If the loads and boundary conditions are defined in the solid model, it is unnecessary to change them when different element sizes are used.

Automatic Mesh Generation

1. NEi Fusion can create mesh and update the boundary conditions (BC) automatically with feature changes in the model.

2. Right-click on **Mesh** in the entity list, and select **Edit** to define mesh parameters.

3. In Mesh window, type in 0.2 in **Element Size (in)** section. NEi Fusion will try to make elements whose edges are close to 0.2 in.

4. Unselect the check mark in **Solid Elements** and select **Shell Elements** because we are going to make plane elements.

5. Click rectangular box under **Quadrilateral** section, and select the back face of the part in the graphic area. The rectangular box will now list **Face <1>**. This means the back face will be meshed using quadrilateral elements. You may need to rotate the part in order to select the back face. In the graphic window, hold the middle mouse button and move the mouse around. You will see dynamic rotation of the part.

6. Click the **checkmark** icon to add the mesh to the finite element model.

7. At the top of the entity list, right-click on Part1 and select **Hide/Display Body**. This will turn off the solid model and only show finite elements.

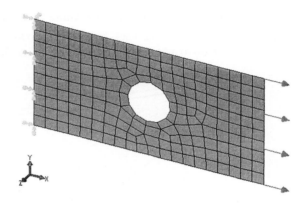

Define Element Properties

1. Right-click on **Physical Properties** in the entity list and select **New**.

2. In the **Physical Property** form, type in **PlaneStress1** for Name section; select **Shell Elements** under Geometry Type; and type in 0.25 for Thickness.

3. Click the **checkmark** icon to add the physical property to the finite element model.

Setup the Analysis

1. In the entity list, right-click on **Analysis 1** and select **Edit**.

2. In Analysis window, type in **First try** in the Name section and **Plate with a hole** in the Title Section.

3. Click the **checkmark** to leave the analysis setup.

D. Running the Analysis

1. Right-click on the analysis name, **First try** and select **Solve in NASTRAN**.

2. This option allows the model to be run in a hidden mode; the default name of the analysis input file would be the same as the saved part file name, **PlateHole.NAS**.

In the hidden mode, the NEi Nastran engine runs the model and displays the progress in the NEi Nastran Output window inside NEi Fusion Modeler. Depending on the system, it may take a few minutes to run. Results are automatically loaded into NEi Fusion Modeler after the model is solved.

E.   Post-Processing the Results

1. If you are in the NEi Nastran Output window, click on the **Model** tab at the bottom of your screen to view the Model from the calculations.

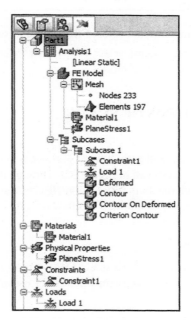

After the solution is complete, NEi Fusion places four different contour plots under **Subcase 1** entity. The tree view looks as shown in the figure after the results are loaded.

2. Right-click on the **Contour on Deformed** plot under **Subcase 1** and select **Edit**.

3. Select **SHELL VON MISES** in the **Result Data** of **Contour Options**. This will plot the von Mises stress contour plot.

4. Type in 10 in Levels section.

5. Unselect Min/Max checkmark.

6. Click the checkmark icon at the top of the Plot window.

7. Right-click on the **Contour on Deformed** plot under **Subcase 1** and select **Display**. The von Mises stress contour plot is displayed on the deformed geometry, as shown in the figure. Compared to the nominal stress of 600 psi, the concentrated stress is about 1790 psi, which is consistent to the traditional stress concentration factor of 3.

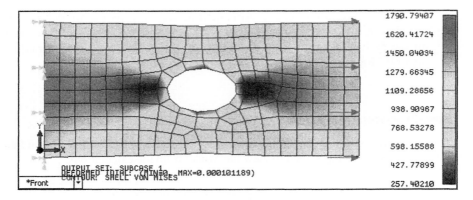

8. In the main menu, select **File > Save** to save the model file.

9. In the main menu, select **File > Exit** to leave the NEi Fusion program. This will end the tutorial.

B.4   STATIC ANALYSIS OF BEAMS[1]

A. Introduction

In this tutorial, we will learn how to perform a linear static analysis of a trailer model composed of beams, as shown in Figure B.3. Through this analysis, the user will learn how to use beam and bar elements in NEi Fusion/NEi Nastran. The units used in this analysis are IPS (inches, pounds, and seconds). The effects of gravity are considered in this model.

B. Opening the Model

1. From the main menu select **File > Open** and open **Tutorial 7.SLDPRT** file located in the ''**C:\Program Files\NEi Fusion\Application\Tutorial**'' folder.

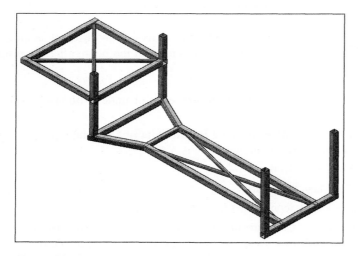

Figure B.3 A trailer model composed of beams

---

[1] Courtesy of Noran Engineering, Inc.

2. Click **File> Save As** and save the part as "**Tutorial 7 Trailer Frame.SLDPRT**" in the working folder. This way the original part can be reused.

3. Switch to the NEi Fusion by clicking ▓ tab in the Tab menu.

C. Define Materials

1. Create a new material by loading AISI 304 Steel from the material library.

2. Right-click on **Materials** and choose **New**, then click on **Load**.

3. Browse to the location of the material library file, "**C:\Program Files\NEi Fusion\ Application**" and open the "**solidworks materials.sldmat**" file.

4. The material tree is populated with the available materials. Under the **Steel** category, select **AISI 304**.

5. Delete the **G** value from the Structural properties field under the tree.

6. Click the **checkmark** icon to load the material.

D. Define Element Properties

1. In the tree view, right-click on **Physical Properties** and choose **New**.

2. Select Line Elements under Geometry Type.

3. The trailer frame is composed of structural members created using the NEi Fusion Weldments Tools. By creating a line element property, you will be able to create one-dimensional elements that have the physical properties based on the Structural Member cross section and the material property. Select a structural member (1) on the model and click on the NEXT ⊕ arrow to create another structural member physical property.

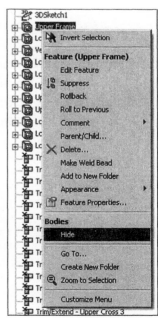

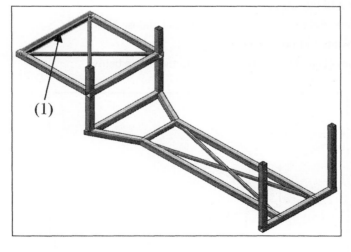

4. Repeat step 3 until all of the structural members have been selected. Click the **checkmark** instead of the **NEXT** arrow to create the final physical property. In total there should be 25 physical properties.

E. Define Constraints

1. To make it easier to define the constraints and mesh the model, you will hide the structural members. Switch to the Feature Manager Design Tree by clicking in the Tab menu.

2. In order to hide structural members, it will be convenient to add the **Hide** feature to the **Structural Member** right-click menu. To do this right click on a structural member "Upper Frame" (it can be distinguished by the icon beside them) and click on double arrows at the bottom to expand the full menu. Click on **Customize Menu**.

3. Check the **Hide** option. Now the Hide feature will be shown to the **Structural Member** right-click menu.

4. Right-click on a **Structural Member** in the tree and select **Hide**. Repeat for other structural members until all are hidden. Now all structural members are displayed as lines.

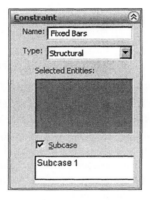

5. Switch to the NEi Fusion tab

6. In the tree view, right-click on **Constraints** and choose **New**.

7. Change the **Name** to **Fixed Bars.**

8. Select the lines (1) and (2) to fix as shown in the figure below.

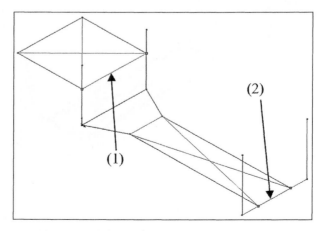

9. Click the **checkmark** to create the fixed constraint.

F.   Define the Loads

1. Right-click on the **Loads** and choose **New**. Change the **Name** to **Gravity Load.**

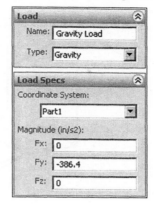

2. Under **Load**, select **Gravity** for **Type**. Type in **-386.4** in the **Fy** field under **Magnitude**.

3. Click the checkmark to create the load.

G.   Mesh the Model

1. Right-click on **Mesh** and choose **Edit**. Type **4** for **Element Size**.

2. Uncheck the **Solid Elements** checkbox and check the **Beam/Bar Elements** checkbox. Click on the **Lines/Arc** box in the **Beam/Bar Elements** form so that it is active.

3. Select all the sketch lines by clicking on the graphics area and drag a box over the entire model. This will select all the lines within the box to create the **Beam/Bar Elements.**

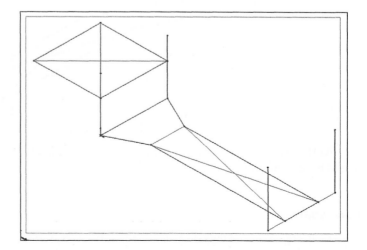

4. Click the checkmark to generate the mesh.

5. The mesh should look like the one shown below.

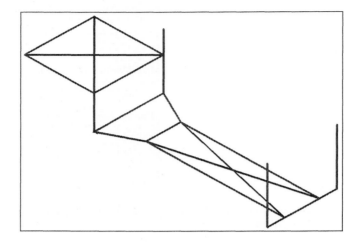

H.   Add Entities to Model

1. Drag and drop constraint **Fixed Bars**, and load **Gravity Load**, one by one onto **Subcases**.

2. Now the model looks like as shown in the image below.

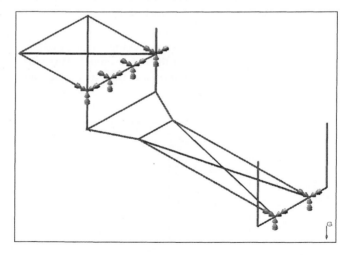

NOTE: By selecting Subcase 1 when defining your loads and constraints, this section could have been skipped.

I. Run the Analysis

1. Right click on **Analysis 1** and select **Solve in Nastran**.

J.   Post-Process Results

1. The results are automatically loaded into NEi Fusion.

2. Right-click on **Contour On Deformed** under Subcase 1 and choose **Edit**.

3. Select BAR VON MISES STRESS for the Results Data under Contour Options. Deselect the Max/Min checkbox.

4. Make sure **Deformed** and **Contour** are selected under **Name**. Click on **Display**. Note the results.

5. Hide the sketches by switching to the Feature Manager Design Tree 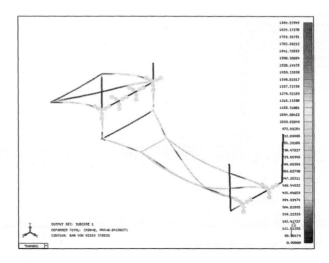, and hiding the sketches similar to how you hid the structural members. The results should look like the image below.

This concludes the tutorial. The major topics that were covered in this tutorial were as follows:
- Running an analysis of a bar/beam structure
- Create line element physical properties based on existing structural elements created with NEi Fusion Weldments Tools
- Using gravity loads

B.5 EXAMPLES IN THE TEXT

NEi Nastran Unser Interface

The finite element modeling and analysis can also be performed using NEi Nastran Editor. The NEi Nastran Editor is a utility that allows the user to manage analysis files, monitor analysis progress, queue multiple analysis jobs, change NEi Nastran settings and options, and access the NEi Nastran help file. NEiNastran Editor can be launched from **Start> All Programs> NEi Nastran> NEi Nastran Editor** in Windows system.

In the large **Editor** View of the NEi Nastran Editor (see Fig. B.4) you should see the **Errors/Warnings** View. In this view you will see a listing of any errors and warnings generated during the analysis. You can see more information by selecting from the several tabs at the bottom of the Editor View.

Click the **Nastran** tab and you will see the Model Input File. This is the file that NEi Fusion prepares when you choose **Solve in Nastran** command. It contains all of the pre-processed model data needed by the NEi Nastran solver to perform the analysis.

The **Analysis** View shows the status information provided by the solver during the analysis.

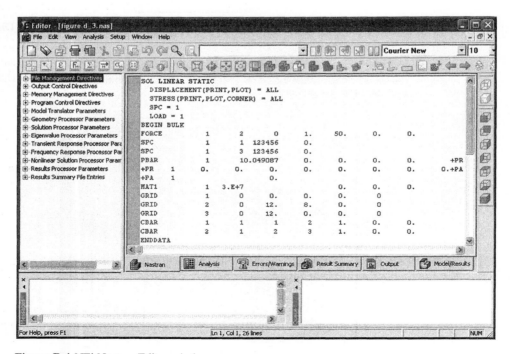

Figure B.4 NEi Nastran Editor window

Click the **Result Summary** tab and the Result Summary File appears in the view. The Results Summary File is a report of any runtime errors or warnings, the model's properties, the number of elements, number of degrees of freedom, maximum aspect ratios of the elements, total mass, epsilon, strain energy, extreme values of various stresses, and other data. The report's contents can be interrogated or printed from the Editor.

The **Output** View has the listing of the output file that is written by NEi Nastran when the analysis is complete. This file contains all of the analysis results in an ASCII format.

To the left of the **Editor** View is the **Options** View. Here the Model Initialization File settings can be viewed or modified. These settings allow you to configure NEi Nastran.

The window on the bottom is the **Messages** Window. The messages displayed here come from the NEi Nastran Editor and solver.

NEi Nastran Model Input File Format

The first step in performing analysis using NEi Nastran is to generate the Model Input File, which defines the geometry, material properties, boundary conditions, and applied loads. The Model Input File is an 80 column ASCII text file and can be created using NEi Nastran Editor or any text editors. The Model Input file also specifies analysis procedures as well as outputs. The Model Input file is divided into two distinct sections: the Case Control Section and the Bulk Data Section. Input in the Case Control Section is referred to as a command and in the Bulk Data Section as an entry. The Case Control section must appear first, followed by Bulk Data section.

The Case Control Section begins with the first command and ends with the command, BEGIN BULK. It defines the subcase structure for the problem, makes selections from the Bulk Data Section, defines the output coordinate system for element and grid point results, and makes output requests for the Model Results Output File.

The Bulk Data section begins with the entry following BEGIN BULK and ends with the entry ENDDATA. It contains all of the details of the structural model and the conditions for the solution. BEGIN BULK and ENDDATA must be present even though no new bulk data is being introduced into the problem or all of the bulk data is coming from an alternate source, such as user-generated input. The format of the BEGIN BULK entry is either a free field or a structured field. Data are separated by comma in free field, while each data is written within eight columns in structured field.

Plane Truss Example 1

In this section, the two–bar truss in Example 2.4 is solved using NEi Nastran Editor. The problem we are analyzing is shown in Figure 2.11 in Chapter 2 and the corresponding NEi Nastran Model Input File is shown in Table B.1. The Case Control section has five commands. The first command ''SOL LINEAR STATIC'' specifies that the linear static analysis will be employed. Next two commands request for outputs. In this case, displacements of all nodes and stresses of all elements will be included. Also these data will be available in the output file as well as database.

The next ''SPC = 1'' command directs NEi Nastran to apply constraints defined by the SPC entry with an identification number (ID) of 1 in the Bulk Data Section. The ''LOAD = 1'' command directs NEi Nastran to apply loading defined by the FORCE entry with an ID of 1 in the Bulk Data Section.

The start of the Bulk Data Section is denoted by the BEGIN BULK delimiter and the end, the ENDDATA delimiter. Both are delimiters are required. The model's geometry is defined via the GRID entry, which corresponds to node. Each grid point coordinate is defined in the default basic coordinate system. The format for GRID entry is:

Table B.1   NASTRAN Commands for Plane Truss Example 1

```
SOL LINEAR STATIC
DISPLACEMENT(PRINT,PLOT) = ALL
STRESS(PRINT,PLOT,CORNER) = ALL
SPC = 1
LOAD = 1
BEGIN BULK
GRID           1         0              0.  0.     0.   0
GRID           2         0             12.  8.     0.   0
GRID           3         0             12.  0.     0.   0
CBAR           1         1              1   2   1.  0.      0.
CBAR           2         1              2   3   1.  0.      0.
PBAR           1         10.04908       7   0.  0.  0.      0.
MAT1           1      3.E+7                         0.  0.  0.
FORCE          1         2              0   1.  50. 0.      0.
SPC            1         1        123456 0.
SPC            1         3        123456 0.
ENDDATA
```

GRID	ID	CP	X	Y	Z	CD			

ID is non-overlapping node number, CP = 0 for the default basic coordinate system, and (X, Y, Z) is the coordinates of the node, and CD is the displacement coordinate system.

Element connectivity is defined via the CBAR entries. The format of CBAR entry is:

CBAR	EID	PID	GA	GB	X1	X2	X3		

EID is the element identification number, PID is property identification number of PBAR, GA and GB are two grid points, and (X1, X2, X3) are vector that defines the local x-y plane in the displacement coordinate system.

The properties of BAR element are defined using PBAR entry. The format of PBAR entry is:

PBAR	PID	MID	A	I1	I2	J	NSM		

PID is the property identification number, MID is the material identification number, A is the cross-sectional area, I1 and I2 are area moment of inertia, J is the torsional constant, and NSM is nonstructural mass per unit length.

The material property of the BAR element is defined using MAT1 entry, which defines the material properties of linear, temperature-independent, isotropic materials. The format of MAT1 entry is:

MAT1	MID	E	G	NU	RHO	A	TREF	GE	

MID is the material identification number, E the Young's modulus, G the shear modulus, NU Poisson's ratio, RHO the density, A the thermal expansion coefficient, TREF the reference temperature, and GE structural damping coefficient.

The applied load can be specified using FORCE entry. The format of FORCE entry is:

FORCE	SID	G	CID	F	N1	N2	N3		

SID is the load set identification number, G the grid point that the force is applied, CID the coordinate system, F the load scale factor, (N1, N2, N3) is the vector of load in CID coordinate system.

The displacement boundary conditions are applied using two SPC entries. The format of SPC entry is:

SPC	SID	G1	C1	D1	G2	C2	D2		

SID is the identification number of single point constraint set, G1 and G2 are Grid points, C1 and C2 are component numbers of global coordinates, and D1 and D2 are enforced displacements. For example, when all six DOFs are fixed, C1 and C2 have the value of 123456.

Once, the Model Input File is prepared, finite element analysis can be performed by selecting **Analysis > Run** command or simply clicking **F5** key. Analysis progress will be shown in Analysis window. Once the analysis is finished, **Errors/Warnings** window will show ant errors or warnings that occur during analysis.

The Output window shows analysis results. There is a quite amount of information in the output. By scrolling down the output window, we can find the nodal displacement and element stress. Since Nodes 1 and 3 are fixed, their displacements will be zero. Node 2 is the only free node, whose displacement is

GRID	T1	T2	T3	R1	R2	R3
2	8.28E-4	−1.81E-4	0.0	0.0	0.0	0.0

Note that T1, T2, and T3 are displacements in X, Y, and Z-coordinates, respectively, and R1, R2, and R3 are rotations.

The axial stresses of two elements can also be found in the Output window.

Element	E1	E2
Axial Stress	1.22E+3	−6.79E+2

Plane Truss Example 2

In this section, a nine-bar truss in Figure B.5 is solved using NEi Nastran Editor. The corresponding NEi Nastran Model Input File is shown in Table B.2. The Case Control section is identical to that of the two-bar truss example.

In the Bulk Data section, the model has six GRID entries, nine CBAR entries. But, their definitions are similar to that of the previous example. We used Young's modulus of $E = 2.9 \times 10^7$ psi. Since 1,200 lb force is applied in the negative y-axis at Node 3, the FORCE entry has −1200 for N2 column. Also, 400 lb force is applied in N1 column of Node 6. Since Node 1 is fixed, all six DOFs are constrained, while Node 4 is allowed to move in the x-coordinate direction.

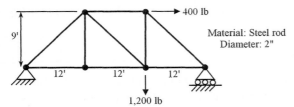

Figure B.5 Nine-bar truss model

After finishing finite element analysis, the displacements and stresses of the model can be found in the Output window. However, NEi Nastran Editor provides graphical post-processing capability.

Select Model/Results window. In the **Options** window (left to the main screen), Results> NEiNastran>Vector Results> Total Translation. This command will produce a contour plot of displacement. In the Toolbar, click "Display Contour" icon and "Display Deformed View" icon. The main window now shows the deformed geometry of the model, as shown in Figure B.6.

Portal Frame Analysis

Consider the plan frame in Figure B.7. All frames have the same material properties. The horizontal frame has different cross-sectional geometry from two vertical frames. All frames are made of the same material with Young's modulus $E = 210$ GPa. Two vertical frames have area $A_1 = A_2 = 4.53 \times 10^{-3}\,\mathrm{m}^2$ and moment of inertia

Table B.2   NASTRAN Commands for Plane Truss Example 2

```
SOL LINEAR STATIC
DISPLACEMENT(PRINT,PLOT) = ALL
STRESS(PRINT,PLOT,CORNER) = ALL
SPC = 1
LOAD = 1
BEGIN BULK
GRID           1        0          0.       0.      0.       0
GRID           2        0        144.       0.      0.       0
GRID           3        0        288.       0.      0.       0
GRID           4        0        432.       0.      0.       0
GRID           5        0        144.     108.      0.       0
GRID           6        0        288.     108.      0.       0
CBAR           1        1          1        2      0.       1.   0.
CBAR           2        1          2        3      0.       1.   0.
CBAR           3        1          3        4      0.       1.   0.
CBAR           4        1          1        5      0.       1.   0.
CBAR           5        1          2        5      1.       0.   0.
CBAR           6        1          3        5      0.       1.   0.
CBAR           7        1          5        6      0.       1.   0.
CBAR           8        1          3        6      1.       0.   0.
CBAR           9        1          4        6      0.       1.   0.
PBAR           1        1    3.14159       0.      0.       0.   0.
MAT1           1    2.9E+7
FORCE          1        3          0       1.      0.   -1200.   0.
FORCE          1        6          0       1.     400.      0.   0.
SPC            1        1     123456       0.
SPC            1        4      23456       0.
ENDDATA
```

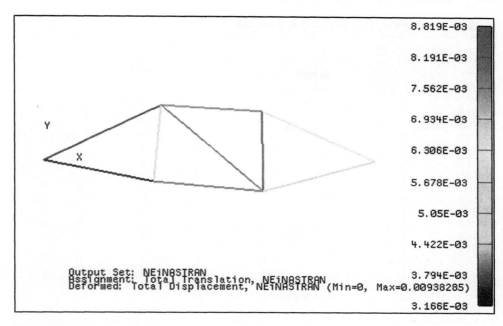

Figure B.6 Displacement contour plot of nine-bar truss

$I_1 = I_2 = 2.51 \times 10^{-5}\,\text{m}^4$. The horizontal frame has area $A_3 = 1.43 \times 10^{-2}\,\text{m}^2$ and moment of inertia $I_3 = 3.309 \times 10^{-4}\,\text{m}^4$. Uniformly distributed load $p = 75\,\text{kN/m}$ is applied at the horizontal frame, and a concentrated force $F = 1\,\text{kN}$ is applied in the horizontal direction at Node 2.

The NEi Nastran commands for the portal frame example are shown in Table B.3. Note that FORCE command is added in the CASE Control section, which will print element forces (bending moment, shear force, and axial force). In the Bulk Data section, the same CBAR elements are used for the frame elements. In fact, CBAR can be used for truss, beam and frame elements. In PBAR entry, not only for the cross-sectional area, but also the moment of inertia column is provided because the element will behave like the combined uniaxial bar and beam element. In addition to the FORCE entry, PLOAD1 entry is provided, which defines distributed loads to CBAR elements at user chosen points along the axis. The format of PLOAD1 entry is:

PLOAD1	SID	EID	TYPE	SCALE	X1	P1	X2	P2	

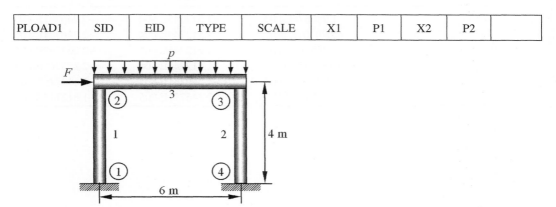

Figure B.7 Plane frame structure under distributed and concentrated load

Table B.3 NASTRAN Commands for Portal Frame

```
SOL LINEAR STATIC
DISPLACEMENT(PRINT,PLOT) = ALL
STRESS(PRINT,PLOT,CORNER) = ALL
FORCE(PRINT,PLOT,CORNER) = ALL
SPC = 1
LOAD = 1
BEGIN BULK
GRID       1       0       0.         0.        0.        0
GRID       2       0       0.         4.        0.        0
GRID       3       0       6.         4.        0.        0
GRID       4       0       6.         0.        0.        0
CBAR       1       1       1          2         1.        0.       0.
CBAR       2       1       3          4         1.        0.       0.
CBAR       3       2       2          3         0.        1.       0.
PBAR       1       1    0.00453    2.51E-5      0.        0.       0.
PBAR       2       1    0.0143     3.309E-4     0.        0.       0.
MAT1       1    2.1E+11
FORCE      1       2          0      1.       1000.       0.       0.
PLOAD1     1       3        FYE      FR         0.     -75000.     1.    -75000.
SPC        1       1     123456      0.
SPC        1       4     123456      0.
ENDDATA
```

SID and EID are load set and element identification numbers, respectively; TYPE is the type of distributed load (FX, FY, FZ, FXE, FYE, FZE, MX, MY, MZ, MXE, MYE, MZE); SCALE is the scale factor; X1 and X2 are locations starting from the first node; and P1 and P2 are load factors at location X1 and X2, respectively. If SCALE = FR (fractional), the Xi values are ratios of the distance along the axis to the total length, and (if X2 ≠ X1) Pi are load intensities per unit length of the element.

The analysis procedure is identical to the previous examples. In the Output window, we can find the bending moment, shear force, and axial force of each element, as shown in Table B.4.

Table B.4 Bending Moment, Shear Force, and Axial Force of the Portal Frame

Element	Node	Bending moment	Shear force	Axial force
1	1	−1.980678E+04	−1.512745E+04	−2.246731E+05
	2	4.070301E+04	−1.512745E+04	−2.246731E+05
2	3	−4.266460E+04	−1.612745E+04	−2.253269E+05
	4	2.184518E+04	−1.612745E+04	−2.253269E+05
3	2	−4.070301E+04	−2.246731E+05	−1.612745E+04
	3	−4.266460E+04	−2.246731E+05	−1.612745E+04

Appendix C

Finite Element Analysis Using ANSYS

C.1 INTRODUCTION

ANSYS is the original (and commonly used) name for ANSYS Mechanical or ANSYS Multiphysics, general-purpose finite element analysis software. ANSYS, Inc actually develops a complete range of CAE products, but is perhaps best known for ANSYS Mechanical & ANSYS Multiphysics. The academic versions of these commercial products are referred to as ANSYS Academic Research, ANSYS Academic Teaching Advanced, Introductory etc. All of these products are general purpose finite element self contained analysis tools incorporating preprocessing (geometry creation, meshing), solver and post processing modules in a unified graphical user interface (GUI). For the remainder of this appendix, when we state "ANSYS", we mean the FEA capability of either the commercial or academic ANSYS Inc. products discussed above. Further we are limiting our discussion to the traditional (or "classic") user interface of ANSYS, not the ANSYS Workbench environment.

One of the advantages of "ANSYS" is the user-programmable capability. The ANSYS Command Language contains several thousand commands relating to creating geometry, mesh, boundary conditions, solver settings & many other features. A subset of these commands are termed ANSYS Parametric Design Language (APDL), and focus on managing parameters, macros, if-then-else branching, do-loops, and scalar, vector and matrix operations. For example, if multiple analyses are required with different values of parameters, do-loops can be used to change parameters, or the Optimization module (/OPT) can be used to automatically manage parameters & populate a response surface.

There are two different modes of performing finite element analysis in ANSYS: batch and interactive modes. The batch mode requires an input file of commands and executes it from the command line, while the interactive mode turns on the graphical user interface and shows the result of each user action (picked from a menu or typed as a command) in a graphics window. The batch mode is best suited for those experienced with the ANSYS Command Language. In this tutorial, only the interactive mode will be explained.

There are two different modes of inputting commands in ANSYS: graphic mode and text mode. In the graphic mode, the user can pick commands and type in required data values. Thus, there is no need to memorize all commands. This mode is beneficial in a sense that the user can see the result of each command in the graphic window. In the text mode, the user type in commands and data in the command area. In this case, the user should remember what command should be used to perform a particular action. However, the text mode has its own advantages. For example, during the design process, the user may want to perform the analysis multiple times with different values of parameters. In such a case, the user can prepare a text file of all commands and then input the file as an input command file. In fact, all actions in the graphic mode are stored as commands in a text file, **jobname.log**. Thus, it would be a good practice to prepare the initial model using the graphic mode, and then, modify the input file for the further modification. In this chapter, we will present two tutorials using the graphic mode, and then will present several examples in Chapters 2–7 using the text mode.

C.2 GETTING START

ANSYS can be started in different ways. In this tutorial, we will start ANSYS in the traditional mode. To access ANSYS program on Windows systems, choose **Start> Programs> ANSYS 11.0> ANSYS**. This command will open ANSYS window as shown in Figure C.1. The ANSYS window is composed of different areas. The graphic area shows geometries and finite elements as well as results of simulation. The main menu contains primary ANSYS functions. The user can directly type in commands in the command area. The standard toolbox collects commands that are frequently used. The utility menu contains utility functions that are available throughout the ANSYS session.

In order to finish ANSYS, choose **File> Exit** command. Then, **Exit from ANSYS** window will show up as shown in the right-lower corner of Figure C.1. Choose a proper option and **OK** to finish the program.

ANSYS has its full documents in help menu. In the utility menu, choose **Help> Help Topics**. Then ANSYS 11.0 Documentation windows will show up. In this documentation, the user can find all manuals and tutorials available in ANSYS.

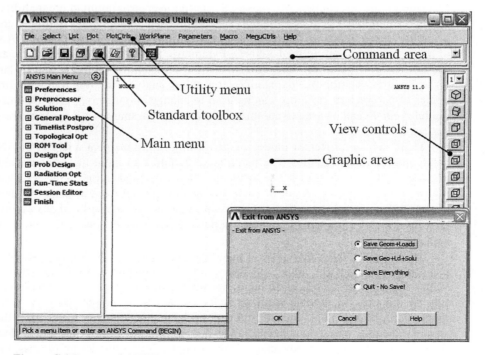

Figure C.1 Layout of ANSYS main window and exit window

C.3   STATIC ANALYSIS OF A CORNER BRACKET[2]

A. Problem Description

This is a simple, single load step, structural static analysis of the corner angle bracket shown in Figure C.2. The upper left-hand pin hole is constrained (welded) around its entire circumference, and a tapered pressure load is applied to the bottom of the lower right-hand pin hole. The objective of the problem is to demonstrate the typical ANSYS analysis procedure. The US Customary system of units is used.

The dimensions of the corner bracket are shown in the figure. The bracket is made of A36 steel with a Young's modulus of 30E6 psi and Poisson's ratio of 0.27. Since the bracket is thin in the z-direction (1/2 inch thick) compared to its x and y dimensions, and since the pressure load acts only in the x-y plane, the problem can be assumed as plane stress. We will use solid modeling to generate the two-dimensional model and automatically mesh it with nodes and elements. (Another alternative in ANSYS is to create the nodes and elements directly.)

If your system includes a Flash player (Macromedia, Inc.), one can view demonstration videos of each step by pointing the web browser to the following URL address: http://www.ansys.com/techmedia/structural_tutorial_videos.html

B. Building the Geometry

In order to perform the finite element analysis, we need to create the geometry, and elements, and to apply boundary conditions. This step is called pre-processing.

Step 1:   Define rectangles

There are several ways to create the model geometry within ANSYS, some more convenient than others. The first step is to recognize that you can construct the bracket easily with combinations of rectangle and circle primitives. Decide where the origin will be located and then define the rectangle and circle primitives relative to that origin. The location of the origin is arbitrary. Here, use the center of the upper left-hand hole. ANSYS does not need to know where the origin is. Simply begin by defining a rectangle relative to that location. In ANSYS, this origin is called the ***global origin***.

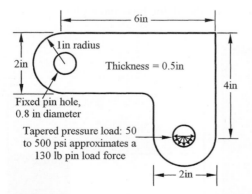

Figure C.2 Corner bracket stress analysis problem

---

[2] By the courtesy of ANSYS, Inc.

1. Choose **Main Menu> Preprocessor> Modeling> Create> Areas> Rectangle> By Dimensions**
2. Enter the following data to create the first rectangle: $X1 = 0$, $X2 = 6$, $Y1 = -1$, $Y2 = 1$ (Note: Press the Tab key between entries)
3. **Apply** to create the first rectangle.

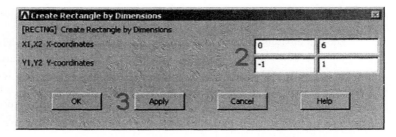

4. Enter the following data to create the second rectangle: $X1 = 4$, $X2 = 6$, $Y1 = -1$, $Y2 = -3$
5. **OK** to create the second rectangle and close the dialog box.
Now the geometry should look like the figure below.

Step 2: Change plot controls and replot

The area plot shows both rectangles, which are *areas*, in the same color. To more clearly distinguish between areas, turn on area numbers and colors. The ''Plot Numbering Controls'' dialog box on the Utility Menu controls how items are displayed in the Graphics Window. By default, a ''replot'' is automatically performed upon execution of the dialog box. The replot operation will repeat the last plotting operation that occurred (in this case, an area plot).

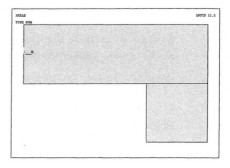

1. Choose **Utility Menu> Plot Ctrls> Numbering**
2. Turn on area numbers.
3. Choose **OK** to change controls, close the dialog box, and replot.

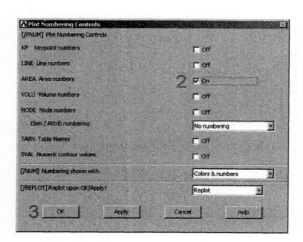

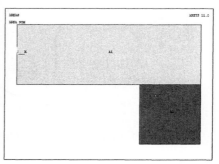

Before going to the next step, save the work you have done so far. ANSYS stores any input data in memory to the *ANSYS database*. To save that database to a file, use the SAVE operation, available as a tool on the Toolbar. ANSYS names the database file using the format *jobname.db*. You can check the current jobname at any time by choosing **Utility Menu> List> Status> Global Status**. You can also save the database at specific milestone points in the analysis (such as after the model is complete, or after the model is meshed) by choosing **Utility Menu> File> Save As** and specifying different jobnames (*model.db*, or *mesh.db*, etc.).

It is important to do an occasional save so that if you make a mistake, you can restore the model from the last saved state. You restore the model using the **RESUME** operation, also available on the Toolbar. (You can also find SAVE and RESUME on the Utility Menu, under File.)

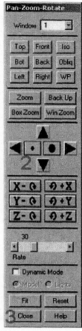

Step 3:   Change working plane to polar and create first circle

The next step in the model construction is to create the half circle at each end of the bracket. You will actually create a full circle on each end and then combine the circles and rectangles. To create the circles, you will use and display the working plane. You could have shown the working plane as you created the rectangles but it was not necessary.

Before you begin however, first "zoom out" within the Graphics Window so you can see more of the circles as you create them. You do this using the "Pan-Zoom-Rotate" dialog box, a convenient graphics control box you'll use often in any ANSYS session.

1. Choose **Utility Menu> PlotCtrls> Pan, Zoom, Rotate**
2. Click on small dot once to zoom out.
3. Close dialog box.
4. Choose **Utility Menu> WorkPlane> Display Working Plane** (toggle on)

Notice the working plane origin is immediately plotted in the Graphics Window. It is indicated by the WX and WY symbols; right now coincident with the global origin X and Y symbols. Next you will change the WP type to polar, change the snap increment, and display the grid.

5. Choose **Utility Menu> WorkPlane> WP Settings**
6. Click on **Polar**.
7. Click on **Grid and Triad**.
8. Enter **.1** for snap increment.
9. **OK** to define settings and close the dialog box.

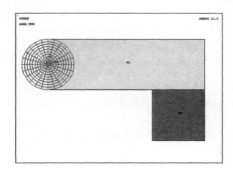

10. Choose **Main Menu> Preprocessor> Modeling> Create> Areas> Circle> Solid Circle**

Be sure to read prompt before picking.

11. Pick center point at: WP X = 0, WP Y = 0, Radius = 1 (in Graphics Window shown below)

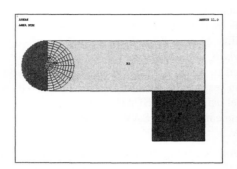

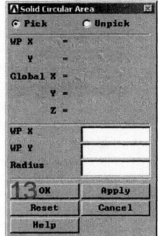

12. Choose **OK** to close picking menu.
13. Toolbar: **SAVE_DB**.

Step 4:   Move working plane and create second circle

To create the circle at the other end of the bracket in the same manner, you need to first move the working plane to the origin of the circle. The simplest way to do this without entering number off-sets is to move the WP to an average keypoint location by picking the keypoints at the bottom corners of the lower, right rectangle.

1. Choose **Utility Menu> WorkPlane> Offset WP to> Keypoints**
2. Pick keypoint at lower left corner of rectangle.
3. Pick keypoint at lower right of rectangle.

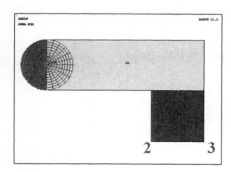

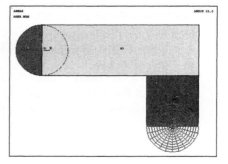

4. OK to close picking menu.
5. Choose **Main Menu> Preprocessor> Modeling> Create> Areas> Circle> Solid Circle**
6. Pick center point at: WP X = 0, WP Y = 0, Radius = 1
7. OK to close picking menu.
8. Toolbar: SAVE_DB.

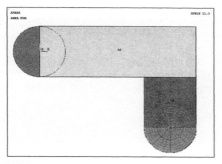

Step 5: Add areas.

Now that the appropriate pieces of the model are defined (rectangles and circles), you need to add them together so the model becomes one continuous piece. You do this with the Boolean add operation for areas.

1. Choose **Main Menu> Preprocessor> Modeling> Operate> Booleans> Add> Areas**
2. Choose **Pick All** for all areas to be added.

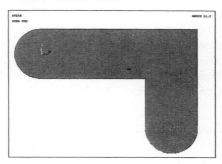

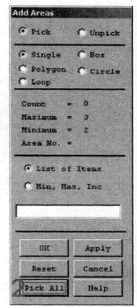

3. Toolbar: SAVE_DB.

Step 6:   Create line fillet.

1. Choose **Utility Menu> PlotCtrls> Numbering**
2. Turn on line numbering.
3. OK to change controls, close the dialog box, and automatically replot.

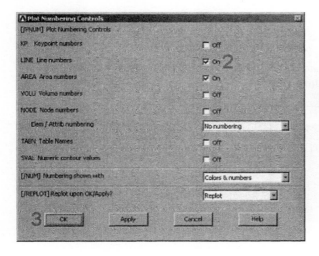

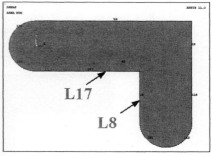

4. Choose **Utility Menu> WorkPlane> Display Working Plane** (toggle off)

5. Choose **Main Menu> Preprocessor> Modeling> Create> Lines> Line Fillet**

6. Pick lines 17 and 8.
7. OK to finish picking lines (in picking menu).
8. Enter 0.4 as the radius.
9. OK to create line fillet and close the dialog box.
10. Choose **Utility Menu> Plot> Lines**

Step 7: **Create fillet area.**

1. Choose **Utility Menu> PlotCtrls> Pan, Zoom, Rotate**
2. Click on Zoom button.
3. Move mouse to fillet region, click left button, move mouse out and click again.
4. Choose **Main Menu> Preprocessor> Modeling> Create> Areas> Arbitrary> By Lines**

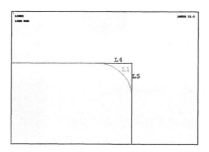

5. Pick lines 4, 5, and 1.
6. OK to create area and close the picking menu.
7. Click on Fit button in Pan, Zoom, Rotate dialog box.
8. Close the Pan, Zoom, Rotate dialog box.
9. Choose **Utility Menu> Plot> Areas**

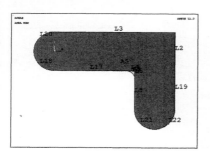

10. Toolbar: SAVE_DB.

Step 8:   Add areas together.

1. Choose **Main Menu**> **Preprocessor**> **Modeling**> **Operate**> **Booleans**> **Add**> **Areas**
2. Pick All for all areas to be added.
3. Toolbar: SAVE_DB.

Step 9:   Create first pin hole.

1. Choose **Utility Menu**> **WorkPlane**> **Display Working Plane** (toggle on)
2. Choose **Main Menu**> **Preprocessor**> **Modeling**> **Create**> **Areas**> **Circle**> **Solid Circle**
3. Pick center point at: WP X = 0, WP Y = 0, Radius = 0.4
4. OK to close picking menu.

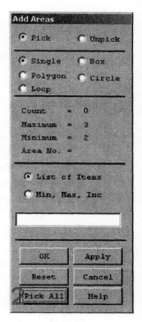

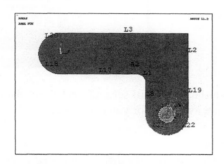

Step 10:   Move working plane and create second pin hole.

1. Choose **Utility Menu**> **WorkPlane**> **Offset WP to**> **Global Origin**
2. Choose **Main Menu**> **Preprocessor**> **Modeling**> **Create**> **Areas**> **Circle**> **Solid Circle**
3. Pick center point at: WP X = 0, WP Y = 0, Radius = 0.4 (in Graphics Window)
4. OK to close picking menu.
5. **Utility Menu**> **WorkPlane**> **Display Working Plane** (toggle off)
6. **Utility Menu**> **Plot**> **Lines**

If you plot areas, it appears that one of the pin hole areas is not there. However, it is there (as indicated by the presence of its lines), you just can't see it in the final display of the screen. That is because the bracket area is drawn on top of it. An easy way to see all areas is to plot the lines instead.

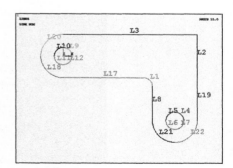

7. Toolbar: SAVE_DB.

Step 11:   Subtract pin holes from bracket.

1. Choose **Main Menu**> **Preprocessor**> **Modeling**> **Operate**> **Booleans**> **Subtract**> **Areas**
2. Pick bracket as base area from which to subtract.

3. Apply (in picking menu).
4. Pick both pin holes as areas to be subtracted.
5. OK to subtract holes and close picking menu.

Step 12: Save the database as model.db.

At this point, you will save the database to a named file—a name that represents the model before meshing. If you decide to go back and remesh, you'll need to resume this database file. You will save it as model.db.

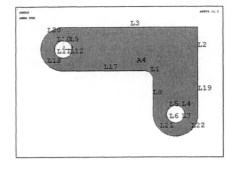

1. Choose **Utility Menu> File> Save As**
2. Enter model.db for the database file name.
3. OK to save and close dialog box.

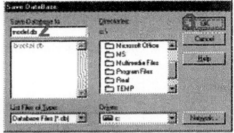

C. Define Materials

Step 13: Set preferences.

In preparation for defining materials, you will set preferences so that only materials that pertain to a structural analysis are available for you to choose.

To set preferences:

1. Choose **Main Menu> Preferences**
2. Turn on structural filtering. The options may differ from what is shown here since they depend on the ANSYS product you are using.
3. OK to apply filtering and close the dialog box.

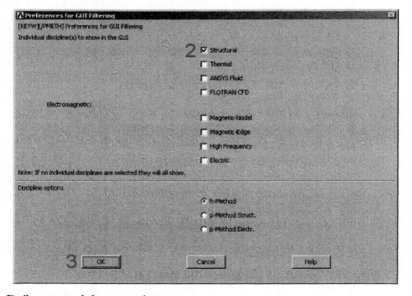

Step 14: Define material properties.

To define material properties for this analysis, there is only one material for the bracket, A36 Steel, with given values for Young's modulus of elasticity and Poisson's ratio.

1. Choose **Main Menu**> **Preprocessor**> **Material Props**> **Material Models**
2. Double-click on Structural, Linear, Elastic, Isotropic.
3. Enter 30e6 for EX and 0.27 for PRXY.
4. OK to define material property set and close the dialog box.
5. In the Define Material Model window, choose **Material**> **Exit**

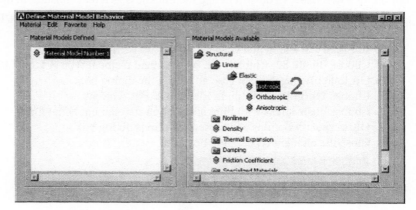

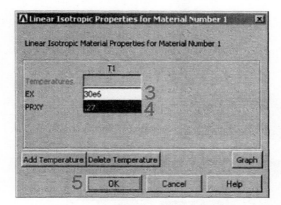

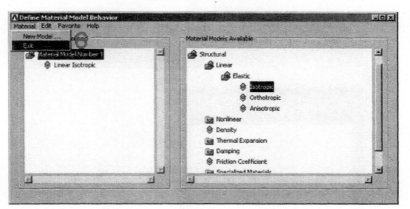

Step 15:   Define element types and options.

In any analysis, you need to select from a library of element types and define the appropriate ones for your analysis. For this analysis, you will use only one element type, **PLANE82**, which is a 2-D, quadratic, structural, higher-order element. The choice of a higher-order element here allows you to have a coarser mesh than with lower-order elements while still maintaining solution accuracy. Also, ANSYS will generate some triangle shaped elements in the mesh that would otherwise be inaccurate if you used lower-order elements (**PLANE42**). You will need to specify plane stress with thickness as an option for **PLANE82**. (You will define the thickness as a real constant in the next step.)

1. Choose **Main Menu**> **Preprocessor**> **Element Type**> **Add/Edit/Delete**
2. Choose **Add**. . . button in Element Types window.
3. In Library of Element Types window, Choose **Solid**.
4. Choose **8node 82**, which is the 8-node quad element type (**PLANE82**).
5. OK to apply the element type and close the dialog box.
6. Choose **Options**. . . button in Element Types window.
7. Choose plane stress with thickness option for element behavior.
8. OK to specify options and close the options dialog box.
9. Close the element type dialog box.

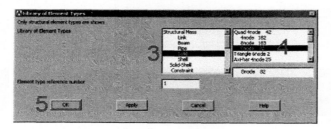

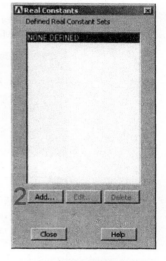

Step 16: Define real constants.

For this analysis, since the assumption is plane stress with thickness, you will enter the thickness as a real constant for **PLANE82**. To find out more information about **PLANE82**, you will use the ANSYS Help System in this step by clicking on a Help button from within a dialog box.

1. Choose **Main Menu> Preprocessor> Real Constants> Add/Edit/Delete**
2. Choose **Add**. . . button for a real constant set.
3. OK for **PLANE82**.
4. Help to get help on **PLANE82**.
5. Hold left mouse button down to scroll through element description.
6. Close the Help window
7. Enter .5 for THK.
8. OK to define the real constant and close the dialog box.
9. Close the real constant dialog box.

D. Generate Mesh

Step 17: Mesh the area.

One nice feature of the ANSYS program is that you can automatically mesh the model without specifying any mesh size controls. This is using what is called a **default mesh**. If you're not sure how to determine the mesh density, let ANSYS try it first! Meshing this model with a default mesh, however, generates more elements than are allowed in the

ANSYS ED program. Instead you will specify a global element size to control overall mesh density.

1. Choose **Main Menu**> **Preprocessor**> **Meshing**> **Mesh Tool**
2. Set Global Size control.
3. Type in 0.5.
4. OK.
5. Choose Area Meshing.
6. Click on Mesh.
7. Pick All for the area to be meshed (in picking menu). Close any warning messages that appear.
8. Close the Mesh Tool.

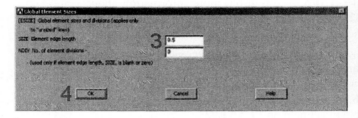

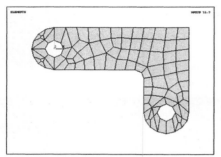

The mesh you see on your screen may vary slightly from the mesh shown here. As a result of this, you may see slightly different results during postprocessing.

Step 18:   Save the database as mesh.db.

Here again, you will save the database to a named file, this time mesh.db.

1. Choose **Utility Menu**> **File**> **Save as**
2. Enter mesh.db for database file name.
3. OK to save file and close dialog box.

E. Apply Loads

Now, we finish preprocessing phase and move on to the solution phase. A new, static analysis is the default, so you will not need to specify analysis type for this problem. Also, there are no analysis options for this problem.

Step 19:   Apply displacement constraints.

You can apply displacement constraints directly to lines.

1. Choose **Main Menu**> **Solution**> **Define Loads**> **Apply**> **Structural**> **Displacement**> **On Lines**

2. Pick the four lines around left-hand hole (Line numbers 10, 9, 11, 12).
3. OK (in picking menu).
4. Click on All DOF.
5. Enter 0 for zero displacement.
6. OK to apply constraints and close dialog box.
7. Choose **Utility Menu> Plot> Lines**

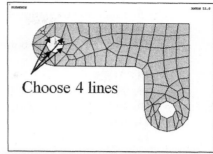

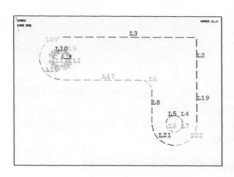

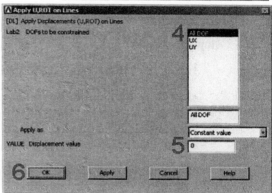

8. Toolbar: SAVE_DB.

Step 20:   Apply pressure load.

Now apply the tapered pressure load to the bottom, right-hand pin hole. ("Tapered" here means varying linearly.) Note that when a circle is created in ANSYS, four lines define the perimeter. Therefore, apply the pressure to two lines making up the lower half of the circle. Since the pressure tapers from a maximum value (500 psi) at the bottom of the circle to a minimum value (50 psi) at the sides, apply pressure in two separate steps, with reverse tapering values for each line. The ANSYS convention for pressure loading is that a positive load value represents pressure into the surface (compressive).

1. Choose **Main Menu> Solution> Define Loads> Apply> Structural> Pressure> On Lines**
2. Pick line defining bottom left part of the circle (line 6).

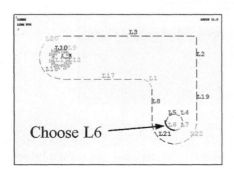

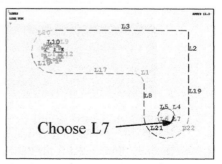

3. Apply.
4. Enter 50 for VALUE.

5. Enter 500 for optional value.
6. Apply.
7. Pick line defining bottom right part of circle (line 7).
8. Apply.
9. Enter 500 for VALUE.
10. Enter 50 for optional value.
11. OK.
12. Toolbar: SAVE_DB.

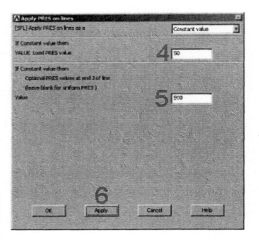

F. Obtain Solution

Step 21:   Solve.

1. Choose **Main Menu> Solution> Solve> Current LS**
2. Review the information in the status window, then choose **File> Close** (Windows), or Close (X11/Motif), to close the window.
3. OK to begin the solution. Choose Yes to any Verify messages that appear.
4. Close the information window when solution is done.

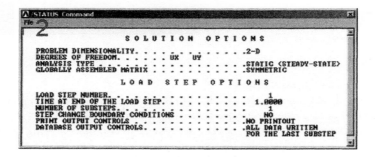

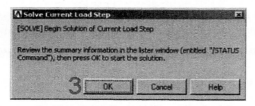

ANSYS stores the results of this one load step problem in the database and in the results file, Jobname.RST (or Jobname.RTH for thermal, Jobname.RMG for magnetic, and Jobname.RFL for fluid analyses). The database can actually contain only one set of results at any given time, so in a multiple load step or multiple substep analysis, ANSYS stores only the final solution in the database. ANSYS stores all solutions in the results file.

G. Review Results

Now, we finish the solution phase and move to the postprocessing phase. The results you see may vary slightly from what is shown here due to variations in the mesh.

Step 22: **Enter the general postprocessor and read in the results.**

   1. Choose **Main Menu**> **General Postproc**> **Read Results**> **First Set**

Step 23: **Plot the deformed shape.**

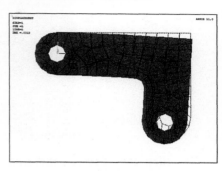

   1. Choose **Main Menu**> **General Postproc**> **Plot Results**> **Deformed Shape**

   2. Choose Def + undeformed.

   3. OK. You can also produce an animated version of the deformed shape:

   4. Choose **Utility Menu**> **Plot Ctrls**> **Animate**> **Deformed Shape**

   5. Choose Def + undeformed.

   6. OK to start animation.

   7. Make choices in the Animation Controller, if necessary, then choose Close.

Step 24: **Plot the von Mises equivalent stress.**

   1. Choose **Main Menu**> **General Postproc**> **Plot Results**> **Contour Plot**> **Nodal Solu**

   2. Choose **Stress** item to be contoured.

   3. Scroll down and choose **von Mises stress**.

   4. OK.

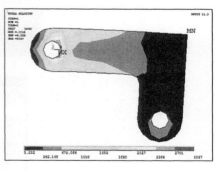

The colors in the contour plot stand for the level of von Mises stress. The maximum stress occurs at the fixed pin hole. The value of stress can be read from the color legend in the bottom of the graphic screen. Note that the color legend can be shown on the right side of graphic screen. You can also produce an animated version of these results:

   5. Choose **Utility Menu**> **Plot Ctrls**> **Animate**> **Deformed Results**

6. Choose **Stress** item to be contoured.
7. Scroll down and choose **von Mises SEQV**.
8. OK.
9. Make choices in the Animation Controller, if necessary, then choose Close.

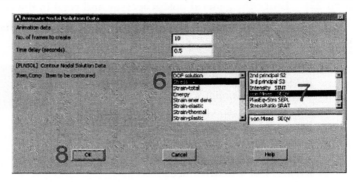

Step 25: List reaction solution.

1. Choose **Main Menu> General Postproc> List Results> Reaction Solu**
2. OK to list all items and close the dialog box.
3. Scroll down and find the total vertical force, FY.
4. Choose **File> Close** (Windows), or Close (X11/Motif), to close the window.

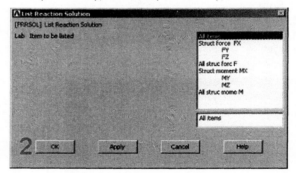

The value of 134.61 is comparable to the total pin load force. Note that the values shown are representative and may vary from the values you obtain. There are many other options available for reviewing results in the general postprocessor. You have finished the analysis. Exit the program in the next step.

Step 26:   Exit the ANSYS program.

When exiting the ANSYS program, you can save the geometry and loads portions of the database (default), save geometry, loads, and solution data (one set of results only), save geometry, loads, solution data, and postprocessing data (i.e., save everything), or save nothing. You can save nothing here, but you should be sure to use one of the other save options if you want to keep the ANSYS data files.

1. Toolbar: Quit.
2. Choose Quit - No Save!
3. OK.

C.4   EXAMPLES IN THE TEXT

The tutorial in the previous section was performed using graphical user interface (GUI) in which the user chooses commands using the menu. Then the GUI interprets these inputs into commands and sends them to ANSYS program. However, it is possible to type in ANSYS commands directly in the Command area. You can prepare commands in text editor and copy and paste them to the Command area. Or, you can save commands as a file and read them using **Utility Menu> File> Read Input From**... In this section, we present various example problems in the text using the command mode.

Example 2.4 – Two–bar Truss Structure

The two–bar truss shown in Figure C.3 has circular cross-sections with diameter of 0.25cm and Young's modulus $E = 30 \times 10^6 \, \text{N/cm}^2$. An external force $F = 50 \, \text{N}$ is applied in the horizontal direction at Node 2. Calculate the displacement of each node and stress in each element. The commands that can perform the finite element analysis is listed in Table C.1.

The detailed explanation of each command can be found by typing `HELP, COMMAND` in the command area. In ANSYS command lines, any text after ''!'' is considered as a comment; i.e., ignored by the program. The above commands are composed of three phases: preprocessing (`/PREP7`), solution (`/SOLUTION`), and postprocessing (`/POST1`). Each phase is ended using `FINISH` command. In this simple example, we generate nodes

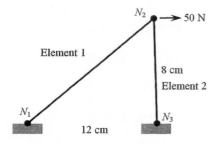

Figure C.3 Two–bar truss

Table C.1 ANSYS Commands for Two–bar Truss Analysis

```
/PREP7                          ! Start preprocessing
ET,1,LINK1                      ! Element type 1 = 2D Truss element
MP,EX,1,3E7                     ! Elastic modulus of material set 1
R,1,3.1416*0.125**2            ! Cross-sectional area of
!                                  real constant set 1
!
N,1,0,0                         ! Define Node 1 (0, 0)
N,2,12,8                        ! Define Node 2 (12, 8)
N,3,12,0                        ! Define Node 3 (12, 0)
!
TYPE,1                          ! Use Element type 1
MAT,1                           ! Use material set 1
REAL,1                          ! Use real constant set 1
E,1,2                           ! Create Element 1 (N1 - N2)
E,2,3                           ! Create Element 2 (N2 - N3)
!
D,1,ALL,0                       ! Fix all DOFs of Node 1
D,3,ALL,0                       ! Fix all DOFs of Node 3
F,2,FX,50                       ! Apply Fx = 50N at Node 2
FINISH                          ! Finish preprocessing
!
!
/SOLUTION                       ! Start simulation
OUTRES,ALL
OUTPR,ALL
SOLVE                           ! Solve the matrix equation
FINISH                          ! Finish simulation
!
/POST1                          ! Start postprocessing
SET,FIRST                       ! Choose the first result set
PLDISP,1                        ! Plot the deformed geometry
PRDISP                          ! Print out the nodal displacement
PRESOL,ELEM                     ! Print out all element results
FINISH                          ! Finish postprocessing
```

and elements directly (N and E commands). Displacement boundary conditions are imposed using D command, and the nodal force is applied using F command. The output from the PRDISP command is shown below:

NODE	UX	UY
1	0.0000	0.0000
2	0.82803E-03	-0.18108E-03
3	0.0000	0.000

Note that the nodal displacements are the same with Example 2.4 in Chapter 2. The output from PRESOL command is shown below:

```
EL= 1        NODES= 1    2           MAT= 1  LINK1
MFORX=       60.093
SAXL=        1224.2      EPELAXL= 0.000041


EL= 2        NODES= 2    3           MAT= 1  LINK1
MFORX=       -33.333
SAXL=        -679.06     EPELAXL=-0.000023
```

`MFORX`, `SAXL`, and `EPELAXL` are, respectively, element force, axial stress, and axial strain. Note that Element 1 is in tension, while Element 2 is in compression. The output from `PLDISP` command is shown below:

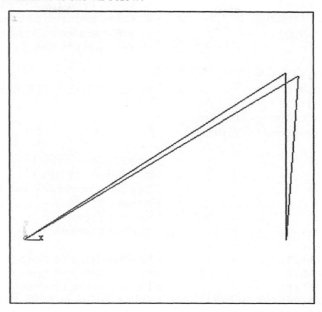

In general, the displacements are relatively small. Thus, most graphical postprocessors amplify the displacement so that the deformed shape of the truss can be shown in the figure.

Example 2.5 – Plane Truss with Three Elements

The plane truss shown in Figure C.4 consists of three members connected to each other and to the walls by pin joints. The members make equal angles with each other and element 2 is vertical. The members are identical to each other with properties: Young's modulus $E = 206 \times 10^9$ Pa, cross-sectional area $A = 1 \times 10^{-4}$ m^2, and length $L = 1$ m. An inclined force $F = 20,000$ N is applied at node 1. Solve for the displacements at node 1 and stresses in the three elements.

One of powerful capabilities in ANSYS is parametric programming. The user can assign a value to a variable and use it as an argument in the following commands. For example, we define a variable `val = 1/sqrt(2)`, and then, use it in the definition of

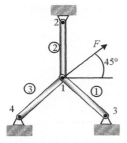

Figure C.4 Plane structure with three truss elements

Table C.2 ANSYS Commands for Three Truss Elements Analysis

```
/PREP7                      ! Start preprocessing
ET,1,LINK1                  ! Element type 1 = 2D Truss element
MP,EX,1,206E9               ! Elastic modulus of material set 1
R,1,0.0001                  ! Cross-sectional area of
!                             real constant set 1
!
ang=3.14159/6
N,1,0,0                     ! Define nodal coordinates
N,2,0,1
N,3,cos(ang),-sin(ang)
N,4,-cos(ang),-sin(ang)
!
TYPE,1                      ! Use Element type 1
MAT,1                       ! Use material set 1
REAL,1                      ! Use real constant set 1
E,1,3                       ! Create Element 1
E,1,2                       ! Create Element 2
E,1,4                       ! Create Element 3
!
D,2,ALL,0                   ! Fix all DOFs of Node 2
D,3,ALL,0                   ! Fix all DOFs of Node 3
D,4,ALL,0                   ! Fix all DOFs of Node 4
F,1,FX,20000/sqrt(2)        ! Apply force at Node 2
F,1,Fy,20000/sqrt(2)        ! Apply force at Node 2
FINISH                      ! Finish preprocessing
!
/SOLUTION                   ! Start simulation
OUTRES,ALL
OUTPR,ALL
SOLVE                       ! Solve the matrix equation
FINISH                      ! Finish simulation
!
/POST1                      ! Start postprocessing
SET,FIRST                   ! Choose the first result set
PLDISP,1                    ! Plot the deformed truss geometry
PRDISP                      ! Print out the nodal displacement
PRESOL,ELEM                 ! Print out all element results
FINISH                      ! Finish postprocessing
```

nodal coordinates, as $N,3,val,-val$. The variable can be an array or matrix. Table C.2 shows the list of ANSYS commands for plane truss. The output from the PRDISP command is shown below:

NODE	UX	UY
1	0.45767E−03	0.45767E−03
2	0.0000	0.0000
3	0.0000	0.0000
4	0.0000	0.0000

Note that the nodal displacements are the same as in Example 2.5 in Chapter 2. The output from PRESOL command is shown below:

```
EL= 1 NODES=       1              3 MAT=         1
MFORX= -3450.9
SAXL=-0.34509E+08     EPELAXL=-0.000168


EL= 2 NODES=       1              2 MAT=         1
MFORX= -9428.1
SAXL=-0.94281E+08 EPELAXL=-0.000458


EL= 3 NODES=       1              4 MAT=         1
MFORX= 12879.
SAXL= 0.12879E+09 EPELAXL= 0.000625
```

The output from `PLDISP` command is shown in Figure C.5.

Example 2.6 – Space Truss Problem

Use the finite element method to determine the displacements and forces in the space truss, shown in Figure C.6. The coordinates of the nodes in meter units are given in the

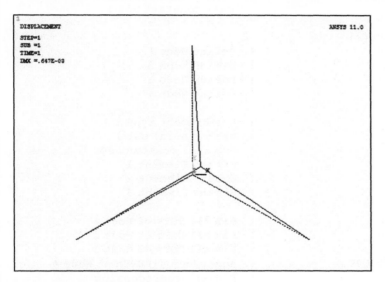

Figure C.5 Deformed shape of the plane structure with three truss elements

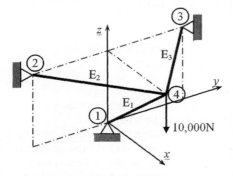

Figure C.6 Three–bar space truss structure

table. Assume Young's modulus E = 70 GPa and area of cross section $A = 1\,cm^2$. The magnitude of the downward force at Node 4 is equal to 10,000 N.

Node	x	y	z
1	0	0	0
2	0	−1	1
3	0	1	1
4	1	0	1

Table C.3   ANSYS Commands for Space Truss Problem

```
/PREP7                    ! Start preprocessing
ET,1,LINK8                ! Element type 1 = 3D Truss element
MP,EX,1,70E9              ! Elastic modulus of material set 1
R,1,0.0001               ! Cross-sectional area of
!                          real constant set 1
!
Val=1/sqrt(2)
N,1,0,0,0                 ! Define Node 1
N,2,0,-1,1               ! Define Node 2
N,3,0,1,1                ! Define Node 3
N,4,1,0,1                ! Define Node 4
!
TYPE,1                    ! Use Element type 1
MAT,1                     ! Use material set 1
REAL,1                    ! Use real constant set 1
E,1,4                     ! Create Element 1
E,2,4                     ! Create Element 2
E,3,4                     ! Create Element 3
!
D,1,ALL,0                 ! Fix all DOFs of Node 1
D,2,ALL,0                 ! Fix all DOFs of Node 2
D,3,ALL,0                 ! Fix all DOFs of Node 3
F,4,Fz,-10000            ! Apply Fz = -10000N at Node 4
FINISH                    ! Finish preprocessing
!
/SOLUTION                 ! Start simulation
OUTRES,ALL
OUTPR,ALL
SOLVE                     ! Solve the matrix equation
FINISH                    ! Finish simulation
!
/POST1                    ! Start postprocessing
SET,FIRST                 ! Choose the first result set
/VIEW,1,1,2,3            ! Set isometric view
PLDISP,1                  ! Plot the deformed truss geometry
PRDISP                    ! Print out the nodal displacement
PRESOL,ELEM               ! Print out all element results
FINISH                    ! Finish postprocessing
```

Table C.3 shows the list of ANSYS Commands for space truss. Note that different element type, `LINK8`, is used for three-dimensional truss. The list of available elements can be found in ANSYS Help. `/VIEW` command controls the viewpoint direction. ANSYS provides various view controls including dynamic view. The output from the `PRDISP` command is shown below:

NODE	UX	UY	UZ
1	0.0000	0.0000	0.0000
2	0.0000	0.0000	0.0000
3	0.0000	0.0000	0.0000
4	0.20203E-02	0.0000	-0.60609E-02

Note that the nodal displacements are the same as in Example 2.6 in Chapter 2. The output from `PRESOL` command is shown below:

```
EL=      1 NODES=   1     4 MAT= 1    LINK8
MFORX= -14142.
SAXL=-0.14142E+09 EPELAXL=-0.002020


EL=      2 NODES=   2     4 MAT= 1    LINK8
MFORX=   7071.1
SAXL= 0.70711E+08 EPELAXL= 0.001010


EL=      3 NODES=   3     4 MAT= 1    LINK8
MFORX=   7071.1
SAXL= 0.70711E+08 EPELAXL= 0.001010
```

The figure below shows the deformed shape of the space truss.

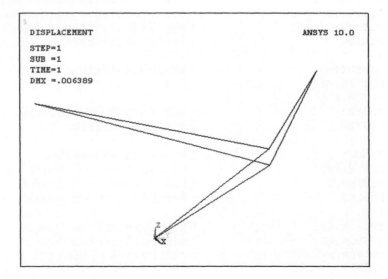

Simply Supported Beam Problem

Consider a simply supported beam, as shown in Figure C.7, loaded by a single load $F = 10$ kN, applied at a point 3.0 m from the left support. Plot the deflection curve, shear

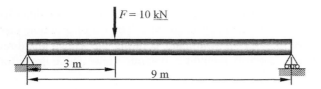

Figure C.7 Simply supported beam

Table C.4   ANSYS Commands for Space Truss Problem

```
/PREP7                        ! Start preprocessing
ET,1,BEAM3                    ! Element type = 2D Beam element
MP,EX,1,210E9                 ! Elastic modulus
R,1,4.53E-3,2.51E-5,0.1       ! Area and Izz
!
N,1,0,0                       ! Define Node 1
N,2,3,0                       ! Define Node 2
N,3,6,0                       ! Define Node 3
N,4,9,0                       ! Define Node 4
!
TYPE,1                        ! Use Element type 1
MAT,1                         ! Use material set 1
REAL,1                        ! Use real constant set 1
E,1,2                         ! Create Element 1
E,2,3                         ! Create Element 2
E,3,4                         ! Create Element 3
!
D,1,UX,0,,,,Uy                ! Fix x- & y-displacement of Node 2
D,4,Uy,0                      ! Fix y-displacement of Node 4
F,2,Fy,-10000                 ! Apply Fy = -10000N at Node 2
FINISH                        ! Finish preprocessing
!
/SOLUTION                     ! Start simulation
OUTRES,ALL
OUTPR,ALL
SOLVE                         ! Solve the matrix equation
FINISH                        ! Finish simulation
!
/POST1                        ! Start postprocessing
SET,FIRST                     ! Choose the first result set
PLDISP,1                      ! Plot the deformed geometry
PRDISP                        ! Print out the nodal displacement
!
ETABLE,MzzI,SMISC,6           ! Bending moment at Node I
ETABLE,MzzJ,SMISC,12          ! Bending moment at Node J
ETABLE,VyI,SMISC,2            ! Shear force at Node I
ETABLE,VyJ,SMISC,8            ! Shear force at Node J
PRETAB,MzzI,MzzJ,VyI,VyJ      ! Print results
PLLS,MzzI,MzzJ                ! Plot bending moment diagram
PLLS,VyI,VyJ                  ! Plot shear force diagram
FINISH                        ! Finish postprocessing
```

force and bending moment diagrams using three equal–length beam elements. The following data apply to the beam: Young's modulus $E = 210$ GPa, cross-sectional area $A = 4.53 \times 10^{-3}$ m^2, and moment of inertia $I = 2.51 \times 10^{-5}$ m^4.

Table C.4 shows the list of ANSYS commands for simply supported beam. For beam analysis, element type BEAM3 is used, which is 2-D beam element. For the beam analysis, both cross-sectional area and the moment of inertia are required as real constants. In order to plot the shear force and bending moment diagram, ETABLE command is used, which defines a table of values per element (the element table) for use in further processing. The element table is organized as a ''worksheet,'' with the rows representing all selected elements, and the columns consisting of result items which have been moved into the table with ETABLE. Each column of data is identified by a user-defined label for listings and displays. PRETAB command prints the element table, and PLLS plots the data in the element table. The output from the PRDISP command is shown below:

NODE	UX	UY	ROTZ
1	0.0000	0.0000	-0.94859E-02
2	0.0000	-0.22766E-01	-0.37943E-02
3	0.0000	-0.19920E-01	0.47429E-02
4	0.0000	0.0000	0.75887E-02

The output from PRETAB command is shown below:

ELEM	MZZI	MZZJ	VYI	VYJ
1	-0.54570E-11	20000.	-6666.7	-6666.7
2	20000.	10000.	3333.3	3333.3
3	10000.	-0.14552E-10	3333.3	3333.3

The above table shows bending moments at the first node (MZZI) and at the second node (MZZJ), as well as the shear force at the first node (VYI) and at the second node (VYJ). The figure below shows the deformed shape of the beam.

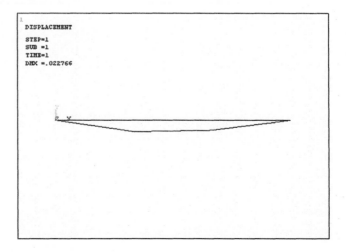

The bending moment and shear force diagrams are shown below:

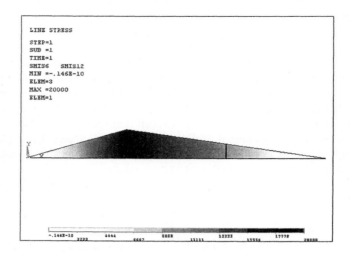

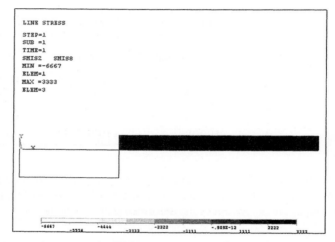

Appendix D

Finite Element Analysis Using MATLAB Toolbox

In this section, a MATLAB[3] toolbox CALFEM[4] is introduced. This toolbox is developed by Lund University in 1999 and can be downloaded free of charge from http://www .byggmek.lth.se/Calfem. CALFEM provides functions that can be used for finite element analysis of a variety of structures. Compared to other finite element programs, this toolbox has many advantages for education. Many commercial programs automatically perform finite element process. Thus, the users do not understand the process of constructing element stiffness matrices, assembling them, and applying boundary conditions. In this toolbox, however, the users must provide every step of finite element process. The toolbox then provides functions that can perform particular step of finite element process.

In general, MATLAB toolboxes are stored in the toolbox folder in MATLAB installation folder. In order to access the toolbox, the user must provide path information using **File, Set Path . . .** command from the main menu. Once the CALFEM folder location is stored in MATLAB paths, the user can access CALFEM functions anywhere. Type ''help springle'' in MATLAB window in order to test accessibility. If the path is properly assigned, then MATLAB should return help content of ''springle'' function, as shown in Figure D.1.

```
Ke=springle(ep)

PURPOSE
  Compute element stiffness matrix for spring element.
INPUT:  ep = [k]; spring stiffness or analog quantity
OUTPUT: Ke :      stiffness matrix, dim(Ke)= 2 x 2
```

Figure D.1 Help content of springle function

D.1 FINITE ELEMENT ANALYSIS OF BAR AND TRUSS

Three Uniaxial Bar Elements

In this section, the uniaxial bar problem in Example 2.3 will be solved using the MATLAB toolbox. The problem consists of four nodes and three elements, as shown in Figure D.2. In one–dimensional problems, each node has a single DOF, and nodes are

---

[3] MATLAB is the trademark of The MathWorks, Inc.

[4] CALFEM A finite element toolbox for MATLAB, Lund University, 1999, Web: http://www.byggmek.lth.se/ Calfem

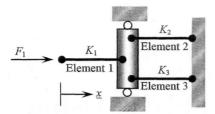

Figure D.2 One–dimensional structure with three uniaxial bar elements

connected by elements. The information of the connectivity is defined by the following array:

$$\texttt{Edof=[1 \quad 1 \quad 2;}$$
$$\texttt{2 \quad 2 \quad 3;} \qquad \text{(P.1)}$$
$$\texttt{3 \quad 2 \quad 4];}$$

The first column of `Edof` array represents element numbers, and the second and third columns represent the DOF of Nodes i and j, respectively. For example, Element 2 (the second row) connects Nodes 2 and 3.

Next, the global stiffness matrix and force vector are defined:

```
K=zeros(4,4);
F=zeros(4,1);                    (P.2)
F(1)=40;
```

Since there are four nodes and each node has a single DOF, the dimension of the global stiffness matrix is 4×4. Initially, components of the stiffness matrix and force vector are set to zero. In the last line of (P.2), the external force 40 N is applied to Node 1. Note that the reaction forces are not specified in the global force vector.

Next, the element stiffness matrices are constructed using the given spring constants. For this purpose, ''`springle`'' function is used:

```
ep1=50; ep2=30; ep3=70;
Ke1=springle(ep1);
                                 (P.3)
Ke2=springle(ep2);
Ke3=springle(ep3);
```

The three variables, `ep1`, `ep2`, and `ep3`, represent the axial rigidity (or, spring constants) of each element. When a uniaxial bar with constant cross-section is used, the axial rigidity, EA/L, can be used as the spring constant. Using these spring constants, the ''`springle`'' function calculates the 2×2 stiffness matrix that corresponds to Eq. (2.6) in Chapter 2. For example, `Ke3` contains the 2×2 stiffness matrix for Element 3. This step must be repeated for all elements. When the material properties are the same, the program can be simplified by using the `for–end` loop command in MATLAB.

Once the element stiffness matrices are obtained, they are assembled to form the global stiffness matrix using ''`assem`'' function:

```
K=assem(Edof(1,:),K,Ke1);
K=assem(Edof(2,:),K,Ke2);        (P.4)
K=assem(Edof(3,:),K,Ke3);
```

In order to assemble the element stiffness matrix into the global stiffness matrix, it is necessary to specify the relation between the local DOFs in the global DOFs. This information is stored in `Edof` array. Thus, the first argument of "`assem`" function is the row of `Edof` array. Then, the first line of (P.4) will copy 2×2 matrix `Ke1` into the global stiffness matrix K in the location that is specified in the first row of `Edof` array. The result is returned to the global stiffness matrix K so that the matrix K contains the accumulated data from all elements.

The global stiffness matrix K in (P.4) is singular because there is a rigid body motion in the system. In order to remove the rigid body motion, the boundary condition must be applied. The displacement boundary condition is specified by

$$
\boxed{\begin{array}{l} \texttt{bc=[3   0;} \\ \texttt{   4   0];} \end{array}} \tag{P.5}
$$

The first column of `bc` array represents the DOF number and the second column is the prescribed value of the DOF. Since Nodes 3 and 4 are fixed, zero values are provided for both nodes.

Using the global stiffness matrix and global force vector, the unknown DOF can be solved using "`solveq`" function:

$$
\boxed{\texttt{Q=solveq(K,F,bc);}} \tag{P.6}
$$

In this particular implementation, the reduced stiffness matrix is not constructed explicitly. The prescribed DOFs are deleted during the matrix solution phase. The results of the matrix solution (nodal displacement) are stored in the variable Q. For this problem, the stored nodal displacements are $Q = \{1.2, 0.4, 0.0, 0.0\}$. Note that Nodes 3 and 4 have zero displacement.

The nodal displacement that is calculated in (P.6) can be used to calculate the element force. For that purpose, the element displacement needs to be extracted from the global displacement Q, as

$$
\boxed{\begin{array}{l} \texttt{ed1=extract(Edof(1,:),Q);} \\ \texttt{ed2=extract(Edof(2,:),Q);} \\ \texttt{ed3=extract(Edof(3,:),Q);} \end{array}} \tag{P.7}
$$

The array `Edof` is used to find the location of those nodes in the element from the global node numbers. For example, `ed2` contains the displacement of Nodes 2 and 3. Using the element nodal displacements and spring constant for the element, the element force can be calculated by calling "`spring1s`" function:

$$
\boxed{\begin{array}{l} \texttt{es1=spring1s(ep1,ed1);} \\ \texttt{es2=spring1s(ep2,ed2);} \\ \texttt{es3=spring1s(ep3,ed3);} \end{array}} \tag{P.8}
$$

The element force corresponds to that in Eq. (2.4). It is positive when the force is tension and negative when compression. For this problem, the element forces are `es1` = $-40\,\text{N}$, `es2` = $-12\,\text{N}$, and `es3` = $-28\,\text{N}$.

Table D.1 provides the list of MATLAB program that can solve the same problem in Example 2.3.

Table D.1   MATLAB Commands for
Spring Finite Elements

```
Edof=[1  1  2;2  2  3;3  2  4];
K=zeros(4,4);
F=zeros(4,1); F(1)=40;
ep1=50;ep2=30;ep3=70;
Ke1=spring1e(ep1);
Ke2=spring1e(ep2);
Ke3=spring1e(ep3);
K=assem(Edof(1,:),K,Ke1);
K=assem(Edof(2,:),K,Ke2);
K=assem(Edof(3,:),K,Ke3);
BC=[3  0;4  0];
Q=solveq(K,F,bc);
ed1=extract(Edof(1,:),Q);
ed2=extract(Edof(2,:),Q);
ed3=extract(Edof(3,:),Q);
es1=spring1s(ep1,ed1);
es2=spring1s(ep2,ed2);
es3=spring1s(ep3,ed3);
```

The above example is the classical finite element analysis procedure. As can be seen, the formulation is based on the general force–displacement equations of a single one–dimensional truss element. The two–force member element or truss element is the simplest type of element used in finite element analysis. The procedure to formulate and solve the force–displacement equation is straightforward, but somewhat tedious. In real life applications, the use of one–dimensional truss element is rare. In the next section, we will expand the procedure to solving two–dimensional truss.

Plane Truss Example

In this section, the two–bar truss structure in Example 2.4 is solved using the MATLAB finite element program. Unlike one–dimensional problem, there are two DOFs at each node. Two DOFs are assigned at each node in a sequence along with the node numbers, as illustrated in Figure D.3.

The element connectivity array now contains four DOFs:

$$Edof=[1 1 2 3 4;$$
$$\quad\quad\quad 2 3 4 5 6];$$

(P.9)

The first element connects Node 1 (DOF 1 and 2) and Node 2 (DOF 3 and 4), while the second elements connects Node 2 (DOF 3 and 4) and Node 3 (DOF 5 and 6).

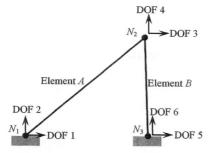

Figure D.3 Degrees–of–freedom of two–bar truss

The size of global stiffness matrix and force vector can be determined based on the total number of DOFs, as

```
K=zeros(6);
F=zeros(6,1);
F(3)=50;
```
(P.10)

Note that the applied force at Node 2 in the x–direction, it must be assigned in the DOF 3.

In the two–dimensional problem, two properties are required for the element:

```
ep=[3e7 .125*.125*pi];
```
(P.11)

The first represents the Young's modulus, while the second for the cross-sectional area. The length of the element will be calculated based on the coordinates of the two nodes.

Unlike one-dimensional problem, the coordinates of nodes are required in order to calculate the length of the element and the coordinate transformation matrix.

```
ex1=[ 0   12 ];
ey1=[ 0   8 ];
```
(P.12)

For example, `ex1` array includes the x–coordinates of two nodes of Element 1, while `ey1` contains y–coordinates.

The element stiffness matrix can be obtained using "`bar2e`" function, as

```
Kel=bar2e(ex1,ey1,ep);
```
(P.13)

The first two arguments are nodal coordinates of x– and y–directions, respectively, and the last argument is the property of the element.

Assembly, boundary conditions, solving the matrix equation, and extracting element displacements are the same with that of one–dimensional case.

Once the element displacements are available, the element forces can be calculated using "`bar2s`" function:

```
N1=bar2s(ex1,ey1,ep,ed1);
```
(P.14)

The first two arguments are nodal coordinates of x– and y–directions, respectively, and the last arguments are the element displacements.

Table D.2   Element for in the Local Coordinates

```
Edof=[1  1  2  3  4;
      2  3  4  5  6];
K=zeros(6);
F=zeros(6,1); F(3)=50;
ep=[3e7   .125*.125*pi];
ex1 =[0   12]; ey1=[0   8];
ex2 =[12  12]; ey2=[8   0];
Ke1=bar2e(ex1,ey1,ep);
Ke2=bar2e(ex2,ey2,ep);
K=assem(Edof(1,:),K,Ke1);
K=assem(Edof(2,:),K,Ke2);
bc=[1  0;2  0;5  0;6  0];
Q=solveq(K,F,bc);
ed1=extract(Edof(1,:),Q);
ed2=extract(Edof(2,:),Q);
N1=bar2s(ex1,ey1,ep,ed1)
N2=bar2s(ex2,ey2,ep,ed2)
plotpar=[1  3  1];scale=1000;
eldraw2(ex1,ey1);
eldraw2(ex2,ey2);
eldisp2(ex1,ey1,ed1,plotpar,scale);
eldisp2(ex2,ey2,ed2,plotpar,scale);
```

One of the most advantageous features of the two–dimensional program is that it can provide the geometry of the elements graphically. First, "`eldraw2`" function plots the undeformed geometry of the element.

$$\boxed{\texttt{eldraw2(ex1,ey1);}} \qquad\qquad (\text{P.15})$$

This function should be called twice for Elements 1 and 2.

Next, "`eldisp2`" function plots the deformed geometry of the element using the element displacement:

$$\boxed{\texttt{eldisp2(ex1,ey1,ed1,plotpar,scale);}} \qquad (\text{P.16})$$

The variable "`plotpar`" specifies the property of the deformed geometry plot. For example, `plotpar=[1 3  1]` plots the deformed geometry using a solid line, magenta color, and circle mark for the node. Since the actual displacement is usually very small, it is often difficult to identify the deformed geometry from the undeformed one. Thus, the "`scale`" is multiplied with the actual displacement in order to magnify the deformation. It is important to specify the value of "`scale`" so that the actual magnitude of displacements can be estimated.

Since the above "`eldraw2`" and "`eldisp2`" functions plot a single element, they must be repeated for all elements in order to plot the entire structure. Table D.2 lists the program for the two–bar truss. Since the deformation is relatively large, a unit scale is used for plotting the deformed geometry. Figure D.4 shows the initial geometry of the truss and deformed geometry with the scale of 1,000. Different values for `plotpar` are used to distinguish the deformed geometry from the undeformed one.

Plane Truss Example 2

Determine the normal stress in each member of the truss shown in Figure D.5. Compare the finite element result with that from the analytical calculation.

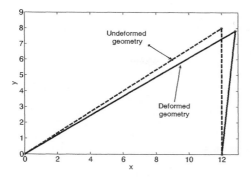

Figure D.4 Deformed geometry of the two-bar truss. Deformation is scaled by 1,000 times.

Prior to carrying out the finite element analysis, it is important to do preliminary analysis to gain some insight into the problem and as a mean to check the finite element analysis results. Since the truss is statically determinate, it is possible to calculate member forces. The free–body–diagram of the structure becomes:

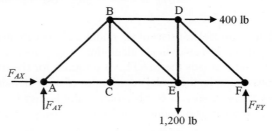

Member BC is a zero–force member. Thus, the force and stress in BC will be zero.

From the equilibrium of moment with respect to Point A, we can calculate the vertical reaction force, as

$$\sum M_A = 36 \times F_{FY} - 24 \times 1200 - 9 \times 400 = 0$$
$$\therefore \quad F_{FY} = 900 \text{ lb}$$

Next, using the joint method, solve for the internal forces in members DF and EF.

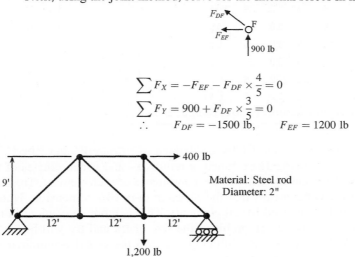

$$\sum F_X = -F_{EF} - F_{DF} \times \frac{4}{5} = 0$$
$$\sum F_Y = 900 + F_{DF} \times \frac{3}{5} = 0$$
$$\therefore \quad F_{DF} = -1500 \text{ lb}, \qquad F_{EF} = 1200 \text{ lb}$$

Figure D.5 Nine-bar truss model

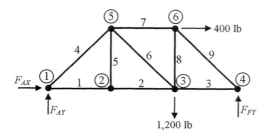

Figure D.6 Finite element model of nine-bar truss

Therefore

$$\sigma_{DF} = -\frac{1500}{\pi} = -477.5 \quad \text{psi}$$

$$\sigma_{EF} = \frac{1200}{\pi} = 382 \quad \text{psi}$$

The finite element model consists of six nodes and nine elements, as shown in Figure D.6. For the comparison with analytical calculation, $\sigma_{DF} = \sigma_9$ and $\sigma_{EF} = \sigma_3$.

Table D.3 summarizes the MATLAB program for the truss structure in Figure D.5. In the previous example, the nodal coordinates of each element are provided using arrays `ex1`, `ey1`, etc. This process is cumbersome and it is easy to make mistakes. An approach that is more convenient is introduced in Table D.3 using "`coordxtr`" function.

```
Coord=[0  0; 12  0; 24  0; 36  0; 12  9; 24  9];
Dof=[1  2; 3  4; 5  6; 7  8; 9  10; 11  12];
Edof=[1   1   2   3   4;
      2   3   4   5   6;
      3   5   6   7   8;
      4   1   2   9  10;
      5   9  10   3   4;
      6   9  10   5   6;
      7   9  10  11  12;
      8  11  12   5   6;
      9  11  12   7   8];
[Ex,Ey]=coordxtr(Edof,Coord,Dof,2);
```

(P.17)

First, the nodal coordinates and DOFs are defined in "Coord" and "Dof" arrays, respectively. Then, the `coordxtr` function returns x– and y–coordinates of the elements. Each row of `Ex` and `Ey` contains the coordinates of each element. The last argument "2" in `coordxtr` means that the problem is in two–dimension. Note that a single array `Ex` and `Ey` is used for all elements. This is more convenient in the programming purpose. For the given problem, the size of `Ex` and `Ey` will be 9×2. Each row represents an element and each column represents the x- and y-coordinates of the nodes.

Table D.3   MATLAB Commands for Nine-bar Truss

```
Coord=[0  0;12  0;24  0;36  0;12  9;24  9];
Dof=[1  2;3  4;5  6;7  8;9  10;11  12];
Edof=[1  1   2   3   4;2   3   4   5   6;3   5   6   7   8;
      4  1   2   9  10;5   9  10   3   4;6   9  10   5   6;
      7  9  10  11  12;8  11  12   5   6;9  11  12   7   8];
[Ex,Ey]=coordxtr(Edof,Coord,Dof,2);
K=zeros(12);
F=zeros(12,1);  F(6)=-1200;F(11)=400;
ep=[10000 pi];
for i=1:9
   Ke=bar2e(Ex(i,:),Ey(i,:),ep);
   K=assem(Edof(i,:),K,Ke);
end
bc=[1  0;2  0;8  0];
Q=solveq(K,F,bc);
Ed=extract(Edof,Q);
for i=1:9
   N(i)=bar2s(Ex(i,:),Ey(i,:),ep,Ed(i,:));
end
eldraw2(Ex,Ey);
plotpar=[1  3  0];scale=1;
eldisp2(Ex,Ey,Ed,plotpar,scale);
```

When we construct the element stiffness matrix, we need to send a single row that corresponds to the element. Thus, the following command assembles all nine elements:

$$
\begin{aligned}
&\texttt{for i=1:9}\\
&\quad\texttt{Ke=bar2e(Ex(i, :),Ey(i, :),ep);}\\
&\quad\texttt{K=assem(Edof(i, :),K,Ke);}\\
&\texttt{end}
\end{aligned}
\tag{P.18}
$$

Applying boundary conditions, solving the matrix equation, and extracting element displacements are the same with that of one–dimensional case.

Once the element displacements are available, the element forces can be calculated using "bar2s" function:

$$
\begin{aligned}
&\texttt{Ed=extract(Edof,Q);}\\
&\texttt{for i=1:9}\\
&\quad\texttt{N(i)=bar2s(Ex(i, :),Ey(i, :),ep,Ed(i, :));}\\
&\texttt{end}
\end{aligned}
\tag{P.19}
$$

In (P.19), the element forces are stored in N array. The element stress can be obtained by dividing the element forces by cross-sectional area. The following stress can be obtained: $\sigma_{DF} = \sigma_9$ and $\sigma_{EF} = \sigma_3$.

$$
\sigma_{EF}=\sigma_3 = \frac{N(3)}{A} = \frac{1200}{\pi \cdot 1^2} = 382\text{psi}
$$

$$
\sigma_{DF}=\sigma_9 = \frac{N(9)}{A} = \frac{-1500}{\pi \cdot 1^2} = -477.5\text{psi}
$$

Thus, the finite element results are identical to the analytical calculations.

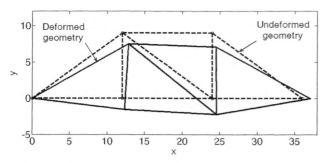

Figure D.7 Deformed geometry of the nine-bar truss model

When `Ex` and `Ey` contains the nodal coordinates of all elements, "`eldraw2`" and "`eldisp2`" functions can plot the entire geometry by a single call.

```
eldraw2(Ex,Ey);
eldisp2(Ex,Ey,Ed,plotpar,scale);
```
(P.20)

The element displacement `Ed` also contains all element displacements. Figure D.7 shows the deformed and undeformed geometry of the truss. Due to relatively large deformation, `scale=1` is used in plotting Figure D.7.

Space Truss Example

In the case of space truss, each node has three DOFs: u_x, u_y, and u_z. Consider the space truss shown in Figure D.8. The Young's modulus $E = 210\,\text{GPa}$, and dimensions of the cross–section are $a = b = 7.5\,\text{mm}$. Solve the problem using MATLAB toolbox and compare the results with analytical calculation.

Let us solve the problem using the conventional vector algebra to solve the space truss problem. The coordinates of the nodes are given as $A(5, 4, 3)$, $B(0, 2, 3)$, $C(5, 0, 0), D(5, 0, 6), E(0, 0, 3)$. Since the space truss consists of two-force members, we can obtain equilibrium relation at Node A using the Free-Body-Diagram as

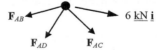

First, we want to calculate the direction of each member force using the position vectors AB, AC, and AD:

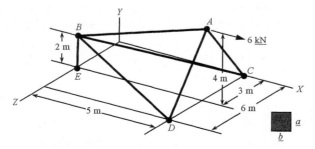

Figure D.8 Space truss model

$$AB = (0 - 5)\mathbf{i}_{-}(2 - 4)\mathbf{j} + (3 - 3)\mathbf{k} = -5\mathbf{i} - 2\mathbf{j}$$
$$AC = (5 - 5)\mathbf{i} + (0 - 4)\mathbf{j} + (0 - 3)\mathbf{k} = -4\mathbf{j} - 3\mathbf{k}$$
$$AD = (5 - 5)\mathbf{i} + (0 - 4)\mathbf{j} + (6 - 3)\mathbf{k} = -4\mathbf{j} + 3\mathbf{k}$$

Unit vectors along AB, AC, and AD become

$$\mathbf{n}_{AB} = \frac{-5\mathbf{i} - 2\mathbf{j}}{\sqrt{(-5)^2 + (-2)^2}} = -0.9285\mathbf{i} - 0.371\mathbf{j}$$

$$\mathbf{n}_{AC} = \frac{-4\mathbf{i} - 3\mathbf{j}}{\sqrt{(-4)^2 + (-3)^2}} = -0.8\mathbf{j} - 0.6\mathbf{k}$$

$$\mathbf{n}_{AD} = \frac{-4\mathbf{i} + 3\mathbf{j}}{\sqrt{(-4)^2 + (3)^2}} = -0.8\mathbf{j} + 0.6\mathbf{k}$$

Thus, the force at each member becomes

$$\mathbf{F}_{AB} = \|\mathbf{F}_{AB}\|\mathbf{n}_{AB} = \|\mathbf{F}_{AB}\|(-0.9285\mathbf{i} - 0.371\mathbf{j})$$
$$\mathbf{F}_{AC} = \|\mathbf{F}_{AC}\|\mathbf{n}_{AC} = \|\mathbf{F}_{AC}\|(-0.8\mathbf{j} - 0.6\mathbf{k})$$
$$\mathbf{F}_{AD} = \|\mathbf{F}_{AD}\|\mathbf{n}_{AD} = \|\mathbf{F}_{AD}\|(-0.8\mathbf{j} + 0.6\mathbf{k})$$

Applying the equilibrium equation at Node A, we have

$$\sum \mathbf{F}_{@A} = 0 = 6000\mathbf{i} + \mathbf{F}_{AB} + \mathbf{F}_{AC} + \mathbf{F}_{AD}$$
$$= 6000\mathbf{i} - 0.9285\|\mathbf{F}_{AB}\|\mathbf{i} - 0.371\|\mathbf{F}_{AB}\|\mathbf{j} - 0.8\|\mathbf{F}_{AC}\|\mathbf{j}$$
$$- 0.6\|\mathbf{F}_{AC}\|\mathbf{k} - 0.8\|\mathbf{F}_{AD}\|\mathbf{j} + 0.6\|\mathbf{F}_{AD}\|\mathbf{k}$$
$$= (6000 - 0.9285\|\mathbf{F}_{AB}\|)\mathbf{i} + (-0.371\|\mathbf{F}_{AB}\| - 0.8\|\mathbf{F}_{AC}\|$$
$$- 0.8\|\mathbf{F}_{AD}\|)\mathbf{j} + (-0.6\|\mathbf{F}_{AC}\| + 0.6\|\mathbf{F}_{AD}\|)\mathbf{k}$$

Also, since the structure is symmetric, $\|\mathbf{F}_{AC}\| = \|\mathbf{F}_{AD}\|$. Therefore,

$$6000 - 0.9285\|\mathbf{F}_{AB}\| = 0, \quad \Rightarrow \|\mathbf{F}_{AB}\| = 6462 \text{ N}$$

$$-0.371\|\mathbf{F}_{AB}\| - 0.8\|\mathbf{F}_{AC}\| - 0.8\|\mathbf{F}_{AD}\| = 0$$
$$\Rightarrow \|\mathbf{F}_{AC}\| = \|\mathbf{F}_{AD}\| = -1500 \text{ N}$$

Thus, stress at member AB, AC, and AD can be calculated as

$$\sigma_{AB} = 6460/(5.63 \times 10^{-5}) = 115 \text{ MPa}$$
$$\sigma_{AC} = \sigma_{AD} = -1500/(5.63 \times 10^{-5}) = -26.7 \text{ MPa}$$

The finite element analysis can be performed using the program listed in Table D.4. Now, each node has three DOFs. Thus, the column size of `Edof` array is seven. In the space truss, the "`Coord`" array contains x-, y-, and z-coordinates of each node. Unlike the plane truss problem, the "`coordxtr`" function extracts x-, y-, and z-coordinates and store them at `Ex`, `Ey`, and `Ez` array. For space truss element, "`bar3e`" function is used to construct the element stiffness matrix. In addition, "`bar3s`" function is used to calculate element force. The space truss can be plotted using "`eldraw3`" function. For plotting the deformed geometry, "`eldisp3`" function can be used.

Table D.4   MATLAB commands for space truss

```
Edof=[1  1  2  3   4   5   6;2  4  5  6   7   8   9;
      3  4  5  6  10  11  12;4  1  2  3   7   8   9;
      5  1  2  3  10  11  12;6  4  5  6  13  14  15];
Coord=[5  4  3;0  2  3;5  0  0;5  0  6;0  0  3];
Dof=[1  2  3;4  5  6;7  8  9;10  11  12;13  14  15];
[Ex,Ey,Ez]=coordxtr(Edof,Coord,Dof,2);
K=zeros(15);F=zeros(15,1); F(1)=6000;
ep=[210e9 .0075*.0075];
for i=1:6
   Ke=bar3e(Ex(i,:),Ey(i,:),Ez(i,:),ep);
   K=assem(Edof(i,:),K,Ke);
end
bc=[7  0;8  0;9  0;10  0;11  0;12  0;13  0;14  0;15  0];
Q=solveq(K,F,bc);
Ed=extract(Edof,Q);
for i=1:6
   N(i)=bar3s(Ex(i,:),Ey(i,:),Ez(i,:),ep,Ed(i,:));
end
eldraw3(Ex,Ey,Ez);
plotpar=[1  3  0];
scale=100;
eldisp3(Ex,Ey,Ez,Ed,plotpar,scale);
```

The element stress can be obtained by dividing the element forces by cross-sectional area. The following stresses can be obtained: $\sigma_{AB} = \sigma_1$ and $\sigma_{AC} = \sigma_3$.

$$\sigma_{AB} = \sigma_1 = \frac{N(1)}{A} = \frac{6462.2}{0.075^2} = 115\text{MPa}$$

$$\sigma_{AC} = \sigma_4 = \frac{N(4)}{A} = \frac{-1500}{0.075^2} = -26.7\text{MPa}$$

Thus, the finite element results are identical to the analytical calculations. Figure D.9 shows the deformed shape of the space truss.

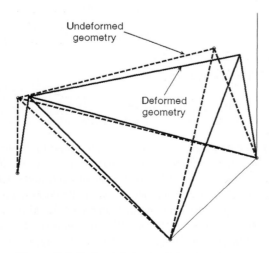

Undeformed geometry

Deformed geometry

Figure D.9 Deformed shape of the space truss

D.2   FINITE ELEMENT ANALYSIS USING FRAME ELEMENTS

Simply-Supported Beam

Consider a simply supported beam, as shown in Figure D.10, loaded by a single load $F = 10\,\text{kN}$, applied at a point 3.0 m from the left support. Calculate the deflection curve, shear force and bending moment using three equal–length beam elements. The following data apply to the beam: Young's modulus $E = 210\,\text{GPa}$, cross-sectional area $A = 4.53 \times 10^{-3}\,\text{m}^2$, and moment of inertia $I = 2.51 \times 10^{-5}\,\text{m}^4$.

In Chapter 4, we distinguished beam and frame elements. However, most finite element programs only have the frame element because it includes the beam element. In addition, they call it a beam element. The beam element in MATLAB toolbox corresponds to the frame element in Chapter 4. The plane beam element in the MATLAB toolbox has three DOFs per each node: axial displacement, lateral displacement, and rotation. However, the simply supported beam will not use any axial component.

With reference to node numbers and corresponding DOFs in Figure D.10, the connectivity of elements is defined as

$$\text{Edof=[1} \quad 1 \quad 2 \quad 3 \quad 4 \quad 5 \quad 6; \\ 2 \quad 4 \quad 5 \quad 6 \quad 7 \quad 8 \quad 9; \\ 3 \quad 7 \quad 8 \quad 9 \quad 10 \quad 11 \quad 12];$$

(P.21)

The first column stands for the node number, next three columns for the DOFs of the first node, and the last three columns for the DOFs of the second node.

Since there are twelve DOFs, the size of the global stiffness matrix will be 12×12, and the global force vector will be 12×1. The concentrated force is applied at DOF 5, which corresponds to the vertical direction of Node 2. In the beam element, three material parameters need to be defined: Young's modulus, cross-sectional area, and the second area moment of inertia.

```
E=2.1e11; A=45.3e-4; I=2510e-8;
ep=[E  A  I];
```

(P.22)

The 6×6 element stiffness matrix is constructed using "**beam2e**" function. Assembly, applying boundary conditions, solving the global matrix equation, and extracting element displacements are identical with the plane truss element. When "**solveq**" function is called with the two return arguments, the second argument contains the reaction force.

```
[Q,R]=solveq(K,f,bc)
```

(P.23)

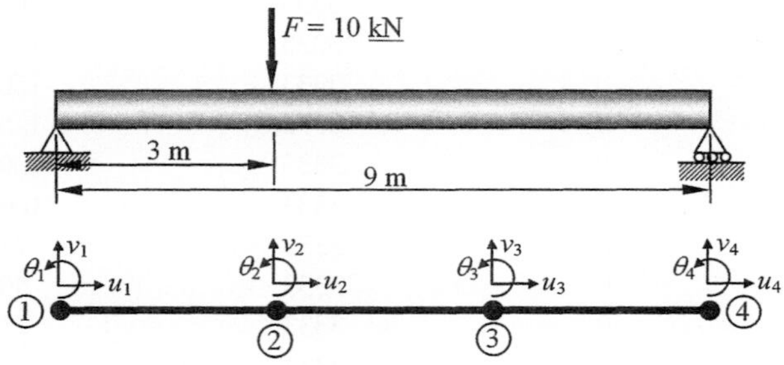

Figure D.10 Simply-supported beam structure

Table D.5 Nodal Displacements and Reaction Forces

DOF	Q (m)	R (N)
u_1	0	0
v_1	0	6,666.7
θ_1	−0.0095	0
u_2	0	0
v_2	−0.0228	0
θ_2	−0.0038	0
u_3	0	0
v_3	−0.0199	0
θ_3	0.0047	0
u_4	0	0
v_4	0	3,333.3
θ_4	0.0076	0

The output of the matrix equation is shown in Table D.5. As expected, there are no axial deformations and axial forces in the beam.

In the beam element, the element forces consist of axial force, bending moment, and transverse shear force. These forces are calculated using "beam2s" function:

```
for I=1:3
  Ed=extract(Edof(I,:),Q);
  es=beam2s(Ex(I,:),Ey(I,:),ep,Ed,[0 0],5)
end
```
(P.24)

The "beam2s" function returns the array "es", which consists of three columns. The first column of es contains the axial force, the second bending moment, and the third the shear force. Since these forces and the moment can change within the element, their values are evaluated at different locations of the element. For example, in (P.24) the last argument in "beam2s" function means that these forces are calculated at five locations that are

Table D.6 Element Force of the Simply-supported Beam

Element	x-coord	Axial force	Shear force	Bending moment
1	0.00	0	−6667	0
	0.75	0	−6667	5000
	1.50	0	−6667	10000
	2.25	0	−6667	15000
	3.00	0	−6667	20000
2	3.00	0	3333	20000
	3.75	0	3333	17500
	4.50	0	3333	15000
	5.25	0	3333	12500
	6.00	0	3333	10000
3	6.00	0	3333	10000
	6.75	0	3333	7500
	7.50	0	3333	5000
	8.25	0	3333	2500
	9.00	0	3333	0

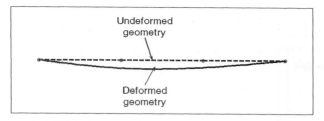

Figure D.11 Deflection curve of the simply-supported beam

equally distributed within the element. These data are stored at each row of "**es**" array. Thus, the "**es**" array in (P.24) is 3×5 matrix. The first row of "**es**" array is the left node, while the last row is the right node. The fifth argument, [0 0], of the "**beam2s**" function is for the distributed load. This argument specifies the distributed load in the local x– and y–directions. Table D.6 shows the axial force, shear force, and bending moment at evenly distributed locations of the beam. Note that the shear force is discontinuous at the Node 2 where concentrated force is applied. As expected, the bending moment changes linearly. In the next example, we will discuss about how to plot axial force, shear force, and bending moment diagrams.

The geometry of the elements and deformed geometry can be plotted using "**eldraw2**" and "**eldisp2**" functions, respectively. Figure D.11 shows the deflection curve of the beam under the concentrated force at Node 2. The complete program list of simply-supported beam problem is shown in Table D.7.

Table D.7   Program list for simply supported beam

```
%Element connectivity
Edof=[1  1  2  3   4   5   6;
      2  4  5  6   7   8   9;
      3  7  8  9  10  11 12];

%Global vector and matrix
K=zeros(12); f=zeros(12,1); f(5)=-10000;
E=2.1e11; A=45.3e-4; I=2510e-8;
ep=[E  A  I];              %Element properties

%Nodal Coordinates
Ex=[0  3; 3  6; 6  9];
Ey=[0  0; 0  0; 0  0];

%Element matrix and assembly
for I = 1:3
   Ke=beam2e(Ex(I,:),Ey(I,:),ep);
   K = assem(Edof(I,:),K,Ke);
end

%Boundary conditions
bc=[1  0; 2  0; 11  0];

%Solution U and reaction force Q
[Q,R]=solveq(K,f,bc);

%Element data
for I = 1:3
   Ed=extract(Edof(I,:),Q);
   es=beam2s(Ex(I,:),Ey(I,:),ep, Ed,[0 0],5)
end
```

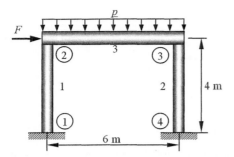

Figure D.12 Plane frame structure under distributed and concentrated load

Portal Frame Analysis

Consider the plan frames in Figure D.12. All frames have the same material properties. The horizontal frame has different cross-sectional geometry from two vertical frames. All frames are made of the same material with Young's modulus $E = 210$ GPa. Two vertical frames have area $A_1 = A_2 = 4.53 \times 10^{-3}$ and moment of inertia $I_1 = I_2 = 2.51 \times 10^{-5}$ m^4. The horizontal frame has area $A_3 = 1.43 \times 10^{-2}$ m^2 and moment of inertia $I_3 = 3.309 \times 10^{-4}$ m^4. Uniformly distributed load $p = 75$ kN/m is applied at the horizontal frame, and a concentrated force $F = 1$ kN is applied in the horizontal direction at Node 2. Plot the axial force, shear force, and bending moment diagram of the frame. Use three frame elements to model the structure.

The "beam2e" function can be used for the frame element. The "beam2e" function also has a capability of calculating the equivalent nodal forces from the distributed load. eq3=[0 -75000] represents the distributed load for Element 3 in the local x- and y-directions. Since the distributed load p is applied downward, it is negative in local y-direction. In this case, the "beam2e" function has additional parameter and additional return value, as

$$
\begin{aligned}
&\texttt{[Ke3,fe3]=beam2e(ex3,ey3,ep3,eq3);} \\
&\texttt{[K,F]=assem(Edof(3,:),K,Ke3,F,fe}
\end{aligned}
\qquad \text{(P.25)}
$$

The above commands calculate the work–equivalent nodal force "fe3" and assembled into the global force vector "F". Since Elements 1 and 2 does not have distributed load, we can use the standard calling convention.

Once the matrix equations are solved for the nodal solutions, the force diagrams can be plotted. For example, the following commands plot the axial force diagram of the three frame elements:

```
[es1  edi1  eci1]=beam2s(ex1,ey1,ep1,Ed(1,:),eq1,20)
[es2  edi2  eci2]=beam2s(ex2,ey2,ep1,Ed(2,:),eq2,20)
[es3  edi3  eci3]=beam2s(ex3,ey3,ep3,Ed(3,:),eq3,20)
magnfac=eldia2(ex1,ey1,es1(:,1),eci1);
magnitude=[3e5  0.5  0];
eldia2(ex1,ey1,es1(:,1),eci1,magnfac);
eldia2(ex2,ey2,es2(:,1),eci2,magnfac);
eldia2(ex3,ey3,es3(:,1),eci3,magnfac,magnitude);
axis([-1.5  7 - 0.5  5.5])
```
(P.26)

Table D.8   Program List for the Plane Frame Structure

```
Edof=[1   1   2   3 4 5 6;
      2 10 11 12 7 8 9;
      3   4   5   6 7 8 9];
K=zeros(12); F=zeros(12,1); F(4)=1000;
A1=45.3e-4; A2=142.8e-4;
I1=2510e-8; I2=33090e-8;
E=2.1e11;

ep1=[E A1 I1]; ep3=[E A2 I2];
ex1=[0 0]; ex2=[6 6]; ex3=[0 6];
ey1=[0 4]; ey2=[0 4]; ey3=[4 4];
eq1=[0 0]; eq2=[0 0]; eq3=[0 -75000];

Ke1=beam2e(ex1,ey1,ep1);
Ke2=beam2e(ex2,ey2,ep1);
[Ke3,fe3]=beam2e(ex3,ey3,ep3,eq3);

K=assem(Edof(1,:),K,Ke1);
K=assem(Edof(2,:),K,Ke2);
[K,F]=assem(Edof(3,:),K,Ke3,F,fe3);

bc=[1 0;2 0;3 0;10 0;11 0;12 0];
Q=solveq(K,F,bc);
Ed=extract(Edof,Q);
[es1 edi1 eci1]=beam2s(ex1,ey1,ep1,Ed(1,:),eq1,20)
[es2 edi2 eci2]=beam2s(ex2,ey2,ep1,Ed(2,:),eq2,20)
[es3 edi3 eci3]=beam2s(ex3,ey3,ep3,Ed(3,:),eq3,20)
%Plot the results
figure(1)
magnfac=eldia2(ex1,ey1,es1(:,1),eci1);
magnitude=[3e5 0.5 0];
eldia2(ex1,ey1,es1(:,1),eci1,magnfac);
eldia2(ex2,ey2,es2(:,1),eci2,magnfac);
eldia2(ex3,ey3,es3(:,1),eci3,magnfac,magnitude);
axis([-1.5 7 -0.5 5.5])
figure(2)
magnfac=eldia2(ex3,ey3,es3(:,2),eci3);
magnitude=[3e5 0.5 0];
eldia2(ex1,ey1,es1(:,2),eci1,magnfac);
eldia2(ex2,ey2,es2(:,2),eci2,magnfac);
eldia2(ex3,ey3,es3(:,2),eci3,magnfac,magnitude);
axis([-1.5 7 -0.5 5.5])
figure(3)
magnfac=eldia2(ex3,ey3,es3(:,3),eci3);
magnitude=[3e5 0.5 0];
eldia2(ex1,ey1,es1(:,3),eci1,magnfac);
eldia2(ex2,ey2,es2(:,3),eci2,magnfac);
eldia2(ex3,ey3,es3(:,3),eci3,magnfac,magnitude);
axis([-1.5 7 -0.5 5.5])
```

First, "beam2s" function is used to calculate the element forces and moment at 20 locations of the each element. Using these element forces, "eldia2" function plots the force diagram on the element geometry. Since the first column of the element force is submitted, the axial force will be plotted, as shown in Figure D.13. The variable "magnitude" is used to scale the magnitude of the diagram. In (P.26), the scale with the size of 3×10^5 will be plotted at the location of (0.5, 0) of the plot.

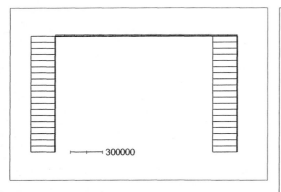

Figure D.13 Axial force diagram for the plane frame structure

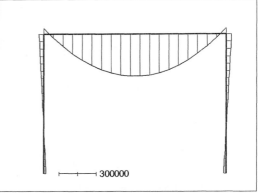

Figure D.14 Bending moment diagram for the plane frame structure

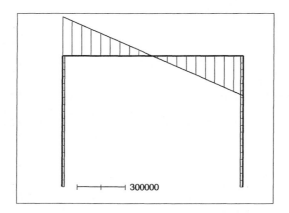

Figure D.15 Shear force diagram for the plane frame structure

Figure D.13, Figure D.14, and Figure D.15 show the axial force, bending moment, shear force diagrams, respectively, plotted using "`eldia2`" function. The complete list of the MATLAB program that can solve the portal frame problem is shown in Table D.8.

D.3   FINITE ELEMENT ANALYSIS USING PLANE SOLID ELEMENTS

Bending analysis of a cantilever beam using CST element

Consider a beam under a pure bending moment, as shown in Figure D.16. Using ten CST elements, calculate the stress of the beam along the neutral axis and top and bottom surfaces using the MATLAB finite element toolbox. For material properties, use $E = 200\text{GPa}$ and $\nu = 0.3$. The thickness of the beam is 0.01 m. Since CST element does not have rotational DOF, the bending moment can be produced by applying an equal and opposite force $F = \pm 100,000\,\text{N}$ at the end of the beam. Compare the numerical results with the elementary theory of beam. Provide an element stress contour plot for σ_{xx}.

The complete list of the MATLAB program that can solve the portal frame problem is shown in Table D.10. Plane solid elements have two DOFs at each node: u_x and u_y.

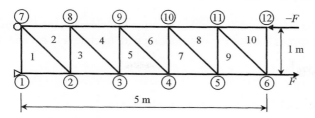

Figure D.16 CST elements for cantilever beam bending analysis

Thus, the plane beam in Figure D.16 has 24 DOFs. First, the coordinates of nodes and element connectivity are produced using the following program:

```
Edof=[ 1    1    2    3    4   13   14;
       2    3    4   15   16   13   14;
       3    3    4    5    6   15   16;
       4    5    6   17   18   15   16;
       5    5    6    7    8   17   18;
       6    7    8   19   20   17   18;
       7    7    8    9   10   19   20;
       8    9   10   21   22   19   20;
       9    9   10   11   12   21   22;
      10   11   12   23   24   21   22];
Coord=[0   0; 1   0; 2   0; 3   0;
       4   0; 5   0; 0   1; 1   1;
       2   1; 3   1; 4   1; 5   1];
dof=[1    2;   3    4;    5    6;    7    8;
     9   10;  11   12;   13   14;   15   16;
    17   18;  19   20;   21   22;   23   24];
[Ex,Ey]=coordxtr(Edof,Coord,dof,3);
```

(P.27)

Since Element 1 connects Nodes 1, 2 and 7, the first row of `Edof` array has the following DOFs: 1, 2, 3, 4, 13, and 14. The last parameter of "`coordxtr`" is 3 because the CST element has three nodes. For material properties, we need to build the stress-strain matrix in Eq. (6.5) for plane stress problems. This can be done using the following program:

```
ep=[1   0.01];
E=200E9;
nu=0.3;
D=hooke(1,E,nu);
```

(P.28)

In arracy `ep`, the first column specifies problem type: (1) plane stress, (2) plane strain, (3) axisymmetric, and (4) three-dimension. The second column stands for the thickness of the plane solid element. The "`hooke`" function calculates the stress-strain matrix D. The first parameter in the "`hooke`" function specifies problem type. For plane problems, the size of stress-strain matrix becomes 3×3.

In order to construct the element stiffness matrix for the i-th element, the "plante" function is called with i-th row of Ex and Ey arrays along with thickness in ep and stress-strain matrix D:

```
for i=1 : 10
   ke=plante(Ex(i,:),Ey(i,:),ep,D);
   K=assem(Edof(i,:),K,ke);
end
```
(P.29)

Applying boundary conditions and solving the matrix equation are identical to the previous examples. After solving for the nodal DOFs, the element stress can be calculated using "plants" function:

```
[es,et]=plants(Ex,Ey,ep,D,ed);
```
(P.30)

Arrays es and et contain element stresses and strains, respectively. The array es is 10×3 dimension. The i-th row represents i-th element resutls. The first column is σ_{xx}, the second column σ_{yy}, and the third column τ_{xy}. The same convention for et. Table D.9 shows element stresses and strains for the cantilevered beam model.

Since element stresses are constant in CST elements, the stress along the neutral axis will be oscillating as illustrated in Figure D.17. From the classical beam theory, the stress along the bottom surface will be 60 MPa, while along the top surface be −60 MPa. The stress results from CST elements are about four times smaller than that of the classical results. This happens due to the inability of the elements in representing linearly varying stress. Also note that σ_{yy} and τ_{xy} are not zero in CST elements. The stress contour can be plotted using "plcontour2" function, which is not a standard function in the MATLAB toolbox. The program list of "plcontour2" can be found in Table D.11.

```
plcontour2(Ex,Ey,es(:,1),ed,500);
```
(P.31)

In (P.31), the displacement is magnified by 500 times.

Table D.9   Element Stresses and Strains for the Cantilevered Beam (stress unit is MPa and strain unit is micro-strain)

Elem	σ_{xx}	σ_{yy}	τ_{xy}	ε_{xx}	ε_{yy}	γ_{xy}
1	15.69	−4.32	4.32	84.9	−45.1	56.1
2	−15.69	−5.00	−4.32	−70.9	−1.5	−56.1
3	14.84	4.16	5.16	68.0	−1.5	67.0
4	−14.84	−4.46	−5.16	−67.5	−0.1	−67.0
5	14.82	4.44	5.18	67.4	−0.1	67.4
6	−14.82	−4.44	−5.18	−67.4	0.1	−67.4
7	14.84	4.46	5.17	67.5	0.1	67.0
8	−14.84	−4.16	−5.17	−78.0	1.5	−67.0
9	15.69	5.00	4.32	70.9	1.5	56.1
10	15.69	4.32	−4.32	−84.9	45.1	−56.1

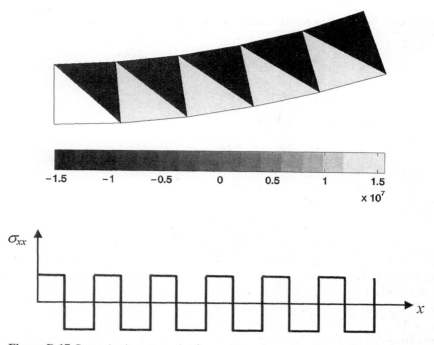

Figure D.17 Stress (σ_{xx}) contour plot for cantilever beam bending analysis and σ_{xx} along the neutral axis

Bending Analysis of a Cantilever Beam Using Rectangular Element

Solve bending of the cantilevered beam in Figure D.16 using five rectangular elements. Compare the numerical results with the elementary theory of beam. Is the rectangular element stiff or soft? Compared to the CST element, is the rectangular element stiffer or softer?

The element numbers and node numbers are shown in Figure D.18. The complete list of program that can analyze the cantilevered bean bending analysis is shown in Table D.13. Compared to CST elements, the main difference is the function that calculates the element stiffness matrix. In the case of rectangular elements, "`plani4e`" function is used to construct the element stiffness matrix:

$$\boxed{\texttt{ke=plani4e(Ex(i, :),Ey(i, :),ep,D);}} \qquad \text{(P.32)}$$

After solving the matrix equation, the element stress are obtained by calling the following function:

```
for i=1 : 5
   es(i, : , : )=plani4s(Ex(i, : ),Ey(i, : ),ep,D,ed(i, : ));
end
```
(P.33)

Note that unlike the CST element, `es` is a three-dimensional array. The first index of `es` is the element number, and the second index is the integration point, and the third index is the stress component. Since the stress is not constant within an element, different stress values are calculated at the integration points.

Table D.10   Program List for the Cantilevered Beam

```
Edof=[ 1    1    2    3    4  13  14;
       2    3    4   15  16  13  14;
       3    3    4    5   6  15  16;
       4    5    6   17  18  15  16;
       5    5    6    7   8  17  18;
       6    7    8   19  20  17  18;
       7    7    8    9  10  19  20;
       8    9   10   21  22  19  20;
       9    9   10   11  12  21  22;
      10   11   12   23  24  21  22];
Coord = [0   0;1   0;2   0;3   0;
         4   0;5   0;0   1;1   1;
         2   1;3   1;4   1;5   1];
dof =[1    2; 3    4; 5    6; 7    8;
      9   10;11   12;13   14;15   16;
     17   18;19   20;21   22;23   24];
[Ex,Ey]=coordxtr(Edof,Coord,dof,3);
%
ep = [1  0.01];
E = 200e9;
nu = 0.3;
D = hooke(1, E, nu);
%
K=zeros(24);
F=zeros(24,1);
F(11)=100000;F(23)=-100000;
%
for i=1:10
   ke = plante(Ex(i,:),Ey(i,:),ep,D);
   K = assem(Edof(i,:),K,ke);
end
%
bc=[1  0;2  0;13  0];
Q=solveq(K,F,bc);
%
ed = extract(Edof,Q);
[es,et]=plants(Ex,Ey,ep,D,ed);
%
plcontour2(Ex,Ey,es(:,1),ed,500);
```

The element stress contour can be plotted by calling the following program:

$$\boxed{\texttt{plcontour(Ex,Ey,es(: , : ,1),ed,200);}} \qquad \text{(P.34)}$$

The function "plcontour" is not a part of standard MATLAB finite element toolbox. Program "plcontour" function is listed in Table D.14.

In the case of the pure bending problem, the axial stress/strain is constant along the same y-location. The beam bending theory yields a 3×10^{-4} strain along the bottom surface of the beam, which is in tension. However, the finite element analysis provides 2.02×10^{-4} strain, which is smaller than the theoretical value. This result shows that the finite element analysis result is stiffer than exact solution. In addition, the beam theory says the shear strain for the pure bending problem is zero. However, the finite element analysis yields linearly varying shear strain between -2×10^{-4} and 2×10^{-4}. Initially

Table D.11   Program List for Plcontour2 Function

```
function plcontour2(Ex, Ey, Es, Ed, scale)
% Plot stress contour for CST or quadrilateral elements
% Input:
% Ex = [x1 x2 x3 (x4);...] Element nodal x-coord.
% Ey = [y1 y2 y3 (y4);...] Element nodal y-coord.
% Es = [s1;s2;s3;(s4)...] Element stress values
figure;
smin = min(min(Es)); smax = max(max(Es));
clength = smax-smin; colmap = colormap;
[csz1 csz2]=size(colormap);
csz1 = csz1 - 1;
[numfele shape] = size(Ex);
if (nargin==5)
  for j=1:shape
    Ex(:,j) = Ex(:,j) + scale * Ed(:,j*2-1);
    Ey(:,j) = Ey(:,j) + scale * Ed(:,j*2);
  end
end
hold on; axis equal; axis off;
for i=1:numfele
  val = (Es(i)-smin)/clength*csz1 + 1;
  color = colmap(floor(val),:);
  fill(Ex(i,:),Ey(i,:),color);
end caxis
([smin smax]); colorbar('horiz'); hold off;
```

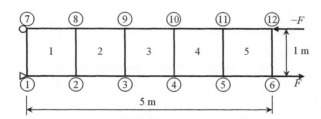

Figure D.18 Rectangular elements for cantilever beam bending analysis

perpendicular corner has to be distorted due to the bending-type deformation. Such a superficial deformation contributes to the stiff result in the bending problem. However, the rectangular element is softer than triangular element. Table D.12 compares finite element analysis results with exact solutions. Note that the finite element solutions are in general stiffer than the exact one. Especially plane solid elements are not accurate for

Table D.12   Comparison Between Finite Element Analysis Results

	CST	Rectangular	Exact
Tip deflection	0.0018 m	0.0051 m	0.0075 m
Max sxx	15.7 MPa	25.6 MPa	60 MPa

Table D.13 Program List for the Cantilevered Beam Bending Analysis

```
Edof=[1  1    2    3    4   15  16  13  14;
       2  3    4    5    6   17  18  15  16;
       3  5    6    7    8   19  20  17  18;
       4  7    8    9   10   21  22  19  20;
       5  9   10   11   12   23  24  21  22];
Coord = [0  0;1  0;2  0;3  0;4  0;5  0;
          0  1;1  1;2  1;3  1;4  1;5  1];
dof = [1    2; 3    4; 5    6; 7    8;
       9   10;11   12;13  14;15  16;
      17   18;19   20;21  22;23  24];
[Ex,Ey]=coordxtr(Edof,Coord,dof,4);
ep = [1  0.01  2];
E = 200e9; nu = 0.3;
D = hooke(1, E, nu);
K=zeros(24);
F=zeros(24,1); F(11)=100000;F(23)=-100000;
for i=1:5
   ke = plani4e(Ex(i,:),Ey(i,:),ep,D);
   K = assem(Edof(i,:),K,ke);
end
bc=[1  0;2  0;13  0];
U=solveq(K,F,bc);
ed = extract(Edof,U);
for i=1:5
  es(i,:,:)=plani4s(Ex(i,:),Ey(i,:),ep,D,ed(i,:));
end
plcontour(Ex,Ey,es(:,:,1),ed,200);
```

Table D.14 Program List for Plcontour Function

```
function plcontour(Ex,Ey,Es,Ed,scale)
% Plot stress contour for CST or quadrilateral elements
% Input:
% Ex = [x1  x2  x3  (x4);...]  Element nodal x-coord.
% Ey = [y1  y2  y3  (y4);...]  Element nodal y-coord.
% Es = [s1  s2  s3  {s4);...]  Nodal stress values
figure;
smin = min(min(Es)); smax = max(max(Es));
[numfele shape] = size(Ex);
if (nargin==5)
  for j=1:shape
    Ex(:,j) = Ex(:,j) + scale * Ed(:,j*2-1);
    Ey(:,j) = Ey(:,j) + scale * Ed(:,j*2);
  end
end
caxis([smin smax]);
hold on; axis equal; axis off;
for i=1:numfele
  color = [];
  for j = 1:shape
    val = Es(i,j);
    color = [color val];
  end
  fill(Ex(i,:),Ey(i,:),color);
end
colorbar('horiz'); hold off;
```

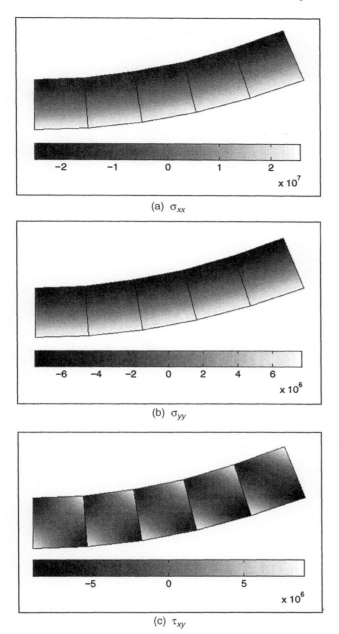

(a) σ_{xx}

(b) σ_{yy}

(c) τ_{xy}

Figure D.19 Stress contour plots for cantilever beam bending analysis

bending type deformation. In such a case, many elements are required through the thickness direction.

Stress contour plots are shown in Figure D.19. Note that σ_{yy} and τ_{xy} components are not zero in the element. This explains the inability of a rectangular element in representing bending type deformation.

Index

A

Allowable stress, 298
Aspect ratio, 268
Assembly, 68, 126, 159, 222
Axial rigidity, 66, 145

B

Basis function, 148
Beam constitutive relation, 145
Bending moment, 166
Boundary condition, 69, 273
  convection, 200
  essential, 108, 114, 188, 212
  natural, 109, 114, 188, 212
  traction, 41
Boundary value problem, 41, 108

C

Characteristic polynomial, 9
Compatibility, 287
Compatibility equation, 43
Completeness, 287
Conductance matrix
  element, 189, 195
  global, 190
Conservation of energy, 186
Constant strain triangular element, 216
Constitutive relation, 35, 212, 213
Constraint, 297, 315
Contour display, 279
Convection, 200, 203
Convection coefficient, 187, 200
Convergence rate, 272
Convergence study, 272
Coordinate transformation, 76, 84
Curvature, 144

D

Deflection curve, 143, 148, 154
Design, 297
  feasible, 298
  parameterization, 304
  variable, 298
Determinant, 5, 25
Direction cosine, 81, 84
Direct stiffness method, 67, 188

Discretization, 117, 261
Displacement field, 41, 42
Displacement variation, 132
Distortion energy theory, 46, 47

E

Eigen value, 8, 9, 25
Eigen vector, 8, 9
Elasticity matrix, 36, 37, 213
Elastic limit, 36
Element
  bar, 66
  beam, 154
  conduction equation, 188
  conforming, 287
  force, 62, 63, 68, 69, 70, 174
  frame, 171
  order, 271
  plane truss, 74
  quadrilateral, 237
  rectangular, 229
  reference, 238
  size, 271
  space truss, 83
  spring, 61, 62
  thermal force, 91
  triangular, 216
  type, 270
Elongation, 67
Energy method, 129
Equilibrium, 18
Equilibrium equation, 40
Error estimation, 291
Euler–Bernoulli beam, 143
Extra zero energy mode, 251

F

Failure envelope, 47
Failure theory, 43
Feasible set, 316
Finite difference method, 309
Finite element analysis procedure, 262
Flexural rigidity, 145
Force, 274
Fourier law, 186
Fully-stressed design, 301

G
Galerkin method, 110, 120, 125, 194
Gauss quadrature, 248
Geometric vector, 1
Goal, 297, 315

H
Heat conduction, 185
Heat conduction equation, 194
Heat flux, 186, 194
Heat source, 194
Hermite polynomial, 155
Hooke's law, 36
  generalized, 36

I
Inter-element displacement compatibility, 219
Internal energy, 187
Interpolation, 216
Interpolation function, 119, 120, 124, 217, 229, 238
Isoparametric mapping, 238, 239
Isotropic, 34

J
Jacobian, 240, 241, 268

L
Lagrange interpolation, 229, 238
Line search, 320
Load factor, 299
Local coordinate, 172

M
Matrix, 2
  addition, 4
  determinant, 5
  diagonal, 3
  equation, 8
  identity, 3
  inverse, 7
  positive definite, 13, 64
  singular, 5, 7
  skew-symmetric, 3
  square, 2
Matrix–matrix multiplication, 6
Matrix–vector multiplication, 6
Maximum principal stress theory, 48
Maximum shear stress theory, 48
Mesh generation, 267

Meshing, 282
Mesh quality, 268
Moment–curvature relation, 145
Moment of inertia, 145

N
Necking, 36
Neutral axis, 143
Norm, 4
Numerical integration, 248

O
Objective function, 315
Optimisation, 313

P
Parameter study, 307
Patch test, 287
Performance, 297
Perimeter, 187, 203
Plane strain, 39, 211, 213
Plane stress, 38, 211, 212
Poisson's ratio, 37
Postprocessing, 278
Potential energy, 44, 132, 133, 147, 160, 215, 222, 223, 224, 233
Preliminary analysis, 262
Preprocess, 265
Principal direction, 34
Principle of minimum potential energy, 132, 133, 162, 215, 225
Principle of virtual work, 129, 130
Proportional limit, 36

Q
Quadratic form, 12
Quilibrium equation, element, 62, 67

R
Rankine yield criterion, 48
Rayleigh–Ritz method, 136, 147, 148
Reaction force, 70
Residual, 109, 115, 194, 204

S
Safety factor, 49, 298
Safety margin, 298, 299
Scalar product, 4
Sensitivity analysis, 307

Shape function, 119, 120, 155, 194, 217, 218, 230, 238
Shear force, 166
Shear modulus, 37
Singularity, 277
Solid model, 267
Solution
  approximate, 124
  derived, 276
  exact, 109
  primary, 276
  trial, 118
St. Vernant principle, 274
Step size, 320
Stiffness matrix
  bar, 67
  beam, 159
  element, 222, 233, 247
  frame, 174
  global, 64, 69, 162
  plane truss, 74, 77, 81
  space truss, 85
  spring, 62
  structural, 64, 68, 159, 162, 222
Strain, 30
  hardening, 36
  matrix, 32
  normal, 31
  principal, 33, 34
  shear, 32
  thermal, 88
  transformation, 33
  vector, 32
Strain–displacement matrix, 218, 232, 245
Strain–displacement vector, 156
Strain energy, 44, 132, 147, 158, 214, 221, 247
  dilatational, 45, 46
Stress
  bending, 167
  component, 20, 21, 22
  dilational, 46
  field, 39, 42
  invariant, 25
  matrix, 22
  maximum shear, 28
  normal, 19
  principal direction, 26
  principal, 24, 45
  shear, 19
  symmetry, 23

  thermal, 87
  transformation, 27, 28
  ultimate, 36
  vector, 24
  volumetric, 46
  von Mises, 46, 47
  yield, 36
Stress averaging, 280
Stress concentration factor, 264
Stress-strain matrix, 36
Stress-strain relation, 35
Superposition, 88
Surface traction, 18, 24, 215, 223
Symmetry, 282

T
Thermal conductivity, 186
Thermal expansion coefficient, 88
Thermal load, 195
Traction boundary condition, 41
Transformation matrix, 75, 76, 84, 173, 222
Transpose, 1, 3
Tresca yield criterion, 48
Trial function, 110, 120

U
Uniaxial tension test, 36
Units, 266

V
Variation, 131
  displacement, 132
Vector, 1
  addition, 4
  magnitude, 2
  product, 5
  scalar multiple, 4
  unit, 5
Virtual displacement, 129, 130
Virtual strain, 130
Virtual work, 130
Von Mises yield criterion, 46

W
Weighted residual, 110
Weight function, 109
Work-equivalent load, 161, 223

Y
Yield stress, 43
Young's modulus, 36, 37, 67